2021 Joint International EUROSOI Workshop and International Conference on Ultimate Integration on Silicon (EuroSOI-ULIS 2021)

Caen, France
1 – 3 September 2021

IEEE Catalog Number:	CFP2149D-POD
ISBN:	978-1-6654-3746-2

**Copyright © 2021 by the Institute of Electrical and Electronics Engineers, Inc.
All Rights Reserved**

Copyright and Reprint Permissions: Abstracting is permitted with credit to the source. Libraries are permitted to photocopy beyond the limit of U.S. copyright law for private use of patrons those articles in this volume that carry a code at the bottom of the first page, provided the per-copy fee indicated in the code is paid through Copyright Clearance Center, 222 Rosewood Drive, Danvers, MA 01923.

For other copying, reprint or republication permission, write to IEEE Copyrights Manager, IEEE Service Center, 445 Hoes Lane, Piscataway, NJ 08854. All rights reserved.

*** This is a print representation of what appears in the IEEE Digital Library. Some format issues inherent in the e-media version may also appear in this print version.

IEEE Catalog Number:	CFP2149D-POD
ISBN (Print-On-Demand):	978-1-6654-3746-2
ISBN (Online):	978-1-6654-3745-5
ISSN:	2330-5738

Additional Copies of This Publication Are Available From:

Curran Associates, Inc
57 Morehouse Lane
Red Hook, NY 12571 USA
Phone: (845) 758-0400
Fax: (845) 758-2633
E-mail: curran@proceedings.com
Web: www.proceedings.com

TABLE OF CONTENTS

INVESTIGATION OF THE INVARIANT DRAIN CURRENT POINT IN DIELECTRIC MODULATED ^{BE}SOI MOSFET BIOSENSOR 1
Leonardo S. Yojo; Ricardo C. Rangel; Katia R. A. Sasaki; Joao A. Martino

ANALYSIS OF ENERGY-DELAY-PRODUCT OF A 3D VERTICAL NANOWIRE FET TECHNOLOGY 5
I. O'Connor; A. Poittevin; S. Le Beux; A. Bosio; Z. Stanojevic; O. Baumgartner; C. Mukherjee; C. Maneux; J. Trommer; T. Mikolajick; G. Larrieu

COMPREHENSIVE KUBO-GREENWOOD MODELLING OF FDSOI MOS DEVICES DOWN TO DEEP CRYOGENIC TEMPERATURES 9
F. Serra Di Santa Maria; L. Contamin; M. Cassé; C. Theodorou; F. Balestra; G. Ghibaudo

THE EFFECTS OF SURFACE PASSIVATION ON THE ELECTROSTATICS OF THE UTB SOI DEVICES 13
Ravi Solanki; Aditya Sankar Medury

IMPROVED INTER-DEVICE VARIABILITY IN GRAPHENE LIQUID GATE SENSORS BY LASER TREATMENT 17
Jorge Ávila; Jose Galdon; Norberto Salazar; Maria-Isabel Recio; Carlos Navarro; Carlos Marquez; Francisco Gamiz

INCORPORATION OF SILICON-CARBIDE (SIC) NANOCRYSTALS IN THE MIM STRUCTURES BASED ON PULSED-DC REACTIVELY SPUTTERED HFO$_X$ LAYERS 21
Robert Mroczynski

MODELING LOW AND HIGH FIELD UNIFORM TRANSPORT IN MONOLAYER MOS$_2$ 25
Alessandro Pilotto; Pedram Khakbaz; Pierpaolo Palestri; David Esseni

MODELING THE PROPAGATION OF AC SIGNAL ON THE CHANNEL OF THE PSEUDO-MOS METHOD 29
Shingo Sato

IMPACT OF HIGH-ASPECT-RATIO ETCHING DAMAGE ON SELECTIVE EPITAXIAL SILICON GROWTH IN 3D NAND FLASH MEMORY 33
Tobias Reiter; Xaver Klemenschits; Lado Filipovic

THE SEMICONDUCTOR MODEL SOLVED BY THE NUMEROV PROCESS OVER A NON-UNIFORM GRID 37
R. Brunetti; M. Rudan

HIGH TEMPERATURE INFLUENCE ON THE TRADE-OFF BETWEEN GM/I$_D$ AND F$_T$ OF NANOSHEET NMOS TRANSISTORS WITH DIFFERENT METAL GATE STACK 41
Vanessa C. P. Silva; Joao A. Martino; Eddy Simoen; Anabela Veloso; Paula G. D. Agopian

ON THE ASYMMETRY OF THE DC AND LOW-FREQUENCY NOISE CHARACTERISTICS OF VERTICAL NANOWIRE PMOSFETS WITH BULK SOURCE CONTACT 45
Eddy Simoen; Anabela Veloso; Philippe Matagne

AVALANCHE TRANSIENT SIMULATIONS OF SPAD INTEGRATED IN 28NM FD-SOI CMOS TECHNOLOGY 49
D. Issartel; S. Gao; S. Hagen; P. Pittet; R. Cellier; D. Golanski; A. Cathelin; F. Calmon

FIRST PRINCIPLES EVALUATION OF TOPOLOGICALLY PROTECTED EDGE STATES IN MOS$_2$ 1T' NANORIBBONS WITH REALISTIC TERMINATIONS 53
Al-Moatasem El-Sayed; Heribert Seiler; Hans Kosina; Markus Jech; Dominic Waldhör; Viktor Sverdlov

IMPACT OF THE BACKGATE ON THE PERFORMANCE OF SOI UTBB NMOSFETS AT CRYOGENIC TEMPERATURES 57
Yi Han; Fengben Xi; Frederic Allibert; Ionut Radu; Slawomir Prucnal; Jin-Hee Bae; Susanne Hoffmann-Eifert; Joachim Knoch; Detlev Grützmacher; Qing-Tai Zhao

CONSTANT-CURRENT TIME DEPENDENT DIELECTRIC BREAKDOWN IN THICK AMORPHOUS SIO$_2$ CAPACITORS 61
Federico Giuliano; Susanna Reggiani; Elena Gnani; Antonio Gnudi; Mattia Rossetti; Riccardo Depetro; Giuseppe Croce

TCAD SIMULATIONS OF HIGH-ASPECT-RATIO NANO-BIOSENSOR FOR LABEL-FREE SENSING APPLICATION 65
Rakshita Pritam Singh Dhar; Naveen Kumar; Cristina Medina-Bailon; César Pascual García; Vihar Petkov Georigiev

CONDUCTANCE DUE TO THE EDGE MODES IN NANORIBBONS OF 2D MATERIALS IN A TOPOLOGICAL PHASE 69

Viktor Sverdlov; Heribert Seiler; Al-Motasem Bellah El-Sayed; Hans Kosina

SYNAPTIC TRANSISTORS BASED ON TRANSPARENT OXIDE FOR NEURAL IMAGE RECOGNITION 73

Q. N. Wang; C Zhao; W. Liu; I. Z. Mitrovic; H. Van Zalinge; Y. N. Liu; C Z Zhao

CURVATURE BASED FEATURE DETECTION FOR HIERARCHICAL GRID REFINEMENT IN TCAD TOPOGRAPHY SIMULATIONS 77

Christoph Lenz; Alexander Toifl; Andreas Hössinger; Josef Weinbub

FEATURE-SCALE MODELING OF LOW-BIAS SF_6 PLASMA ETCHING OF SI 81

Luiz Felipe Aguinsky; Georg Wachter; Frâncio Rodrigues; Alexander Scharinger; Alexander Toifl; Michael Trupke; Ulrich Schmid; Andreas Hössinger; Josef Weinbub

EFFECT OF TEMPERATURE ON PERFORMANCE OF HZO-BASED FD-SOI NCFET 85

Vullakula Rama Seshu; Rameez Raja Shaik; K. P. Pradhan

ON THE BREAKDOWN VOLTAGE TEMPERATURE DEPENDENCE OF HIGH-VOLTAGE POWER DIODE PASSIVATED WITH DIAMOND-LIKE CARBON 89

Luigi Balestra; Susanna Reggiani; Antonio Gnudi; Elena Gnani; Jagoda Dobrzynska; Jan Vobecký

IN-DEPTH CRYOGENIC CHARACTERIZATION OF 22 NM FDSOI TECHNOLOGY FOR QUANTUM COMPUTATION 93

Hung-Chi Han; Farzan Jazaeri; Antonio D'Amico; Zhixing Zhao; Steffen Lehmann; Claudia Kretzschmar; Edoardo Charbon; Christian Enz

IMPACT OF DIFFERENT TYPES OF PLANAR DEFECTS ON CURRENT TRANSPORT IN INDIUM PHOSPHIDE (INP) 97

Christian Dam Vedel; Enrico Brugnolotto; Søren Smidstrup; Vihar P. Georgiev

TEMPERATURE INCREASE IN MRAM AT WRITING: A FINITE ELEMENT APPROACH 102

Tomáš Hadámek; Mario Bendra; Simone Fiorentini; Johannes Ender; Roberto L. De Orio; Wolfgang Goes; Siegfried Selberherr; Viktor Sverdlov

JUNCTIONLESS NANOWIRE TRANSISTORS BASED WILSON CURRENT MIRROR CONFIGURATION 106

André B. Shibutani; Michelly De Souza; Renan Trevisoli; Rodrigo T. Doria

IMPROVING THE PHOTON DETECTION PROBABILITY OF SPAD IMPLEMENTED IN FD-SOI CMOS TECHNOLOGY WITH LIGHT-TRAPPING CONCEPT 110

S. Gao; D. Issartel; R. Orobtchouk; F. Mandorlo; D. Golanski; A. Cathelin; F. Calmon

CHARACTERIZATION AND LAMBERT – W FUNCTION BASED MODELING OF FDSOI FIVE-GATE QUBIT MOS DEVICES DOWN TO CRYOGENIC TEMPERATURES 114

E. Catapano; A. Aprà; M. Cassé; F. Gaillard; S. De Franceschi; T. Meunier; M. Vinet; G. Ghibaudo

DEVICE SIMULATIONS OF ION-SENSITIVE FETS WITH ARBITRARY SURFACE CHEMICAL REACTIONS 118

Leandro Julian Mele; Pierpaolo Palestri; Luca Selmi

RANDOM TELEGRAPH NOISE REAL TIME TESTING BASED ON DOWNSAMPLING FOR MASS DATA EXTRACTION 122

Maximilian Juettner; Michael Otto; Jan Hoentschel

CHARGE PUMPING-BASED METHOD FOR TRAPS DENSITY EXTRACTION IN JUNCTIONLESS TRANSISTORS 126

E. T. Fonte; R. Trevisoli; R. T. Doria

FIELD-EFFECT PASSIVATION OF LOSSY INTERFACES IN HIGH-RESISTIVITY RF SILICON SUBSTRATES 130

Martin Rack; Lucas Nyssens; Massinissa Nabet; Dimitri Lederer; Jean-Pierre Raskin

A THEORETICAL STUDY OF ELECTRON MOBILITY DISTRIBUTION IN FDSOI MOSFET 134

N. D. Akhavan; G. A. Umana-Membreno; R. Gu; J. Antoszewski; L. Faraone; S. Cristoloveanu

TCAD NEGATIVE CAPACITANCE FERROELECTRIC DEVICE MODELING FOR RADIATION DETECTION APPLICATIONS 138

Arianna Morozzi; Michael Hoffmann; Stefan Slesazeck; Roberto Mulargia

PERFORMANCE OF STACKED SOI NANOWIRES IN A WIDE TEMPERATURE RANGE 142

Jaime C. Rodrigues; Genaro Mariniello; Mikael Cassé; Sylvain Barraud; Maud Vinet; Olivier Faynot; Marcelo A. Pavanello

LOW TEMPERATURE INVESTIGATION OF N-CHANNEL GAA VERTICALLY STACKED SILICON NANOSHEETS 146

Bogdan Cretu; Anabela Veloso; Eddy Simoen

TCAD BASED MODELING OF SUB-SURFACE LEAKAGE IN SHORT CHANNEL BULK MOSFETS 150

Harshit Kansal; Aditya Sankar Medury

AN ARTIFICIAL SYNAPTIC THIN-FILM TRANSISTOR BASED ON 2D MXENE–TIO₂ 154
Y. X. Cao; C. Zhao; I. Z. Mitrovic; Y. N. Liu; L. Yang; H Van Zalinge; C. Z. Zhao

SILICON AND HAFNIA THIN FILM TRANSFER ON C-PLANE SAPPHIRE: EFFECT OF SUBSTRATE THICKNESS ON THE FERROELECTRIC HAFNIA PROPERTIES 158
Valentin Antonov; Sergey Tarkov; Vladimir Popov; Andrey Miakonkikh; Andrey Lomov; Konstantin Rudenko

OPERATIONAL TRANSCONDUCTANCE AMPLIFIER DESIGN WITH GATE-ALL-AROUND NANOSHEET MOSFET USING EXPERIMENTAL LOOKUP TABLE APPROACH 162
Júlia C. S. Sousa; Welder F. Perina; Eddy Simoen; Anabela Veloso; Joao A. Martino; Paula G. D. Agopian

IN-SITU RECOVERY OF ON-MEMBRANE PD-SOI MOSFET FROM TID DEFECTS AFTER GAMMA IRRADIATION 166
Amor Sedki; Valeriya Kilchytska; Farès Tounsi; Nicolas André; Laurent A. Francis; Denis Flandre

NEW 10V TO 1V LEVEL SHIFTER BASED ON NEW N/PMOS HIGH VOLTAGE IN FDSOI TECHNOLOGY 170
P. Galy; Sebastien Haendler

USE OF CMOS IMAGE SENSOR FOR EARLY DETECTION OF ISCHEMIC AND HAEMORRHAGIC STROKE 174
G. Pignataro; P. Cepparulo; O. Cuomo; A. Cusano; R. Rao; M. Ruvo; F. Palma

HIGH PERFORMANCE SILICON-BASED SUBSTRATE USING BURIED PN JUNCTIONS TOWARDS RF APPLICATIONS 178
Maxime Moulin; Martin Rack; Thibaud Fache; Zdenek Chalupa; Christophe Plantier; Yves Morand; Joris Lacord; Frédéric Allibert; Fred Gaillard; Jose Lugo; Louis Hutin; Jean-Pierre Raskin

SI/SI₀.₇GE₀.₃ A2RAM NANOWIRES FABRICATION AND CHARACTERIZATION FOR 1T-DRAM APPLICATIONS 182
J. Lacord; F. Tcheme Wakam; Z. Chalupa; J.-M. Hartmann; P. Besson; V. Loup; C. Vizioz; L. Brevard; F. Aussenac; X. Mescot; K. Lee; M. Bawedin

THERMAL STABILITY OF FERROELECTRICITY IN HAFNIA-ZIRCONIA-ALUMINA BURIED OXIDE STACKS 186
Fedor Tikhonenko; Valentin Antonov; Vladimir Popov; Andrey Miakonkikh; Konstantin Rudenko

SI THICKNESS INFLUENCE ON SUBTHRESHOLD CURRENTS AT HIGH TEMPERATURES IN FDSOI CMOS 189
Mattias Ekström; Laura Zurauskaite; Per-Erik Hellström

RF PERFORMANCES AT CRYOGENIC TEMPERATURES OF INDUCTANCES INTEGRATED IN A FDSOI TECHNOLOGY 193
Quentin Berlingard; Jose Lugo-Alvarez; Lauriane Contamin; Cédric Durand; Philippe Galy; Andre Juge; Silvano De Franceschi; Maud Vinet; Tristan Meunier; Mikaël Cassé; Fred Gaillard

COMBINED EFFECTS OF BTI, HCI AND OFF-STATE MOSFETS AGING ON THE CMOS INVERTER PERFORMANCE 197
A. Crespo-Yepes; C. Nasarre; N. Garsot; J. Martin-Martinez; R. Rodriguez; E. Barajas; X. Aragones; D. Mateo; M. Nafria

CHARGE-BASED MODELING OF FIELD EFFECT TRANSISTORS, MAKE IT EASY 201
Jean-Michel Sallese

CONDUCTANCE MODULATION IN AL/SIO₂/N-SI MIS RESISTIVE SWITCHING STRUCTURES 205
Piotr Wisniewski; Jakub Jasinski; Andrzej Mazurak

Author Index

Investigation of the Invariant Drain Current Point in Dielectric Modulated [BE]SOI MOSFET Biosensor

Leonardo S.Yojo[1,*], Ricardo C. Rangel[1,2], Katia R. A. Sasaki[1] and Joao A. Martino[1]

[1] LSI/PSI/USP, University of Sao Paulo, Sao Paulo, Brazil
[2] FATEC-SP, Faculdade de Tecnologia de Sao Paulo, Sao Paulo, Brazil
*Email: leonardo.yojo@usp.br

Abstract—The study of a dielectric-modulated biosensor built on the [BE]SOI MOSFET structure showed a crossing point of the transfer characteristics curves for target biomaterials with different dielectric constants (CID – Current Invariant Dielectric). The CID corresponds to an insensitive region for biosensing application, although it can be interesting as a reference signal measurement. It was evaluated that this behavior arises from the threshold shift and the transconductance increase as a function of the biomaterial dielectric constant. The gate voltage at the crossing point presented higher variation as a function of the metal gate-semiconductor workfunction difference, when compared to the influence of the programming gate voltage and the gate oxide charge concentration.

Keywords— [BE]SOI MOSFET; bio-FET; dielectric-modulated biosensor

I. Introduction

The biosensors are devices compound by a receptor, responsible for capturing a biochemical signal, and a transducer, that translates the received signal to an electric output [1]. There are several types of biosensors, that can differ from each other depending on the target material or sensing principle. Among these variety of biosensors, the ones constructed on field-effect-transistors (FET) are gaining relevance due to the benefits of this technology such as fast response, reduced production cost and small size factor [2].

The Back Enhanced Silicon-on-Insulator ([BE]SOI) Metal-Oxide-Semiconductor (MOS) FET [3] is a reconfigurable device, which means it can work as an n-type or p-type transistor, depending on the back biasing. In previous works it was analyzed the advantages of this structure for the biosensing application through numerical simulations [4]. A peculiar characteristic was observed when the influence of the dielectric constant (k) value of the biomaterial target was varied, which was the presence of a crossing (CID – Current Invariant Dielectric) in the drain current vs. gate voltage ($I_{DX}V_{GF}$) curves in the studied conditions. This work aims to investigate the causes behind the appearance of this invariant point.

II. Device Characteristics

Fig. 1a shows the biosensor [BE]SOI MOSFET structure implemented in the Synopsys Sentaurus TCAD [5]. It is a planar device built on Silicon-on-Insulator wafer. The silicon channel is not intentionally doped, which means an intrinsic doping level (acceptors with concentration of $10^{15}/cm^3$). The drain and source contacts are made of a mid-gap metal to provide a suitable current of either electrons or holes. There are two underlap regions between the gate electrode and the

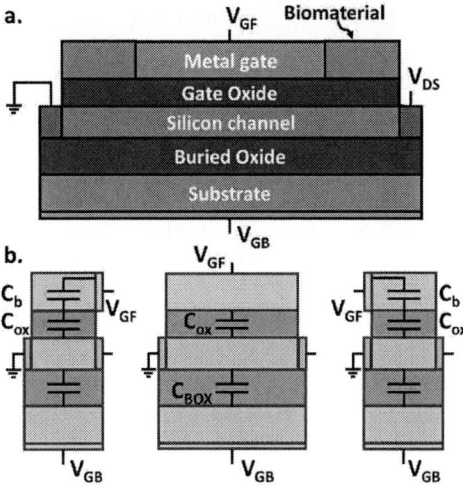

Fig. 1   Schematic profile of the biosensor BESOI MOSFET (a) and of two laterals and middle transistors (b).

source/drain contacts, where the biomaterial to be sensed is deposited. The dielectric constant of this material has influence on the drain current of the transistor. Both gate oxide and silicon channel have thickness of 10nm, the buried oxide is 200nm thick and the metal gate 50nm. The length of each underlap is 100nm and the length of the metal gate is 1μm. Temperature dependence mobility, SRH recombination and non-local tunneling Schottky contacts models were considered.

The operation principle of the [BE]SOI MOSFET relies on the interaction of both electric fields due to the front and back biasing. The back gate (V_{GB}) is responsible for the selection of the transistor type (programming gate) through the buried oxide capacitance C_{BOX}. For a positive enough value, electrons are induced in the channel at the silicon/buried oxide interface, thus the device behaves as a n-type. The opposite case occurs for negative enough V_{GB}, in which holes are induced instead. The front gate voltage (V_{GF}) enables the drain current control (controlling gate). Depending on its value the channel under the front gate electrode can be either totally depleted (off state) or not (on state). Furthermore, it is possible to look at this device as a series connection of three transistors, as shown in Fig. 1b, where the influence of the controlling gate electric field on the channel occurs through the gate oxide capacitance (C_{ox}) on the middle transistor; and through the biomaterial capacitance (C_b) in series with C_{ox} on the lateral transistors by a fringing electric field.

The authors would like to acknowledge the research-funding agencies FAPESP (under grant 2018/01568-4), CAPES and CNPq.

978-1-6654-3746-2/21 $31.00 © 2021 IEEE

III. THE CROSSING POINT

Fig. 2 shows the transfer characteristics as a function of the k. The n-type transistor was simulated with V_{GB}=30V (a high enough value due to the thick buried oxide) and V_{DS}=100mV, while the p-type was biased with V_{GB}=-30V and V_{DS}=-100mV. In each of the following results, three k values were considered (k=1,10,20) to analyze the impact of this parameter [4]. It is possible to note the presence of an intersection between the curves, almost a point, for both p- and n- types. The biasing of the device near this region is undesirable at first, as the drain current remains roughly constant and its sensitivity to the k value would be severely reduced (although it can be interesting as reference signal measurement, for instance). To understand its cause, it was analyzed the drain current for the simplification proposed in Fig. 1b, in which the same curves were simulated for the middle and lateral transistors separately, as shown in Fig. 3. These curves only show the behavior of the structures and the current level cannot be related to Fig.2, as a constant $|V_{DS}|$=100mV was considered in this case. As expected, the middle transistor current is the same for all k's, as it is influenced only by the dielectric constant of the gate oxide. However, the threshold voltage (V_T) of the complete device is determined by the middle transistor. The lateral transistors current presented the crossing point as a function of k. The increase of k in the biomaterial causes an increase of the transconductance of the lateral transistors, as the stronger controlling gate electric field enables a higher carrier concentration in the channel when the device is conducting. Also, the change in the k value causes a shift of V_T, considering a SOI MOSFET model for the lateral transistor [6]. The presence of the crossing point arises from these two phenomena that counterbalances each other, in a similar way to the known ZTC (Zero Temperature Coefficient) effect [3].

In the sequence, the influence of metal gate-semiconductor workfunction difference (Φ_{MS}); programming gate voltage (V_{GB}); and gate oxide charge concentration (Q_{ox}) on the V_{GF} at the crossing point was evaluated, as these parameters have direct influence on the transconductance or V_T. Fig.4 shows the behavior of the I_DxV_{GF} curves as a function of Φ_{MS}, in which only a threshold voltage shift is seen. This shift is directly mirrored in the crossing point as a function of the Φ_{MS} curve, as shown in Fig. 5 for both p- and n-type for Φ_{MS} from

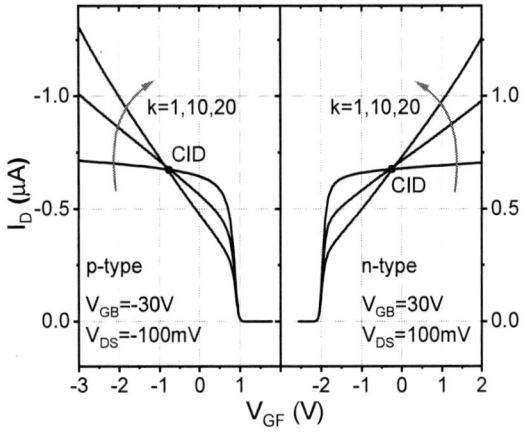

Fig. 2   Biosensor ^{BE}SOI MOSFET drain current as a function of the controlling gate voltage.

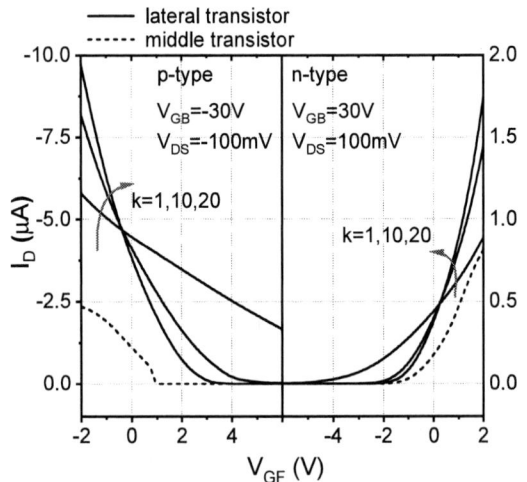

Fig. 3   Drain current as a function of the controlling gate voltage for the middle and lateral transistors of Fig1b.

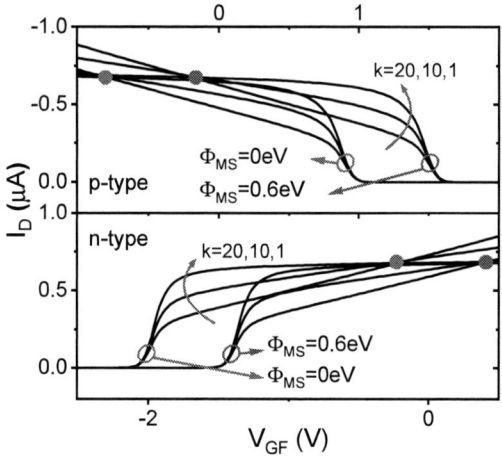

Fig. 4   Drain current for different metal gate/semiconductor workfunction.

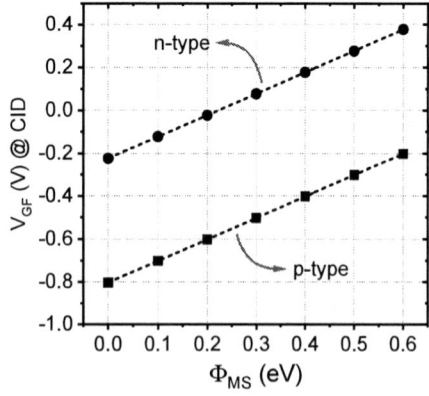

Fig. 5   Gate voltage at CID point as a function of the metal gate/semiconductor workfunction.

978-1-6654-3746-2/21 $31.00 © 2021 IEEE

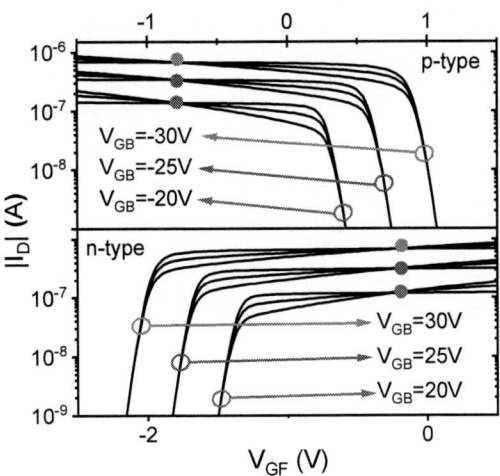

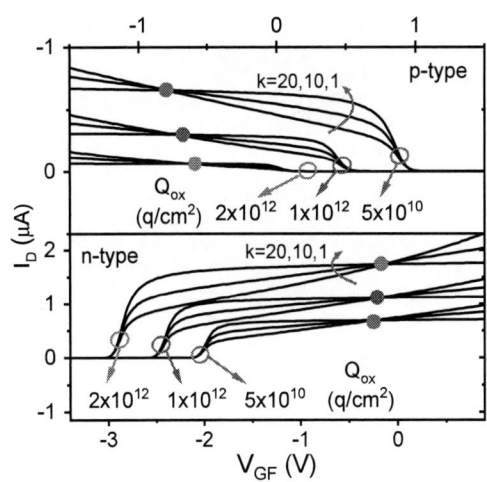

Fig. 6   Drain current for different programming gate voltages (V_{GB}).

Fig. 8   Drain current for different gate oxide charge concentrations (Q_{ox}).

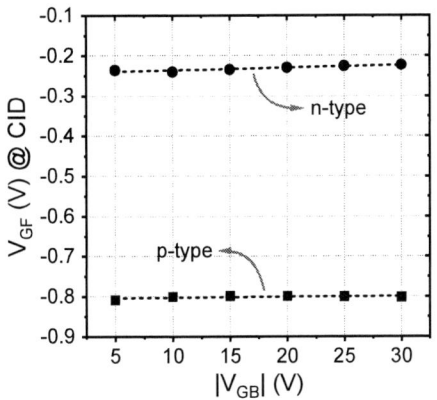

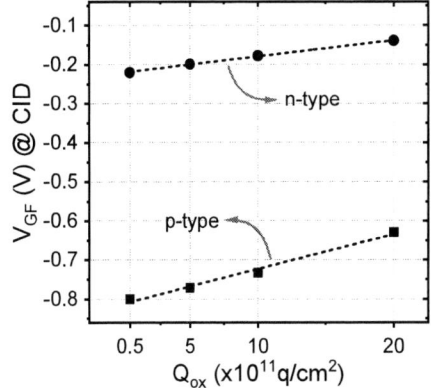

Fig. 7   Gate voltage at CID point as a function of the programming gate voltage (V_{GB}).

Fig. 9   Gate voltage at CID point as a function of the gate oxide charge concentrations (Q_{ox}).

0eV to 0.6eV. It is also possible to observe that the vertical difference between the lines correspond approximately to $2\varphi_F$ (fermi potential) because the channel is in inversion in the n-type transistor, while in the p-type, it is in accumulation regime (disregarding V_{DS} influence once the curves were simulated in the linear region).

Fig. 6 shows the $I_D x V_{GF}$ for different V_{GB}'s values. In this case, according to the operation principle described above, the higher V_{GB} promotes a higher carrier concentration in the channel and, consequently, a higher V_{GF} needed to deplete the channel in the off state (i.e., the threshold voltage is shifted). As result, little variation of the crossing point as a function of V_{GB} is observed in Fig. 7 for $|V_{GB}|$ from 5V to 30V.

The Q_{ox} influence on the $I_D x V_{GF}$ curves is shown in Fig. 8, where it is possible to observe a change in the current level and V_T too. As it was considered positive charges, different tendencies were observed for the p-type (higher Q_{ox}

contributed to lower the hole accumulation) and n-type (higher Q_{ox} contributed to increase the electron induction in the channel). The crossing point as a function of Q_{ox} is presented in Fig. 9 for Q_{ox} from $5x10^{10}q/cm^2$ to $2x10^{12}q/cm^2$.

IV. CONCLUSION

This work studied the presence of an intersection region in the transfer characteristics of the dielectric based ^{BE}SOI MOSFET biosensor. The analysis through numerical simulations showed that this effect arises from the counterbalance of the two phenomena: the increase of the transconductance and the threshold voltage shift as a function of the biomaterial dielectric constant. This investigation supports further studies of the ^{BE}SOI MOSFET usage as a biosensor, as the biasing of the device near this crossing point corresponds to a condition where the drain current is insensitive to the biomaterial, although it can be interesting as a reference signal measurement.

978-1-6654-3746-2/21 $31.00 © 2021 IEEE

REFERENCES

[1] D. R. Thévenot, K. Toth, R. A. Durst and G. S. Wilson., "Electrochemical biosensors: recommended definitions and classification," Pure Appl. Chem., 71(12), pp.2333-2348, 1999. DOI: 10.1351/pac199971122333

[2] D. Sadighbayan, M. Hasanzadehon and E. Ghafar-Zadeh, "Biosensing based on field-effect transistors (FET): Recent progress and challenges," TrAC, Trends Anal. Chem., 133, 2020. DOI: 10.1016/j.trac.2020.116067

[3] K. R. A. Sasaki, R.C. Rangel, L.S. Yojo and J.A. Martino, "Tradeoff between the transistor reconfigurable technology and the zero-temperature-coefficient (ZTC) bias point on BESOI MOSFET," Microelectron. J., 94, 2019. DOI: 10.1016/j.mejo.2019.104658

[4] L. S. Yojo, R.C. Rangel, K. R. A. Sasaki and J.A. Martino, "Study of ^{BE}SOI MOSFET Reconfigurable Transistor for Biosensing Application" ECS J. Solid State Sci. Technol., 10(2), 2021. DOI: 10.1149/2162-8777/abe3cc

[5] Synopsys TCAD, Sentaurus Device User Guide, (Version O-2018.06).

[6] J. P. Colinge, Silicon-on-Insulator Technology, 3rd ed., Springer, 2004.

Analysis of Energy-Delay-Product of a 3D Vertical Nanowire FET Technology

I. O'Connor, A. Poittevin, S. Le Beux[*], A. Bosio
Lyon Institute of Nanotechnology, CNRS UMR 5270, Ecole Centrale de Lyon Ecully, France
[*]Dept. of Electrical and Computer Engineering Concordia University, Canada
ian.oconnor@ec-lyon.fr

J. Trommer, T. Mikolajick[**]
NaMLab gGmbH, 01187 Dresden, Germany,
[**]Institute for Semiconductors and Microsystems and cfaed, TU Dresden, 01187 Dresden, Germany.
Jens.Trommer@namlab.com

Z. Stanojevic, O. Baumgartner
Global TCAD Solutions, Vienna, Austria
z.stanojevic@globaltcad.com

G. Larrieu
LAAS-CNRS, Université de Toulouse, Toulouse, France
guilhem.larrieu@laas.fr

C. Mukherjee, C. Maneux
IMS Laboratory, University of Bordeaux
CNRS UMS 5218,
351, Cours de la Libération - 33405 Talence, France
chhandak.mukherjee@ims-bordeaux.fr

Abstract— To sustain transistor scaling beyond lateral 7nm devices, gate-all-around (GAA) junction-less vertical nanowire field effect transistors (VNWFET) are a promising alternative. This work analyses the energy-delay-product (EDP) for a junction-less 3D vertical gate-all-around nanowire FET technology, with a physical channel length of 14nm. Comparisons with the EDP of a baseline 7nm FinFET technology are carried out. The analysis motivates a new 3D neural network compute cube (N²C²) concept. Our results show that a 10x gain in EDP can be achieved for a physical VNWFET gate length of 14nm.

Keywords—Vertical junctionless NWFET, logic circuit simulation.

I. INTRODUCTION

Emerging computing paradigms for the Internet of Things (IoT), in particular edge computing and edge artificial intelligence (AI), target real-time operations including data creation, decision, and action where milliseconds matter. Existing solutions are computation-intensive and energy-hungry requiring server-based implementations, which introduce latency and jitter, as well as raising data protection and privacy concerns. Deterministic, secure, and real-time operation is important for self-driving cars, robotics, industry 4.0, augmented reality, and many other areas. However, despite recent improvements in algorithm efficiency, energy-efficiency of the hardware has become a challenge. Embedded lightweight energy-efficient hardware remains elusive.

Today's 2D electronic architectures suffer from "unscalable" interconnects and are thus still far from being able to compete with biological neural systems in terms of real-time information-processing capabilities with comparable energy consumption. Recent advances in materials science, device technology, and synaptic architectures have the potential to fill this gap with novel disruptive technologies that go beyond conventional technology. A promising solution comes from vertical nanowire field-effect transistors (VNWFETs) [1-2] that unlock the full potential of truly unconventional 3D networks through a unique integration approach [1] termed Logic Element Gate Overstacking (LEGO). This technology has motivated the concept of a flexible 3D neural network compute cube (N²C²) with high integration density and performance. While significant gains in silicon area and energy-efficiency are expected through the combination of extremely small elementary device footprint and intrinsically

3D device and circuit functionality, this work is the first attempt to quantify the gains in terms of energy-efficiency of 3D logic blocks based on VNWFET devices.

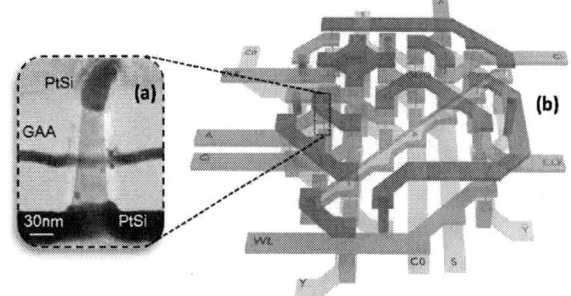

Fig 1: (a) STEM image showing the cross section of a junctionless vertical nanowire transistor, reproduced from [1], (b) 3D logic cell schematic of a 1-bit full adder using vertically stacked VNWFETs.

The rest of this paper is organized as follows: section II recalls the energy-delay product (EDP) metric and develops analytical equations for on-current and load capacitance based on geometrical and material parameters to compare EDP gains between VNWFETs and FinFETs. Section III applies this equation to explore EDP gains for both technologies and finally, Section IV concludes the work.

II. ANALYSIS OF EDP IMPROVEMENT

The VNWFET technology under study [3, 4] is composed of a homogenous highly doped nanowire channel, patterned into boron doped ($2 \times 10^{19} \text{cm}^{-3}$) Si-substrate, working as a junction-less device [5]. The current between the silicided source/drain contacts is controlled by a gate-all-around structure having a physical channel length of 14nm (Fig. 1 (a)). Leveraging vertical integration of stacked VNWFETs, much higher compactness and flexibility of vertical device dimensions can be achieved (Fig. 1 (b)) and a significant area gain compared to a 7nm FinFET technology can be expected.

Energy-Delay Product (EDP) is a useful metric to compare the speed of energy-efficient circuits. While the Power-Delay Product (PDP), or the switching energy, measures the energy per function, it does not satisfactorily capture the speed. In a classical CMOS circuit, considering the PDP for a 0-to-1-to-0 computation cycle, the EDP can be written as

$$EDP = PDP * t_P = C_L V_{DD}^2 * t_P = \frac{C_L^2 V_{DD}^3}{I_{sat}} \qquad (1)$$

978-1-6654-3746-2/21 $31.00 © 2021 IEEE

where t_p represents the propagation time, C_L represents the load capacitance on the gate output, V_{DD} represents the supply voltage and I_{sat} represents the saturation current of the transistor through which the current is drawn from the voltage supply or sunk to the ground to change the output voltage state (we assume that the transistor through which the current is flowing is primarily in the saturation region).

The ratio between the EDP of the FinFET technology and the VNWFET technology can therefore be expressed as:

$$G_{Evf} = \frac{EDP_f}{EDP_v} = \frac{C_{Lf}^2 V_{DDf}^3}{I_{satf}} \frac{I_{satv}}{C_{Lv}^2 V_{DDv}^3} \approx \frac{C_{Lf}^2}{C_{Lv}^2} \frac{I_{satv}}{I_{satf}} \qquad (2)$$

where the subscripts f and v stand for the FinFET and VNWFET, respectively. We assume that supply voltage values for both technologies are identical (i.e. $V_{DDf} = V_{DDv}$). Hence, for a given value n of G_Evf, the EDP of the VNWFET technology is n times lower than that of the FinFET technology.

A. On-current (I_{sat}) considerations

We also assume that material current density and carrier mobility are close enough for both technologies such that their influence can be ignored in this comparison. This then implies that the transistor saturation current can be approximated as

$$I_{sat} = \kappa \frac{W_g}{L_g} \qquad (3)$$

where κ is a constant incorporating technology parameters and operating conditions, W_g represents the effective width of the transistor channel under the gate orthogonal to current flow, and L_g represents the length of the channel under the gate, i.e. the distance between source and drain. This also assumes that the transistor channel occupies all the space available in the given geometry – this implies that the channel is fully depleted in the FinFET and the radius of the nanowire is sufficiently small in the VNWFET to allow junctionless transport.

We can calculate W_g according to the geometry of both FinFET and VNWFET devices:

$$\left.\begin{array}{l} W_{gf} = 2(h_f + w_f) \\ W_{gv} = 2\pi r_v \end{array}\right\} \qquad (4)$$

where W_{gf} and W_{gv} represent the effective widths of the FinFET and VNWFET channels, respectively; h_f and w_f represent the height and width of a fin in the FinFET device; and r_v represents the radius of the VNWFET nanowire channel. The geometries and parameters of both devices are shown in Figs. 2 (FinFET) and 3 (VNWFET). Thus, the ratio between the saturation currents of both devices can be written as:

$$\frac{I_{satv}}{I_{satf}} = \frac{W_{gv}}{L_{gv}} \frac{L_{gf}}{W_{gf}} = \frac{\pi r_v}{(h_f + w_f)} \frac{L_{gf}}{L_{gv}} \qquad (5)$$

In nanowire-based devices, the ratio of channel length L_{gv} to channel diameter $2r_v$ should be kept constant at 2:1 to preserve desirable behavior in the off state [6]. It is also important to avoid degradation of both ballistic and dissipative currents, which occurs with decreasing device size. In fact, ballisticity (i.e. the ratio of dissipative to ballistic current) also degrades for very small devices, with channel lengths below around 10nm. Both constraints can be combined resulting in:

$$L_{gv} \geq 10nm, r_v = \frac{L_{gv}}{4} \text{ subject to } r_v \geq 2.5nm \qquad (6)$$

B. Load capacitance (C_L) considerations

In terms of capacitance, we will, in a first approach, consider that load capacitance is composed of the input gate capacitance C_{in} of F (fanout) logic gates on the output node, as well as interconnect capacitance C_w. Hence,

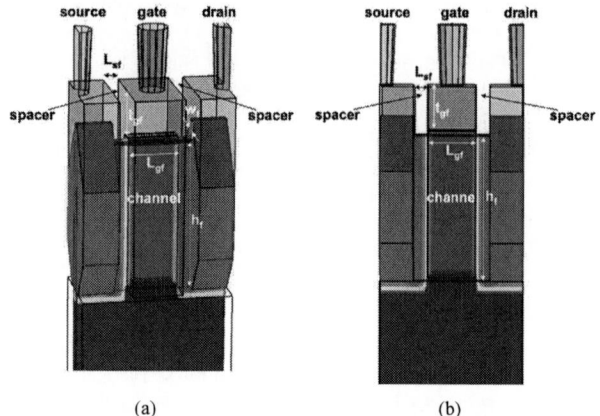

Fig. 2: FinFET geometry (a) 3D view (b) lateral view perpendicular to the gate electrode; color scheme: red (resp. blue) indicates a high (resp. low) electron density for an n-channel device (vice versa for p-channel devices).

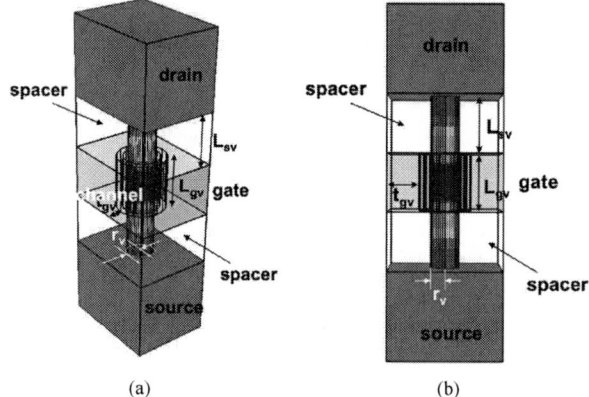

Fig 3: VNWFET geometry (a) 3D view (b) lateral view; color scheme: red (resp. blue) indicates a high (resp. low) electron density for an n-FET (vice versa for p-FETs).

$$C_L = F(C_{in} + C_w) = F(C_{gc} + C_{gs} + C_w) \qquad (7)$$

where the input gate capacitance is composed of the direct gate-channel capacitance C_{gc} linked to the gate-channel area, and the gate-source capacitance C_{gs} linked to the spacer geometry.

1) Gate-channel capacitance

The gate-channel capacitance can be expressed in terms of channel geometry for both devices as:

$$\left.\begin{array}{l} C_{gcf} = \frac{\varepsilon_{ox} L_{gf} 2(h_f + w_f)}{EOT} \\ C_{gcv} = \frac{\varepsilon_{ox} L_{gv} 2\pi r_v}{EOT} \end{array}\right\} \qquad (8)$$

Here, C_{gcf} and C_{gcv} are the gate-channel capacitance for FinFET and VNWFET devices, respectively; ε_{ox} and EOT are the dielectric permittivity for standard SiO$_2$ and the equivalent oxide thickness of the gate dielectric material (i.e. converted to the thickness it would have using SiO$_2$ as the gate dielectric), respectively. Assuming that the dielectric material is identical for both technologies, EOT writes:

$$EOT = t_{high-K} \frac{\varepsilon_{ox}}{\varepsilon_{high-k}} \tag{9}$$

where ε_{high-k} is the permittivity of the hi-k gate dielectric. A typical value for EOT is around 0.89nm in current technologies where HfO$_2$ gate dielectrics are used.

2) Gate-source capacitance

The gate-source capacitance can be expressed in terms of spacer geometry for both devices as:

$$\left.\begin{array}{l} C_{gsf} = \dfrac{\varepsilon_s\, 2t_{gf}(2t_{gf}+h_f+w_f)}{L_{sf}} \\[2mm] C_{gsv} = \dfrac{\varepsilon_s\left(4(t_{gv}+r_v)^2-\pi r_V^2\right)}{L_{sv}} \end{array}\right\} \tag{10}$$

where C_{gsf} and C_{gsv} represent the gate-source capacitance for FinFET and VNWFET devices, respectively; t_{gf} and t_{gv} represent the gate material thickness for FinFET and VNWFET devices, respectively and can be typically assumed to be between 10nm-20nm; L_{sf} and L_{sv} represent the spacer length (between gate and source) for FinFET and VNWFET devices, respectively, and can be typically assumed to be around 8nm; ε_s represents the spacer material dielectric permittivity. We assume that the spacer material (typically Si$_3$N$_4$) is identical for both technologies and that there are no fabrication issues for different dielectric materials for the gate (high-k) and for the spacer. We also assume that the gate material surrounds the channel with uniform thickness (overlap) equal to t_{gf} for the FinFET (although the lateral thickness is defined by the FinFET pitch), and is a square centered around the nanowire with minimum overlap equal to t_{gv} for the VNWFET. Note that this expression does not consider fringing capacitances, which is a reasonable assumption since this is not the dominant component. Hence one can write:

$$\left.\begin{array}{l} C_{inf} = \dfrac{\varepsilon_{ox}L_{gf}2(h_f+w_f)}{EOT} + \dfrac{\varepsilon_s\, 2t_{gf}(2t_{gf}+h_f+w_f)}{L_{sf}} \\[2mm] C_{inv} = \dfrac{\varepsilon_{ox}L_{gv}2\pi r_v}{EOT} + \dfrac{\varepsilon_s\left(4(t_{gv}+r_v)^2-\pi r_v^2\right)}{L_{sv}} \end{array}\right\} \tag{11}$$

3) Wire capacitance

Wire capacitance is considered for local (gate-to-gate) interconnect, and is expressed as:

$$C_w = \frac{\varepsilon_{ox}w_m L_{gg}}{t_{ox}} \tag{12}$$

where w_m and L_{gg} represent the width and length of local (gate-to-gate) interconnect respectively; and t_{ox} represents the metal-substrate oxide thickness for interlayer dielectric SiO$_2$. L_{gg} can be directly linked to circuit compactness since it represents the lateral distance (pitch) between two gates. For a given improvement in compactness A_c between VNWFET and FinFET (i.e. $A_c = A_f/A_v$),

$$\frac{L_{ggf}}{L_{ggv}} = \sqrt{A_c} \tag{13}$$

Assuming identical values for ε_{ox}, t_{ox} and w_m between the FinFET and VNWFET technologies, we can also write:

$$\frac{C_{wf}}{C_{wv}} = \sqrt{A_c} \tag{14}$$

4) Overall expressions for the load capacitances

Leveraging (14) one can write the expression for the load capacitances in both cases as

$$\left.\begin{array}{l} C_{Lf} = F\left[C_{gcf} + C_{gsf} + C_{wf}\right] \\[2mm] C_{Lv} = F\left[C_{gcv} + C_{gsv} + \dfrac{C_{wf}}{\sqrt{A_c}}\right] \end{array}\right\} \tag{15}$$

where C_{Lf} and C_{Lv} represent the load capacitances on FinFET and VNWFET logic gate outputs, respectively.

Thus, the ratio between the load capacitances of both devices can be written as:

$$\frac{C_{Lf}^2}{C_{Lv}^2} = \frac{\left[C_{gcf}+C_{gsf}+C_{wf}\right]^2}{\left[C_{gcv}+C_{gsv}+\dfrac{C_{wf}}{\sqrt{A_c}}\right]^2} \tag{16}$$

And finally, the ratio between the EDPs can be expressed as,

$$G_{Evf} = \frac{C_{Lf}^2}{C_{Lv}^2}\frac{I_{Satv}}{I_{satf}} = \frac{\left[C_{gcf}+C_{gsf}+C_{wf}\right]^2}{\left[C_{gcv}+C_{gsv}+\dfrac{C_{wf}}{\sqrt{A_c}}\right]^2}\frac{I_{Satv}}{I_{satf}} \tag{17}$$

Which can be further re-written in terms of geometric parameters as,

$$G_{Evf} = \frac{\left[\dfrac{\varepsilon_{ox}L_{gf}2(h_f+w_f)}{EOT} + \dfrac{\varepsilon_s\, 2t_{gf}(2t_{gf}+h_f+w_f)}{L_{sf}} + C_{wf}\right]^2}{\left[\dfrac{\varepsilon_{ox}L_{gv}2\pi r_v}{EOT} + \dfrac{\varepsilon_s\left(4(t_{gv}+r_v)^2-\pi r_V^2\right)}{L_{sv}} + \dfrac{C_{wf}}{\sqrt{A_c}}\right]^2}\frac{\pi r_v}{(h_f+w_f)}\frac{L_{gf}}{L_{gv}} \tag{18}$$

5) Impact on 3D design

For this analysis, we assume that A_c is a constant and of the order of 5 according to [7], although it is anticipated that it will vary (negatively) with an increasing number of fins per FinFET / nanowires per VNWFET. This hypothesis is explored quantitatively in the next section and will require further analysis once an automated 3D place and route tool is available. This tends also to support the view that

a) VNWFET-based design performance improves for lower numbers of nanowires per VNWFET – not only for the pitch overhead but also because the on-current varies sublinearly with the nanowire number per VNWFET.

b) Computing should be kept local to enable short interconnects and limit energy consumption in parasitic elements (particularly relevant for mobile low power applications).

The proposed N^{2}C^2 concept (fig. 4) enabled by the VNWFET technology is exactly this – a regular 3D matrix of individual computing functions where intra-cube interconnect is short due to both the limited complexity of the N^{2}C^2 circuit and the limited number of targeted nanowires per VNWFET. While the benefit of local computing is known to limit the impact of interconnect delays, N^{2}C^2 goes beyond the current planar state-of-the-art by extending this principle to 3 dimensions. Fig. 4 illustrates the N^{2}C^2 concept as a scalable and flexible assembly of high-expressivity logic cells based on VNWFETs. By focusing on commonly used functions in deep learning and convolutional neural network architectures such as Multiply-Accumulate, combined with logic design styles suited to multiple transistors in series such as Pass-Transistor logic, we target compact and low-EDP neural network compute cube hardware capable of interfacing seamlessly with other, identical cubes through a versatile 3D interconnect framework. This will facilitate the development of physically regular and logically flexible 3D matrices for AI accelerators, including levers to explore hardware/software co-design and approximation techniques to further improve energy-efficiency both at design time and at run-time.

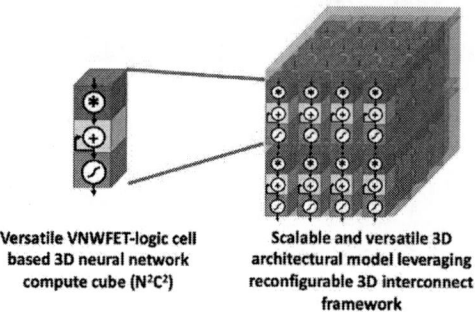

Versatile VNWFET-logic cell based 3D neural network compute cube (N²C²)

Scalable and versatile 3D architectural model leveraging reconfigurable 3D interconnect framework

Fig 4: N²C² concept

III. EDP GAIN ANALYSIS

For a FinFET aspect ratio (h_f/w_f) value of 60nm/7nm with 20nm physical FinFET gate length L_{gf}, and by varying the value of physical VNWFET gate length L_{gv} between 10nm-20nm for varying nanowire radius r_v, gate material thickness t_{gv}, spacer length L_{sv} and compactness A_c, EDP- and PDP-gain values have been calculated, as shown in Fig. 5.

Among the chosen radii, the lowest permissible value is 2.5nm (according to (6)) and the highest is 25nm which corresponds to the largest diameter available for the VNWFET technology. For low r_v, the on-current ratio I_{satv}/I_{satf} is lower than 1, but is offset significantly by the $(C_{lf}/C_{lv})^2$ ratio; while for high r_v, the on-current ratio approaches 2,5 while the $(C_{lf}/C_{lv})^2$ ratio tends towards unity. The results show that 10x gain in EDP between VNWFET and 7nm FinFET technology can be achieved for L_{gv}=14nm and a nanowire radius of 3.5nm. If the smallest fabricated device dimensions are considered, i.e. L_{gv}=14nm and r_v=11nm, an EDP gain of 4.3 has been predicted. Considering the worst-case scenario, for the largest already-fabricated devices (r_v=25nm and L_{gv}=14 nm [4]), a 2x EDP gain can still be observed. This analysis demonstrates that the VNWFET can be designed to be more energy-efficient compared to a 7nm FinFET.

IV. CONCLUSIONS

It is well known that transistor energy efficiency improves as gate length decreases. This is especially true for GAA NWFETs, among which the disruptive vertical device implementations allow the transition from 2D to truly 3D architectures. In this work, we developed a comparative analysis of logic energy-efficiency through first-order equations for EDP of VNWFET and FinFET technologies. Leveraging a set of realistic geometric and material parameter values, our work shows that a 10x gain in EDP over a baseline 7nm FinFET technology can be achieved for a physical VNWFET gate length of 14nm and even for actual fabricated devices, 4.3x gain in EDP has been predicted. These results pave the way towards 3D neural network compute cube required for dense and energy efficient non Von Neumann computing.

ACKNOWLEDGMENT

This work was supported by the LEGO project (Grant ANR-18-CE24-0005-01) and by the project FVLLMONTI funded by European Union's Horizon 2020 research and innovation program under grant agreement N°101016776.

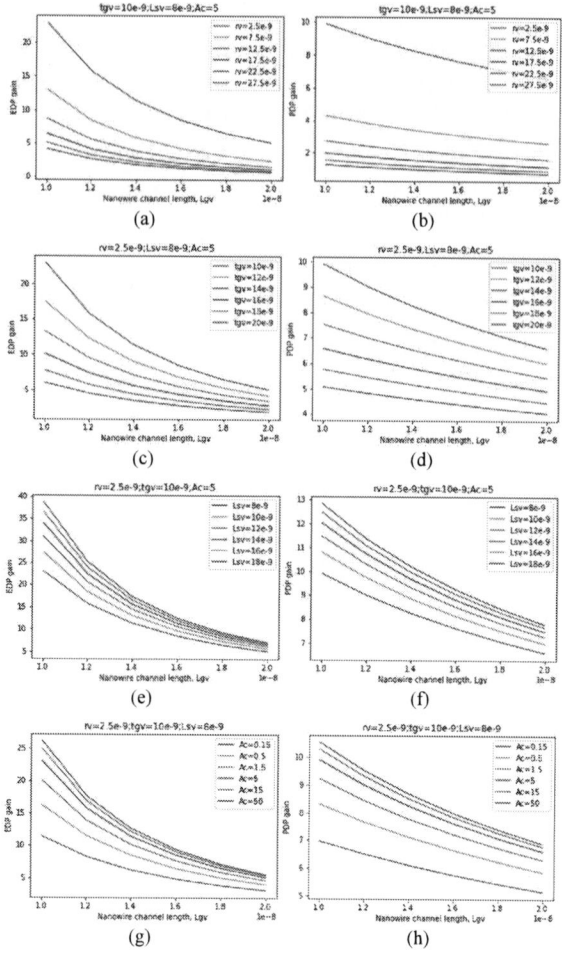

Fig. 5: EDP- and PDP-gain versus VNWFET gate length: a),b) for varying nanowire radius r_v; c),d) for varying gate material thickness t_{gv}; e),f) for varying spacer length L_{sv}; g),h) for varying compactness A_c.

REFERENCES

[1] G. Larrieu and X. L. Han " Vertical nanowire array-based field effect transistors for ultimate scaling", *Nanoscale*, vol. 5, pp. 2437-2441, 2013. DOI: 10.1039/C3NR33738C

[2] A. Veloso et al., "Vertical nanowire FET integration and device aspects", ECS Transactions, vol. 72 (4), pp. 31-42, 2016. DOI: 10.1149/07204.0031ecst

[3] Y. Guerfi and G. Larrieu "Vertical Silicon Nanowire Field Effect Transistors with Nanoscale Gate-All-Around", *Nanoscale Research Letters*, vol. 11, pp. 210, 2016. DOI: 10.1186/s11671-016-1396-7

[4] G. Larrieu, Y. Guerfi, X. L. Han and N. Clément "Sub-15 nm gate-all-around field effect transistors on vertical silicon nanowires", *Solid-State Electronics*, vol. 130, pp. 9-14, 2017. DOI: 10.1016/j.sse.2016.12.008

[5] J. P. Colinge *et al.* "Nanowire transistors without junctions", *Nat. Nanotechnol.* vol. 5, pp. 225-229, 2010. DOI: 10.1038/nnano.2010.15

[6] Z. Stanojevic, O. Baumgartner, M. Karner, F. Mitterbauer, H. Demel, and C. Kernstock, "Simulation Study on the Feasibility of Si as Material for Ultra-Scaled Nanowire Field-Effect Transistors," *Proc. EUROSOI-ULIS*, Vienna, Austria, pp. 25-27, 2016, DOI: 10.1109/ULIS.2016.7440074.

[7] C. Mukherjee, M. Deng, F. Marc, C. Maneux, A. Poittevin, I. O'Connor, S. Le Beux, A. Kumar, A. Lecestre, G. Larrieu, "3D logic cells design and results based on Vertical NWFET technology including tied compact model", *IFIP/IEEE International Conference on Very Large Scale Integration (VLSI-SoC)*, 5-7 October 2020, Salt Lake City (UT), USA. DOI: arXiv:2005.14039v1.

Comprehensive Kubo-Greenwood modelling of FDSOI MOS devices down to deep cryogenic temperatures

F. Serra di Santa Maria[1], L. Contamin[2], M. Cassé[2], C. Theodorou[1], F. Balestra[1], G. Ghibaudo[1]

[1] IMEP-LAHC, Univ. Grenoble Alpes, Minatec, 38016 Grenoble, France,

[2] CEA-LETI, Univ. Grenoble Alpes, Minatec, 38054 Grenoble, France.

Email: francesco.serra-di-santa-maria@grenoble-inp.fr, gerard.ghibaudo@grenoble-inp.fr.

Abstract— **A comprehensive Kubo-Greenwood modelling of FDSOI MOS devices is carried out down to deep cryogenic temperatures. It is found that a single set of mobility parameters is only needed to fit the device characteristics versus temperature for long channel devices. Instead, in short channels, the neutral scattering mobility component μ_N is found to decrease at small gate length due to the increased presence of neutral defects close to source/drain ends whatever the temperature.**

Keywords—Kubo-Greenwood, mobility, modeling, MOSFET, FDSOI, cryogenic temperature.

I. Introduction

The Cryogenic electronics is still a key research topic as allowing circuit performance improvements in terms of operation speed, turn-on behavior, thermal noise reduction, punch-through current decrease etc. [1-5]. It finds application in high speed computing, sensing and detection, space electronics and recently in readout CMOS electronics for quantum computing [6,7]. In this context, the characterization and modelling of MOSFETs down to cryogenic temperatures is still a key issue. Besides, Kubo-Greenwood formalism has proven a powerful approach for the modelling of transport in MOS inversion layers, enabling detailed mobility calculations [8-10] and recently subthreshold slope calculations [11].

In this work, we propose, for the first time, to apply the Kubo-Greenwood approach for the drain current modelling as a function of gate voltage in 28nm FDSOI MOSFETs down to deep cryogenic temperatures. We first validate the single 2D subband approximation for the description of the MOS inversion charge with gate voltage down to liquid helium temperatures. Then, we show that the drain current transfer characteristics can be modeled within the Kubo-Greenwood formalism down to very low temperatures, using calibrated scattering limited mobility laws (Phonon, Coulomb, Neutral, Surface roughness).

II. Experiments Details

The measurements were performed on 28nm FDSOI MOSFETs with silicon film thickness t_{si}=7nm and buried oxide (BOX) thickness t_{box}=25nm from STMicroelectronics. NMOS transistors were processed from (100) handle substrate, with <100>-oriented channel, and a high-k/metal gate Gate-First architecture [12]. Low-V_{th} transistors were available with un-doped channel through a doped back plane (NWELL doping $N_A=10^{18}cm^{-3}$) below the BOX. Thin gate oxide (with equivalent oxide thickness EOT=1.1nm) devices with gate length L varying from 30nm up to 10µm and with gate width W=1µm or 10µm were tested using a cryogenic probe station down to 4.2K.

The gate-to-channel capacitance $C_{gc}(V_g)$ was measured with an HP 4284 LCR meter at 1MHz frequency and 10mV AC level using the standard split C-V technique. The drain current $I_d(V_g)$ MOSFET transfer characteristics were recorded in linear region (V_d=30-50mV) with an HP4156 parameter analyzer. All the measurements were made at zero back bias.

III. Kubo-Greenwood Transport and FDSOI MOSFET Modelling

The main equations used for the Kubo-Greenwood transport and FDSOI MOSFET modelling are recalled in the following sub-sections.

A. Kubo-Greenwood transport modelling

The inversion layer density n within a single subband approximation is related to the Fermi level E_f by [8,11],

$$n = kT.A_{2d}.ln\left(1 + e^{\frac{E_f}{kT}}\right) \qquad (1)$$

where A_{2D}=g.m_d*/(π.h) is the 2D density of states with g the subband degeneracy factor, m_d* the DOS effective mass and $\hbar$ the reduced Planck constant and kT is the thermal energy, T being the temperature. Note that E_f is referred to the subband edge E_c, stated here to zero.

The inversion layer sheet conductivity σ, which is a function of Fermi level and temperature, can be computed by integration over energy E of the so-called energy conductivity function $\sigma_E(E)$ as [8,11],

$$\sigma(E_f,T) = \int_0^{+\infty} \sigma_E(E)\left(-\frac{\partial f}{\partial E}\right)dE \qquad (2)$$

where $f = 1/\left(1 + e^{\frac{E-E_f}{kT}}\right)$ is the Fermi function and,

$$\sigma_E(E) = q.E.A_{2d}.\mu(E) \qquad (3)$$

with E being the carrier kinetic energy and $\mu(E)$ the energy mobility function.

When several scattering take place in the electronic transport, the energy mobility function $\mu(E)$ can be evaluated by the Matthiessen rule as,

$$\mu(E) = \left(\frac{1}{\mu_{ph}} + \frac{1}{\mu_N} + \frac{1}{\mu_C} + \frac{1}{\mu_{SR}}\right)^{-1} \qquad (4)$$

which accounts for the phonon (μ_{ph}), neutral (μ_N), Coulomb (μ_C) and surface roughness (μ_{SR}) limited mobility components, respectively.

The phonon limited mobility (in cm²/Vs) in an inversion layer has been expressed by [13],

978-1-6654-3746-2/21 $31.00 © 2021 IEEE

$$\mu_{ph}(T,F) = 1180.\left[\left(\frac{T}{300}\right)^{2.11} + \left(\frac{T}{300}\right)^{1.7}.\left(\frac{F}{F_0}\right)^{\alpha(T)}\right]^{-1} \quad (5)$$

with F being the effective electric field (defined below), $\alpha(T) = 0.2.\left(\frac{300}{T}\right)^{0.1}$ and $F_0 = 7 \times 10^4 \left(\frac{V}{cm}\right)$ a critical field.

The Coulomb scattering limited mobility (in cm^2/Vs) is proportional to the carrier kinetic energy E and takes the form [8],

$$\mu_C(E) = 1650.\left(\frac{E}{E_{coul}}\right) \quad (6)$$

with here E_{coul}=0.032eV.

The surface roughness limited mobility (cm^2/Vs) is proportional to the square of the reciprocal effective electric field and is well approximated by [14]:

$$\mu_{SR}(T,F) = \frac{8.8 \times 10^{14}}{F^2}.exp\left[-\left(\frac{T}{850}\right)^2\right]. \quad (7)$$

Finally, the neutral scattering limited mobility is constant with energy and temperature and reads [15],

$$\mu_N = Constant. \quad (8)$$

where the constant is proportional to the reciprocal neutral defect number.

B. FDSOI MOSFET modelling

For the modelling of an FDSOI MOSFET device, we consider that the 2D inversion layer is located at the front oxide/silicon channel interface, such that the gate charge conservation equation yields,

$$V_g = V_{fb} + V_s + \frac{Q_i}{C_{ox}} + \frac{C_{it}.(V_s - V_0)}{C_{ox}} + \frac{C_b.(V_s - V_b)}{C_{ox}} \quad (9)$$

where $Q_i(E_f,T)$=q.n(E_f,T) is the inversion charge, V_s is the front surface potential, V_{fb} is the flat band voltage, V_g is the front gate voltage, V_b is the back bias, C_{ox} the front oxide capacitance, C_{it} (=q.N_{it}) is the front interface trap capacitance, N_{it} being the interface trap density, and, $C_b = (C_{si}C_{box})/(C_{si} + C_{box})$ is the substrate coupling capacitance, C_{si} being the silicon capacitance and C_{box} the BOX oxide capacitance.

The surface potential $V_s(V_g,V_b)$ can be obtained by solving Eq. (9) for given front gate voltage and back bias. Thus, the Fermi level E_f can be calculated for any bias as:

$$E_f(V_g, V_b) = q.\left[V_s(V_g, V_b) - V_0\right] \quad (10)$$

where V_0 is a reference potential depending on channel doping level.

The effective electric field F entering the phonon and surface roughness limited mobility components of Eqs (5) and (7) is evaluated by the usual expression accounting for the inversion charge and back electric field as,

$$F(V_g, V_b) = \frac{\frac{Q_i(V_g, V_b, T)}{2} + C_b.\left[V_s(V_g, V_b) - V_b\right]}{\varepsilon_{si}} \quad (11)$$

where ε_{si} is the silicon permittivity.

The drain current is then calculated in linear operation region, i.e. for small drain voltage V_d, as,

$$I_d(V_g, V_d, V_b, T, F) = \frac{W}{L}.\sigma\left[E_f(V_g, V_b), T, F\right].V_d. \quad (12)$$

In short channel devices, the source/drain series resistance R_{sd} effect is accounted for through Ohm's law as,

$$I_d(V_g, V_d, V_b, T, F) = \frac{\frac{W}{L}.\sigma[E_f(V_g,V_b),T,F].V_d}{1+R_{sd}.\frac{W}{L}.\sigma[E_f(V_g,V_b),T,F]}. \quad (13)$$

IV. RESULTS AND DISCUSSION

A. Long channel devices

$C_{gc}(V_g)$ characteristics were measured on large area MOSFETs with W=L=10µm. The inversion charge was obtained after integration of the $C_{cg}(V_g)$ curves starting from V_g=0 as is usual in split C-V technique. Experimental and modeled $C_{gc}(V_g)$ and $Q_i(V_g)$ characteristics are shown in Fig. 1 for various temperatures. They clearly reveal the adequacy of the single subband approximation of Eq. (1) and of the MOSFET approach of Eq. (9) for the modelling of the gate charge control in FDSOI MOS transistors down to liquid helium temperature. The model parameters are indicated in the caption of Fig. 1.

In Fig. 2 are reported the experimental and Kubo-Greenwood modeled $I_d(V_g)$, $g_m(V_g)$ and Y-function $Y(V_g) = I_d/\sqrt{g_m}$ [16] characteristics for such a long channel device and for various temperatures, showing very good agreement. The best Kubo-Greenwood model fits have been achieved after proper calibration of the phonon, Coulomb and Surface roughness mobility law parameters, whose values are those displayed in Eqs (5), (6) and (7). It should be mentioned that these values are close to the original ones in [8,13,14]. The value of μ_N, indicated in Fig. 1 caption, has been set to a large value in the case of long channel device.

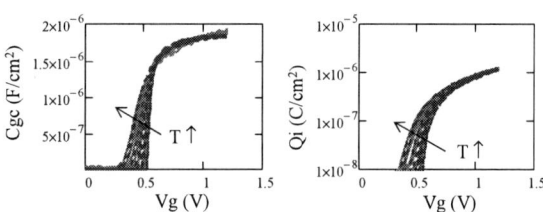

Fig. 1. Experimental (red solid lines) and modeled (blue dashed lines) $C_{gc}(V_g)$ and $Q_i(V_g)$ characteristics for various temperatures T(K)=4.2, 10, 20, 50, 100, 150, 200, 250 and 300 (W=L=10µm, model parameters: C_{ox}=2.1 µF/cm², C_{box}=0.14 µF/cm², C_{si}=1.52 µF/cm², C_{it}=0.16 µF/cm2, V_0= 0.5 V).

The variations of the effective mobility, μ_{eff}=σ/(qn), with inversion layer density deduced from the Kubo-Greenwood modeling are shown in Fig. 3a for various temperatures. They reveal the onset of a bell-shaped mobility behavior at very low temperatures as is usual [2,8], due to the dominance of Coulomb and surface roughness scattering processes. Figure 3b displays the change in the temperature dependence of μ_{eff} for various carrier densities taken at weak, intermediate and strong inversion.

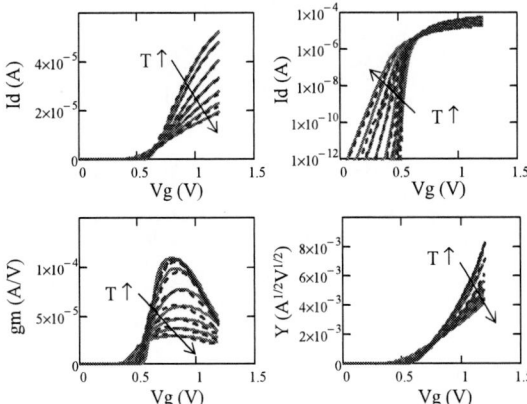

Fig. 2. Experimental (red solid lines) and modeled (blue dashed lines) $I_d(V_g)$, $g_m(V_g)$ and $Y(V_g)$ characteristics for various temperatures T(K)=4.2, 10, 20, 50, 100, 150, 200, 250 and 300 (μ_N=3000cm²/Vs, V_d=50mV, W=L=10µm, N_{it}=2-8×10¹²/eVcm²).

At low carrier density, μ_{eff} increases with temperature due to Coulomb scattering predominance before to decrease at higher temperature due to enhanced phonon scattering. At high carrier density, μ_{eff} always decreases with temperature due to phonon diffusion supremacy. These features are also illustrated in Fig. 4, where the variations of μ_{eff} with carrier density are compared to the various mobility law components. As expected, at very low temperature, Coulomb scattering is dominating, whereas, at high temperature, phonon scattering is prevailing.

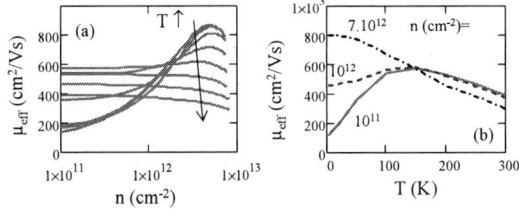

Fig. 3. a) Variations of μ_{eff} with 2D carrier density n for various temperatures T(K)=4.2, 10, 20, 50, 100, 150, 200, 250 and 300 and b) with temperature T for various 2D carrier densities as obtained from Kubo-Greenwood modeling (W=L=10µm).

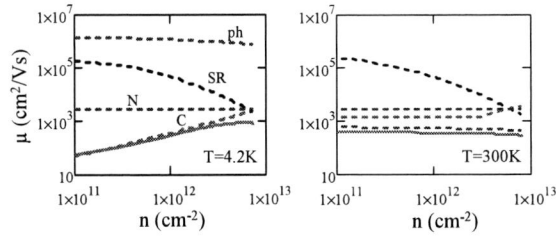

Fig. 4. Variations of μ_{eff} (red solid line) and of various scattering mobility component with 2D carrier density n for T=4.2K and T=300K as obtained from Kubo-Greenwood modeling (W=L=10µm).

B. Short channel devices

In Figs 5 and 6 are displayed the experimental and Kubo-Greenwood-modeled $I_d(V_g)$, $g_m(V_g)$ and $Y(V_g)$ characteristics for MOS devices with gate length ranging from 30nm up to 1µm and for various temperature, showing again very good agreement. In this case, the best Kubo-Greenwood model fits have been obtained by keeping the same mobility parameters as for long devices (Fig. 2), at the exception of the neutral mobility component μ_N, and of the reference potential V_0, which were adjusted versus gate length for each temperature as shown in Fig. 7. It is found that V_0 follows the same trend vs L as the threshold voltage exhibiting a slight roll-off due to short channel effect (not shown). However, as can be seen from Fig. 7b, μ_N is strongly degraded as the channel length is reduced, regardless of temperature, due to the increased influence of neutral defects close to source and drain in agreement with the data of [17] (square symbols in Fig. 7b) and with the theoretical analysis of [18].

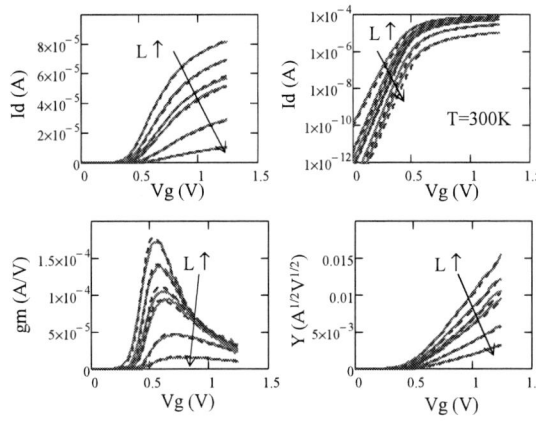

Fig. 5. Experimental (red solid lines) and modeled (blue dashed lines) $I_d(V_g)$, $g_m(V_g)$ and $Y(V_g)$ characteristics for various gate lengths L(nm)=30, 60, 90, 120, 300 and 1000 and T=300K (V_d=30mV, W=1µm).

It should also be noted that, when fitting the transfer characteristics for short channel devices, the source/drain series resistance (R_{sd}) effects were taken care using the Y-function, which is independent of R_{sd} [16]. To this end, the mobility parameter μ_N was first tuned to adjust the experimental $Y(V_g)$ curves. Then, R_{sd} was adjusted using Eq. (13) to fit the experimental $I_d(V_g)$ and $g_m(V_g)$ characteristics. Typical values for R_{sd} were found in the range 230-250Ω.µm for temperatures varying from 25K up to 300K.

978-1-6654-3746-2/21 $31.00 © 2021 IEEE

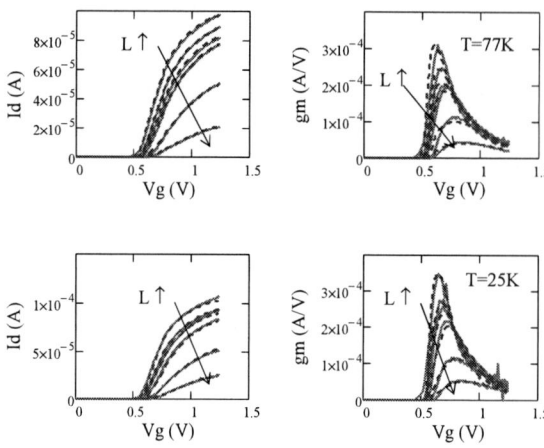

Fig. 6. Experimental (red solid lines) and modeled (blue dashed lines) I_dV_g) and $g_m(V_g)$ characteristics for various gate lengths L(nm)=30, 60, 90, 120, 300 and 1000 and for T=77K and T=25K (V_d=30mV, W=1μm).

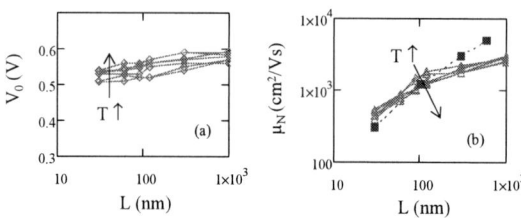

Fig. 7. Variations of (a) parameter V_0 and (b) neutral mobility component μ_N with gate length L for various temperatures T(K)= 25, 77, 100, 150, 200, 250 and 300 as obtained from Kubo-Greenwood modeling (W=1μm). The square symbols are results taken from [17].

V. CONCLUSIONS

A comprehensive Kubo-Greenwood modelling of FDSOI MOS devices has been performed down to deep cryogenic temperatures. Interestingly, a single set of mobility parameters was only needed to fit the data versus temperature for long channel devices. Instead, in short channel MOSFETs, the neutral scattering mobility component μ_N was found to be nearly temperature independent and to significantly decrease at small gate length due to the enhanced influence of neutral defects close to source/drain ends. Therefore, such a Kubo-Greenwood modelling can provide a physical insight into the scattering processes limiting the transport in FDSOI MOSFETs down to very low temperatures and could serve as a good basis for developing compact models.

VI. ACKNOWLEDGMENT

The authors are grateful to EU H2020 RIA project SEQUENCE (Grant No. 871764) and to ERC Synergy QuCube (Grant No. 810504) for financial support.

VII. REFERENCES

[1] E. A. Gutiérrez, J. Deen, and C. Claeys, Low temperature electronics: physics, devices, circuits, and applications. Academic Press, 2000.

[2] F. Balestra and G. Ghibaudo, Device and Circuit Cryogenic Operation for Low Temperature Electronics. 2001.

[3] T. Wada, H. Nagata, H. Ikeda, Y. Arai, M. Ohno and K. Nagase, Development of Low Power Cryogenic Readout Integrated Circuits Using Fully-Depleted-Silicon-on-Insulator CMOS Technology for Far-Infrared Image Sensors, J. Low Temp. Phys., 167, 602 (2012).

[4] R. M. Incandela, L. Song, H. Homulle, E. Charbon, A. Vladimirescu, F. Sebastiano, Characterization and Compact Modeling of Nanometer CMOS Transistors at Deep-Cryogenic Temperatures, IEEE Journal of the Electron Devices Society, 6, 996 (2018).

[5] A. Beckers, F. Jazaeri, C. Enz, Characterization and Modeling of 28-nm Bulk CMOS Technology Down to 4.2 K, IEEE J. Electron Devices Soc., 6, 1007 (2018).

[6] J.M. Hornibrook, J. I. Colless, I. D. Conway Lamb, S. J. Pauka, H. Lu, A. C. Gossard, J. D. Watson, G. C. Gardner, S. Fallahi, M. J. Manfra, and D. J. Reilly, Cryogenic control architecture for large-scale quantum computing, Phys. Rev. Applied, 3, 024010 (2015).

[7] R. Maurand, X. Jehl, D. Kotekar-Patil, A. Corna, H. Bohuslavskyi, R. Laviéville, L. Hutin, S. Barraud, M. Vinet, M. Sanquer and S. De Franceschi, A CMOS silicon spin qubit, Nature Commun., 7, 13575 (2016).

[8] G. Ghibaudo, Transport in the inversion layer of a MOS transistor. Use of Kubo-Greenwood formalism, Journal Phys. C: Solid State Physics, 19, 767 (1985).

[9] J. Dura, F. Triozon, S. Barraud, D. Munteanu, S. Martinie, and J. L. Autran, Kubo-Greenwood approach for the calculation of mobility in gate-all-around nanowire metal-oxide-semiconductor field-effect transistors including screened remote Coulomb scattering—Comparison with experiment, Journal of App. Physics, 111, 103710 (2012).

[10] O. Bonno, S. Barraud, D. Mariolle, and F. Andrieu, Effect of strain on the electron effective mobility in biaxially strained silicon inversion layers: An experimental and theoretical analysis via atomic force microscopy measurements and Kubo-Greenwood mobility calculations, Journal of App. Physics, 103, 063715 (2008).

[11] G. Ghibaudo, M. Aouad, M. Casse, S. Martinie, F. Balestra, On the modelling of temperature dependence of subthreshold swing in MOSFETs down to cryogenic temperature, Solid-State Electronics, 170, 107820 (2020).

[12] N. Planes et al., 28nm FDSOI technology platform for high-speed low-voltage digital applications, in Symposium on VLSI Technology, Digest of Technical Papers, 33, pp. 133–134 (2012).

[13] F. Gámiz and J. A. López-Villanueva, A comparison of models for phonon scattering in silicon inversion layers, Journal of Appl. Phys., 77, 4128, 1995.

[14] S. Villa, A.L. Lacaita, L.M. Perron, R. Bez, A physically-based model of the effective mobility in heavily-doped n-MOSFETs, IEEE Trans Electron Devices, 45, 110 (1998).

[15] C. Erginsoy, Neutral impurity scattering in semiconductors, Phys. Rev., 79, 1013 (1950).

[16] G. Ghibaudo, A new method for the extraction of MOSFET parameters, Electronics Letters, 24, 543 (1988).

[17] M. Shin, M. Mouis, A. Cros, E. Josse, G.-T. Kim and G. Ghibaudo, Low temperature characterization of mobility in 14nm FD-SOI CMOS devices under interface coupling conditions, Solid State Electronics, 108, 30 (2015).

[18] G. Ghibaudo, Mobility characterization in advanced FD-SOI CMOS devices (part III pages 307-322) in "Semiconductor-On-Insulator Materials for NanoElectronics Applications", Springer, Berlin (2010).

978-1-6654-3746-2/21 $31.00 © 2021 IEEE

The effects of surface passivation on the electrostatics of the UTB SOI devices

Ravi Solanki and Aditya Sankar Medury
Department of Electrical Engineering and Computer Science
Indian Institute of Science Education and Research
Bhopal, India
ravisolanki.vnit@gmail.com and adityam@iiserb.ac.in

Abstract—The surface passivation conditions at top and bottom channel-oxide interface modify channel band structure and band gap of the (Ultra-Thin Body) UTB device. In this work, the channel band structure is calculated using the $sp^3d^5s^*$ tight-binding method with various passivation energies and passivation strategies. The interaction between quantum confinement and the surface passivation effect is investigated. The effect of surface passivation energy on electrostatics parameters, such as threshold voltage and integrated charge density as a function of gate voltage, is shown over a wide range of SOI channel thicknesses. A significant reduction in threshold voltage is seen at lower passivation energies. The thin channel is shown to be more sensitive to the surface passivation energy below threshold voltage as compared to thicker channels, while, the importance of properly passivating both the top and the bottom interface, simultaneously, to reduce the off-state current is also shown.

Index Terms—Tight binding; UTB; passivation; electrostatics

I. Introduction

THE Ultra-Thin Body (UTB) Silicon-on-Insulator (SOI) field effect transistors with double gates are known to reduce the short channel effects due to better control of the channel electrostatics [1]. Due to its channel thickness comparable to the electron De Broglie wavelength, which is ≈ 5 nm for silicon, quantum confinement effects (QCE) starts dominating the physics of the UTB device [2] [3]. The increase in threshold voltage of UTB device compared to bulk device is a clear manifestation of QCE controling the device electrostatics [4]. Along with the confinement effects, the atomic fluctuations and properties of surfaces and oxide-channel interfaces also become prominent at such length scale [5].

Generally, the UTB device simulation utilizes the $sp^3d^5s^*$ tight-binding (TB) method applied on the channel material [6]. As the unpassivated edge generates dangling bonds which results in energy states within the band gap, the atoms at the channel-oxide interface are assumed to be passivated with hydrogen to remove the states in band gap [7]. In the more realistic simulations, instead of hydrogen passivation, the SiO_2 is included in the simulation along with the channel by utilizing computationally extensive DFT calculations [8].

Meanwhile, the effect of surface passivation for InAs nanowires with various passivation species like H, NH_2 and

We acknowledge financial support from DST, (SERB), Government of India (Grant No: ECR/2017/000011)

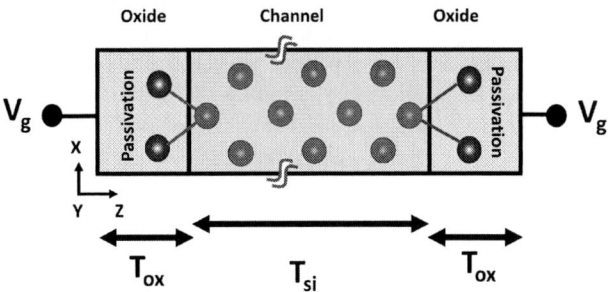

Fig. 1: The schematic of Si-(100) UTB double gate device. The channel atoms at oxide/channel interface is passivated with species through sp^3 bonding.

F is studied using the Density Functional Theory (DFT) and the change of band gap and effective mass of the channel with these different passivation species is shown in [9]. However, the effect of surface passivation on the electrostatics of the device is not studied, which needs to be evaluated to use the device for any electronic applications.

In this work, we explore the interaction of the two prominent effects in the UTB double gate device, namely, surface properties and the confinement energy on the electrostatics. We have seen the effect of various surface passivation on the band structure of UTB device with channel thickness of 2 nm (higher confinement) and 10 nm (lower confinement) and analyzed its effect on integrated charge and threshold voltage of the device.

II. Simulation Approach

We consider the double gate UTB device as shown in Fig. (1) with Si-(100) channel of thickness T_{si} and oxide thickness of T_{ox}. The 10 orbital $sp^3d^5s^*$ tight-binding (TB) model is used to calculate the band structure of the ultra-thin channel material for device simulations. The TB hamiltonian is formed using orbital energies at the atomic sites (on-site energies) and the interactions energies of different orbitals with nearest neighbouring atoms, which at room temperature are taken from [10]. The on-site energies are modified by adding the on-site's potential energy due to the application of the gate voltage. The channel surface atoms are implicitly passivated

978-1-6654-3746-2/21 $31.00 © 2021 IEEE

with species, for example hydrogen, through sp^3 bonding. This is done by altering the on-site and interaction energies of the surface atoms with a passivation potential ΔE_{edge}. The passivation potential can be related to the electron affinity of the different passivation species making sp^3 bonding. The TB Hamiltonian finally obtained is then solved self-consistently with the Poisson's equation to get the electrostatics parameters such as channel potential and charge density [6].

III. RESULTS AND DISCUSSION

The silicon (100) channel is considered as a typical case for studying the effect of surface passivation along with the confinement energy. For the case of (100) channel, edge atoms will be bonded with two passivation species with a passivation potential ΔE_{edge}. The device structure with 7 atomic layers in channel is shown in Fig. 1 with passivated surface atoms. The device have finite thickness T_{si} along the z direction, while assumed to have infinite dimensions in other two axis. In this section the effect of ΔE_{edge} on the electrostatics is discussed in detail.

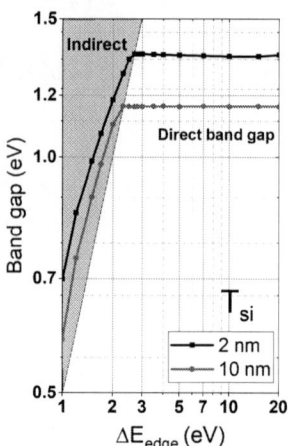

Fig. 3: Variation of band gap with passivation energy ΔE_{edge} and T_{si} of 2 nm and 10 nm.

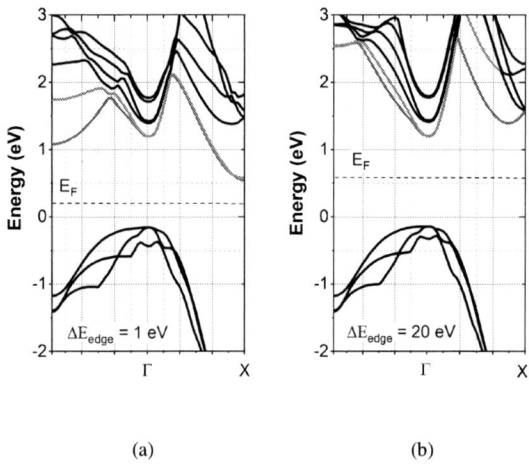

Fig. 2: The band structure of (100) silicon UTB channel with various surface passivation energies (ΔE_{edge}).

Firstly, the effect of two extreme values of ΔE_{edge} on the band structure of channel material with T_{si} of 2 nm is shown in Fig. 2. With the weaker passivation potential of 1 eV (Fig. 2(a)), the conduction band minima can be seen near X point, while, it appears at Γ point for ΔE_{edge} of 20 eV (Fig. 2(b)). This indicates the transition from indirect (ID) to direct (D) band gap semiconductor. Also, the change in band gap can be noticed between the case of ΔE_{edge} of 1 eV and 20 eV.

The dependence of the band gap on the ΔE_{edge} for various values of T_{si} (2 nm and 10 nm) is shown in Fig. 3. It can be seen from the figure that the change in band gap with ΔE_{edge} can only be seen in the indirect regime, whereas, in direct regime band gap is independent of ΔE_{edge}. This shows that

the valley minima near X point is significantly affected by ΔE_{edge}, while the valley minima at Γ point remains unaffected by ΔE_{edge}.

Comparing the point of transition from indirect to direct band gap behavior for T_{si} of 2 nm (with greater confinement) and 10 nm (with weaker confinement), it can be seen that due to the greater structural confinement in thinner channels, the transition occurs at slightly higher ΔE_{edge} value, compared to the case of thicker channels.

The effect of ΔE_{edge} value on the integrated charge density for T_{si} of 2 nm and 10 nm is shown in Fig. 4(a) and (b), respectively. In Fig. 4(a), the significant change in order of magnitude of integrated carrier density with ΔE_{edge} can be seen below threshold voltage (V_{TH}) in the regime of indirect band gap that is ΔE_{edge} between 1 eV and 2 eV. After which there is no change in band gap with ΔE_{edge} in direct regime results in no change in integrated carrier density for ΔE_{edge} of 5 eV and 20 eV.

Comparing the change of integrated charge density between the ΔE_{edge} of 2 eV and 20 eV, for T_{si} 2 nm, with the change of integrated charge density for T_{si} 10 nm with same range of ΔE_{edge}, it can be seen the thinner channel is more affected by variation of ΔE_{edge}. This is due to the larger edge influence at the center for thinner channel as compared to the thicker channel.

It can also be seen that with lesser passivation potential ($\Delta E_{edge} = 1$ eV), the off current for T_{si} 2 nm and 10 nm will be nearly same, whereas, with $\Delta E_{edge} = 20$ eV, the off current will be significantly lower for T_{si} 2 nm. The reason behind this observation is higher structural confinement in thin channel which results in higher band gap, consequently, lower off current. The change in band gap with channel thickness will also be seen in the threshold voltage.

978-1-6654-3746-2/21 $31.00 © 2021 IEEE

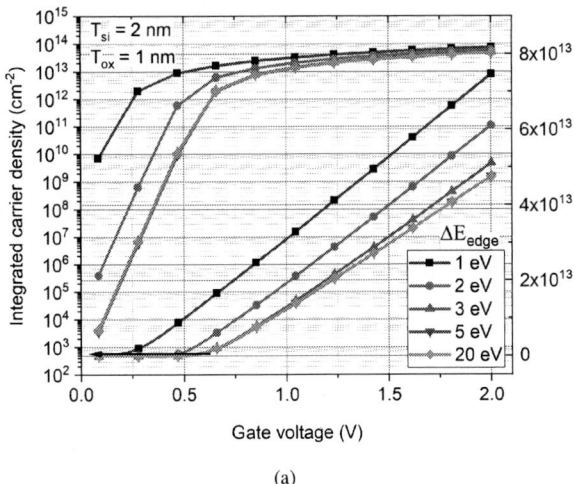

(a)

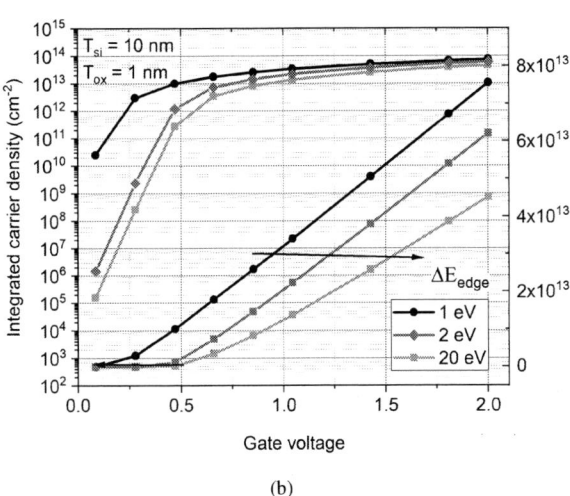

(b)

Fig. 4: The integrated charge density for Si-(100) UTB channel for T_{si} (a) 2 nm and (b) 10 nm and 1 nm SiO$_2$ oxide with various surface passivation energies (ΔE_{edge}).

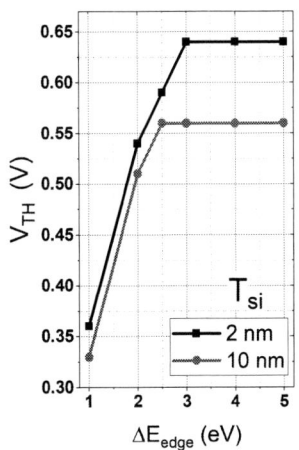

Fig. 5: Variation of V_{TH} with ΔE_{edge} and T_{si}.

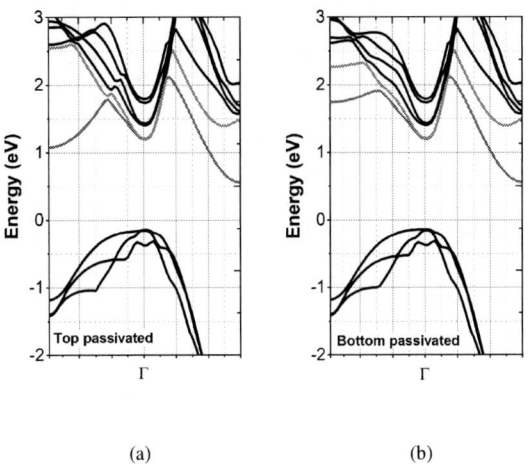

(a)   (b)

Fig. 6: Band diagram for (a) top surface with $\Delta E_{edge} = 20$ eV. (b) bottom surface with $\Delta E_{edge} = 20$ eV.

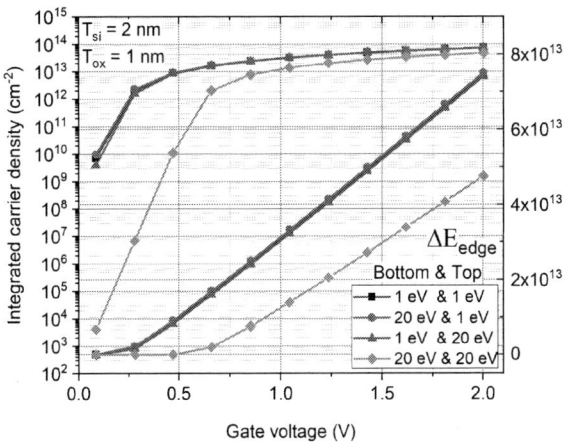

Fig. 7: The integrated charge density for $T_{si} = 2$ nm and with various surface passivation conditions.

The variation of threshold voltage V_{TH} with ΔE_{edge} for two channel thicknesses 2 nm and 10 nm is shown in Fig. 5. It can be seen that for lower value of ΔE_{edge}, the V_{TH} values for 2 nm and 10 nm T_{si} is nearly equal. This manifest in nearly same integrated charge density below V_{TH} for the case of 2 nm and 10 nm for lower values of ΔE_{edge}. For the case of higher values of ΔE_{edge}, there is significant variation in V_{TH} between T_{si} of 2 nm and 10 nm. Therefore, at higher ΔE_{edge} value, significant change in integrated charge density below V_{TH} can be seen.

It can also be seen from Fig. 5 that compared to $T_{si} = 10$ nm, $T_{si} = 2$ nm have much larger variation of V_{TH} as a function of ΔE_{edge}. Again, this is due to a larger edge influence at the center for thinner channels.

Now, we will see the effect of top and bottom surface passivation conditions on the band structure and electrostatics

978-1-6654-3746-2/21 $31.00 © 2021 IEEE

of the device. The band structure for only top passivated (ΔE_{edge} at top is 20 eV and 1 eV at bottom surface) in Fig. 6(a) and for only bottom passivated (ΔE_{edge} at top is 1 eV and 20 eV at bottom surface) in Fig. 6(b) is shown. From the figures it can be seen that the valley near L point (left side of Γ) is significantly affected by the passivation condition, however, the valley at Γ and at X point remains unaffected. As the valley at X is global minima of the band structure that will govern the properties of device, that remains unaffected by the passivation condition. The effect of passivation condition on the integrated carrier density can be seen in Fig. 7. It can be seen that any one interface poorly passivated = 1 eV (ΔE_{edge} = 1 eV) can degrade the device performance significantly in the subthreshold regime, while increase the on current at the same time.

IV. CONCLUSION

In this work, the effect of various passivation conditions at channel oxide interface along with the confinement on the electrostatics of the UTB device is studied. The band structure is calculated using tight-binding approach along with edge passivation with sp^3 bonded species. From the results presented in this work, we firstly show that the transition from direct to indirect band gap is seen when passivation energy on one or both surfaces is reduced. We also show that the passivation energy has a greater effect on the threshold voltage in the strong confinement regime compared to weaker confinement.

REFERENCES

[1] Y. K. Choi, K. Asano, N. Lindert, V. Subramanian, T.-J. King, J. Bokor, and C. Hu, "Ultra-thin body SOI MOSFET for deep-sub-tenth micron era," IEEE IEDM, pp. 919–921, April 1999.

[2] F. Stern and W.E. Howard,"Properties of semiconductor surface inversion layers in the electric quantum limit",Physical Review, vol. 163(3), pp. 816, Nov. 1967.

[3] F. Stern, "Self-consistent results for n-type Si inversion layers." Physical Review B, vol. 5(12), pp. 4891, Jun. 1972.

[4] YK Choi, D Ha, TJ King, C. Hu, "Threshold voltage shift by quantum confinement in ultra-thin body device", Device Research Conference. Conference Digest (Cat. No. 01TH8561) IEEE, pp. 85-86, Jun. 2001.

[5] HE Jung and M Shin, "Effects of Si/SiO$_2$ interface stress on the performance of ultra-thin-body field effect transistors: a first-principles study.", Nanotechnology, vol. 29(2), pp. 025201, Dec. 2017.

[6] A. Rahman, Exploring new channel materials for nanoscale CMOS devices: A simulation approach. PhD diss., Purdue University, 2005.

[7] S. Lee, F. Oyafuso, P. Von Allmen, G. Klimeck, "Boundary conditions for the electronic structure of finite-extent embedded semiconductor nanostructures" Phys. Rev. B., vol. 69(4), pp. 045316, Jan. 2004.

[8] S. Markov, C. Yam, G. Chen, B. Aradi, G. Penazzi, T. Frauenheim, "Towards atomic level simulation of electron devices including the semiconductor-oxide interface", In 2014 International Conference on Simulation of Semiconductor Processes and Devices (SISPAD) IEEE, pp. 65-68, Sep. 2014.

[9] P. Razavi and J. C. Greer, "Influence of Surface Passivation on Indium Arsenide Nanowire Band Gap Energies," J. Electron. Mater., vol. 48(10) , pp. 6654–6660, Oct. 2019.

[10] TB Boykin , G Klimeck, F Oyafuso, " Valence band effective-mass expressions in the sp^3d^5s* empirical tight-binding model applied to a Si and Ge parametrization", Physical Review B.,vol. 69(11), pp. 115201.Mar. 2004.

Improved inter-device variability in graphene liquid gate sensors by laser treatment

Jorge Ávila*, Jose Galdon*, Norberto Salazar, Maria-Isabel Recio, Carlos Navarro, Carlos Marquez and Francisco Gamiz

Nanoelectronics Research Group (CITIC-UGR), Department of Electronics,
University of Granada, 18071, Granada, Spain
*These authors contributed equally to this work.
jorgeavila@ugr.es, jcgaldon@ugr.es

Abstract—We investigate the influence of a visible laser treatment on the electrical performance of CVD-grown graphene-based liquid gate sensors. This method allows to treat locally the graphene sheet, improving the electrical characteristic for biochemical sensing applications. Optimizing the laser exposure, the Dirac point (minimum conductivity voltage) was shifted around 300 mV to lower voltages, together with a decrease of the inter-device electrical variability. These results open the door to use the laser treatment to increase the sensibility and reproducibility of liquid gate graphene-based devices as sensors or biosensors.

I. INTRODUCTION

Graphene, a monolayer of carbon atoms packed into a two-dimensional honeycomb lattice, has received an exceptional attention due to its excellent electrical, optical, thermal and bendability properties [1]–[3]. In addition, graphene is very sensitive to changes in the surroundings while its reduced thickness improves the electrostatic control of the channel, thus making it particularly attractive for sensor applications [4]. However, in the case of biosensors, they should operate at or near physiological conditions, so it is essential to use graphene in aqueous solutions. Electrochemical studies have shown that the ions accumulate at the surface of graphene when a gate voltage is applied between the electrodes, without charge transferred across the interface [5]. These studies suggest that graphene operates nearly as an ideal polarizable electrode [6], [7]. The graphene–electrolyte interface is typically modeled as an electrical double layer capacitance (EDLC), constituted by two layers of ions that are created at the surface of graphene. The first layer is composed of ions of opposite charges to those present in the graphene, and the second layer is composed of positive and negative charged ions that progressively reach the potential of the solution far from the graphene surface. The EDLC can be modulated applying a voltage at a reference electrode immersed in the electrolyte solution, controlling the number of free carriers in graphene, and therefore its conductivity. The minimum of free carriers, and thus the minimum of conductivity, is reached when the valence and the conduction band meet at a point called the Dirac point, the gate bias at which the Fermi level reaches the Dirac point is the charge neutrality point or Dirac voltage.

Electrochemical gated graphene sensors employing electrolytes such as ionic liquids and aqueous solutions have been extensively reported, showing excellent performances [8], [9]. Nonetheless, due to the high sensitivity of graphene and the complex nature of the electrolyte solutions, the reproducibility between devices can be challenging. For example, some factors that affect the inter-device reproducibility are the different doping levels of the graphene sheet due to graphene–substrate interactions or fabrication residues [10]. Moreover, common reliability disturbances such as interface states, traps and mobile charges also affect the sensing capability of the fabricated devices. Different strategies have been explored in order to clean of residues and improve the quality of the graphene sheet, as thermal annealing, plasma, UV-Vis light, electrical or mechanical treatments [11]. In this work, we have explored the ablation of the graphene layer using a visible laser as a straightforward and rapid treatment to improve the inter-device electrical variability in graphene-based liquid gate sensors. This strategy can modify locally the graphene surface without compromising the rest of the device, being useful if sensible materials are presented and to avoid damaging the contacts. Employing this method, we have observed a reduced Dirac voltage and an improved inter-device electrical variability in graphene-based sensors.

II. EXPERIMENTAL SETUP

Graphene was synthesized through low-pressure chemical vapor deposition (LPCVD) and polycrystalline copper foil was used as catalytic substrate. The reaction was made at 1 Torr using methane (CH_4) as a carbon precursor. The foils were first heated up to 1000 °C in a H_2 environment to reduce the native copper oxide on the copper foil surface. Then, a H_2/CH_4 gas mixture (50 sccm : 30 sccm) was added during the graphene growth at 1000 °C for 30 minutes. The cooling down step was made by opening the furnace, during this step the gas composition was remained the same. Then, graphene layers were transferred to cleaned quartz substrates using the PMMA based technique [12]. The device processing is illustrated in Figure 1: Initially, reactive-ion etching (RIE) at 10 W and 30 sccm of O_2 was used to pattern the graphene employing a solid mask. Up to six sensors were fabricated on each substrate. Then, the Cr (5 nm)/Au (100 nm) electrodes were deposited by physical vapor deposition. Finally, the laser treatment was performed on the graphene layer using a computer-

978-1-6654-3746-2/21 $31.00 © 2021 IEEE

controlled laser engraver (Laserbot from Makeblock). The laser wavelength was 445 nm, and the power was set to 25 % with respect to the maximum power (1600 mW). To evaluate the electrochemical characteristics of the devices, a phosphate buffered saline solution (PBS) at a 1X concentration, was placed covering the graphene to perform the liquid gate. PBS is a buffer solution commonly used in biochemical research, as the osmolarity and ion concentrations of the solutions match those of the human body. Hence, using this buffer solution, we evaluated the sensor in near human physiological conditions. The static DC characteristics were acquired using a Keithley SCS 4200 and an Agilent B1500 systems. The low-frequency noise characterization was carried out using a low-noise-current amplifier connected to a software-based spectrum analyzer [13].

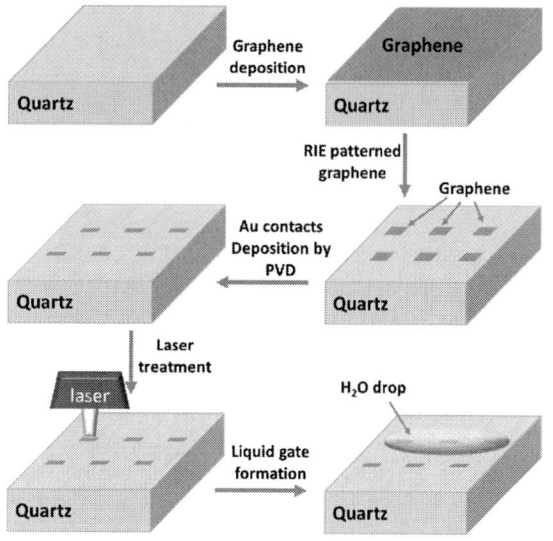

Fig. 1. Scheme of the fabrication flow of the liquid gate graphene devices including the laser treatment.

III. RESULTS AND DISCUSSION

Before considering the double layer capacitance effect in the devices, the laser radiation on the graphene sheet has been evaluated. The graphene layer resistance was measured, without a liquid gate, at a fixed drain-source voltage ($V_D = 0.1\,V$) before and after each laser exposure (Figure 2.a). An increase in the resistance of the device was observed following an exponential trend with the number of laser exposure cycles. The resistance value saturated around 2.8 kΩ after 6 to 8 cycles. In order to evaluate the structural repercussion of the laser ablation and determine the origin of the increased electrical resistance of the graphene, Raman spectroscopy analysis was carried out before and after the laser treatment (Figure 2.b). Prior to laser exposure, the Raman spectrum of the graphene device shows the signature for pristine single-layer graphene, with a G peak at $\approx 1590\,cm^{-1}$ and a 2D peak at $\approx 2680\,cm^{-1}$. After the laser treatment, an increase of the D

peak ($\approx 1340\,cm^{-1}$), as well as the emergence of the D' peak ($\approx 1620\,cm^{-1}$) can be observed.

These results corroborate that the laser treatment has an effect on the structural order of the graphene. The emergence of the D' peak could be related to the formation of sp^3-type defects [14], [15]. These new defects have an adverse impact on the electrical behavior, increasing the graphene resistance. Nevertheless, the resistance stabilized at higher number of laser treatments, indicating that possibly the effect of the laser in the graphene structure saturates. A possible explanation is that these defects are formed mainly in the grain boundaries, where dangling bonds are receiving enough energy to react with the ambient atmosphere, generating some oxide species [16]. These species would hinder the electric transport between grains, increasing the electric resistance. Despite deteriorating the sheet resistance of the graphene layers, these defects may have different implications when the graphene-electrolyte interface is formed.

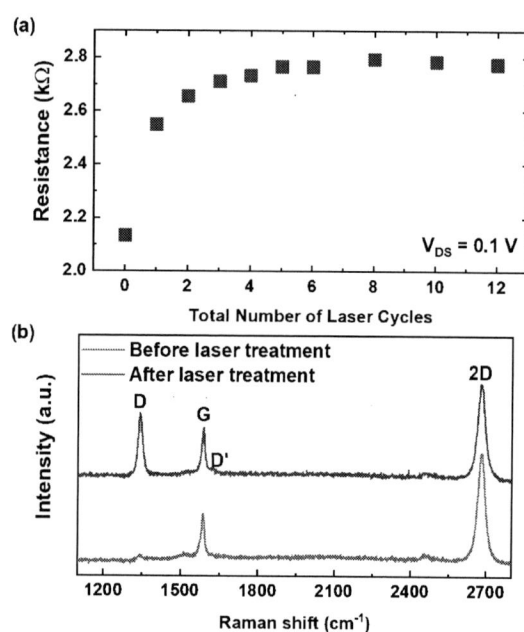

Fig. 2. a) The graphene device resistance measured at 0.1 V drain-source bias without liquid gate for different numbers of laser cycles and, b) the graphene Raman spectra before and after the laser treatment. The Raman laser wavelength is 532 nm.

After the deposition of the liquid gate and the formation of the electrical double layer capacitance, we measured the transfer characteristic curve ($R = V_D/I_D$) of a device using two different metals as the gate electrode (Figure 3). A proper modulation of the resistance as a function of the liquid gate voltage is observed, together with a shift of the Dirac point depending on the metal electrode. This shift from 0.83 V (gold) to 0.27 V (silver) is mainly due to the difference in the work function of the employed metals. In order to work in aqueous solutions, low voltages are desirable to keep the working conditions below the water electrolysis potential

978-1-6654-3746-2/21 $31.00 © 2021 IEEE

(1.2 V), avoiding secondary reactions, keeping a low gate-drain current [17]. The lower Dirac voltage for the silver gate electrode can be interesting for low energy consumption applications. However, silver can easily be oxidized in contact with the aqueous solution. Hence, it is preferable the use of a gold electrode as a more electrochemical stable option despite the higher voltages. In the following devices, gold was used as the gate electrode.

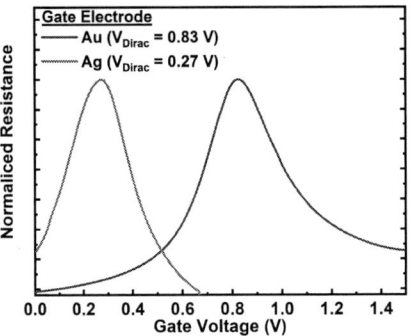

Fig. 3. Resistance versus gate voltage characteristic curves for a device measured employing gold(blue line) or silver (red line) as the gate electrode. The drain source bias was set at 0.1 V. The Dirac voltage is defined as the voltage when the resistance reaches the maximum.

Figure 4.a shows the transfer characteristic curves of a device when using the liquid gate for successive laser treatment cycles. Note that in all the cases, there is a modulation of the channel conductivity as a function of the liquid gate voltage. As observed, the device resistance initially decreases with the successive laser cycles, but after 4 cycles it starts to saturate. At the same time, the Dirac point shifts to lower voltages from 1 V before the laser treatment, down to 0.66 V after 8 cycles (inset in Figure 4.a). Then, at higher number of laser cycles, the rise of the resistance is accelerated and the Dirac voltage shifts to higher voltages. This indicates that a high number of laser cycles is drastically damaging the sensor. The increase in the resistance is in agreement with the Raman interpretation, indicating that after the laser exposure the graphene conductivity is hindered, harming the electrical performance. However, at low number of laser cycles, the sensing capabilities are improved with a lower resistance and reduced Dirac voltages compared to the device without laser treatment. When comparing the Dirac voltages of several devices before and after the laser treatment (for a total number of 4 laser cycles) a significant reduction of the electrical variability among devices is observed (Figure 4.b). The laser treatment consisting of 4 cycles reduces the standard deviation from 65 to 24 mV, together with an improved average Dirac voltage of 810 mV. Note that the inter-device variability is a critical constraint for sensing applications. A possible explanation of the variability improvement observed could reside in a laser-induced transformation of the graphene structure. Induced defects and oxidized species at grain boundaries can

improve the electrical double layer capacitance, which finally improves the liquid gate/interface, making it more reliable for sensing and less variable among devices [18], [19].

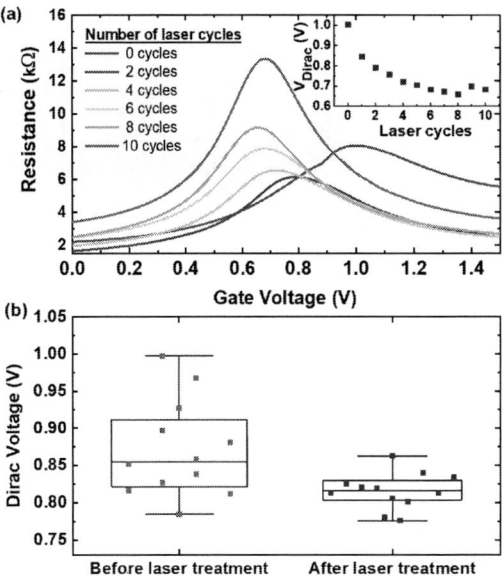

Fig. 4. a) Resistance versus gate voltage characteristic curves for a device measured at different number of laser cycles. The drain source bias was set at 0.1 V. The Dirac voltage is defined as the voltage when the resistance reaches the maximum. b) Dirac voltages distribution of 16 devices measured before and after a laser treatment of 4 cycles.

Low-frequency noise characterization sheds light on the defect implication in the graphene electrical performance after the laser ablation. As figure 5 shows, the graphene layer presents a normalized power spectral density (PSD) of the noise formed by a flicker or 1decade/1decade (1/f) contribution at low frequency and a Lorentzian contribution with center frequency around 10^4 Hz. This result indicates that graphene sheets are affected by carrier number fluctuations at low frequencies and by capture and emission processes of carriers at higher frequencies [20]. According to the similar spectrum observed before and after the laser irradiation (even lightly lower), these fluctuations are not enhanced by the laser treatment, suggesting a graphene resistance increased without a degradation of the interface after the laser exposure. This result is in agreement with the lower variability among devices observed in Figure 4.b after the laser irradiation.

IV. CONCLUSION

Despite presenting a slight increase in the graphene resistance after laser exposure, the graphene-based liquid gate sensors show a reduced Dirac point and inter-device variability. These results indicate that there is a trade-off between some grade of graphene degradation and an improvement at the liquid gate/graphene interface. Moreover, the laser exposure does not affect the power spectral density of the current

978-1-6654-3746-2/21 $31.00 © 2021 IEEE

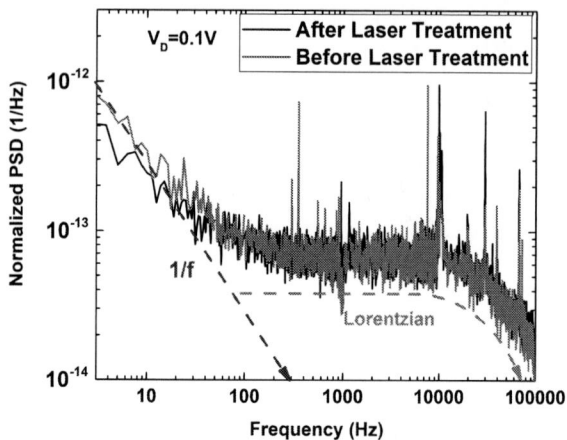

Fig. 5. Normalized power spectral density of the noise for a device before and after the laser treatment.

discarding an interface degradation. Finding the origin of this phenomenon seems critical to optimize the device performance for sensing using liquid gates. Hence, the use of the laser treatment could be an advantageous technique to increase the sensibility and reproducibility of these graphene-based devices for applications as sensors or biosensors.

ACKNOWLEDGMENT

This work has received funding from the European Union's Horizon 2020 research and innovation programme under the Marie Skłodowska-Curie grant agreement No 895322, from the Spanish Program (TEC2017-89800-R). SUPERA COVID-19 Fund and CRUE-Santander, Regional Program FEDER UGRVID (CV20-36685), P18-RT-4826 project and UGR-MADOC CEMIX 2D-EDEX are also thanked for financial support.

REFERENCES

[1] K. S. Novoselov, A. K. Geim, S. V. Morozov, D. Jiang, Y. Zhang, S. V. Dubonos, I. V. Grigorieva, and A. A. Firsov, "Electric Field Effect in Atomically Thin Carbon Films," *Science*, vol. 306, no. 5696, pp. 666–9, oct 2004.

[2] A. K. Geim and K. S. Novoselov, "The rise of graphene." *Nature materials*, vol. 6, no. 3, pp. 183–191, 2007. [Online]. Available: http://www.nature.com/doifinder/10.1038/nmat1849

[3] K. S. Novoselov, V. I. Fal ko, L. Colombo, P. R. Gellert, M. G. Schwab, and K. Kim, "A roadmap for graphene," *Nature*, vol. 490, no. 7419, pp. 192–200, 2012.

[4] C. W. Lee, J. M. Suh, and H. W. Jang, "Chemical Sensors Based on Two-Dimensional (2D) Materials for Selective Detection of Ions and Molecules in Liquid," *Frontiers in Chemistry*, vol. 7, no. November, pp. 1–21, nov 2019. [Online]. Available: https://www.frontiersin.org/article/10.3389/fchem.2019.00708/full

[5] A. T. Valota, I. A. Kinloch, K. S. Novoselov, C. Casiraghi, A. Eckmann, E. W. Hill, and R. A. W. Dryfe, "Electrochemical Behavior of Monolayer and Bilayer Graphene," *ACS Nano*, vol. 5, no. 11, pp. 8809–8815, nov 2011. [Online]. Available: https://pubs.acs.org/doi/10.1021/nn202878f

[6] M. Dankerl, M. V. Hauf, A. Lippert, L. H. Hess, S. Birner, I. D. Sharp, A. Mahmood, P. Mallet, J.-Y. Veuillen, M. Stutzmann, and J. A. Garrido, "Graphene Solution-Gated Field-Effect Transistor Array for Sensing Applications," *Advanced Functional Materials*, vol. 20, no. 18, pp. 3117–3124, sep 2010. [Online]. Available: https://onlinelibrary.wiley.com/doi/10.1002/adfm.201000724

[7] X. Du, H. Guo, Y. Jin, Q. Jin, and J. Zhao, "Electrochemistry Investigation on the Graphene/Electrolyte Interface," *Electroanalysis*, vol. 27, no. 12, pp. 2760–2765, dec 2015. [Online]. Available: https://onlinelibrary.wiley.com/doi/10.1002/elan.201500302

[8] F. Chen, Q. Qing, J. Xia, J. Li, and N. Tao, "Electrochemical gate-controlled charge transport in graphene in ionic liquid and aqueous solution," *Journal of the American Chemical Society*, vol. 131, no. 29, pp. 9908–9909, 2009.

[9] N. Liu, R. Chen, and Q. Wan, "Recent advances in electric-double-layer transistors for bio-chemical sensing applications," *Sensors (Switzerland)*, vol. 19, no. 15, 2019.

[10] A. Pirkle, J. Chan, A. Venugopal, D. Hinojos, C. W. Magnuson, S. McDonnell, L. Colombo, E. M. Vogel, R. S. Ruoff, and R. M. Wallace, "The effect of chemical residues on the physical and electrical properties of chemical vapor deposited graphene transferred to SiO 2," *Applied Physics Letters*, vol. 99, no. 12, p. 122108, sep 2011. [Online]. Available: http://aip.scitation.org/doi/10.1063/1.3643444

[11] B. Zhuang, S. Li, S. Li, and J. Yin, "Ways to eliminate PMMA residues on graphene superclean graphene," *Carbon*, vol. 173, pp. 609–636, mar 2021. [Online]. Available: https://linkinghub.elsevier.com/retrieve/pii/S000862232031126X

[12] G. Borin Barin, Y. Song, I. de Fátima Gimenez, A. G. Souza Filho, L. S. Barreto, and J. Kong, "Optimized graphene transfer: Influence of polymethylmethacrylate (PMMA) layer concentration and baking time on graphene final performance," *Carbon*, vol. 84, no. C, pp. 82–90, apr 2015. [Online]. Available: https://linkinghub.elsevier.com/retrieve/pii/S0008622314011269

[13] J. Chroboczek, "Automatic, wafer-level, low frequency noise measurements for the interface slow trap density evaluation," in *International Conference on Microelectronic Test Structures, 2003.* IEEE, 2003, pp. 95–98. [Online]. Available: 10.1109/ICMTS.2003.1197409

[14] A. Eckmann, A. Felten, A. Mishchenko, L. Britnell, R. Krupke, K. S. Novoselov, and C. Casiraghi, "Probing the Nature of Defects in Graphene by Raman Spectroscopy," *Nano Letters*, vol. 12, no. 8, pp. 3925–3930, aug 2012. [Online]. Available: https://pubs.acs.org/doi/10.1021/nl300901a

[15] T. Huang, J. Long, M. Zhong, J. Jiang, X. Ye, Z. Lin, and L. Li, "The effects of low power density co2 laser irradiation on graphene properties," *Applied surface science*, vol. 273, pp. 502–506, 2013.

[16] D. L. Duong, G. H. Han, S. M. Lee, F. Gunes, E. S. Kim, S. T. Kim, H. Kim, Q. H. Ta, K. P. So, S. J. Yoon, S. J. Chae, Y. W. Jo, M. H. Park, S. H. Chae, S. C. Lim, J. Y. Choi, and Y. H. Lee, "Probing graphene grain boundaries with optical microscopy," *Nature*, vol. 490, no. 7419, pp. 235–239, 2012. [Online]. Available: http://dx.doi.org/10.1038/nature11562

[17] H. Ohta, Y. Sato, T. Kato, S. Kim, K. Nomura, Y. Ikuhara, and H. Hosono, "Field-induced water electrolysis switches an oxide semiconductor from an insulator to a metal," *Nature Communications*, vol. 1, no. 8, 2010.

[18] B. Standley, A. Mendez, E. Schmidgall, and M. Bockrath, "Graphene–Graphite Oxide Field-Effect Transistors," *Nano Letters*, vol. 12, no. 3, pp. 1165–1169, mar 2012. [Online]. Available: https://pubs.acs.org/doi/10.1021/nl2028415

[19] Y. Yang and R. Murali, "Binding mechanisms of molecular oxygen and moisture to graphene," *Applied Physics Letters*, vol. 98, no. 9, p. 093116, feb 2011. [Online]. Available: http://aip.scitation.org/doi/10.1063/1.3562317

[20] T. A. Oproglidis, T. A. Karatsori, C. G. Theodorou, D. Tassis, S. Barraud, G. Ghibaudo, and C. A. Dimitriadis, "Origin of Low-Frequency Noise in Triple-Gate Junctionless n-MOSFETs," *IEEE Transactions on Electron Devices*, vol. 65, no. 12, pp. 5481–5486, dec 2018. [Online]. Available: https://ieeexplore.ieee.org/document/8494709/

978-1-6654-3746-2/21 $31.00 © 2021 IEEE

Incorporation of silicon-carbide (SiC) nanocrystals in the MIM structures based on pulsed-DC reactively sputtered HfO$_x$ layers

Robert Mroczyński[*]

Warsaw University of Technology, Institute of Microelectronics and Optoelectronics, Koszykowa 75,
00-662 Warsaw, Poland
[*]Author's email address: robert.mroczynski@pw.edu.pl

Abstract – **Colloidal cubic SiC-NCs have been introduced into MIM structures with HfO$_x$ layers. The examined electrical performance of NCs-MIM structures has demonstrated improved switching characteristics, good retention, and a high HRS/LRS ratio compared to reference structures. The presented results have shown the feasibility of applying SiC-NCs in RRAM devices.**

Keywords – **silicon carbide nanocrystals; memory; metal-insulator-metal; electrical characterization; high-*k* dielectric**

I. INTRODUCTION

After the demonstration of the first practical integration of resistive random-access memory (RRAM) in conventional, complementary metal-oxide-semiconductor (CMOS) technology in the form of one transistor-one resistor (1T1R) structure, the investigations and feasibility studies of the metal-insulator-metal (MIM) structures have noticeably increased [1]. Compared to typical non-volatile semiconductor memory (NVSM) devices, RRAM has been considered for future memory applications with its simple structure, fast switching properties, low power consumption level, or excellent scalability [2-4]. Many different architectures based on several dielectric and conductive materials have been proposed so far [5-10]. However, the bottleneck of all these devices is large operation voltage or deteriorated retention. Several concepts have been presented to overcome those limitations, among others the introduction of nanocrystals (NCs)/nanoparticles (NPs) into the resistive material stack [11]. The literature primarily reports the application of silicon (Si) NCs or metal NPs in various semiconductor devices and structures [12-15]. However, the noticeable growth of interest in other types of semiconductor NCs or all-inorganic perovskite NCs can be noticed recently [16]. Our previous works demonstrated the feasibility of applying Si and silicon-carbide (SiC) NCs in Metal-Insulator-Semiconductor (MIS) structures for possible applications in memory devices [17-18]. In this work, we report the technology of the MIM structures with SiC-NCs embedded in HfO$_x$ ensembles. We present the preliminary results of the electrical characterization of fabricated test structures.

The novelty of these structures is the application of colloidal cubic SiC-NCs fabricated employing the reactive bonding method followed by electroless wet chemical etching [19]. The advantage of the proposed application of SiC-NCs is the much easier introduction of NCs into gate-stack as the embedding process is performed using the same tools as standard photoresist spinning-on. Most importantly, the technology can also be characterized as the extremely low temperature one, which is compatible with CMOS technology's back-end-of-the-line (BEOL) conditions [20]. The results of the electrical characterization of the fabricated test structures are described. The comparative electrical measurements of the MIMs with and without NCs proved the essential difference resulting from the presence of the nanocrystalline film. The results presented in this work indicate the feasibility of applying the studied structures in non-volatile resistive memory devices (e.g., RRAMs).

II. EXPERIMENTAL

In this study, highly antimony-doped silicon (Si) substrates with a resistivity of 0.001 Ωcm were used. The processing sequence of SiC-NCs MIM devices is described below. Si substrates were cleaned using a modified RCA method (Piranha + SC1 + SC2 + HF dipping). In the first step, a 10 nm bottom hafnium oxide (HfO$_x$) layer was deposited. In the next step, the SiC-NCs spinning off on top of the hafnia surface was performed, followed by 10 nm top HfO$_x$ layer deposition. Two types of colloidal nanocrystals were used, i.e., with the dimensions of 1−3 nm (SiC-I) and 4−6 nm (SiC-II) (Fig. 1). After forming SiC-NCs embedded in dielectric layer ensembles, the 500 nm aluminum (Al) contact pads were created through the standard lithography process in ultraviolet (UV) radiation @380 nm. The bottom surface of Si substrates was also covered by Al film to improve the contact properties. The HfO$_x$ and Al layers were deposited utilizing the pulsed-DC reactive magnetron sputtering technique as it allows for a reliable formation of different materials with the possibility to easy tailoring of electrical properties [21]. The Post-Metallization Annealing (PMA) at 300°C in a vacuum atmosphere was performed as the ultimate step of test devices fabrication.

978-1-6654-3746-2/21 $31.00 © 2021 IEEE

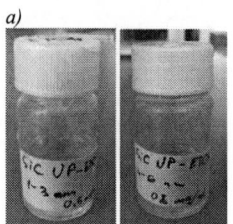

Average diameter	Density of SiC nanocrystals in methanol solvent
1−3 nm (SiC-I)	0.5 mg/mL
4−6 nm (SiC-II)	0.8 mg/mL

Figure 1. Colloidal silicon-carbide nanocrystals dispersed in methanol (a) and specifics of the liquid solution (b).

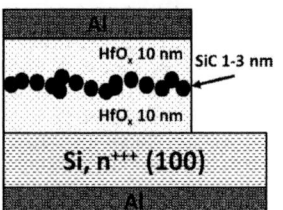 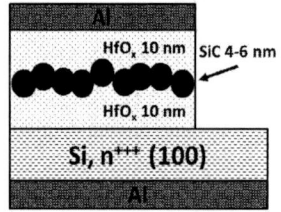

Figure 2. The test structures investigated in this work; a schematic cross-section of MIM stacks with the SiC-NCs dimensions of 1−3 nm (a) and 4−6 nm (b).

For the sake of comparison, reference MIM structures (without the introduction of SiC-NCs) were also fabricated. The schematics of the MIM stacks studied in this work have been presented in Fig. 2. The electrical measurements were conducted with the Keithley 4200 semiconductor characterization system and SUSS PM-8 probe station. The results presented in this study were obtained for MIM devices with a top electrode area of 1.8×10^{-4} cm^2. The current compliance (I_{cc}) 10^{-6} or 10^{-7} A was set to protect the devices from the complete breakdown phenomenon.

III. RESULTS AND DISCUSSION

The comparison of representative current-voltage (I-V) characteristics of the investigated in this study MIM structures was depicted in Fig. 3. The presented data distinctly show that the application of NCs results in significantly lower gate-voltage (V_g) values necessary to switch the device into a low-resistant state (LRS) or high-resistant state (HRS) compared to reference devices.

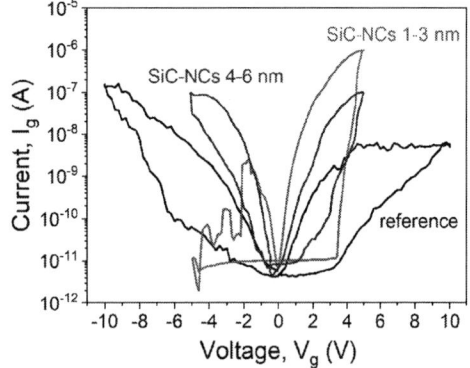

Figure 3. Comparison of representative resistive switching curves of MIM devices investigated in this work.

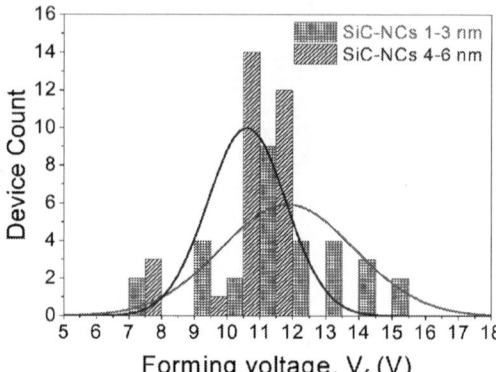

Figure 4. Forming voltage values distribution; data for both types of structures was approximated using the normal distribution.

As can be drawn from Fig. 3, the V_{set} is ~3.4 V, and ~3 V, whereas the V_{reset} is ~−2.1 V and ~−5 V for MIM structures with SiC-I and SiC-II NCs, respectively. In MIM structures without NCs, the reset/set processes occur in the range of 9-10 V. The significantly decreased set/reset phenomenon compared to reference devices may be because the inserted nanoparticles enhance the local electric field [22]. In this case, SiC-NCs effectively support the formation of current paths and the leakage filament formation under the influence of electrical polarization, thus affecting the switching properties of the memory cell. Such a finding is of great importance for the reduction of the device's power consumption level.

Fig. 4 shows the distribution of the forming voltage (V_f) values of examined devices. In each type of NCs-MIM device, around 30 structures were tested. The forming voltage lies within the range of 9−15 V and 11−12 V for smaller and larger SiC-NCs, respectively. The possible explanation of this behavior relies on the fact that the density of smaller nanocrystals across the gate area is noticeably higher than larger nanoparticles. It increases the number of possibilities of forming leakage paths at different voltage values as the soft-breakdown is a very stochastic phenomenon [23].

Fig. 5 exhibits the representative I-V curves of SiC-NCs MIM devices with smaller and larger SiC-NCs after several duty cycles. In this work, around 30 devices of each type were examined. The obtained curves show repetitive behavior. It is also particularly evident that the LRS/HRS ratio of MIM structures with SiC-I is significantly higher than SiC-II MIM structures.

The latter observation is more pronounced by the observing data presented in Fig. 6. This picture illustrates the comparison of set/reset states retention of both investigated types of MIM devices. The resistance (R) values were calculated using I-V characteristics measured at +/−2 V. The I-V curves were measured in a limited range of V_g after a particular time to limit the undesirable switch of the device to the other state.

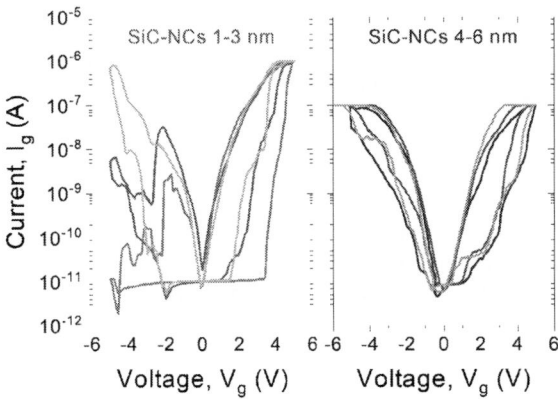

Figure 5. Representative basic resistive switching curves of MIM devices with SiC-I (1–3 nm) and SiC-II (4–6 nm) NCs.

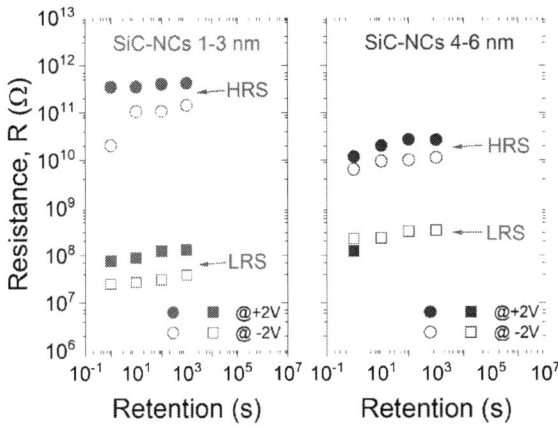

Figure 6. Retention characteristics of studied in this work RRAM structures with SiC-NCs; resistance values were taken at +/−2 V.

For structures with SiC-NCs 1−3 nm, the HRS/LRS ratio is about four orders of magnitude compared to two orders of magnitude obtained in the case of SiC-NCs 4−6 nm. The performed investigation of the HRS/LRS states retention also proves the excellent stability, thus the minor degradation of investigated structures. The approximation of achieved R values up to ten years also revealed good stability.

In the last part of the examination, the statistical stability of the retention of MIM devices with SiC-NCs was investigated. Fig. 7 depicts the cumulative retention plot expressed by the change of resistance of examined MIM devices within the time. The shift in resistance was calculated as the difference between the values measured after the set/reset operation within the range $1-10^3$ s.

Several conclusions can be drawn after analyzing the data presented in Fig. 7. It is noticeable that the stability of HRS/LRS states is more repeatable in the case of MIM structures with SiC-I NCs. It is especially evident for HRS state, where the resistance value has changed a maximum tenfold for around 90% of MIM devices. Still, the minimum LRS/HRS ratio is two orders of magnitude, proving good retention.

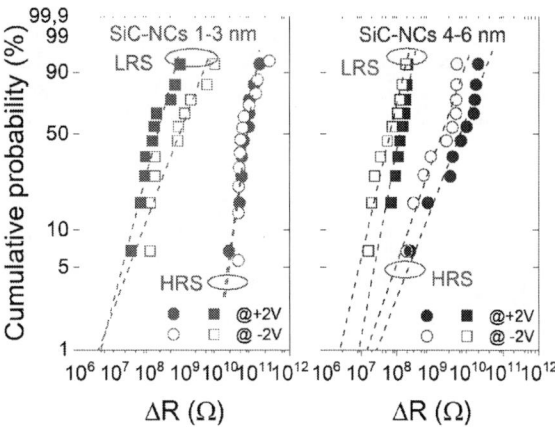

Figure 7. Cumulative plot of retention expressed as resistance changes within the time; resistance values were taken for all devices at +/−2 V.

In the case of structures with SiC-II NCs, a higher degradation of retention and lower LRS/HRS ratio has been observed compared to samples with SiC-I NCs. In the worst case, the LRS/HRS ratio difference lies in the range of one order of magnitude. However, more than 50% of devices yield in the HRS/LRS ratio around 10^2, which confirms relatively good memory characteristics, as was presented in Fig. 6.

IV. CONCLUSION

In summary, the feasibility of the introduction of colloidal cubic SiC nanocrystals in MIM structures was demonstrated. A good overall performance of MIM devices has been achieved by introducing the NCs into the HfO_x oxide sputtered using the pulsed-DC reactive magnetron process. The preliminary results have shown the significantly narrower set/reset operation and larger HRS to LRS ratio of the SiC-NCs devices compared to the reference structures. Moreover, the comparison between the performance of MIM structures with SiC-NCs demonstrated that the improved switching properties, larger HRS to LRS ratio, retention of HRS/LRS states, and repeatability are characteristic for devices with a smaller dimension of nanocrystals, i.e. 1−3 nm. The potential explanation of this behavior lies in the difference in nanoparticles' density across the structure area. The latter finding pointed out a next step towards optimizing the proposed technology: control and tailoring the nanocrystals distribution over the active area of MIM structure. The presented results have shown the potential application of examined structures in RRAM devices.

ACKNOWLEDGMENT

This work has been supported in part by the National Centre for Research and Development (NCBiR) under grant No. TECHMATSTRATEG1/347012/3/NCBR/ 2017 (HYPERMAT) in the course of "Novel technologies of advanced materials – TECHMAT-

STRATEG" and by Materials Technologies project granted by Warsaw University of Technology under the program Excellence Initiative: Research University (ID-UB) entitled "Study on the charge transport mechanisms and filament formation in Metal-Insulator-Metal (MIM) structures".

REFERENCES

[1] I. G. Baek, M. S. Lee, S. Seo, M. J. Lee, D. H. Seo, D.-S. Suh, J. C. Park, S. O. Park, H. S. Kim, I. K. Yoo, U.-In. Chung, and J. T. Moon, in Tech. Dig. IEEE Int. Electron Devices Meeting, 587–590 (2004).

[2] D. Ielmini, Semicond. Sci. Technol. 31(6), 063002 (2016), DOI: 10.1088/0268-1242/31/6/063002.

[3] H.-S. Philip Wong, H.-Y. Lee, S. Yu, Y.-S. Chen, Y. Wu, P.-S. Chen, B. Lee, F. T. Chen, and M.-J. Tsai, Proc. of IEEE 100(6), 1951-1970 (2012), DOI: 10.1109/JPROC.2012.2190369.

[4] Y. Yang, P. Gao, S. Gaba, T. Chang, X. Pan, and W. Lu, Nat. Commun. 3(1), 732 (2012), DOI: 10.1038/ncomms1737.

[5] H.-D. Kim, H.-M. An, K. C. Kim, Y. Seo, K.-H. Nam, H.-B. Chung, E. B. Lee, and T. G. Kim, Semicond. Sci. Technol. 25, 065002 (2010), DOI: 10.1088/0268-1242/25/6/065002.

[6] D. Panda and T.-Y. Tseng, Ferroelectrics 471, 23 (2014), DOI: 10.1080/00150193.2014.922389.

[7] A. L. Lacaita and D. J. Wouters, Phys. Status Solidi A 205, 2281 (2008), DOI: 10.1002/pssa.200723561.

[8] I. Valov and M. N. Kozicki, J. Phys. D: Appl. Phys. 46, 074005 (2013), DOI: 10.1088/0022-3727/46/7/074005.

[9] J. C. Scott and L. D. Bozano, Adv. Mater. 19, 1452 (2007), DOI: 10.1002/adma.200602564.

[10] T. W. Hickmott, J. Appl. Phys. 33, 2669 (1962), DOI: 10.1063/1.1754187.

[11] M. Lanza, K. Zhang, M. Porti, M. Nafria Z.Y. Shen, L.F. Liu, J.F. Kang, D. Gilmer, and G. Bersuker, App. Phys. Lett. 100, 123508 (2012), DOI: 10.1063/1.3697648.

[12] F. Priolo, T. Gregorkiewicz, M. Galli, and T.F. Krauss, Nature Nanotechnology 9, 19–32 (2014), DOI:10.1038/nnano.2013.271.

[13] Conibeer, G., Perez-Wurfl, I., Hao, X., D. Di, and D. Lin, Nanoscale Res. Lett. 7, 193 (2012), DOI: 10.1186/1556-276X-7-193.

[14] R. J. Walters, P. G. Kik, J. D. Casperson, H. A. Atwater, R. Lindstedt, M. Giorgi, and G. Bourianoff, Appl. Phys. Lett. Vol. 85(13) (27), 0003-6951 (2004), DOI: 10.1063/1.1795364.

[15] S. Huang, S. Banerjee, R.T. Tung, and S. Oda, J. Appl. Phys. 93 (1), 576–581 (2003), DOI: 10.1063/1.1529094.

[16] C. de Weerd, L. Gomez, A. Capretti, D.M. Lebrun, E. Matsubara, J. Lin, M. Ashida, F.C.M. Spoor, L.D.A. Siebbeles, A.J. Houtepen, K. Suenaga, Y. Fujiwara, and T. Gregorkiewicz, Nature Communications 9:4199 (2018), DOI: 10.1038/s41467-018-06721-02

[17] A. Mazurak, R. Mroczyński, J. Jasiński, D. Tanous, B. Majkusiak, S. Kano, H. Sugimoto, M. Fujii, and J. Valenta, Microelectron. Eng. 178, 298-303 (2017), DOI: 10.1016/j.mee.2017.05.050.

[18] A. Mazurak, R. Mroczyński, D. Beke, and A. Gali, Nanomaterials 10(12) 1-11, 2387 (2020), DOI: 10.3390/nano10122387.

[19] D. Beke, Z. Szekrényes, I. Balogh, Z. Czigány, K. Kamarás, and A. Gali, J. Mater. Res. 28(1), 44-49 (2013), DOI: 10.1557/jmr.2012.223.

[20] I. G. Baek, D. C. Kim, M. J. Lee, H.-J. Kim, E. K. Yim, M. S. Lee, J. E. Lee, S. E. Ahn, S. Seo, J. H. Lee, J. C. Park, Y. K. Cha, S. O. Park, H. S. Kim, I. K. Yoo, U.-In. Chung, J. T. Moon, and B. I. Ryu, in Tech. Dig. IEEE Int. Electron Devices Meeting 750–753 (2005).

[21] R. Mroczyński and R.B. Beck, Microelectron. Reliab. 52(1), 107-111 (2012), DOI: 10.1016/j.microrel.2011.08.010.

[22] Q. Wu, W. Banerjee, J. Cao, Z. Ji, L. Li, and M. Liu, Appl. Phys. Lett. 113, 023105 (2018), DOI: 10.1063/1.5030780.

[23] F. Stellari, E.Y. Wu, T. Ando, E. Cartier, M.M. Frank, C. Cabral, P. Song, and D. Pfeiffer, IEEE Electron Device Letters, 42(6), 828-831 (2021), DOI: 10.1109/LED.2021.3071168.

Modeling Low and High Field Uniform Transport in Monolayer MoS$_2$

Alessandro Pilotto, Pedram Khakbaz, Pierpaolo Palestri, and David Esseni

DPIA, University of Udine, Via delle Scienze 206, 33100 Udine, Italy. e-mail: pilotto.alessandro@spes.uniud.it

Abstract—We have developed a multi-valley Monte Carlo simulator to study uniform transport in MoS$_2$ monolayers. At low electric field, our solver is in excellent mutual agreement with a numerical solution of the linearized Boltzmann Transport Equation. We have then explored high field transport and analyzed the influence of different scattering mechanisms on the electron saturation velocity. Although scattering with neutral defects and Coulomb centers strongly affects the mobility, the effect on the saturation velocity is only modest. On the other hand, scattering with surface optical phonons has a significant influence on the saturation velocity, which we physically interpreted by inspecting the energy and momentum distributions of carriers.

Index Terms—MoS$_2$, Monte Carlo, High Field Transport

I. Introduction

MoS$_2$ is arguably the most technologically mature 2D semiconductor. The sub-nanometer thickness, extreme flexibility and large piezo-resistance [1] qualify MoS$_2$ as a promising baseline material for tactile sensing in soft robotics and electronic skin [2], [3], for fast thermal sensors [4], as well as for nanoelectronics devices. While linear transport and low field mobility have been addressed in several contributions [5], [6], high field transport has received a more limited scrutiny [7]–[9].

Even if the high field saturation velocity is not the most appropriate figure of merit for the performance of nanoscale FETs [10], the velocity versus field curves in MoS$_2$ are relevant for many experimentally characterized transistors (which are not at all nanoscale) and, moreover, for sensor applications that do not require ultra-scaled devices.

We here report on the development of a Monte Carlo (MC) solver for the multi-valley Boltzmann Transport Equation (BTE) in MoS$_2$, that so far has been used and validated for a uniform field regime, but in the near future will be extended to describe MoS$_2$ based transistors. The paper proceeds as follows. Section II provides a brief description of the employed model. In Section III, the MoS$_2$ electron mobility, calculated from MC simulations and by including different scattering mechanisms is compared with the results of the linearized-BTE (LBTE) solver described in [1]. Section IV reports MC simulations of velocity versus field curves in MoS$_2$, and in Section V we finally offer some concluding remarks.

This work was supported by the Italian MIUR through the PRIN 2017 Project under Grant 2017SRYEJH.

II. Model Description

We have employed a non-parabolic description of the band structure based on the Effective Mass Approximation (EMA) close to the K and Q conduction band minima. The effective masses m_l, m_t and the non parabolic factor α were extracted from DFT calculations [11], and their values are reported in Fig. 1a.

Scattering with elastic and inelastic MoS$_2$ phonons (Ph) and with the Surface Optical (SO) phonons originating in the top (TOX) and bottom (BOX) oxides were accounted for according to the formulation discussed in [1], [10]. Deformation potentials and phonon energies for MoS$_2$ were taken from [12]. The treatment of scattering with Coulomb Centers (CC) and with Neutral Defects (ND) can also be found in [1], [10].

Static screening for scattering with ND and CC has been included by means of the dielectric function approach [10].

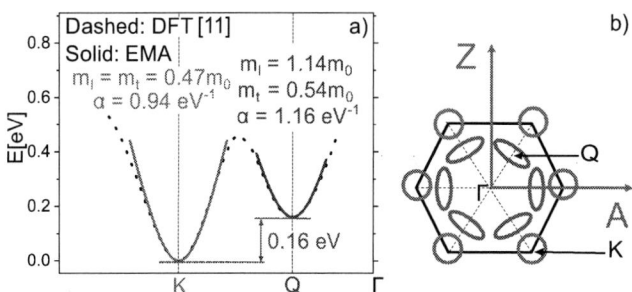

Fig. 1. Comparison between the MoS$_2$ energy dispersion close to the K (red) and Q (blue) conduction band minima calculated with DFT simulations [11] (dotted) and then fitted with the non-parabolic EMA model (solid). The minimum in Q is plotted along the K-Γ (transverse) direction. b) Sketch of the first Brillouin zone of MoS$_2$: A and Z denote the armchair and zigzag directions, respectively.

III. Mobility Extraction: MC vs LBTE

As a first validation of the multi-valley MC simulator developed in this paper, in Fig. 2 we compare the MoS$_2$ electron mobility obtained either with MC simulations or with the LBTE solver in [1]. Mobility is analyzed at room temperature, for different values of the electron density N_s and considering different combinations of scattering mechanisms (Ph, Ph and ND, Ph and CC, Ph and SO). The defects concentrations are $N_d = 2.4 \times 10^{13}$ cm^{-2} and $N_c = 3 \times 10^{12}$ cm^{-2} for ND and CC scattering, respectively. The agreement between the two modeling approaches is good for all the considered cases.

978-1-6654-3746-2/21 $31.00 © 2021 IEEE

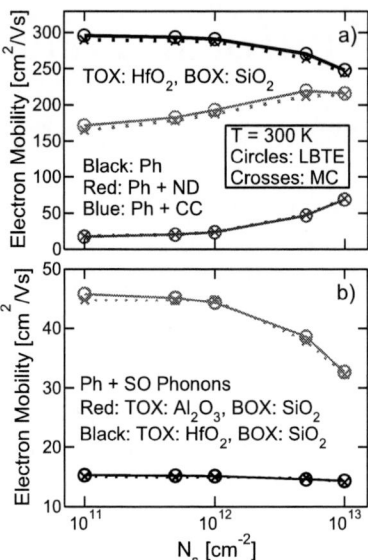

Fig. 2. Comparison between the MoS_2 electron mobility at $T = 300$ K as a function of N_s computed by using either the LBTE (circles) or the MC simulations (crosses). (a) Phonon scattering and scattering with either ND ($N_d = 2.4 \times 10^{13}$ cm^{-2}) or CC ($N_c = 3 \times 10^{12}$ cm^{-2}); (b) scattering with SO phonons is included.

IV. MC SIMULATIONS OF $v(F)$ CURVES

In this Section, we focus on MC simulations of the velocity versus field (i.e. $v(F)$) curves computed by including different scattering mechanisms.

A. Intrinsic Phonons

Figure. 3a shows the $v(F)$ curves at different temperatures computed with the MC simulator by using the EMA parameters reported in [12] instead of those in Fig. 1a, and by accounting only for the scattering with MoS_2 intrinsic phonons. The electron density is $N_s = 10^{12}$ cm^{-2}. We notice that the MC and the results of [12] are in good mutual agreement at all temperatures. However, the EMA parameters in [12] are not consistent with from DFT calculations reported in Fig. 1a. Figure 3a can thus be seen as a validation of the phonon scattering modeling for a given set of EMA parameters. In order to investigate the influence of the bandstructure description, Fig. 3b compares the results of the MC simulations at room temperature by using the EMA parameters of Fig. 1a, the MC results in [12] (based on analytical bands), and the full band MC simulations in [7]. As it can be seen, our results predict a higher low-field mobility and higher peak velocity ($v_p \simeq 4.5 \times 10^6$ cm/s), but essentially the same saturation velocity as in [12], whereas a closer agreement over the whole range of applied electric fields is observed with simulations from [7]. We verified that the discrepancy with [12] in Figure 3b at low and medium field is mainly due to the energy separation between the K and Q valleys, which is 160 meV in our simulations (see Fig. 1a) and 70 meV in [12].

Figures 4a and 4b report respectively the average energy (referred to the valley's minimum) and the valleys' occupation versus the electric field, and for simulations including only intrinsic phonons scattering. By inspecting Fig. 4a and Fig. 1a

we can conclude that the EMA energy model employed in this work is an adequate approximation: in fact the EMA is in good agreement with DFT calculations in the entire energy range relevant for low and high field simulations. The set of EMA parameters in Fig. 1a will be used throughout the rest of the paper.

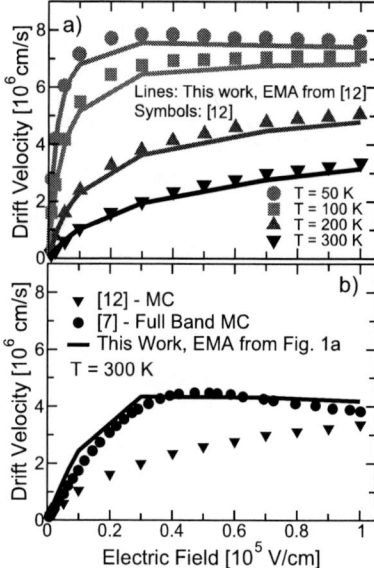

Fig. 3. a) Comparison between the $v(F)$ curves from [12], and the results of this work by using the phonon scattering parameters and the EMA parameters as in [12]. b) Comparison between the $v(F)$ curves at room temperature from [12], [7] and the results of this work obtained by using the phonon scattering parameters from [12], but the EMA parameters from Fig. 1a.

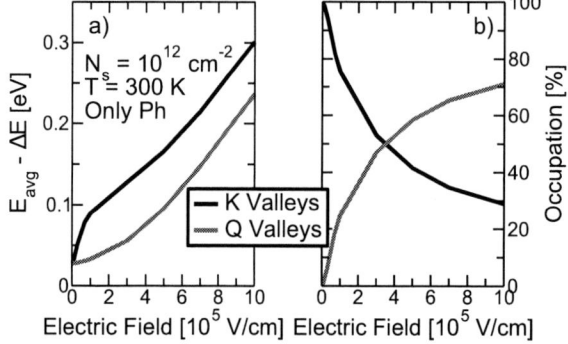

Fig. 4. Average electron energy (referred to the valley minimum) versus the longitudinal electric field. b) Valleys' occupation versus the electric field. Results are reported in black for the K valley and in red for the Q valley. Only Ph scattering is active in the simulations.

B. Intrinsic Phonons and Scattering with Defects

In Fig. 5 we report the $v(F)$ curves computed with MC simulations when only intrinsic phonon scattering is considered (black line), compared with the results obtained by including also scattering with ND ($N_d = 2.4 \times 10^{13}$ cm^{-2}) or CC ($N_c = 3 \times 10^{12}$ cm^{-2}). We notice that the scattering with impurities has a strong effect on the mobility, which degrades from values above 200 cm^2/(Vs) down to about 9 cm^2/(Vs) when CC is included. The drastic mobility reduction also increases the electric field required to reach velocity saturation up to almost 10^6 V/cm, which is much larger than the maximum

978-1-6654-3746-2/21 $31.00 © 2021 IEEE

field (10^5 V/cm) that it has been experimentally explored in MoS$_2$ devices [7]. Figure 5 also shows that, however, the saturation velocity value that is eventually reached at high fields is essentially the same for the different sets of scattering mechanisms.

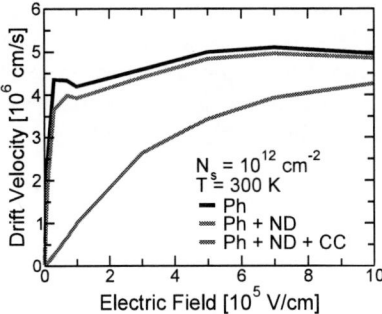

Fig. 5. Comparison between the room temperature $v(F)$ curves computed with the MC simulator, with the EMA parameters of Fig. 1a and for $N_s = 10^{12}$ cm^{-2}. Results are shown for intrinsic phonon scattering only (black), then by adding the scattering by neutral defects (red), and finally also Coulomb centers (green). Bottom Oxide is SiO$_2$.

C. Intrinsic Phonons and SO Phonons

Figure 6 compares the velocity versus electric field curves obtained with the MC simulator when only scattering with intrinsic phonons is included with the ones obtained by considering also scattering with SO phonons originating in the top and bottom oxides. We notice that, in the latter case, the mobility is reduced, as previously observed in Fig. 2b, but v_{sat} is larger than the counterpart obtained neglecting the SO phonons (see also Fig. 5). The velocity saturation in Fig. 6, however, is reached only for an electric field that is larger than 10^5 V/cm, which is difficult to probe in actual experiments [7].

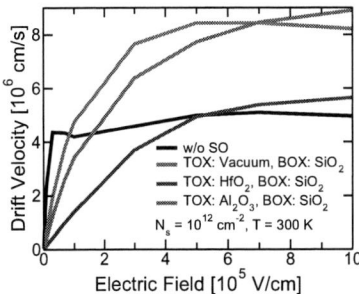

Fig. 6. Comparison between the room temperature $v(F)$ curves computed with the MC simulator when scattering with SO phonons from TOX and BOX is neglected (black) or is instead accounted for. Vacuum (red), HfO$_2$ (blue) and Al$_2$O$_3$ (green) are the dielectric options here considered for the TOX. The BOX is always SiO$_2$. The electron density is $N_s = 10^{12}$ cm^{-2}.

In order to better understand the counterintuitive influence of the SO scattering on the saturation velocity, we plot in Fig. 7 the comparison between the valleys' occupation as a function of the applied electric field obtained from the MC simulations when scattering with SO phonons is neglected (plot a) or instead accounted for (plot b, considering vacuum as TOX). Figure 7b shows that in the presence of SO phonons and for longitudinal electric fields up to $4 \div 5 \times 10^5$ V/cm the electron population in the 6 Q valleys is below 15 %, so that carriers mostly occupy the K valleys. Figure 7a shows

that, when SO phonon scattering is instead neglected, a large fraction of electrons (52÷58 %) resides in the Q valleys, that feature a much lower mobility. This is confirmed also by the energy distribution functions at $F_x = 5 \times 10^5$ V/cm reported in Fig. 8, where we notice that the presence of scattering with SO phonons results in a more peaked $f(E)$ at low energies.

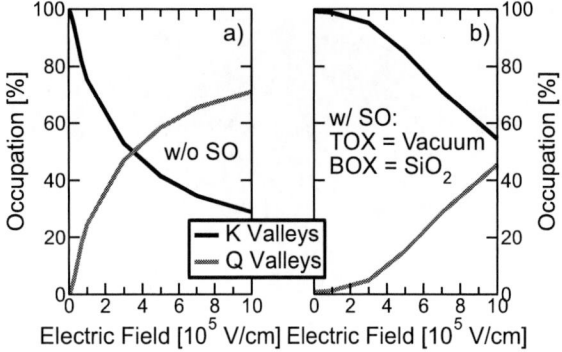

Fig. 7. Comparison between valleys' occupation vs. electric field when (a) scattering with SO phonons is turned off and when (b) scattering with SO phonons is considered: TOX is vacuum and BOX is SiO$_2$.

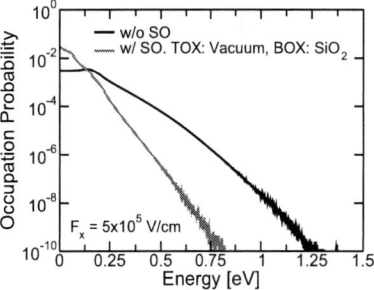

Fig. 8. Comparison between the energy distribution functions extracted from MC simulations at $F_x = 5 \times 10^5$ V/cm when scattering with SO phonons is off (black) and when scattering with SO phonons is considered: TOX is vacuum and BOX is SiO$_2$ (red).

The reason of the increase in the saturation velocity observed by including SO phonons is thus due to the anisotropic and inelastic nature of the SO scattering mechanism. In fact, as it can be deduced from the occupation functions in the reciprocal space reported in Fig. 9, the presence of SO phonons tends to reduce the energy of the electrons that populate a valley (see Fig. 8) without effectively randomizing their momentum. Therefore, for a given electric field, the $f(\vec{k})$ is more elongated in the direction of the electric field when SO phonons are active. In other words, SO phonons tend to induce an electron relaxation that prevents electrons from occupying the Q valleys, whereas it produces only a small $\vec{k}$ randomization.

It is interesting to note that the effects of SO phonons on v_{sat} shown in Fig. 6 are quite different from those obtained from MC simulations in [9] that, instead, reported a $v_{sat} = 5.8 \times 10^6$ cm/s at $F_x \geq 5 \times 10^4$ V/cm when a SiO$_2$ substrate is considered and CC is active (with $N_c = 10^{12}$ cm^{-2} and $N_s = 5 \times 10^{11}$ cm^{-2}). This discrepancy may be ascribed to a difference in the electron mobility, that originates from different scattering rates with intrinsic phonons. In fact, the authors in [9] reported a phonon-limited mobility for monolayer MoS_2 in the order of 2000 cm^2/Vs, while in this work we found about 300 cm^2/Vs (see Fig. 2). Moreover, Fig. 4a of [9] shows that, for all the simulated electric fields, intrinsic phonons are the dominant scattering mechanism whereas in the MC model, the scattering rate of SO phonons largely exceeds the rate of acoustic and optical intrinsic phonons (see Fig. 10).

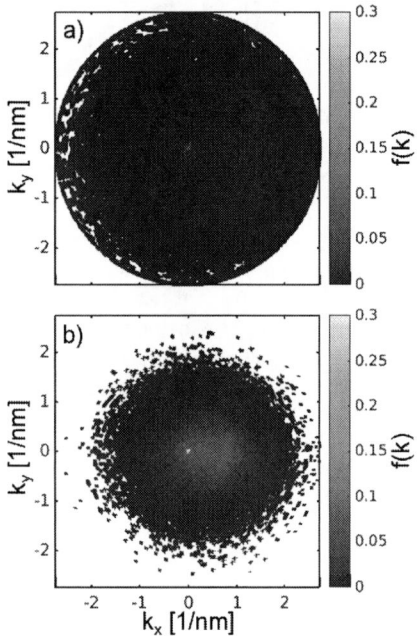

Fig. 9. Occupation functions in the $\vec{k}$-space for the K valleys of MoS_2 at F_x = 5e5 V/cm when (a) scattering with SO phonons is turned off and when (b) scattering with SO phonons is considered, TOX is vacuum and BOX is SiO_2.

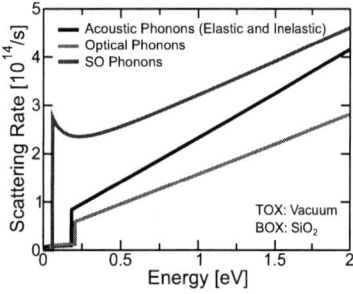

Fig. 10. Total scattering rate for emission of acoustic (black), optical (red) and SO phonons (blue) as a function of energy for electrons in K valleys. The curve for acoustic phonons includes both elastic and inelastic transitions.

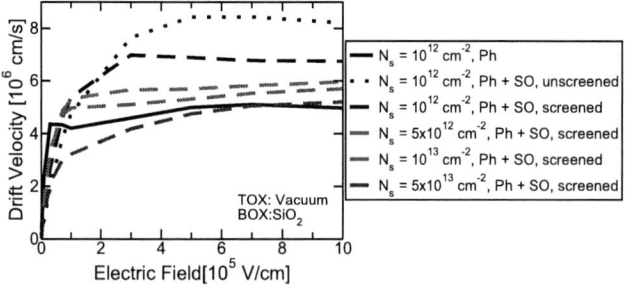

Fig. 11. Comparison between the room temperature $v(F)$ curves for electrons in MoS_2 computed with MC simulations by considering only Ph (black solid line) or Ph and SO scattering for different values of N_s. Static screening is either turned off (black dotted line) or on (dashed lines).

Finally, in this work we have assumed that scattering with SO phonons is not screened by the free electrons. This leads to an overestimation of the scattering rate, but more accurate results could only be achieved by employing a dynamic screening approach [10] that, on the other hand, would dramatically increase the computational burden of MC simulations. Preliminary results with static screening suggest that the inclusion of screening in the treatment of the scattering with SO phonons may be in fact important for an accurate evaluation of the value of v_{sat} in MoS_2 (see Fig. 11).

V. CONCLUSIONS

Monte Carlo simulations of high field transport in monolayer MoS_2 have shown that the saturation velocity is only slightly affected by scattering with Coulomb centers and neutral defects while the effect of SO phonons is more subtle.

The analysis also pointed out that a multi-valley non-parabolic description of the band structure is accurate. This paves the way towards the development of a Monte Carlo simulator for MOSFETs based on MoS_2.

REFERENCES

[1] M. Hosseini, M. Elahi, M. Pourfath and D. Esseni, "Strain-Induced Modulation of Electron Mobility in Single-Layer Transition Metal Dichalcogenides MX_2 (M = Mo , W; X = S , Se)," in IEEE Transactions on Electron Devices, vol. 62, no. 10, pp. 3192-3198, Oct. 2015, doi: 10.1109/TED.2015.2461617.

[2] D. Rus, M. Tolley, "Design, fabrication and control of soft robots," in Nature, vol. 521, pp. 467-475, May 2015, doi: 10.1038/nature14543.

[3] A. Chortos, J. Liu, Z. Bao, "Pursuing prosthetic electronic skin," in Nature Materials, vol. 15, pp. 937–950, Oct. 2016, doi: 10.1038/nmat4671.

[4] A. I. Khan, P. Khakbaz, K. A. Brenner, K. K. H. Smithe, M. J. Mleczko, D. Esseni, E. Pop, "Large temperature coefficient of resistance in atomically thin two-dimensional semiconductors," in Applied Physics Letters, vol. 116, no. 20, p. 203105, May 2020, doi:10.1063/5.0003312.

[5] K. Kaasbjerg, K. S. Thygesen, K. W. Jacobsen, "Phonon-limited mobility in n-type single-layer MoS_2 from first principles," in Physical Review B, vol. 85, no. 11, pp. 115317-1-115317-16, Mar. 2012, doi: 10.1103/PhysRevB.85.115317.

[6] M. Hosseini, M. Elahi, M. Pourfath, D. Esseni, "Strain induced mobility modulation in single-layer MoS_2," in Journal of Physics D: Applied Physics, vol. 48, no. 37, p. 375104, Aug. 2015, doi: 10.1088/0022-3727/48/37/375104.

[7] C. Zhang, L. Cheng, Y. Liu, "Understanding high-field electron transport properties and strain effects of monolayer transition metal dichalcogenides," in Physical Review B, vol. 102, no. 11, pp. 115405-1-115405-7, Sep. 2020, doi: 10.1103/PhysRevB.102.115405.

[8] K. K. H. Smithe, C. D. English, S. V. Suryavanshi, E. Pop, "High-Field Transport and Velocity Saturation in Synthetic Monolayer MoS_2," in Nano Letters, vol. 18, no. 7, pp. 4516-4522, Jun. 2018, doi: 10.1021/acs.nanolett.8b01692.

[9] D. K. Ferry, "Electron transport in some transition metal dichalcogenides: MoS_2 and WS_2," in Semiconductor Science and Technology, vol. 32, no. 8, p. 085003, Jul. 2017, doi: 10.1088/1361-6641/aa7472.

[10] D. Esseni, P. Palestri, L. Selmi, "Nanoscale MOS transistors: semiclassical transport and applications," Cambridge University Press, 2011.

[11] P. Giannozzi et al., "QUANTUM ESPRESSO: a modular and open-source software project for quantum simulations of materials," in Journal of Physics: Condensed Matter, vol. 21, no. 39, p. 395502, Sep. 2009, doi: 10.1088/0953-8984/21/39/395502.

[12] X. Li, J. T. Mullen, Z. Jin, K. M. Borysenko, M. Buongiorno Nardelli, K. W. Kim, "Intrinsic electrical transport properties of monolayer silicene and MoS_2 from first principles," in Physical Review B, vol. 87, no. 11, pp. 115418-1-115418-9, Mar. 2013, doi: 10.1103/PhysRevB.87.115418.

Modeling the propagation of ac signal on the channel of the pseudo-MOS method

Shingo Sato
Faculty of Engineering Science
Kansai University
Suita, Japan
satos@kansai-u.ac.jp

Abstract— **This paper presents an analytical model in which a small ac signal propagates on the channel of a pseudo-metal oxide semiconductor structure. The analytical impedance model with a modified Bessel function can be derived by modeling the carrier conduction on the channel using the transmission line model. Possible measurement conditions to confirm the influence of channel conduction are also clarified.**

Keywords-component; pseudo-MOS method; channel sheet resistance; transmission line model; impedance

I. INTRODUCTION

In recent times, the technology node of the complementary metal oxide semiconductor (MOS) manufacturing process has been significantly scaled down to below 10 nm [1]. To further improve the drivability of scaled transistors, various device structures fabricated on stacked wafers, such as gate-all-around transistors and nano-sheet transistors [2], have been investigated. To maintain the quality of stacked wafers, such as silicon-on-insulator (SOI) wafers, reliable and simple inspection methods are required. A powerful method is the pseudo-MOS method [3]. It has been utilized to characterize the electrical properties of SOI wafers without fabricating any device structures. First, the pseudo-MOS method was developed as a dc measurement method. Then, it was extended to the ac measurement method to extract the universal mobility of carriers [4]. In some studies, the ac response of pseudo-MOS structures has been explored [4,5]. However, a detailed analytical model including the influence of channel resistance has not yet been constructed. This paper presents an analytical model that expresses the ac response (impedance) of a pseudo-MOS structure.

II. EXPERIMENTAL SET UP AND MODELING

A. Experimental Set Up

Figure 1 shows a typical configuration for measuring the ac response of a pseudo-MOS structure. Four probe stations provided by HiSOL, Inc. and a home-made conduction vacuum chuck are used in the measurement process to characterize the electrical properties of the SOI wafer with the pseudo-MOS method. The metal probes are made of tungsten carbide. The probe radius, pressure, and pitch are 40μm, 85 g and 1.27 mm, respectively.

Furthermore, a 4-in-bonded SOI wafer with Si island having an area size (S) of 25 mm² is used. The nominal thicknesses of the SOI and buried oxide (BOX) layer are 2.0 μm and 1.0 μm, respectively. The polarity and resistivity of the SOI layer and substrate are p-type and 1~10 Ωcm.

The dc voltage, which is called the gate voltage (V_G), and the superimposed ac signal are supplied from an Agilent 4294A impedance analyzer. The voltages are applied to one of the metal probes loaded on the SOI layer, and the ac response (impedance) of the pseudo-MOS structure is

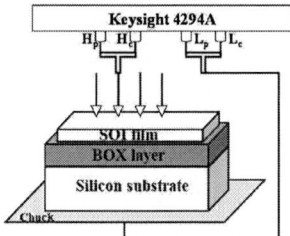

Figure 1 Typical configuration for measuring ac characteristics of stacked wafer with pseudo-MOS method

measured. The frequency of the superimposed ac signal is swept from 40 Hz to 2 MHz.

B. Modeling

Figure 2 shows the (a) top and (b) half side views of the SOI wafer with electrical components for modeling the propagation of the ac signal on the channel created by V_G. It is assumed that an ac current propagates on the channel in parallel with the SOI/BOX interface, passes through the BOX layer, and reaches the substrate chuck, as a result of applying a small voltage. The small ac signal spreads concentrically around the loaded metal probe. These propagation mechanisms are modeled by using a transmission line model [6]. The differential equation for the potential on the channel, $v(r)$, and the total impedance, Z, between the probe and the chuck are expressed by Eqs. (1) and (2), respectively, with ρ as the channel sheet resistance and Z_u as the impedance for the MOS structure without the SOI layer per unit area.

$$\frac{d^2v}{dr^2} + \frac{1}{r}\frac{dv}{dr} - \frac{\rho}{Z_u}v = 0 \tag{1}$$

$$Z = Z_C + Z_{sub} + Z_{MOS}\frac{\sqrt{\dfrac{\rho}{\pi Z_{MOS}}}}{2I_1\left(\sqrt{\dfrac{\rho}{\pi Z_{MOS}}}\right)} \tag{2}$$

,where Z_{MOS} ($=Z_u/S$), Z_C, and Z_{sub} are impedances for the MOS structure without the SOI layer and the contact

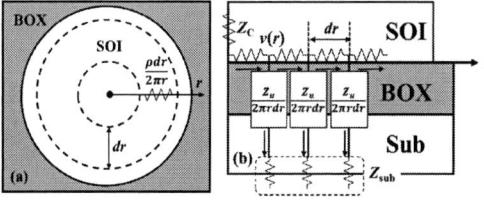

Figure 2 (a) Top and (b) half side views of the SOI wafer for modeling the propagation of the ac signal on the channel of pseudo-MOS method

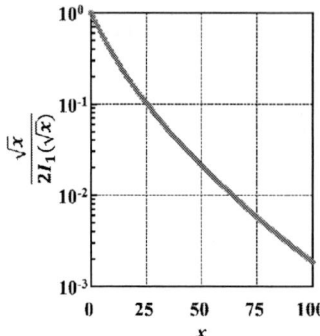

Figure 3 characteristic of last term in Eq. (2), normalized with Z_{MOS}.

resistance from the metal probe and substrate chuck, respectively. $I_n(x)$ is a 1st kind modified Bessel function with order n.

Figure 3 depicts the characteristic of the last term in Eq. (2), normalized with Z_{MOS}. This term can be regarded as a correction term for Z_{MOS}, owing to channel conduction. We can confirm that it monotonically decreases as x, which corresponds to $\rho/\pi Z_{MOS}$, increases. The following equation (Eq.(3)) can be obtained by expanding Eq. (2) using a Taylor series under $\rho \ll \pi Z_{MOS}$.

$$Z = Z_C + Z_{sub} + Z_{MOS}\left(1 - \frac{\rho}{8\pi Z_{MOS}} + \frac{\rho^2}{96\pi^2 Z_{MOS}^2} + \cdots\right)$$
$$= Z_C + Z_{sub} + Z_{MOS} - \frac{\rho}{8\pi} + \frac{\rho^2}{96\pi^2 Z_{MOS}} - \cdots \quad (3)$$

This result clearly indicates that the proposed impedance model reduces the sum of the contact resistances (Z_C and Z_{sub}) and impedance for the MOS structure (Z_{MOS}) in the limit of $\rho = 0$, which is the same as replacing the SOI layer with metal. The proposed model shows that the influence of the contact resistances from both the metal probe and the chuck must be considered, as already reported in [5,7], and ρ must be considerably smaller than Z_{MOS} for eliminating channel conduction in parallel with the BOX layer because of the fourth term in Eq.(3), even when the contact resistance can be eliminated. When the real part of ($Z_C+Z_{sub}+Z_{MOS}$) is smaller than $\rho/8\pi$, the series resistance, R_S, in the series RC circuit model has a negative value, and the absolute value of phase $|\theta|$ is over 90°.

III. RESULT AND DISCUSSION

Figure 4 shows the frequency dependence of $|Z|$, θ, Real(Z), and Imag(Z) for the proposed model with $Z_C=Z_{sub}=0$ and $Z_{MOS}=(j\omega C_{BOX})^{-1}$ as a parameter of ρ, where ω and C_{BOX} are the angular frequency and capacitance of the BOX layer, respectively. The Si island size, S, is assumed to be 25 mm². The black line represents the typical characteristics of Z_{MOS}, which has f^{-1} dependence due to the BOX layer capacitance, as shown in Fig. 3(a). We can confirm from Figs. 4(a) and (b) that the impedance of the proposed model becomes smaller than that of Z_{MOS} and its phase is below -90° in the high-frequency region. Real(Z) in the low-frequency region shown in Fig. 4(c) has a value of $-\rho/8\pi$, as expected from Eq. (3). Furthermore, the wavering characteristics that oscillate between positive and negative as a function of the applied

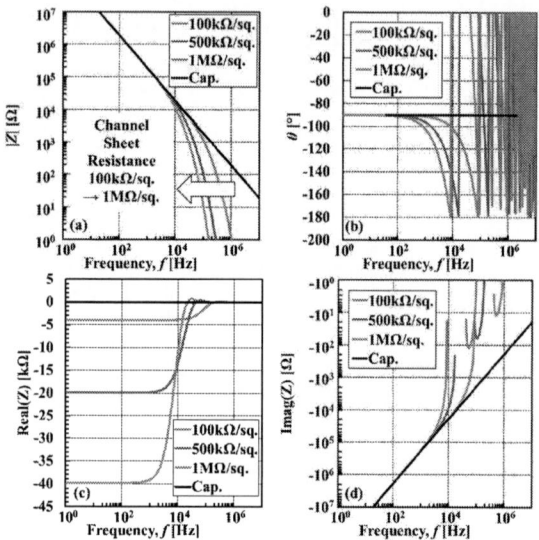

Figure 4. Frequency dependence of proposed model as a parameter of ρ. (a) impedance $|Z|$, (b) phase θ, (c) Real(Z) and (d) Imag(Z)

frequency in Real(Z) and Imag(Z) can be observed in the high-frequency region presented in Figs. 4(c) and (d).

Although some ac measurement results using the pseudo-MOS method have already been reported [4,5], such impedance characteristics have not been observed to date. One possible reason for this is the contact resistance Z_C and Z_{sub} [5,7-9]. The influence of channel conduction, which results from the signal propagation on the channel, is observed in the high-frequency region, as shown in Fig. 4. However, the pure resistance component is independent of the applied frequency. Thus, the contact resistance masks the effect of the channel conduction observed at high frequencies.

To prove this assumption, the resistance components that are expected to be crucial in the pseudo-MOS method as a function of V_G are analyzed from the measurement, as shown in Fig. 5. The blue and orange symbols indicate Z_C+Z_{sub} and ρ respectively. The green lines indicate the Z_{MOS} for each applied frequency. The ρ, threshold, and flatband voltages, V_{TH} and V_{FB}, are extracted by employing the dual-configuration Kelvin pseudo-MOSFET method [9]. The

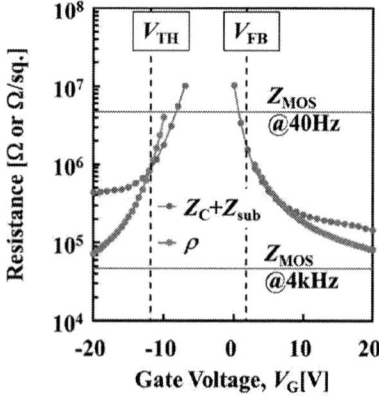

Figure 5 V_G dependence of resistance components in pseudo-MOS method

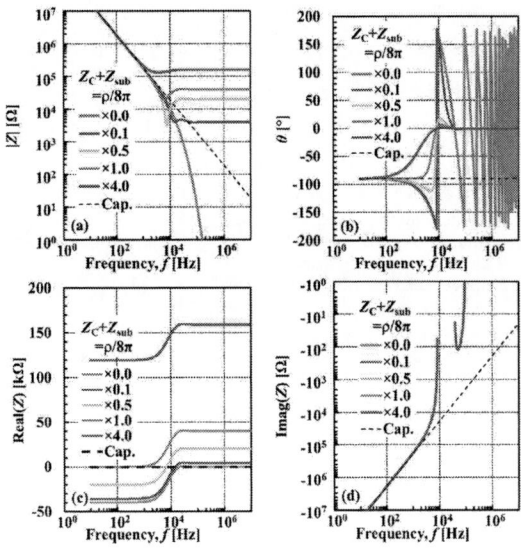

Figure 6. Frequency dependence of proposed model with $\rho = 1\text{M}\Omega/\text{sq.}$ as a parameter of contact resistance $Z_C + Z_{\text{sub}}$.
(a) impedance $|Z|$, (b) phase θ, (c) Real(Z) and (d) Imag(Z)

$Z_C + Z_{\text{sub}}$ is extracted from the Cole-Cole plot [5,10]. As the effect of channel conduction is probably enhanced when $\rho/\pi Z_{\text{MOS}}$ becomes large, as expected from Fig. 3, it is more apparent as the gate voltage is close to V_{TH} and V_{FB}. In addition, it can be easily observed in the high-frequency measurement depicted in Fig. 4, because the Z_{MOS} becomes small, owing to the impedance characteristic of the BOX layer capacitance. However, the contact resistance, $Z_C + Z_{\text{sub}}$, has a larger value in any V_G than $Z_{\text{MOS}}@4\text{kHz}$. It is possible to observe it based on the fourth term in Eq.(3), even in low-frequency measurements. The real part of ($Z_C + Z_{\text{sub}} + Z_{\text{MOS}}$) must be smaller than $\rho/8\pi$. However, this will most likely not be satisfied in the standard configuration of the ac pseudo-MOS method for the SOI wafer, as shown in Fig. 5 and in [5]. These results give us a definite reason for why we could not observe the influence of the channel conduction thus far; it was due to the large contact resistance.

Figure 6 shows the frequency dependence of the proposed model with ρ of 1 MΩ/sq. for (a) impedance $|Z|$, (b) phase θ, (c) Real(Z), and (d) Imag(Z) as a parameter of the contact resistance, $Z_C + Z_{\text{sub}}$. The $Z_C + Z_{\text{sub}}$ is parameterized from 0 to $4.0 \times \rho/8\pi$. In Fig. 6(a), the f^0 dependence of the impedance can be observed by adding a finite $Z_C + Z_{\text{sub}}$. As the resistance value increases, the impedance characteristic including channel conduction, which is smaller than that of the BOX layer, becomes invisible and is no longer observed with a high contact resistance of $4.0 \times \rho/8\pi$. In most cases, $Z_C + Z_{\text{sub}}$ is higher than ρ as shown in Fig. 5 and in [5]. Again, it can be concluded that $Z_C + Z_{\text{sub}}$ would mask the effect of channel conduction. In contrast, there is no influence on Imag(Z), as shown in Fig. 6(d), because $Z_C + Z_{\text{sub}}$ is assumed to be a multiplier of $\rho/8\pi$, which is a real number. As $Z_C + Z_{\text{sub}}$ is considered, Real(Z), shown in Fig. 6(c), increases for all the applied frequencies and turns from negative to positive values at low frequencies. When $Z_C + Z_{\text{sub}}$ is equal to $\rho/8\pi$, Real(Z) is at a low frequency of exactly 0, as expressed by Eq. (3). As shown in Fig. 6(b), θ at low frequencies is almost equal to -90° because $|\text{Imag}(Z)|$ is much larger than $|\text{Real}(Z)|$, as shown in Figs. 6(c) and (d). As

the applied frequency increases, $|\text{Imag}(Z)|$ decreases and is comparable to $|\text{Real}(Z)|$. In response to this, $|\theta|$ increases and finally reaches 180° without $Z_C + Z_{\text{sub}}$. When the positive $Z_C + Z_{\text{sub}}$ is added to Real(Z), $|\theta|$ decreases from 180° as Real(Z) is close to 0. At $Z_C + Z_{\text{sub}}$ of $\rho/8\pi$, $|\theta|$ is no more than 90° because Real(Z) is exactly 0. When $Z_C + Z_{\text{sub}}$ is over $\rho/8\pi$, $|\theta|$ monotonically decreases and approaches 0°, which is the pure resistance value.

To observe the influence of the channel conduction in the measurement, it is necessary to decrease $Z_C + Z_{\text{sub}}$ as much as possible. As shown in Fig. 7, the contact resistance from the metal probe, Z_C, is eliminated in the measurement. Figures 7(a) and (b) show the measurement results of impedance $|Z|$ and its phase θ without Z_C, as a function of the measured frequency.

As shown in Fig. 7, the proposed model using channel sheet resistance extracted with the dual-configuration Kelvin pseudo-MOSFET method [9] (ρ is ~80 kΩ/sq. for both carriers) is presented together with the solid line, with a Z_{sub} value of 2.2 kΩ. It can be observed that the measured impedance, $|Z|$, after neglecting Z_C becomes smaller than theoretical impedance $|Z|$ for the capacitor structure of $(\omega C_{\text{BOX}})^{-1}$, and θ approaches -100° around a few kHz. This characteristic indicates that the effect of the channel conduction implied from the proposed model can be successfully observed by suppressing the contact resistance.

However, the proposed model does not have a good correlation with the measurement result yet and shifts to the high-frequency region. In contrast, the measurement result for the hole has a small shift for the proposed model in comparison with that for an electron. There are two possible explanations for this shift. The first is the depletion layer effect for both the SOI layer and the substrate. The proposed model does not consider the effects of the depletion layer. The fact that $|Z|$ for holes has a smaller shift than that for electrons indicates that the depletion layer in the SOI layer may have a large impact on the amount of the frequency shift. The second is the influence of interface traps. The frequency shift in the model considering the effect of channel conduction has already been expected in conventional MOSFETs [11]. An improved model that includes both effects and a more detailed analysis is needed.

The θ in measurements over a few kilohertz approaches 0° once, and then it becomes close to -90° again. This characteristic indicates that the other resistance, which is probably Z_{sub}, as indicated in Fig. 6, and the capacitance component would be expected in the measurement configuration. It is well-known that the substrate of the wafer

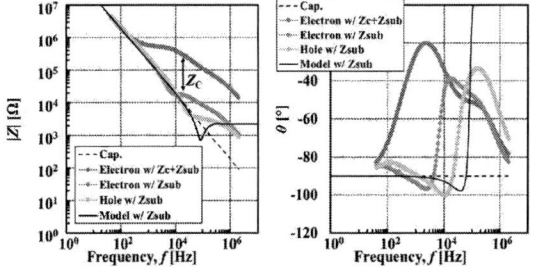

Figure 7. The measurement results of (a)impedance $|Z|$ and its phase θ without Z_C for both carriers. The proposed model with ρ of ~80kΩ/sq. and Z_{sub} of 2.2kΩ is also shown.

without chemical treatment and any metal deposition has native oxides with high resistance values [7]. To account for θ in the high-frequency region, Z_{sub} must be considered. In addition, the capacitance component in a higher frequency region is expected to be affected by the parasitic capacitance of the native oxide or equipment [5,7].

IV. CONCLUSION

This paper presented an analytical model for propagating a small ac signal on the channel of the pseudo-MOS method. We clarified that the reason why the effect of channel conduction has not yet been observed is because of the contact resistances from the metal probe and chuck. We confirmed that the impedance characteristics of the proposed model and the measurement results without the contact resistance from the metal probe have a phase characteristic, over a pure capacitor of -90°. However, there is a frequency shift between the measurement results and the proposed model. This could be caused by the depletion layer or the interface trap. Another contact resistance, which may be caused by the substrate chuck, must be considered for the proposed model to account for all of the measurement results. The proposed model indicates that the contact resistance must be suppressed below $\rho/8\pi$ to observe the influence of channel conduction in the pseudo-MOS method.

ACKNOWLEDGMENT

This work was financially supported in part by the Kyoto Technoscience Center 2019 and JSPS KAKENHI Grant number JP21K04160.

REFERENCES

[1] G. Yeap, S. S. Lin, Y. M. Chen, H. L. Shang, P. W. Wang, et. al., "5nm CMOS Production Technology Platform Featuring Full-fledged EUV, and High Mobility Channel FinFETs with Densest 0.021μm2 SRAM Cells for Mobile SoC and High Performance Computing Applications," Tech. Digest of 2019 IEEE International Electron Devices Meeting (IEDM), San Francisco, CA, USA, 2019, pp.36.7.1–36.7.4.

[2] N. Loubet, T. Hook, P. Montanini, C. -W. Yeung, S. Kanakasabapathy, et. al., "Stacked Nanosheet Gate-All-Around Transistor to Enable Scaling Beyond FinFET," in Proc. Symp. VLSI Technol., Kyoto, Japan, Jun. 2017, pp. T230–T231, doi: 10.23919/VLSIT.2017.7998183.

[3] S. Cristoloveanu, D. Munteanu and M. S. T. Liu, "A Review of the Pseudo-MOS Transistor in SOI wafers: Operation, Parameter Extraction, and Applications," IEEE Trans. Electron Devices, vol. 47, no. 5, pp. 1018–1027, 2000, doi: 10.1109/16.841236

[4] A. Diab, I. Ionica, G. Ghibaudo and S. Cristoloveanu, "RC Model for Frequency Dependence of Split C−V Measurements on Bare SOI Wafers," IEEE Electron Device Lett., vol. 34, no. 6, pp. 792–794, 2013, doi: 10.1109/LED.2013.2257663

[5] S. Sato, G. Ghibaudo, L. Benea, I. Ionica, Y. Omura and S. Cristoloveanu, "Impact of Contact and Channel Resistance on the Frequency-dependent Capacitance and Conductance of Pseudo-MOSFET," Solid-State Electron., vol. 159, pp. 197–203, 2019, doi: 10.1016/j.sse.2019.03.059.

[6] P. M. D. Chow and K. L. Wang, "A New AC technique for Accurate Determination of Channel Charge and Mobility in Very Thin Gate MOSFET's," IEEE Trans. Electron Devices, vol. 33, no. 9, pp. 1299–1304, 1986, doi: 10.1109/T-ED.1986.22662.

[7] I. Yarita, S. Sato and Y. Omura, "Impact of Native Oxide Growth on the Capacitance-Voltage Characteristic of Pseudo-MOS Structure," ECS Trans., vol. 77, no. 11, pp.1887–1892, 2017, doi: 10.1149/07711.1887ecst

[8] M. Alepidis, A. Bouchard, C. Delacour, M. Bawedin and I. Ionica, "Out-of-Equilibrium Body Potential Measurement on Silicon-on-Insulator With Deposited Metal Contacts," IEEE Trans. Electron Devices, vol. 67, no. 11, pp. 4582–4586, 2020, doi: 10.1109/TED.2020.3023872

[9] D. Mori, I. Nakata, M. Matsuda and S. Sato, "Detailing Influence of Contact Condition and Island Edge on Dual-Configuration Kelvin Pseudo-MOSFET Method," IEEE Trans. Electron Devices, vol. 68, no. 6, pp. 2906–2911, 2021, doi: 10.1109/TED.2021.3074115.

[10] K. S. Cole and R. H. Cole, "Dispersion and Absorption in Dielectrics I. Alternating Current Characteristics," J. Chem. Phys., vol. 9, no. 4, pp. 341–351, 1941, doi: 10.1063/1.1750906.

[11] S. B. F. Sicre and M. M. D. Souza, "Dit Extraction From Conductance-Frequency Measurements using a Transmission-Line Model in Weak Inversion of poly/TiN/HfO2 nMOSFETs," IEEE Trans. Electron Devices, vol. 59, no. 3, pp. 827–834, Mar. 2012, doi: 10.1109/TED.2011.2179657.

Impact of High-Aspect-Ratio Etching Damage on Selective Epitaxial Silicon Growth in 3D NAND Flash Memory

Tobias Reiter
Institute for Microelectronics
TU Wien
Vienna, Austria
reiter@iue.tuwien.ac.at

Xaver Klemenschits
Institute for Microelectronics
TU Wien
Vienna, Austria
klemenschits@iue.tuwien.ac.at

Lado Filipovic
Institute for Microelectronics
TU Wien
Vienna, Austria
filipovic@iue.tuwien.ac.at

Abstract—A physical process model for inductive plasma dry etching is presented and applied to simulate vertical channel hole etching, a critical fabrication step in modern 3D NAND flash memory. A specialized advection algorithm is subsequently applied to simulate the selective epitaxial growth (SEG) of silicon on the bottom source line. The induced etching damage on the bottom silicon substrate, which is included in the etching model, is shown to heavily reduce the quality of the SEG. The removal of this damaged layer is shown to result in highly crystalline epitaxially grown silicon.

Index Terms—Plasma Etching; Plasma-induced damage; Process Simulation; Selective Epitaxial Growth; 3D NAND;

I. INTRODUCTION

Modern three-dimensional flash memory (3D NAND) employs a large number of stacked control gates and insulating layers, comprised of alternating silicon dioxide (SiO_2) and silicon nitride (Si_3N_4) thin films, leading to a recent increase in storage size up to 1.33 TB [1] on a single die [2]. The rapid scaling drives new challenges in the etching of these stacks in order to meet the demands of further increasing memory capacity by adding more stacked layers, thereby dramatically increasing aspect-ratios of the feature sizes. The specific challenge is the dry etching of high-aspect-ratio (HAR) channel holes through these stacks using high energy plasmas. The plasma can often introduce damage in the silicon layers by implanting ions in the silicon substrate which forms the source line [3]. This heavily impacts the subsequent selective epitaxial silicon growth (SEG) of the source contact, severely decreasing its quality and leading to the formation of voids. Recent studies have shown that removing the damaged silicon layer using a post etch plasma treatment can effectively remove impurities, leading to well formed SEG silicon on the bottom [4].

We propose a physical model which is capable of simulating the channel hole dry etch process including the resulting source material damage through ion implantation, leading to undesirable voids in the source contact after SEG. The model uses a level set powered topography simulator in combination with top-down Monte Carlo ray tracing to accurately describe the surface kinetics for multiple materials and 3D geometries, combined and implemented in the ViennaTools software ecosystem [5].

This paper is organized as follows. In Section II we describe the physical process model, starting with the basic idea of level set based surface representations, the applied surface kinetics, describing the surface evolution, the flux calculations using Monte Carlo ray tracing, the modeled ion damage, and the selective epitaxial growth. Section III shortly introduces the plasma chemistry which determines our various model parameters, after which the simulation setup is described in Section IV. The obtained results are then presented in Section V and a conclusion is given in Section VI.

II. MODEL

A. Level Set Surface Representation

To accurately represent the substrate surface and its time evolution during the etching process, a level set based description is used [6]. In this method, the surface is described implicitly by a level set function $\phi(\vec{x})$, defined at every point $\vec{x}$ in space. This function is obtained using signed distance transforms, containing the surface S as the zero level set:

$$S = \{\vec{x}\colon \phi(\vec{x}) = 0\}. \tag{1}$$

In order to propagate the surface in an advection step, the level set equation

$$\frac{\partial \phi(\vec{x}, t)}{\partial t} + v(\vec{x})|\nabla \phi(\vec{x}, t)| = 0 \tag{2}$$

is solved given the scalar velocity field $v(\vec{x})$ which describes the surface normal velocity at each point. This is achieved by applying finite difference schemes on a regularly discretized grid, used to store the level set values.

B. Surface Kinetics

To analyze the surface kinetics which describe the etching process, a model based on the theory of active surface sites is used [7]. Three different types of particle species are considered:

978-1-6654-3746-2/21 $31.00 © 2021 IEEE

1) A reactive etchant forming volatile etch products which dissociate thermally and thus etch the substrate;
2) A passivating species which forms protective polymer layers on the surface;
3) Energetic ions which physically sputter the film and enhance the dissociation of the volatile etch products.

The rates of particle types impinging on the surface can be summed to give coverages ϕ_x of different particle types at all surface sites, φ_x, where x represents etchant (e), polymer (p) or etchant on polymer (pe). Since the etching time is usually much larger than the surface reaction time scales, we can assume that the coverages reach a steady state on the surface and can be expressed by the following equations [7]:

$$\frac{d\varphi_e}{dt} = J_e S_e \left(1 - \varphi_e - \varphi_p\right) - k_{ie} J_i Y_e \varphi_e - k_{ev} J_{ev} \varphi_e \approx 0, \quad (3)$$

$$\frac{d\varphi_p}{dt} = J_p S_p - J_i Y_p \varphi_p \varphi_{pe} \approx 0, \quad (4)$$

$$\frac{d\varphi_{pe}}{dt} = J_e S_{pe} \left(1 - \varphi_{pe}\right) - J_i Y_p \varphi_{pe} \approx 0. \quad (5)$$

Here, J_x and S_x represent the different particle fluxes and sticking probabilities, respectively. Y_e is the ion-enhanced etching yield for etchant particles, Y_p is the ion-enhanced etching yield on polymer, Y_{sp} gives the physical ion sputtering yield and k_{ie} and k_{ev} are the stoichiometric factors for ion-enhanced etching and evaporation respectively. Solving these equations for the coverages, one can determine etch or deposition rates on the surface. If deposition of polymer dominates, the surface velocity is positive and given by:

$$v = \frac{1}{\rho_p} \left(Y_p J_i \varphi_{pe} - J_p S_p\right), \quad (6)$$

where ρ_p is the atomic polymer density. In this case, polymer material is deposited on the surface, which acts as passivation layer for the chemical etching process. If etching dominates, the negative surface velocity of the substrate is given by:

$$v = \frac{1}{\rho_m} \left[J_i Y_e \varphi_e + J_i Y_{sp} \left(1 - \varphi_e\right) + J_{ev} \varphi_e\right], \quad (7)$$

where ρ_m is the atomic density of the etched material and depends on which layer in the stack is being etched. Together, these equations describe the temporal evolution of the surface, given the particle fluxes J_x at each location on the etched substrate.

C. Flux Calculation

The fluxes are calculated using a top-down Monte Carlo ray tracing approach, where a large number of particles are launched from a source plane and traced towards the substrate surface in order to determine the point of impact on the surface. Particles are initialized at random positions on the source plane with random directions according to a particle-specific distributions. Both polymer and etchant particle directions are described by a cosine distribution, while ions are represented with a power cosine distribution, thus being more directional towards the surface.

The ion yield efficiencies on the point of impact are dependent on the ion energy E, as well as the incident angle α between the ion particle and the surface normal. The initial energy for each ion particle is assigned based on the process used. In general, the ion yield efficiency for physical sputtering and ion-enhanced etching upon impact on the surface can be expressed as

$$Y = A(\sqrt{E} - \sqrt{E_{th}})f(\alpha), \quad (8)$$

where A is a process-dependent constant, describing the particle yield per unit of energy, E_{th} is the minimum energy ions must have to etch the substrate, referred to as threshold energy, and $f(\alpha)$ is a function of the incident angle. For the physical ion sputtering this function can be fitted using the expression

$$f(\alpha) = \left(1 + B_{sp}\left(1 - \cos^2(\alpha)\right)\right)\cos(\alpha), \quad (9)$$

while for the ion enhanced etching, the expression

$$f(\alpha) = \cos(\alpha) \quad (10)$$

is applied, as presented in [7]. If the energy is below the threshold, the ion is reflected specularly, until it reaches a point with a larger incident angle or it leaves the simulation domain.

Polymer and etchant particles, on the other hand, reflect diffusely with varying sticking probabilities at each discretized surface point.

D. Ion Damage

In HAR etching processes, the use of high energy ions is necessary to achieve the desired channel aspect ratios. However, this introduces the problem of high energy ions being able to penetrate into the bottom substrate, consequently leading to a disordered silicon layer on the bottom of the channel hole, thereby destroying the crystal purity. To model this damaged layer in the crystalline silicon substrate, the energy of traced ions is recorded on the substrate surface at each flux calculation step. Since the highest ion energies occur at normal incidence, the recorded energies can be used to model the ion damage in the material directly below the surface. At each surface point, the impinging ion energy E is assumed to decrease through the substrate due to scattering, following an exponential attenuation given by

$$E(d) = E_i e^{-d/\lambda}, \quad (11)$$

where E_i is the initial ion energy upon surface impact and d is the normal distance to the surface inside the material, equivalent to the penetration depth. Given the observed thickness of the damaged layer d_{th} in [4] and the threshold energy for ion enhanced etching E_{th}, the attenuation length λ is determined as

$$\lambda = d_{th} / \ln\left(\frac{E_i}{E_{th}}\right). \quad (12)$$

Subsequently, an ion damage coefficient $D(d)$ is defined and stored for each surface point of the geometry. The coefficient is proportional to the ion energy at a depth d in the substrate:

$$D(d) \propto E(d) - E_{th}. \tag{13}$$

In order for silicon to grow epitaxially, the material must not be damaged and hence, the ion damage coefficient at the surface must fulfill

$$D \leq 0. \tag{14}$$

To remove the damaged layer in a post etch treatment process simulation, the surface is etched using low energy ions. Assuming these low energy ions do not cause any additional implantations in the substrate, the surface is etched until the damage coefficient D at the bottom drops below 0, thereby forming a suitable interface for the subsequent selective epitaxial growth.

E. Selective Epitaxial Growth

In selective epitaxial growth, silicon is grown only on a clean crystalline silicon surface, with growth rates depending on the surface orientation. For the simulation of SEG, the specialized numerical method presented by Toifl et al. [8] is applied. In this approach an additional top layer level set is advected with growth rates as proposed in [9], using the Stencil-Local-Lax-Friedrichs (SLLF) numerical integration scheme.

III. PLASMA CHEMISTRY

Etching is modeled as a fluorocarbon based etch process, including Ar and O_2 as inhibitor, as is commonly used for the etching SiO_2/Si_3N_4 layers. The fluorocarbon radicals CF_x^+ together with Ar^+ provide ion bombardment, while the O_2 forms a passivation layer on the substrate. Corresponding parameters are extracted from [10] and summarized in Table I. For the post etch treatment, the chemistry described in [4] is used, where an inductance coupled plasma using low energy CL_2 and NF_3/CH_2F_2 gases is proposed.

IV. SIMULATION SETUP

The structure of the simulation domain is based on the multi-material representation, presented by Ertl and Selberherr [11]. In this approach, the stacked material layers are represented by individual level sets which are connected by sequential union operations. This multi-material representation enables the resolution of thin layer regions (i.e. thickness below a single grid spacing) and thus an additional layer, placed on top of all layers, is used capture any deposited polymer. The etching simulation and post etch treatment process consist of multiple steps of alternating ray tracing and surface advection steps. In the ray tracing step, the required fluxes at each surface point are calculated and used to find the resulting surface velocity which is applied for the subsequent level set advection.

TABLE I
PARAMETERS USED FOR THE SIMULATION OF THE VERTICAL CHANNEL HOLE ETCHING PROCESS. VALUES ARE EXTRACTED FROM [7] AND [10].

Parameter	Value	Description
J_{ion}^{src}	$1 \times 10^{17} \text{cm}^{-2}\text{s}^{-1}$	Source ion flux
J_{etch}^{src}	$5 \times 10^{17} \text{cm}^{-2}\text{s}^{-1}$	Source etchant flux
J_{poly}^{src}	$1 \times 10^{17} \text{cm}^{-2}\text{s}^{-1}$	Source polymer flux
A_{sp}	$0.00339 \text{ eV}^{-1/2}$	Yield coefficient in (8) for physical ion sputtering
A_{ie}^e	$0.0361 \text{ eV}^{-1/2}$	Yield coefficient in (8) for ion enhanced etching (substrate)
A_{ie}^p	$4 A_{ie}^e$	Yield coefficient in (8) for ion enhanced etching (polymer)
B_{sp}	9.3	Sputtering yield angular factor in (9)
E_{sp}^{th}	18 eV	Threshold energy for physical ion sputtering
$E_{ie.e}^{th}$	4 eV	Threshold energy for ion enhanced etching (substrate)
$E_{ie.p}^{th}$	4 eV	Threshold energy for ion enhanced etching (polymer)
k_{ie}, k_{ev}	1	Stoichiometric factors
S_e	0.9	Etchant sticking probability
S_p	0.26	Polymer sticking probability
S_{pe}	0.6	Polymer on etchant sticking probability
ρ_{SiO_2}	$2.2 \times 10^{22} \text{ cm}^{-3}$	SiO_2 density
$\rho_{Si_3N_4}$	$10.3 \times 10^{22} \text{ cm}^{-3}$	Si_3N_2 density
ρ_{Si}	$5.02 \times 10^{22} \text{ cm}^{-3}$	Si density
ρ_p	$2.0 \times 10^{22} \text{ cm}^{-3}$	Polymer density

V. RESULTS

The geometric profile of the stacked sheets after the etching simulation is shown in Figure 1. The etch process is performed until the via is slightly over-etched in the silicon substrate. On the sidewall one can observe the thin passivation layer in red.

The damage coefficient on a 3D clip of the surface after the etching process is shown in Figure 2. Due to the high directional trajectories of ions and the angle-dependent etching yield, the resulting ion damage is predominately confined to the bottom regions of the via. In the post etch treatment, the surface is etched with low energy ions, until the damaged layer has been removed and the crystalline silicon material underneath is revealed, so SEG can commence properly.

Next, SEG of silicon at the bottom of the channel hole is carried out with the resulting profiles depicted in Figure 3. When the damaged layer is present on the silicon substrate a large void is observed, leading to an ill-formed contact to the silicon source line (Figure 3a). Such defects will heavily impact the bottom select gate characteristics and reduce the overall quality of the memory stack. However, when the silicon substrate is cleaned prior to SEG, the desired SEG growth is obtained with the grown layer providing full contact with the silicon source line, as shown in Figure 3b.

VI. CONCLUSION

A physical process model is applied to simulate vertical channel hole etching in 3D NAND flash memory layers. The model, based on a level set surface representation and Monte

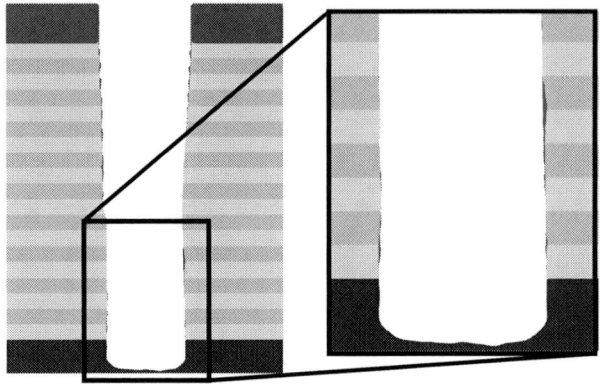

Fig. 1. Final 2D profile slice of the HAR via etching simulation. The alternating SiO_2/Si_3N_4 layers (yellow/cyan) are etched down to a bottom Si layer (blue). The thin sidewall layers (red) represent the deposited polymer formed during the etching process.

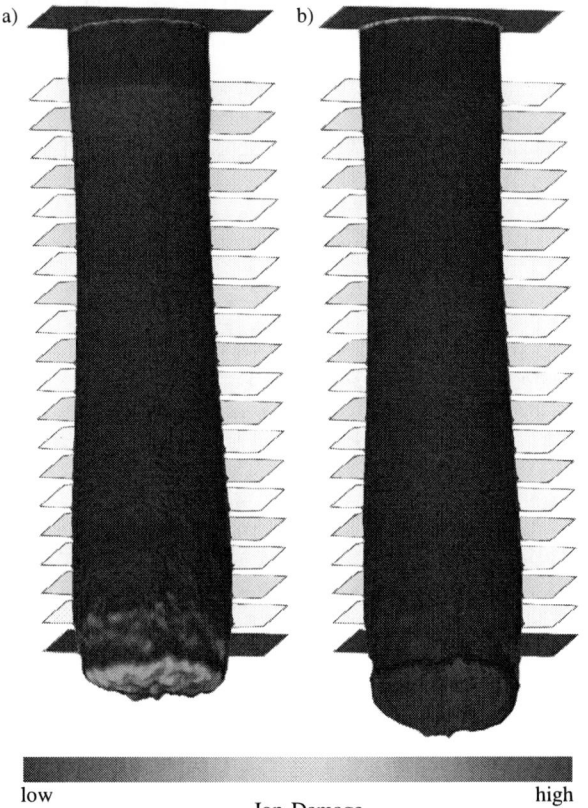

low   Ion Damage   high

Fig. 2. Clipped 3D surface representation of the ion damage (a) prior to the post etch treatment, and (b) after the post etch treatment.

Carlo ray tracing for flux calculation, is able to simulate the damage induced by ion-enhanced etching. The damage caused on the bottom silicon substrate is modeled using a surface damage coefficient. In a post etch treatment process, the damaged substrate is etched until the crystalline silicon at the bottom is exposed and the subsequent selective epitaxial growth of silicon provides full contact with the source line,

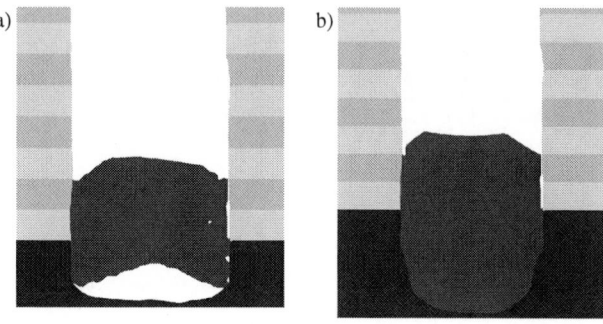

Fig. 3. Results for the selective epitaxial growth of silicon (red) after the etching process, on the bottom of a cropped 2D slice of the trench. In (a) a damaged layer on the surface is covering parts of the Si layer (blue, on the bottom), hence leading to an ill-formed SEG. In (b) the damaged layer has been removed and the SEG covers the entire bottom, leaving no voids.

leaving no undesired voids. This is modeled using the surface damage coefficient which is decreased according to an exponential attenuation during the etching of the bottom substrate.

Due to the physical nature of the presented model, relevant physical effects and underlying etch mechanics are represented appropriately. Therefore, the proposed model allows to analyze the physical behavior of the etch process in order to optimize the fabrication conditions during channel hole etching and the post etch treatment fabrication steps. Therefore, the model serves as a basis for an enhanced understanding of source contact formation in 3D NAND memory cells.

ACKNOWLEDGMENT

This work was supported in part by the Austrian Research Promotion Agency FFG (Bridge Young Scientists) under Project 878662 "Process-Aware Structure Emulation for Device-Technology Co-Optimization".

REFERENCES

[1] Y. Li, "3D NAND memory and its application in solid-state drives: Architecture, reliability, flash management techniques, and current trends," IEEE J. Solid-State Circuits 12, 2020, pp. 56–65
[2] H. Kim, S. Ahn, Y.G. Shin, K. Lee, E. Jung, "Evolution of NAND flash memory: From 2D to 3D as a storage market leader," IEEE Int. Memory Workshop (IMW), 2017, pp. 1–4
[3] N. Yabumoto et al., "Surface damage on Si substrates caused by reactive sputter etching," Jpn. J. Appl. Phys. 20, 1981, pp. 893–900
[4] L. Luo et al., "An effective process to remove etch damage prior to selective epitaxial growth in 3D NAND flash memory," Semicond. Sci. Technol. 34, 2019, pp. 095004-1–095004-5
[5] X. Klemenschits et al., "ViennaTools", https://github.com/ViennaTools, 2021
[6] J.A. Sethian, "Level set methods and fast marching methods", 2nd ed., Cambridge University Press, 1999
[7] A.L. Magna, G. Garozzo, "Factors affecting profile evolution in plasma etching of SiO_2," J. Electrochem. Soc. 150, 2003, pp. 178–185
[8] A. Toifl et al., "The level-set method for multi-material wet etching and non-planar selective epitaxy," IEEE Access 8, 2020, pp. 115406–115422
[9] D. Dutartre, A. Talbot, N. Loubet, "Facet propagation in Si and SiGe epitaxy or etching," ECS Trans. 3, 2006, pp. 473–487
[10] E. Gogolides, P. Vauvert, G. Kokkoris, G. Turban, A.G. Boudouvis, "Etching of SiO_2 and Si in fluorocarbon plasmas: A detailed surface model accounting for etching and deposition," J. Appl. Phys. 88, 2000, pp. 5570–5584
[11] O. Ertl and S. Selberherr, "A fast level set framework for large three-dimensional topography simulations," Comput. Phys. Commun. 180, 2009, pp. 1242–1250

The Semiconductor Model Solved by the Numerov Process
Over a Non-Uniform Grid

R. Brunetti, M. Rudan[1]

FIM Department, University of Modena and Reggio Emilia, Via Campi 213/A, I-41125 Modena, Italy

[1]"E. De Castro" Advanced Research Center on Electronic Systems (ARCES) and Department DEI

University of Bologna, Viale Risorgimento 2, I-40136 Bologna, Italy

Abstract – **The Numerov Process (NP) provides the solution of some classes of ODEs with an accuracy much superior to that of the standard finite-difference or box-integration methods. The original formulation of NP requires a uniform grid, which is a drawback for applications to, e.g., the semiconductor-device equations. Purpose of this work is showing how a method for extending NP to a non-uniform grid is applied to the solution of the drift-diffusion model. The method keeps the fifth-order accuracy of the original NP. In the multi-dimensional case, the variable transformation illustrated in the paper is found beneficial also when standard solution schemes are used; in fact, it makes the current-density vector well defined within each grid element.**

Keywords: **Semiconductor Model, Numerov Process, Non-uniform Grid.**

I. Introduction

The Numerov Process (NP) [1] is a highly-accurate method for solving ODEs of the form $-z'' = S(z, \xi)$, where $z(\xi)$ is the unknown function and S (which may be non linear) is prescribed. In the field of device modeling and design, the accuracy of NP makes its application advantageous for the solution of the semiconductor-device equations (details on the method and its accuracy are found, e.g., in [2], [3], [4]). As it stands, the semiconductor-device model does not have the form specified above; however, as briefly outlined in the next section, it is reduced to it by suitable transformations of the unknowns.

The application of NP in the original form requires the use of a uniform grid; although the better accuracy of the method may compensate for this aspect, the possibility of using a non-uniform grid is in any case important in semiconductor-device analysis: in fact, a typical behavior of the functions describing a semiconductor device (namely, the electric potential and carrier concentrations) is that of being almost constant over some regions of the device, and of exhibiting sharp variations in others. A scheme that extends NP to the case of a non-uniform grid has recently been proposed in [5]; it is applied here to semiconductor-device analysis. The variable transformation of the unknowns is applicable also in the multi-dimensional case; although the multi-dimensional extension of NP to a general grid is still wanting, it is shown here that the variable transformation illustrated is beneficial also when standard solution schemes are used; in fact, it eliminates a long-unsolved difficulty connected to the calculation of the current-density vector within each grid element.

II. Model Equations

In the one-dimensional model of semiconductor devices, one of the equations (the Poisson equation $-\varepsilon \, \mathrm{d}^2\varphi/\mathrm{d}x^2 = \varrho$) is suitable as it stands for the application of NP, because the first derivative $\mathrm{d}\varphi/\mathrm{d}x$ is absent. The normalized form is obtained by letting $\varphi = (k_B \, T/q) \, u$, with u the normalized electric potential and $k_B \, T/q$ the thermal voltage. Defining $\xi = x/L$, where $L = [\varepsilon \, k_B \, T/(q^2 \, n_{\mathrm{int}})]^{1/2}$ is the intrinsic Debye length, ε the semiconductor permittivity, n_{int} the intrinsic concentration at T, and using $\varrho = q \, (p - n + N)$, provides $-u'' = (p - n + N)/n_{\mathrm{int}}$. In the above, p (n) is the hole (electron) concentration and $N(x)$ the net doping concentration; apices indicate derivation with respect to ξ.

In contrast, the continuity and transport equations for electrons and holes are not suitable for NP, but become such after a transformation of the unknown function. Considering the electrons first, the continuity equation reads $\partial n/\partial t - (1/q) \, \mathrm{d}J_n/\mathrm{d}x = W_n$, with J_n the electron-current density and W_n the net generation rate of the conduction band [6]; in turn, the drift-diffusion transport equation reads $J_n = q \, D_n \, (\mathrm{d}n/\mathrm{d}x - n \, \mathrm{d}u/\mathrm{d}x)$. Combining the continuity and transport equations provides a second-order equation in the unknown n, where also the first derivative of n appears. The transformation yielding the NP-suitable form is based on the new unknown $\chi = (n/n_{\mathrm{int}}) \exp(-u/2)$; the second-order equation thus found reads

$$-\chi'' = -\left[\frac{(u')^2}{4} + \frac{u''}{2}\right] \chi - Q \exp(-u/2), \quad (1)$$

with $Q = L^2 \, (\partial n/\partial t - W_n)/(n_{\mathrm{int}} \, D_n)$. As anticipated, the first derivative of the unknown function is not present in (1). The continuity and transport equations for holes are treated in a similar manner, using the new unknown $\omega = (p/n_{\mathrm{int}}) \exp(u/2)$; one finds

$$-\omega'' = -\left[\frac{(u')^2}{4} - \frac{u''}{2}\right] \omega - R \exp(u/2), \quad (2)$$

with $R = L^2 \, (\partial p/\partial t - W_p)/(n_{\mathrm{int}} \, D_p)$. Finally, using the new unknowns in the Poisson equation yields

$$-u'' = \omega \exp(-u/2) - \chi \exp(u/2) + N/n_{\mathrm{int}} . \quad (3)$$

978-1-6654-3746-2/21 $31.00 © 2021 IEEE

III. Factorization of the Transport Equations

Taking the steady-state case ($\partial n/\partial t = \partial p/\partial t = 0$), and assuming that the dominant generation-recombination phenomenon is trap-assisted thermal recombination, it follows $-W_n = -W_p = U = (p\,n - p_B\,n_B)/[\tau_n\,(p + p_B) + \tau_p\,(n + n_B)]$ with τ_p, τ_n the carrier lifetimes [6]. For non-degenerate dopant concentrations it is $p_B \simeq n_B \simeq n_{\text{int}}$ whence, letting $\eta = \exp(u/2)$,

$$\frac{D_n}{L^2}\,Q = \frac{D_p}{L^2}\,R = \frac{\omega\,\chi - 1}{\tau_n\,[1 + \omega/\eta] + \tau_p\,[1 + \chi\,\eta]}\,. \quad (4)$$

The model made of (1–4) is a non-linear system in the unknowns χ, ω, u, whose solution can be tackled with the standard iteration scheme: namely, u, Q, R are taken from the previous iteration when (1, 2) are solved, whereas χ, ω are taken from the previous iteration when (3) is solved. Within the iteration process, the Poisson equation (3) is linearized with respect to u (Sec. V).

The form of (1, 2) is made less unwieldy by introducing the operators $\hat{b} = u'/2 + \mathrm{d}/\mathrm{d}\xi$, $\hat{b}^\dagger = u'/2 - \mathrm{d}/\mathrm{d}\xi$; it is readily found that (1, 2) factorize into two first-order equations: $\hat{b}\,f = Q/\eta$, $-\hat{b}^\dagger\,\chi = f$ for electrons, and $-\hat{b}^\dagger\,g = R\,\eta$, $\hat{b}\,\omega = g$ for holes (factorization is achievable also for the standard form of the transport equations; however, the standard form makes the application of NP impossible). Given a non-uniform grid with nodes labeled $0 = \xi_0 < \xi_1 < \cdots < \xi_N < \xi_{N+1} = H$, and elements $h_{i+1} = \xi_{i+1} - \xi_i$, $i = 0, 1, \ldots, N$, the factorized equations are integrated over each element with no loss of generality (the details, along with the treatment of the boundary conditions, are given in [5]); the problem is therefore transformed into the calculation of integrals: considering, e.g., element h_{i+1} ($\xi_i \leq \xi \leq \xi_{i+1}$), they are

$$I(\xi, \xi_i) = \int_{\xi_i}^{\xi} Q\,\mathrm{d}\theta\,, \qquad Y(\xi, \xi_i) = \int_{\xi_i}^{\xi} f/\eta\,\mathrm{d}\theta\,. \quad (5)$$

IV. The Numerov Process Over a Non-Uniform Grid

The above derivation shows that the quality of the solution depends on the accuracy with which integrals (5), or the equivalent for R, g, are calculated. Integrals of the same type, still containing the unknowns, are found when solving the Poisson equation (3). For a given element h_{i+1}, the integrals of the form $\int_{\xi_i}^{\xi_{i+1}} \sigma\,\mathrm{d}\xi$ are calculated by Simpson's integration, whose accuracy is $\mathrm{O}(h_{i+1}^5)$ [7]; Simpson's integration requires the knowledge of the nodal values σ_i, σ_{i+1} (which are unknowns proper) and of the midpoint value $\sigma_{i+1/2}$ (which, instead, is not available). The usefulness of the Numerov Process becomes evident at this stage: one observes in fact that the functions whose midpoint values are sought depend in all cases on the solutions of equations of the form $-z'' = S(z, \xi)$ (namely, such that the first derivative of the unknown is absent, compare with (1–3)), and that NP applies because the three nodes ξ_i, $\xi_{i+1/2}$, ξ_{i+1} are equally spaced. The NP formula provides a linear interpolation of $\sigma_{i+1/2}$ in terms of σ_i, σ_{i+1} (details are in

[5]), which allows one to complete the calculation of the integral; since the accuracy of NP is also $\mathrm{O}(h_{i+1}^5)$, the order of accuracy of the Simpson integration is preserved. A last consideration is that the procedure depicted above applies to each element independently; therefore, the scheme is applicable regardless of the size of the elements, namely, it holds for a general, non uniform grid.

V. Reconstruction

As anticipated, the Poisson equation (3) is linearized; this is done by letting $u = u_{\text{old}} + \delta u$, to yield

$$-\delta u'' = c\,\delta u + s\,, \qquad c = -(1/2)\,[\omega/\eta + \chi\,\eta]_{\text{old}}\,, \quad (6)$$

$$s = [u'' + \omega/\eta - \chi\,\eta + N/n_{\text{int}}]_{\text{old}}\,. \quad (7)$$

The discretization procedure transforms the model equations into algebraic systems, whose unknowns are the vectors corresponding to χ, ω, or δu. Suffix "old" reminds one that $\eta(u)$, χ, and ω take the values available at the end of the last iteration; thus, prior to convergence the right hand side of (7) differs from zero.

Coefficients c and s playing the same role as those of (6) and (7) are defined also for (1) and (2); specifically,

$$c = -(u')^2/4\,, \qquad s = -Q/\eta - u''\,\chi/2 \quad (8)$$

for the electron equation (1), and

$$c = -(u')^2/4\,, \qquad s = -R\,\eta + u''\,\omega/2 \quad (9)$$

for the hole equation (2), with Q, R given by (4). As above, the quantities entering the definition of c and s are "old", namely, they are taken from the last iteration.

The solution of the algebraic equations provides the updated nodal values of the unknowns; after this, it is necessary to reconstruct the updated midpoint values. Indicating χ, ω, or δu with ψ, the midpoint values read, for $i = 2, N$ [5]

$$\psi_{i-1/2} = a_i^{i-1}\,\psi_{i-1} + a_i^i\,\psi_i + h_i^2\,b_i^{i-1/2}\,, \quad (10)$$

where $a_i^{i-1} = (48 + h_i^2\,c_{i-1})/D_i$, $a_i^i = (48 + h_i^2\,c_i)/D_i$, and

$$b_i^{i-1/2} = \frac{s_{i-1} + 10\,s_{i-1/2} + s_i}{D_i}\,, \quad (11)$$

$$D_i = 96 - 10\,h_i^2\,c_{i-1/2}\,. \quad (12)$$

It is worth observing that c in (6), (8), and (9) is never positive; as a consequence, definition (12) provides a strictly-positive denominator D_i. From the numerical standpoint, the non-positive definitenesss of c brings the advantage that c adds weight to the main diagonal of the algebraic system to be solved.

From (8), (9) it also follows that the nodal and midpoint values of u' and u'' must be used as well in the calculation. The "old" second derivative is always available for both nodal and midpoint values; as for the first derivative one applies NP to find, at the midpoint,

$$u'_{i-1/2} = \frac{u_i - u_{i-1}}{h_i} + \frac{h_i}{24}\left(u''_{i-1} - u''_i\right). \quad (13)$$

For the first derivative at the nodes, instead, one must combine NP with the Simpson integration, because the nodes involved are not equally spaced; one finds

$$u'_{i-1} = \frac{u_i - u_{i-1}}{h_i} - \frac{h_i}{6}\left(u''_{i-1} + 2\,u''_{i-1/2}\right), \quad (14)$$

$$u'_i = \frac{u_i - u_{i-1}}{h_i} + \frac{h_i}{6}\left(u''_i + 2\,u''_{i-1/2}\right). \quad (15)$$

Note that subtracting (14) from (15) yields Simpson's formula $u'_i - u'_{i-1} = h_i\left(u''_{i-1} + 4\,u''_{i-1/2} + u''_i\right)/6$. Also, letting $i \leftarrow i+1$ in (14), and adding term-by-term to (15) the resulting relation, provides a more symmetric expression of u'_i.

VI. Results

The procedure has been applied to a one-dimensional structure of the $n^+ - n$ type, with a $G = LH = 10\ \mu$m length and a donor-doping concentration of the form

$$N = N(G) + \frac{N(0) - N(G)}{1 + \exp[k\,(x - G/2)]}, \quad (16)$$

where $N(0) = 10^{22}$ m^{-3}, $N(G) = 10^{21}$ m^{-3}, and $k = 300$ m^{-1}. Being the structure unipolar, it suffices to solve (1) and (3), after letting $\omega = 0$ in the latter. Fig. 1 shows the electron concentration $n(x)$ when $V(0) - V(G) = 1$ V; the figure zooms in on the central part of the curve, where the variation of n is maximal and the nodal distribution denser. For comparison, a simulation on a 2000-node, uniform grid was carried out first; subsequent simulations were then carried out using non-uniform grids, whose node spacing was denser in the central part of the curve, and coarser elsewhere. The figures in the inset of Fig. 1 show the total number of nodes of each grid. No appreciable difference was detected between the results from the 2000-node grid and the others, down to 120 nodes. Some discrepancies, visible in the figure, appeared with the 60-node grid. A similar behavior was found after reversing the bias, namely, $V(0) - V(G) = -1$ V.

VII. Multi-Dimensional Case

The natural evolution of this research line is extending the method to more than one dimension. By way of example, the multi-dimensional form of (1) reads

$$-\nabla^2\chi = -\left[\frac{|\nabla u|^2}{4} + \frac{\nabla^2 u}{2}\right]\chi - Q\exp(-u/2), \quad (17)$$

also expressible in factorized form by means of the operators $\hat{b} = \nabla u/2 + \nabla$ and $\hat{b}^\dagger = \nabla u/2 - \nabla$. The extension of NP to the multi-dimensional case using tensor-product grids has been achieved for uniformly-spaced grids (in [8] for the

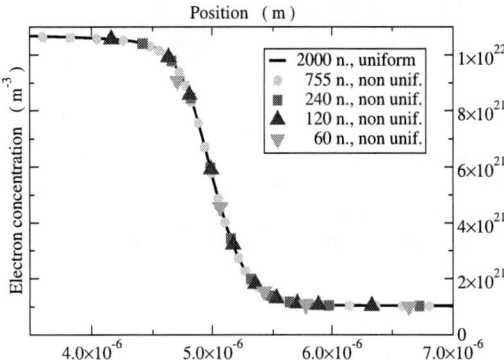

Fig. 1. Electron concentration calculated over different, non-uniform grids, compared with that obtained with a denser, uniform grid ($V(0) - V(G) = 1$ V, $n_{\text{int}} = 10^{16}$ m^{-3}). For the sake of clarity, only some of the markers are shown.

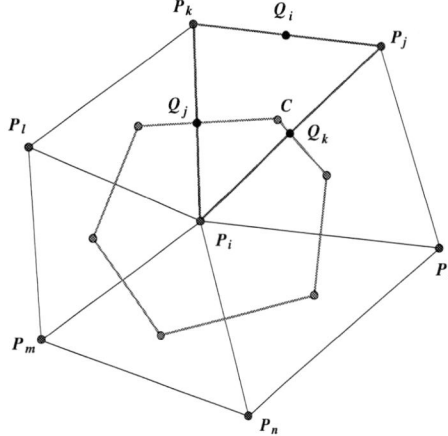

Fig. 2. Example of triangle-based grid used in the simulation of semiconductor devices in two dimensions.

Schrödinger equation, and in [3], [4] for the semiconductor-device model), whereas that to non-uniform, tensor-product grids has not been achieved yet.

In semiconductor-device analysis, due to the complex form of the integration domain and to the presence of internal interfaces between different materials, a type of grid more general than the tensor-product one is often adopted, whose elements are simplexes: considering for instance a two-dimensional case, the grid is made of triangular elements like those shown in Fig. 2. The triangles correspond to the elements like h_i or h_{i+1} of the one-dimensional case; the domain with the red boundary is the Wigner-Seitz cell of node P_i, and its sides are the axes of the triangles' sides emanating from P_i; such a domain corresponds to

978-1-6654-3746-2/21 $31.00 © 2021 IEEE

the $(h_i + h_{i+1})/2$ cell of the one-dimensional case.

The typical method for solving the multi-dimensional form of the Poisson equation (6), (7) on this grid consists in applying the Gauss theorem over the Wigner-Seitz cell Ω_i, and in approximating the area integral at the right hand side by replacing the integrand with a constant. In turn, the flux across the cell boundary is approximated by replacing δu with a linear function over each triangle; this procedure, also called *Box-Integration Method* (BIM), has become popular for solving the drift-diffusion model since the late 70s [9], [10], and has subsequently been extended to the higher moments of the BTE [11], [12], [13], [14].

When BIM is applied to the transport part of the model, the vector equation like, e.g., $\boldsymbol{J}_n/(q\,D_n) = \operatorname{grad} n - n \operatorname{grad} u$, is projected onto the sides of each triangle, and the projection is approximated with a constant over the side itself; the resulting expression is integrated over each side, this providing the so-called *exponential-fitting scheme*; finally, the result is used in the calculation of the flux across the cell boundary $\partial\Omega_i$.

A drawback of the exponential-fitting scheme is that the current density is not well defined over each triangle: one finds in fact a different vector depending on which pair of projections one uses to reconstruct the current from the scalar components. This is due to the fact that, apart from the special case of a purely-diffusive transport, the current density is not a gradient; therefore, forcing its components to be constant along arbitrarily-oriented directions introduces an inconsistency. In the practical applications, the impossibility of reconstructing the current-density vector introduces an error in the calculation of the currents at the device contacts, and in the analysis of physical effects driven by the current density like, e.g., impact ionization.

The variable transformation leading to the form (17) of the transport model eliminates the drawback: the equation to be solved is in fact identical to the linearized Poisson equation; therefore, it can be solved with BIM with respect to χ (given u), using the same approximations. The dependence of the electron concentration on the coordinates is extracted from $n = n_{\text{int}}\, \chi \exp(u/2)$ only after the solution is achieved; also, combining the latter with $\boldsymbol{J}_n/(q\,D_n) = \operatorname{grad} n - n \operatorname{grad} u$ yields the current density in terms of χ and u:

$$\boldsymbol{J}_n/(q\,D_n\, n_{\text{int}}) = \exp(u/2)\,(\operatorname{grad}\chi - \chi\,\operatorname{grad} u/2)\,. \quad (18)$$

The above equation defines uniquely $\boldsymbol{J}_n$, whose projections turn out to be quite different from constants. This result shows that the variable transformation $n = n_{\text{int}}\,\chi \exp(u/2)$, which in one dimension makes it possible to implement NP thanks to the elimination of the first derivative, is also beneficial when the multi-dimensional version of the transport model in semiconductors is considered: thanks again to the elimination of the first derivatives,

all physical quantities of interest are in fact well defined over the grid elements.

VII. Conclusions

The main conclusions drawn from this work are:

1) A suitable transformation of the unknowns makes it possible to apply the Numerov Process to the mathematical model of semiconductor devices.
2) In one dimension, NP is extendable to the case of a non-uniform grid by combining it with the Simpson integration.
3) The strategy can be applied to higher-order transport models (e.g., hydrodynamic) because the structure of the equations is basically the same.
4) The variable transformation of the mathematical model of semiconductor holds also in the multi-dimensional case, and can be used over tensor-product grids.

Finally, in the multi-dimensional case the transformation of the unknowns is applicable also when standard solution schemes (e.g., BIM) are used over grids made of simplexes; in this case, it has the advantage of eliminating a long-unsolved difficulty connected to the calculation of the current-density vector within each grid element.

References

[1] B. Numerov, *Publ. Obs. Central Astrophys.*, vol. 2, pp. 188–288, 1923 (in Russian).

[2] F. Buscemi, E. Piccinini, R. Brunetti, and M. Rudan, in *IEEE International Conference on Simulation of Semiconductor Processes and Devices (SISPAD)*, Yokohama, September 2014, pp. 161–164.

[3] N. Speciale, R. Brunetti, and M. Rudan, in *2019 International Conference on Simulation of Semiconductor Processes and Devices (SISPAD 2019)*, D. Esseni, P. Palestri, and D. Rideau, Eds. Udine, Italy: IEEE, Sept. 4–6 2019, pp. 37–40.

[4] ——, *Advances in Science, Technology and Engineering Systems Journal*, vol. 5, no. 6, pp. 1414–1421, 2020.

[5] R. Brunetti, N. Speciale, and M. Rudan, *J. of Computational Electronics*, vol. 20, no. 4, pp. 1105–1113, April 2021.

[6] M. Rudan, *Physics of Semiconductor Devices*. Springer, 2018.

[7] W. H. Press, B. R. Flannery, S. A. Teukolsky, and W. T. Wetterling, *Numerical Recipes — The Art of Scientific Computing*. New York: Cambridge University Press, 1988.

[8] T. Graen and H. Grubmüller, *Computer Physics Communications*, vol. 198, pp. 169–178, 2016.

[9] P. E. Cottrell and E. M. Buturla, in *Proc. of the* NASECODE I *Conference*, B. T. Browne and J. J. H. Miller, Eds. Dublin: Boole Press, 1979, pp. 31–64.

[10] E. M. Buturla, P. E. Cottrell, B. M. Grossman, and K. A. Salsburg, *IBM J. of Res. and Dev.*, vol. 25, no. 4, pp. 218–231, July 1981.

[11] M. Rudan and F. Odeh, *COMPEL*, vol. 5, no. 3, pp. 149–183, 1986.

[12] M. Rudan, F. Odeh, and J. White, *COMPEL*, vol. 6, no. 3, pp. 151–170, 1987.

[13] A. Forghieri, R. Guerrieri, P. Ciampolini, A. Gnudi, M. Rudan, and G. Baccarani, *IEEE Trans. on CAD of ICAS*, vol. CAD-7, no. 2, pp. 231–242, 1988.

[14] A. Gnudi, F. Odeh, and M. Rudan, *European Trans. on Telecommunications and Related Technologies*, vol. 1, no. 3, pp. 307–312 (77–82), 1990.

978-1-6654-3746-2/21 $31.00 © 2021 IEEE

High Temperature Influence on the Trade-off between gm/I$_D$ and f$_T$ of nanosheet NMOS Transistors with Different Metal Gate Stack

Vanessa C. P. Silva[1,*], Joao A. Martino[1], Eddy Simoen[2], Anabela Veloso[2] and Paula G. D. Agopian[1,3]

[1] LSI/PSI/USP, University of Sao Paulo, Sao Paulo, Brazil
[2] imec, Leuven, Belgium
[3] UNESP, Sao Paulo State University, Sao Joao da Boa Vista, Brazil
*e-mail: vcpsilva@usp.br

Abstract— **This work presents an experimental analysis of the trade-off between transistor efficiency (gm/I$_D$) and unit gain frequency (f$_T$) of nanosheet field effect transistors (NSFETs) with different metal gate (MG) stack, considering the influence of high temperature (T), until T=200 °C. The results are very promising for both MG stacks. The MG stack (n*) presents a high f$_T$ about 260 GHz (T=25 °C and L=28 nm) and a gm/I$_D$ about 37 V^{-1} (T=25 °C and L=200 nm). The MG stack (m*) also presents very good characteristics, like a f$_T$ about 252 GHz (T=25 °C and L=28 nm) and a gm/I$_D$ about 35 V^{-1} (T=25 °C and L=200 nm). From the analyses as a function of the inversion coefficient (IC), it was possible to determine that the optimal operation point occurs in the transition from moderate to strong inversion for L=28 nm and it is in strong inversion for long channel devices. In all cases, although the intrinsic voltage gain (A$_V$) is degraded moving away from weak inversion, the degradation was not very pronounced up to the optimal operation point and considering the temperature variation, the A$_V$ presents a greater stability at the optimal point than in weak inversion.**

Keywords-Nanosheets (NS), MOSFET, Analog operation, f$_T$, Transistor Efficiency.

I. INTRODUCTION

Gate-all-around nanowire or nanosheet field effect transistors (GAA NWFETS or NSFETs) have emerged as the leading alternative to meet the industry's requirements for sub 5 nm technology nodes, representing the ultimate scaled architecture of a transistor for complementary metal oxide semiconductor (CMOS). NSFETs can achieve higher drive currents due to their larger effective widths, and the GAA configuration provides improved electrostatic control, which allows further gate length scaling [1].

Due to their superior characteristics the NSFETs are analyzed in this work, to evaluate the optimal application point for analog circuits, from a trade-off analysis between transistor efficiency and unit gain frequency and its behavior with the influence of high temperature and channel length. The influence of the metal gate stack on the A$_V$ is also presented in this work.

II. DEVICE CHARACTERISTICS

In this work GAA NSFETs consisting of 22 parallel fins in the device layout, each corresponding to two vertically stacked nanosheets, have been analyzed. Each transistor has the following characteristics: effective oxide thickness (EOT) about 0.9 nm, channel width (W$_{NS}$) about 15 nm, channel height (h$_{NS}$) about 11 nm, and three channel lengths (L) 28, 70 and 200 nm have been investigated. Each transistor was analyzed for two different metal gate (MG) stacks: one composed of TiN and Al- based layer (m*), with a total thickness of about 7.5 nm – MG stack (m*), and the other composed of TiN and Al- based layer (n*) with a total thickness

of about 4.7 nm – MG stack (n*). Figure 1-A and Figure 1-B present the Transmission Electron Microscopy (TEM) cross section images for MG stack (m*) and (n*), respectively [2].

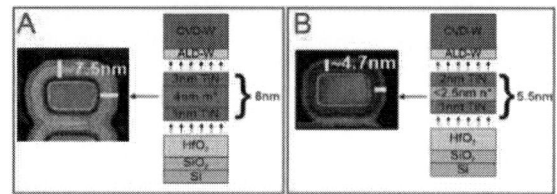

Figure 1 – TEM images of the different MG stacks: (A) with an Al-based layer (m*) and (B) with an Al-based layer (n*) [2].

III. RESULTS AND ANALYSIS

The first analysis is the drain current (I$_{DS}$) as a function of drain voltage (V$_{GT}$) with a gate voltage overdrive (V$_{GT}$) of 200 mV. The presented data was experimentally measured for L=28 nm, varying the temperature (T) from 25 °C to 200 °C and for both MG stack, as presented in Figure 2.

The behavior when analyzing the temperature is the same for both MG stacks, there is a reduction in I$_{DS}$ with the increase of T, which is expected, due the mobility degradation at high temperatures. The MG stack (m*) presents an I$_{DS}$ with a level about 30 μA lower than the MG stack (n*).

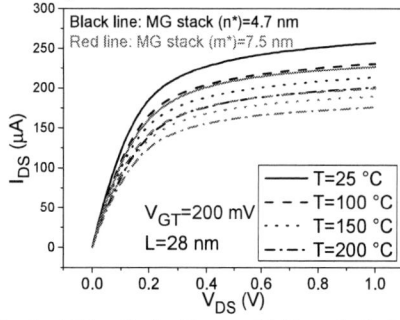

Figure 2 – I$_{DS}$ X V$_{DS}$, for L=28 nm and MG stacks (m* and n*).

A. Inversion Coefficient

The following results will be presented as a function of the Inversion Coefficient (IC) which numerically represents the channels' level of inversion, which can be Weak Inversion (WI: IC < 0.1), Moderate Inversion (MI: 0.1 < IC < 10) and Strong Inversion (SI: IC >10). The IC is denoted as the ratio between the drain current (I$_{DS}$) and the transition current (I$_{DSt}$)- the current between the transition from WI to SI [3], in our study I$_{DSt}$ was extracted as the correspondent current for the threshold voltage (V$_T$) from each transistor. The use of IC for the analysis allows a conscious choice of the region of operation

978-1-6654-3746-2/21 $31.00 © 2021 IEEE

for MOSFET devices, enabling to achieve higher figures-of-merit (FoM) for a given situation [3], [4].

B. Trade-off Between Transistor Efficiency and Unit Gain Frequency

The transistor efficiency (gm/I_D) for the MG stack (n*) is presented in Figure 3-A and for the MG stack (m*) in Figure 4-A, both for L=28, 70 and 200 nm, varying the temperature from 25 to 200 °C. The channel length influence on gm/I_D is low and the obtained values for WI, did not change too much indicating a slight degradation of the subthreshold swing (SS) with L reduction. Focusing on the temperature influence on gm/I_D, it affects all three inversion regions. In WI it degrades SS and consequently transistor efficiency, while in SI, it degrades the mobility and consequently the transconductance (gm). Both analyzed devices show a similar degradation of gm/I_D with increasing temperature, while long devices with MG stack (n*), in WI present a degradation from 37 V^{-1} (at 25 °C) to 23 V^{-1} (at 200 °C), the degradation presented for MG stack (m*) devices was from 35 V^{-1} (at 25 °C) to 22 V^{-1} (at 200 °C).

For both MG stacks the gm/I_D achieves its maximum value at WI (IC<0.1), where this parameters is inversely proportional to the SS, remaining almost constant, and starts to reduce in MI while IC increases, due to the mobility degradation and series resistance that starts to rule the gm/I_D [5].

The unit gain frequency (f_T) is another important analog parameter and can be calculated through $f_T = gm_{sat}/2\pi C_{gg}$, where gm_{sat} is the transconductance when biased at saturation region (V_{DS}=700mV) and C_{gg} is the total gate capacitance. For extracting this parameter, the capacitance was experimentally measured for a transistor with the same characteristics of the ones analyzed in this work, with a channel length of approximately 1 μm.

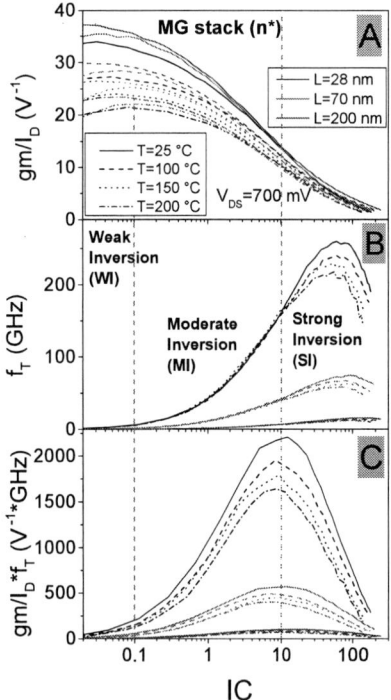

Figure 3 – gm/I_D (A), f_T (B) and gm/I_D*f_T (C) as a function of Inversion Coefficient for MG stack (n*).

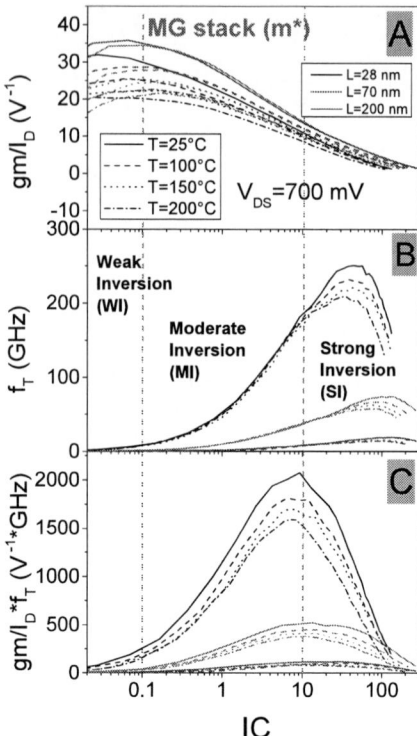

Figure 4 – gm/I_D (A), f_T (B) and gm/I_D*f_T (C) as a function of Inversion Coefficient for MG stack (m*).

Figure 3-B and Figure 4-B present the f_T as a function of IC for L=28, 70 and 200 nm, and for T from 25 °C to 200 °C for MG stacks (n*) and (m*) devices, respectively. The influence of L is noted in this case since it affects C_{gg} and gm_{sat} (the increase of L increases C_{gg} and degrades gm_{sat}) and the temperature mainly affects the carrier mobility and consequently gm_{sat} values. Both devices present the same behavior. Although the temperature causes a degradation of approximately 20% in the f_T value, the most important factor in this analysis is the channel length, which in turn causes a strong reduction of f_T as L increases, as presented in Table 1.

For both cases (MG stacks n* and m*) the f_T increase with IC, achieving its maximum at SI, where gm_{sat} plays a role and the maximum value is also obtained in SI, and after the peak there is a drop caused by the velocity saturation [3].

It is very challenging to cope with the trade-off between gm/I_D and f_T, since gm/I_D characterizes the DC performance, the f_T characterizes the frequency performance, and they behave differently when analyzing the IC levels. The f_T achieves its maximum value at SI, while for gm/I_D the maximum value is obtained at WI.

The compromise between these two parameters (gm/I_D*f_T) is a helpful tool for a design project, where it helps to find an optimal point of application for a given situation. Figure 3-C and Figure 4-C present gm/I_D*f_T for MG stack n* and m*, respectively. The best application point is obtained at the transition from MI to SI, for both MG stacks and L=28 nm. For L=70 and 200 nm, this point is in SI, as presented in Figure 5, where the IC values extracted for the maximum value of gm/I_D*f_T for both MG stacks are represented. There is a reduction of IC when increasing T, because while I_{DSt} increases due to V_T reduction, I_{DS} in strong inversion reduces due to the

978-1-6654-3746-2/21 $31.00 © 2021 IEEE

mobility degradation, and both contribute to the reduction in IC at the optimal application point. The IC values, for L=70 and 200 nm, are higher than 10, i.e., the optimal application point is in SI.

Table 1 – Unit gain frequency values for MG stacks n* and m* and their respective degradation (in percentage) with temperature.

L (nm)	f_T (GHz) – MG stack (n*)			f_T (GHz) – MG stack (m*)		
	25°C	200°C	%	25°C	200°C	%
28	260	219	16	252	209	17
70	75	59	21	74	58	22
200	16	13	25	19	14	26

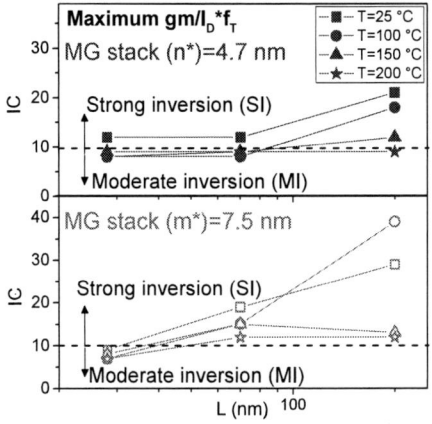

Figure 5 – Corresponding Inversion Coefficient values for maximum gm/I$_D$*f$_T$, for 2 different MG stacks.

C. Intrinsic Voltage Gain

The intrinsic voltage gain (A$_v$) is also an important analog parameter and can be obtained through A$_V$=(gm/I$_D$)*V$_{EA}$, V$_{EA}$ being the Early voltage. For a first order analysis, V$_{EA}$ is considered constant for all IC levels, since it did not present a significant variation with IC [3], [6].

Figure 6-A and Figure 6-B present A$_V$ as a function of IC for MG stack (n*) and (m*), respectively, for L=28, 70 and 200 nm, and a temperature range from 25 °C to 200 °C. The L influences the A$_V$ due to the channel length modulation, as presented in ref [7], i.e., longer transistors present a higher A$_V$. The influence of temperature also degrades A$_V$ since it affects SS and mobility, from 25 to 200 °C. Temperature degrades A$_V$ from 59 dB to 56 dB (WI; IC=0.1) and from 51 dB to 49 dB (SI; IC=10), both for L=200 nm and MG stack (n*). The increase in IC causes a degradation in A$_V$, for all L and T. Comparing the maximum A$_V$ value at IC=0.1 with A$_V$ in the optimal application point (at IC=10), it degrades from 59 dB to 51 dB (for L=200 nm and MG stack n*).

For both MG stacks the influence of temperature in intrinsic voltage gain reduces while increasing IC. Although SI presents a lower A$_V$ in SI, it is still the best application region, even in terms of A$_V$ due to its higher stability when increasing temperature.

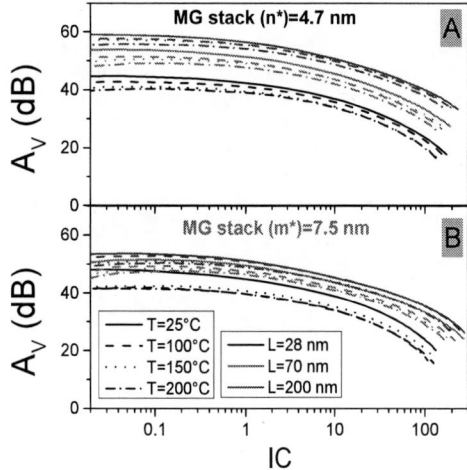

Figure 6 – Intrinsic voltage gain as a function Inversion Coefficient for metal gate stack n* (A) and m* (B).

IV. CONCLUSIONS

A trade-off analysis between gm/I$_D$ and f$_T$ was presented for NSFETs with different MG stack and channel length. Additionally, the influence of temperature was also analyzed. The influence of channel length is mainly observed for f$_T$, where for L=28 nm it reaches about 260 GHz and for L=200 nm reaches about 16 GHz, at T=25 °C and MG stack (n*).

The A$_V$ was also analyzed in this work, and L also affects this parameter, causing a degradation from 59 dB (L=200 nm) to 44 dB (L=28 nm) (T=25 °C and MG stack (n*)). This degradation with L, that is expected, is mainly explained by the channel length modulation.

Both MG stacks present the same behavior for increasing IC and temperature. When increasing T, gm/I$_D$, f$_T$ and A$_V$ degrade, due to the influence of T on the carrier mobility.

Making the compromise between gm/I$_D$ and f$_T$ it was demonstrated that the optimal application point is near the transition between Moderate Inversion to Strong Inversion for L=28 nm and at Strong Inversion for L=70 and 200 nm. Even for A$_V$ it is a good operation point since the temperature dependence reduces when increasing IC. For T from 25°C to 200 °C, at IC=0.1 (WI) A$_V$ degrades about 6% (L=200 nm, MG stack m*) and at IC=10 (SI) A$_V$ degrades about 2% (L=200 nm, MG stack m*).

ACKNOWLEDGMENT

The authors acknowledge CNPq and CAPES for the financial support. The devices have been processed in the frame of imec's Core Partner Program on Logic Devices.

REFERENCES

[1] A. V. De Oliveira, A. Veloso, C. Claeys, N. Horiguchi, and E. Simoen, "Low-Frequency Noise in Vertically Stacked Si n-Channel Nanosheet FETs," *IEEE Electron Device Lett.*, vol. 41, no. 3, pp. 317–320, Mar. 2020.

[2] A. Veloso *et al.*, "Scaled, novel effective workfunction metal gate stacks for advanced low-V T, gate-all-around vertically stacked nanosheet FETs with reduced vertical distance between sheets," in *International Conference on Solid State Devices and Materials*, 2019.

[3] D. M. Binkley, "Tradeoffs and Optimization in Analog CMOS Design," in *2007 14th International Conference on Mixed Design of Integrated Circuits and Systems*, 2007, pp. 47–60.

[4] E. Afacan and G. Dundar, "Inversion coefficient optimization assisted analog circuit sizing tool," *SMACD 2017 - 14th Int. Conf. Synth. Model. Anal. Simul. Methods Appl. to Circuit Des.*, 2017.

[5] D. Boudier, B. Cretu, E. Simoen, A. Veloso, and N. Collaert, "Detailed characterisation of Si Gate-All-Around Nanowire MOSFETs at cryogenic temperatures," *Solid. State. Electron.*, vol. 143, pp. 27–32, May 2018.

[6] P. Jespers, *The gm/ID Methodology, A Sizing Tool for Low-voltage Analog CMOS Circuits*. Boston, MA: Springer US, 2010.

[7] V. C. P. Silva, W. F. Perina, J. A. Martino, E. Simoen, A. Veloso and P. G. D. Agopian, "Analog Figures of Merit of Vertically Stacked Silicon Nanosheets nMOSFETs With Two Different Metal Gates for the Sub-7 nm Technology Node Operating at High Temperatures," in IEEE Transactions on Electron Devices, vol. 68, no. 7, pp. 3630-3635, July 2021.

On the Asymmetry of the DC and Low-Frequency Noise Characteristics of Vertical Nanowire pMOSFETs with Bulk Source Contact

Eddy Simoen, Anabela Veloso, and Philippe Matagne

Imec, UPM, Kapeldreef 75, B-3001 Leuven, Belgium
eddy.simoen@imec.be

Abstract—**In this work, the impact of switching the source and drain on the low-frequency noise of Gate-All-Around (GAA) Vertical Nanowire (VNW) pMOSFETs with a bulk silicon source contact is investigated. As shown, significantly lower 1/f noise is observed in reverse operation compared with the normal forward configuration. Considering the mobility fluctuations origin of the 1/f noise, this indicates a pronounced impact of the choice of the source/drain contact on the conduction in the silicon nanowire. On the other hand, little effect has been found from the doping density in the NWs.**

Keywords-vertical nanowires; gate-all-around; low-frequency noise; forward and reverse operation

I. INTRODUCTION

Gate-All-Around (GAA) architectures provide the ultimate control of the short-channel effects, thereby optimizing the device performance. GAA Vertical Nanowire (VNW) transistors are the most promising devices from a viewpoint of minimum footprint and 3D integration [1]. As a bonus, it turns out that VNW FETs exhibit superior low-frequency (LF) noise behavior, compared with FinFET and planar architectures, for frequencies f below 1 kHz [2], [3]. It has been found that the dominant 1/f or flicker noise in VNW FETs is govened by so-called mobility ($\Delta\mu$) fluctuations [2]-[5], resulting in an input-referred voltage noise power spectral density (PSD) that reduces markedly going from weak to strong inversion.

A potential drawback or attention point for vertical NWFETs lies in the asymmetry of the connections to the top and bottom part of the pillars. Particularly, in the case where the silicon substrate is employed as the source contact [1], [6], this can result in pronounced asymmetry in the DC characteristics [6]. An example is shown in Fig. 1 for a VNW pMOSFET fabricated on a silicon wafer, with 50 nm epitaxial in-situ highly B-doped Si source (S) and drain (D) contacts to a 100 nm undoped Si NW. The source is biased through the backside bulk contact. Clearly, a pronounced threshold voltage (V_T) shift exists between the Forward (F) and Reverse (R) operation. In the former case, the back

substrate contact is used as source, while it becomes the drain in the R case. To the Authors' best knowledge, such asymmetry study has not been carried out yet with respect to the LF noise, which is the goal of the present work.

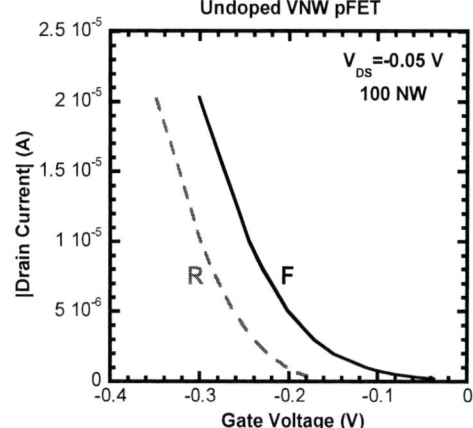

Figure 1 Transfer characteristic of a GAA VNW pMOSFET in linear operation (V_{DS}=-0.05 V) in Forward (F) and Reverse (R) operation. 10 by 10 undoped pillars are connected in parallel (=100 NWs).

II. EXPERIMENTAL DETAILS

The VNW pFETs studied have been fabricated on 300 mm diameter low-resistivity p-type Si wafers, with the substrate as source contact, using the process flow described elsewhere [1]. The intrinsic or B-doped silicon pillars have a diameter of $\geq$26 nm and a gate length (defined vertically) of about 100 nm. The source and drain regions of the pillars are composed of ~50 nm thick in-situ highly B-doped Si. On top of the pillars, additional highly B-doped SiGe (25 % Ge) was grown for obtaining an increased drain contact area. The gate stack consists of IL-SiO₂/HfO₂/TiN/W with an Equivalent Oxide Thickness of ~1 nm.

Transistors with arrays of 10×10 NWs in parallel (= 100 NWs per device) have been studied for their LF

978-1-6654-3746-2/21 $31.00 © 2021 IEEE

noise performance, as described previously [3]. The LF noise spectra were measured in linear operation at a V_{DS}=-50 mV, stepping the gate voltage V_{GS} from weak to strong inversion and for F and R operation. The input-referred voltage noise PSD (S_{VG}) was derived from the measured drain current noise PSD (S_I) by dividing with g_m^2 (g_m the transconductance measured in each gate bias point).

III. RESULTS AND DISCUSSION

Figure 2 represents the frequency-multiplied noise spectra of a GAA VNW pFET with intrinsic (undoped) channel in F and R operation, for two drain currents. The undoped channel most likely results in an Inversion Mode (IM) type of device, as the background doping of the epitaxial silicon pillars is expected to be n-type.

The frequency multiplied spectra exhibit 1/f or Generation-Recombination (GR) behavior at high or low drain current I_D, respectively, for frequencies below about 100 Hz to 1 kHz. Beyond 1 kHz, the spectra become white, which is most likely related to the substrate source contact [7]. Indeed, replacing the backside bulk contact by a top source contact on a Silicon-on-Insulator (SOI) wafer removes the white noise, while flicker noise persists over the whole frequency range up to 10 kHz. More importantly, a pronounced asymmetry in the noise Power Spectral Density (PSD) is observed, with lower values for the R operation, i.e., using the top contact as the injecting source.

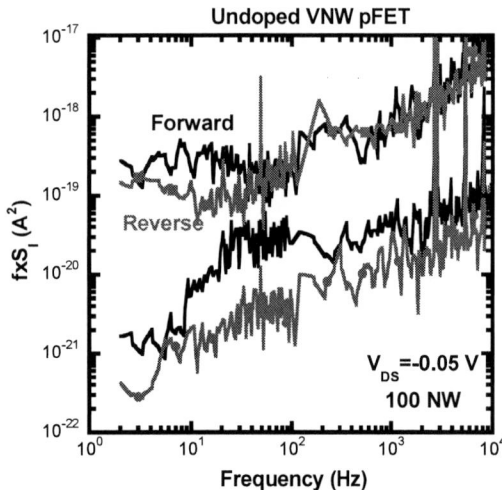

Figure 2   Frequency normalized current noise spectra for an undoped VNW pMOS in F and R operation at V_{DS}=-0.05 V. The top spectra correspond with a higher I_D.

According to the normalized current noise PSD at f=10 Hz in Fig. 3, the difference amounts to about one order of magnitude in weak inversion and lowers for higher $|I_D|$. This is confirmed by the input-referred voltage noise PSD (S_{VG}) in Fig. 4. From the reduction

of S_{VG} with more negative gate voltage (V_{GS}), one can derive that the 1/f noise in weak inversion in this case is rather due to mobility fluctuations [3]. This is also derived from the fact that the (g_m/I_D) function, shown in Fig. 3 is not proportional with S_I/I_D^2 [8]-[11]. At the same time, a higher g_m is observed in R operation, indicating better conduction (higher mobility) when using a top source contact.

At high currents, the impact of the access resistance starts to dominate the 1/f noise magnitude. This may explain the reversal of the noise PSD trend at higher absolute I_D for R operation in Figs 3 and 4.

Figure 5 represents the normalized drain current noise PSD at 10 Hz versus $|I_D|$ for a GAA NW pFET with an in-situ B-doped channel with a concentration of 5×10^{18} cm^{-3}. In other words, the FETs correspond with a junctionless (JL) p$^+$-p-p$^+$ architecture. A similar trend as in Fig. 2 can be derived between F and R operation. This is confirmed by the S_{VG} data in Fig. 6.

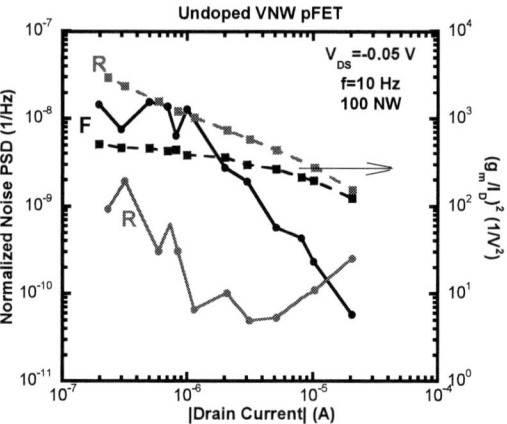

Figure 3 Normalized drain current noise PSD at 10 Hz for an undoped VNW pFET in F and R operation at V_{DS}=-0.05 V.

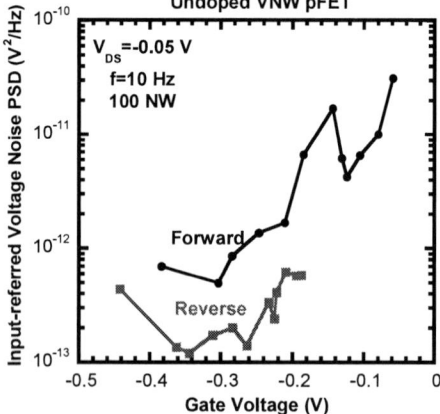

Figure 4   Input-referred voltage noise PSD at 10 Hz for an undoped VNW pFET in F and R operation at V_{DS}=-0.05 V.

978-1-6654-3746-2/21 $31.00 © 2021 IEEE

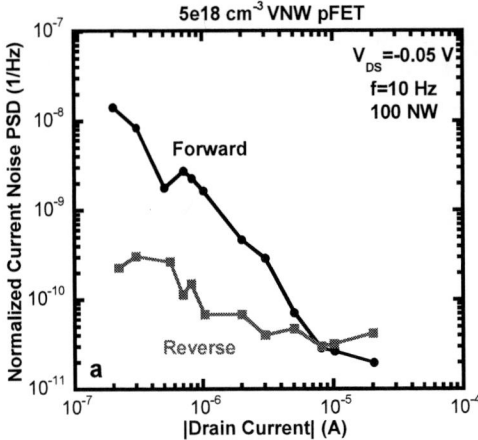

Figure 5   Normalized drain current noise PSD at 10 Hz for a 5×10^{18} cm^{-3} in-situ B-doped GAA VNW pFET in F and R operation at V_{DS}=-0.05 V.

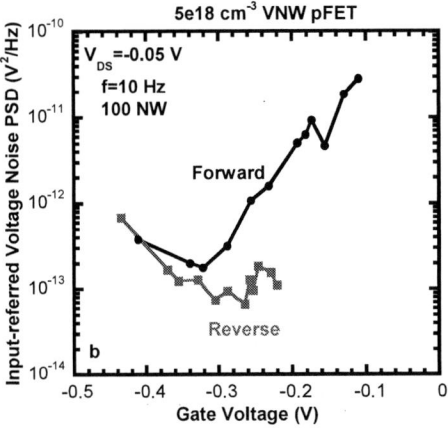

Figure 6   Input-referred voltage noise PSD at 10 Hz for an 5×10^{18} cm^{-3} in-situ B-doped GAA VNW pFET in F and R operation at V_{DS}=-0.05 V.

Another striking feature is that the noise PSD for the undoped IM and the doped JL GAA VNW pFETs are of the same order of magnitude for both F and R operation. This is a bit surprising as in the past, it has been demonstrated that JL architectures yield a better 1/f noise performance [12]-[15]. Minimum noise PSD has been reported for an intermediate B doping density around 5×10^{18} cm^{-3} [7], [12].

In Fig. 7, the average input-referred S_{VG} in linear operation and 10 Hz is represented for a set of undoped and B-doped VNW pFETs. Again, rather similar values are obtained for both types of devices. It is clear that in this case, the pillar doping has a marginal impact on the 1/f noise PSD in both F and R

operation. At the same time, it is evident that the silicon substrate as a source contact has a major effect on the LF noise behavior of these devices. Choosing a small-area source contact in R operation yields better 1/f noise and DC performance, which is promising as in a 3D integration scheme, there will be a local source contact fabricated.

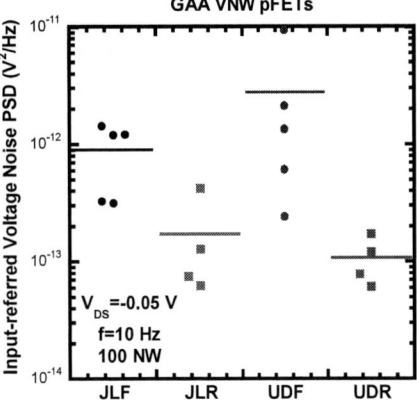

Figure 7 Average input-referred voltage noise PSD for undoped and junctionless VNW pFET at f=10 Hz and $V_{GS} \sim V_T$. V_{DS}=-0.05 V, for F and R operation.

Finally, a few words should be spent regarding the $\Delta\mu$ origin of the flicker noise. Normally, Δn fluctuations dominate the 1/f noise in advanced MOSFETs, indicating that carrier exchange with traps in the gate stack is the underlying mechanism. The fact that no Δn fluctuations are found here can be explained by two reasons: one, the density of oxide traps is much smaller in the GAA VNW pFETs or, the carriers are concentrated in the core of the nanowires away from the traps in the gate stack. While this should be supported by 3D devices simulations, we strongly believe that the second mechanism is responsible for the observations. If true, the data also suggests a strong impact on the carrier distribution in the NW by the type of source contact, i.e., bulk versus top.

IV.   CONCLUSIONS

It has been shown that the LF noise at low f (<100 Hz) and drain currents exhibits a pronounced asymmetry when swapping the source and drain contacts. The fact that the wire doping does not impact significantly this behavior indicates the dominant role of the type of source contact on the transport behavior in the silicon nanowires.

ACKNOWLEDGMENT

This work has been performed in the frame of imec's GAA Nanowire Core Partner Program.

REFERENCES

[1] A. Veloso, G. Eneman, T. Huynh-Bao, A. Chasin, E. Simoen, E. Vecchio, K. Devriendt, S. Brus, E. Rosseel, A. Hikavyy, R. Loo, V. Paraschiv, B. T. Chan, D. Radisic, W. Li, J. J. Versluijs, L. Teugels, F. Sebaai, P. Favia, H. Bender, E. Vancoille, J. Scheerder, C. Fleischmann, N. Horiguchi, and P. Matagne, "Vertical nanowire and nanosheet FETs: device features, novel schemes for improved process control and enhanced mobility, potential for faster & more energy efficient circuits," International Electron Devices Meeting, IEEE Xplore, pp. 230-233, Dec. 2019.

[2] T. Imamoto, Y. Ma, M. Muraguchi, and T. Endoh, "Low-frequency noise reduction in vertical MOSFETs having tunable threshold voltage fabricated with 60 nm CMOS technology on 300 mm wafer process," Jpn. J. Appl. Phys., vol. 54, p. 04DC11, 2015.

[3] E. Simoen, A. Vinicius Oliveira, A. Veloso, A. Chasin, R. Ritzenthaler, H. Mertens, N. Horiguchi, and C. Claeys, "Low frequency noise performance of horizontal, stacked and vertical silicon nanowire MOSFETs," Solid-State Electron., vol. 184, p. 108087, 2021.

[4] N. Clément, X. L. Han, and G. Larrieu, "Electronic transport mechanisms in scaled gate-all-around silicon nanowire transistor arrays," Appl. Phys. Lett., vol. 103, pp. 263504/1-5, Dec. 2013.

[5] C. Mukherjee, C. Maneux, J. Pezard, and G. Larrieu, "1/f noise in 3D vertical gate-all-around junction-less silicon nanowire transistors," in Proc. ESSDERC 2017, IEEE Xplore, pp. 34-37, Sep. 2017.

[6] M. Liu, F. Lentz, S. Trellenkamp, J.-M. Hartmann, J. Knoch, D. Grützmacher, D. Buca, and Q.-T. Zhao, "Diameter scaling of vertical Ge Gate-All-Around nanowire pMOSFETs," IEEE Trans. Electron Devices, vol. 67, pp. 2988-2994, Jul. 2020.

[7] E. Simoen, P. Matagne, A. Veloso, and C. Claeys, "Impact of processing factors on the low-frequency noise of gate-all-around silicon vertical nanowire FETs", Inv. Paper to be published in the Proc. of the ECS Fall Meeting Symposium on VLSI Technology, Orlando (Fl.), Oct. 2021.

[8] G. Ghibaudo, O. Roux, Ch. Nguyen-Duc, F. Balestra, and J. Brini, "Improved analysis of low frequency noise in field-effect MOS transistor," Phys. Status Solidi (a), vol. 124, pp. 571-581, Apr. 1991.

[9] E. Simoen and C. Claeys, "On the flicker noise in submicron silicon MOSFETs," Solid-State Electronics, vol. 43, pp. 865-882, May 1999.

[10] G. Ghibaudo and T. Boutchacha, "Electrical noise and RTS fluctuations in advanced CMOS devices," Microelectron. Reliab., vol. 42, pp. 573-582, Apr.-May 2002,

[11] E. Simoen, H.-C. Lin, A. Alian, G. Brammertz, C. Merckling, J. Mitard, and C. Claeys, "Border traps in Ge/III-V channel devices: analysis and reliability aspects," IEEE Trans. Device and Mater. Reliability, vol. 13, pp. 444-455, Dec. 2013.

[12] A. Veloso, G. Hellings, M. J. Cho, E. Simoen, K. Devriendt, V. Paraschiv, E. Vecchio, Z. Tao, J. J. Versluijs, L. Souriau, H. Dekkers, S. Brus, J. Geypen, P. Lagrain, H. Bender, G. Eneman, P. Matagne, A. De Keersgieter, W. Fang, N. Collaert, and A. Thean, "Gate-all-around NWFETs vs. triple-gate FinFETs: Junctionless vs. extensionless and conventional junction devices with controlled EWF modulation for multi-VT CMOS," in Tech. Dig. of the Symposium on VLSI Technol., IEEE Xplore, pp. T138-T139, Jun. 2015.

[13] P. Singh, N. Singh, J. Miao, W.-T. Park, and D. L. Kwong, "Gate-all-around junctionless nanowire MOSFET with improved low-frequency noise behavior," IEEE Electron Device Lett., vol. 32, pp. 1752-1754, Dec. 2011.

[14] C.-H. Park, M.-D. Ko, K.-H. Kim, S.-H. Lee, J.-S. Yoon, J.-S. Lee, and Y.-H. Jeong, "Investigation of low-frequency noise behavior after hot-carrier stress in an n-channel junctionless nanowire MOSFET," IEEE Electron Device Lett., vol. 33, pp. 1538-1540, Nov. 2012.

[15] E. Simoen, A. Veloso, P. Matagne, N. Collaert, and C. Claeys, "Junctionless versus inversion mode gate-all-around nanowire transistors from a low-frequency noise perspective," IEEE Trans. Electron Devices, vol. 65, pp. 1487-1492, Apr. 2018.

Avalanche Transient Simulations of SPAD integrated in 28nm FD-SOI CMOS Technology

D. Issartel
Univ Lyon
INSA Lyon, CNRS, INL, UMR5270,
Villeurbanne, France
dylan.issartel@insa-lyon.fr

P. Pittet
Univ Lyon
UCBL, CNRS, INL, UMR5270,
Villeurbanne, France
patrick.pittet@univ-lyon1.fr

A. Cathelin
STMicroelectronics
Crolles, France
andreia.cathelin@st.com

S. Gao
Univ Lyon
INSA Lyon, CNRS, INL, UMR5270,
Villeurbanne, France
shaochen.gao@insa-lyon.fr

R. Cellier
Univ Lyon
CPE Lyon, CNRS, INL, UMR5270,
Villeurbanne, France
remy.cellier@cpe.fr

F. Calmon
Univ Lyon
INSA Lyon, CNRS, INL, UMR5270,
Villeurbanne, France
francis.calmon@insa-lyon.fr

S. Hagen
Univ Lyon
INSA Lyon, CNRS, INL, UMR5270,
Villeurbanne, France
sarah.hagen@insa-lyon.fr

D. Golanski
STMicroelectronics
Crolles, France
dominik.golanski@st.com

Abstract—**This article presents a study of Single Photon Avalanche Diodes (SPAD) implemented in 28nm Fully Depleted Silicon-On-Insulator (FD-SOI) CMOS technology based on transient TCAD simulations. The integration of SPAD in this technology is currently being studied. This work allows for a better understanding of the mechanism behind the quite high Dark Count Rate (DCR) measured at relative low excess bias voltages with the previous FD-SOI SPAD design. In this study, TCAD transient simulation methodology is introduced to better understand SPAD behavior during the avalanche process. TCAD simulations revealed that Shallow Trench Isolation (STI) structures in the active area have a negative effect on avalanche quenching, because of slower carrier evacuation with possible occurrence of secondary avalanches in series. Based on this analysis, we propose a new SPAD architecture to achieve a lower DCR.**

Keywords—*SPAD, FD-SOI CMOS, TCAD Simulation, avalanche process*

I. INTRODUCTION

Single Photon Avalanche Diodes (SPAD) are photodetectors appreciated for weak intensity light detection, thanks to high avalanche gain and fast response time [1-2]. SPAD are used in various application domains such as self-driving cars, facial recognition, visible light communication, cryptography, and quantum communications [3].

SPAD pixels include addressing, quenching and recharging electronics. Avalanche cycles occurring in SPAD consist mainly of an exponential current increase followed by i) a quenching step where the avalanche is stopped allowing for current reduction and ii) a recharging step for the restoration of the initial biasing condition. Quenching can be implemented either by connecting a high resistance in series with the diode (passive quenching) or by using more complex circuits (active quenching), subjects of varied research [4]. In addition, in the last few years, another active field of research was focused on the integration of SPAD in advanced CMOS technologies [5], and recently in a Silicon-On-Insulator (SOI) platform [6]. Two-Tier 3D integration of SPAD arrays and associated electronics has been developed to get higher fill factors. However, 3D SPAD architecture

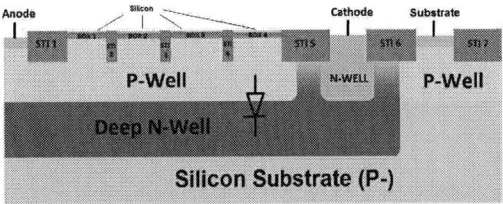

Figure 1. Schematic view of the SPAD implemented in FD-SOI CMOS technology – half structure with cylindrical symmetry is considered (not to scale) [10].

relies on complex Wafer-to-Wafer-Bonding technique [7]. In this paper, we present a detailed study of the photodetection process with an analysis of the avalanche cycle of SPAD designed and fabricated in 28nm Fully Depleted Silicon-On-Insulator (FD-SOI) CMOS technology from STMicroelectronics [8] (Fig. 1). In section II, the layout of the studied SPAD cell is described, while in section III, process and device level simulation setups developed for this analysis are presented. The results of the study are reported and discussed in sections IV and V for SPAD with the initial and proposed optimized designs, respectively.

II. SPAD IMPLEMENTED IN FD-SOI CMOS TECHNOLOGY

FD-SOI CMOS technology is based on two silicon layers electrically insulated by a thin oxide layer: The Buried OXide (BOX). The addition of this BOX layer enables the reduction of junction parasitic capacitances and a better electrostatic control of the channel of transistors with lower short channel effects and leakage currents.

In this technology, electronic circuits are implemented in the thin upper silicon layer while the silicon substrate contains P and N wells (PW, NW), allowing for body-biasing of transistors, and deep N-wells (DNW), which permit isolation between different PW.

Since associated electronics can be integrated in the upper silicon layer, the design of SPAD FD-SOI in this process has been proposed by authors in previous work [8]. Thus, it allows intrinsic 3D implementation of the SPAD and of its electronics. Indirect SPAD event detection through the BOX for Filling Factor (FF) optimization was also

978-1-6654-3746-2/21 $31.00 © 2021 IEEE

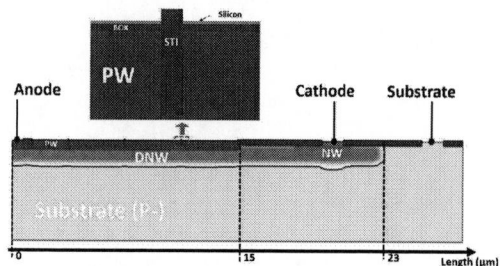

Figure 2. TCAD view of the SPAD in FD-SOI CMOS technology after the process simulations (top: zoom on a STI trench).

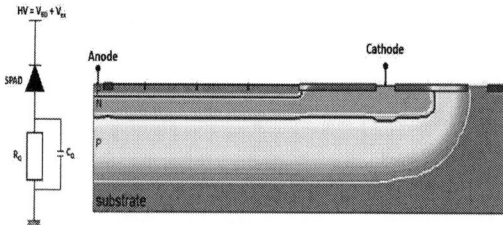

Figure 3. TCAD view of the SPAD with electrical connections for the mixed-mode transient simulations.

presented [8]. The proposed SPAD cell implements the PW-DNW junction as shown in Fig. 1.

This cell has been fabricated in a standard 28 nm FD-SOI CMOS process and characterized. This cell shows a relatively high Dark Count Rate (DCR) for low excess bias voltages (less than 1V) as compared to the state of the art. We have shown recently that a decrease of the DNW doping level can be used to increase the breakdown voltage (V_{BD}) by several volts [9], and may reduce the DCR. We have also worked around the Shallow Trench Isolation (STI) blocks on the edges of the active zone to reduce premature edge breakdown [10]. With such optimizations, it was possible to reduce DCR but not by enough. Deeper TCAD simulations were performed for a better understanding of the origins of this high DCR in order to make further cell optimizations.

III. TCAD SIMULATION METHODOLOGY

Setup process and device level simulations for transient analysis using the TCAD *Synopsys Sentaurus* software workbench [11] are presented in this section. The cross section of the studied structure is shown on Fig. 2 (cylindrical symmetry).

We describe and parametrize the main steps of the manufacturing process in the simulation framework according to confidential information provided by STMicroelectronics.

For the device's electrical simulation, we started by calibrating our simulations using the static I(V) characteristic of the diode [9]. In particular, the avalanche model and its parameters were properly determined to obtain a simulated breakdown voltage, V_{BD}, close to the measured one. The Van Overstraeten model [12], which gives consistent results, i.e. comparable with measurements, was chosen. In addition, the lifetimes of carriers were adjusted in order to obtain a dark current in good agreement with measurements.

Then, transient electrical studies were carried out in two steps. The first one was an initialization phase in which i) the structure was biased to an excess voltage (V_{ex}) of approximately 1V and ii) the avalanche model was disabled (quasi-static simulation). From this unstable equilibrium point, we moved to the second phase, the transient simulation

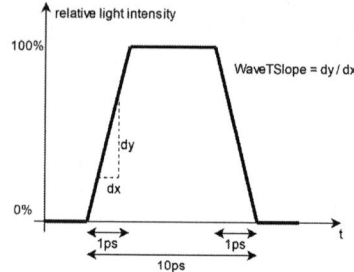

Figure 4. Shape of the light pulse used in transient simulations to launch the avalanche.

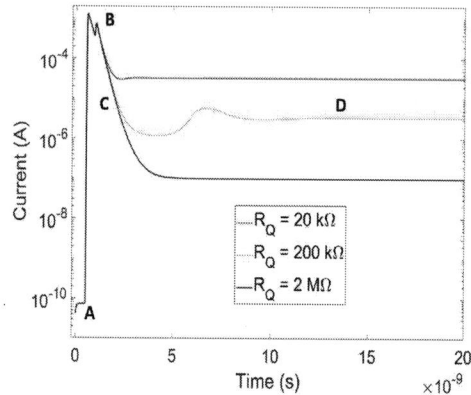

Figure 5. SPAD current curves obtained with transient simulations on the reference structure for various values of R_Q.

with a localized carrier photo-generation process to trigger the avalanche process [13]. The electrical connections are shown in Fig. 3. The quenching circuit implements a 200kΩ external resistance (R_Q) connected to the anode of the SPAD. The parasitic capacitance of 3fF of the pad (C_Q), is also shown in Fig. 3.

A short light pulse (10ps) localized in the multiplication region (as shown in Fig. 4) was used for the photo-generation of electron-hole pairs inside the Space Charge Region (SCR). The transient simulation allows for a dynamic analysis of the avalanche process by monitoring some internal device data (e.g. the current density) and by tracking the current over time.

IV. UNEXPECTED BEHAVIOR OF THE REFERENCE SPAD STRUCTURE

As previously mentioned, our study on transient simulations, which was carried out on the reference SPAD structure (see Fig. 1 and Fig. 2), revealed unexpected behavior. As shown in Fig. 5, the initial current (marker A) was approximately at 0.1nA (dark current). After the triggering of the avalanche, the current exhibited a steep increase and reached ~1mA (marker B). Then, we observed a current decrease resulting from the passive quenching (region C). However, the avalanche current did not fully subside, and the diode current remained several orders of magnitude higher than its initial level (region D). The same behavior was observed for quenching resistances R_Q of 20kΩ, 200kΩ and 2MΩ.

Another representation of the SPAD avalanche cycle is given in Fig. 6. It plots the current as a function of the SPAD voltage. During phase 1 (region ①), the number of impact ionizations increases in time and corresponds to the steep current observed in Fig. 5. The current decreases in phase 2

978-1-6654-3746-2/21 $31.00 © 2021 IEEE

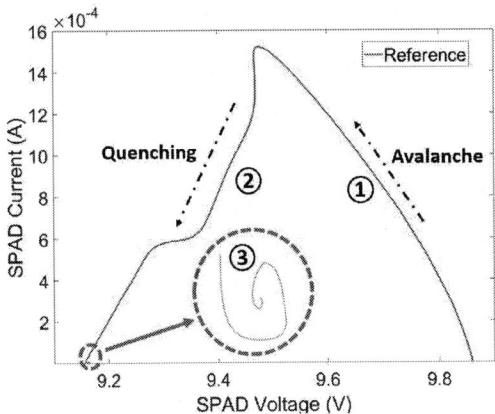

Figure 6. Quenching cycle of the reference structure during transient simulations ($R_Q = 200k\Omega$).

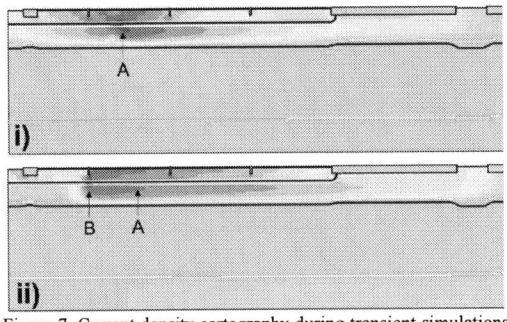

Figure 7. Current density cartography during transient simulations for the reference SPAD structure : i) just after the light pulse, ii) few nanoseconds after i).

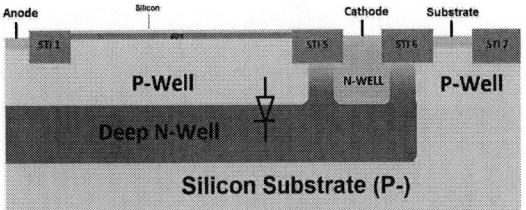

Figure 8. Schematic view of the "Without STI trenches" SPAD structure (not to scale).

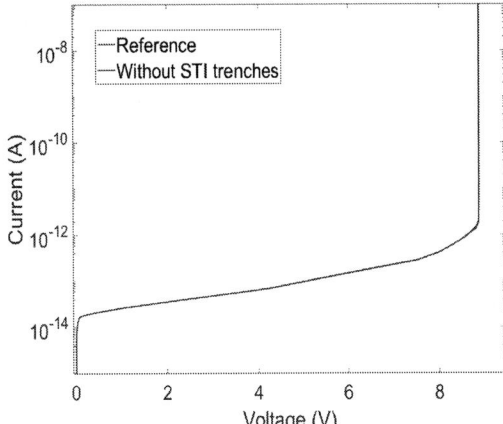

Figure 9. I(V) curves obtained with static simulations for the reference and the "Without STI trenches" SPAD structures.

thanks to passive quenching (region ②). The cycle seems to halt after this quenching phase (at marker ③), but by examining this area up close, we observe some repetitions of the cycle corresponding to the triggering of secondary avalanche in series.

Cartography studies of the current density in the structure explain this behavior. As shown in Fig. 7, following the rise of the light pulse, the current density increases due to the avalanche process (marker A, Fig. 7i). After a short-elapsed time (a few nanoseconds), the current density spreads and a second avalanche appears. The latter is located below the STI trench closest to the illumination area (marker B, Fig. 7ii). Thereafter, a high current density remains present under this STI block, which acts as a funnel limiting the current flow and increasing the current density magnitude locally. It would then seem that these STI blocks, located close to the active zone, caused multiple avalanches in series. To tackle this issue, several methods were considered, such as the displacement or enlargement of the anode contact. The next section studies the most simple and direct way to address this issue, i.e. by removing the STI trenches.

V. SIMULATIONS OF THE MODIFIED SPAD STRUCTURE WITHOUT STI TRENCHES

In the physical design of SPAD in FD-SOI CMOS technology, these STI trenches, which border transistor active areas are required according to the Design Rules Manual (DRM). However, we designed a new SPAD cell with a drastic reduction of the number of STI blocks placed over the active area of the SPAD (as shown in Fig. 8). Static characteristic of the SPAD was determined by simulations to evaluate the impact of this modification on the electrical

behavior of the SPAD: as shown in Fig. 9, no significant modification of the dark current and V_{BD} were observed. Next, we carried out a study of the electrical field in the SCR and found that the three electrical field peaks due to the STI blocks included in the initial SPAD cell were no longer present for this new structure (as shown in Fig. 10).

For transient simulations, we observed that the current increased during the avalanche, collapsed during the quenching phase, and recovered a value very close to its initial level (as shown in Fig. 11). In addition, since STI trenches were no longer present, the secondary avalanches did not occur, the charge density spreading along the p-region, and the current flowing to the anode. Finally, the behavior of the quenching cycle corresponds to the one expected with three phases (as shown in Fig. 12): increase of the current during the avalanche (marker ①), decrease during the quenching (marker ②), and recharging to return to the metastable initial state (marker ③).

VI. CONCLUSION

In this study, we carried out transient simulations of avalanches for SPAD integrated in FD-SOI CMOS technology which took into account the entire technological process. We showed that STI trenches localized close to the SPAD's active area (imposed by the design rules) play a negative role during the avalanche quenching, because they prevent a fast carrier evacuation on the anode side, leading to possible multiple secondary avalanches in series. Finally, the removal of these STI blocks allows for recovery of a correct avalanche quenching process. Based on these results, we have designed optimized SPAD structures without STI trenches in the vicinity of the SPAD active zone. This new design has been launched in fabrication and will be characterized. While waiting to receive the new test-chips for

978-1-6654-3746-2/21 $31.00 © 2021 IEEE

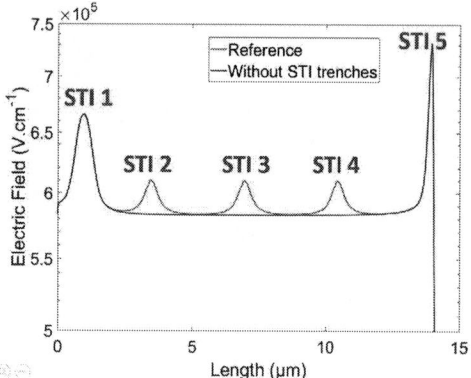

Figure 10. Shape of the electricic field in the SCR obtained with static simulations for the reference and the "Without STI trenches" SPAD structures.

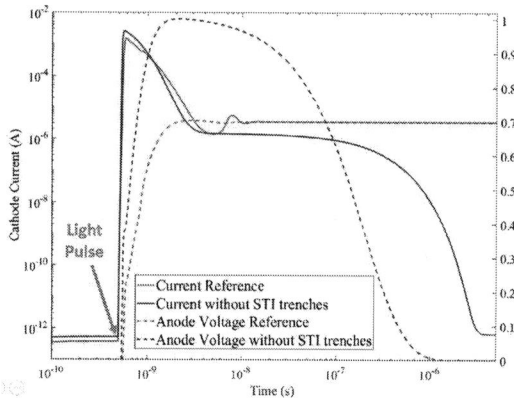

Figure 11. I(t) and V(t) curves obtained with transient simulations on the reference and "Without STI trenches" SPAD structures.

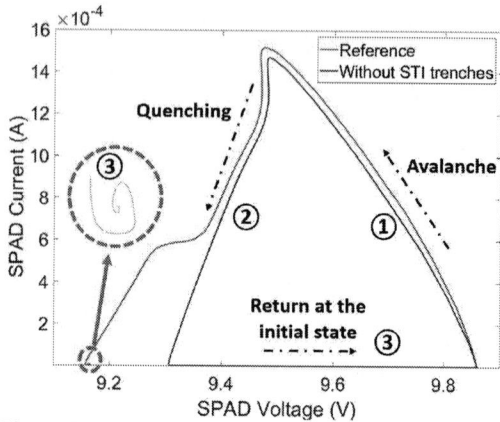

Figure 12. Quenching cycles obtained with transient simulations on the reference and the "Without STI trenches" SPAD structures.

full characterization, future work will focus on employing avalanche transient simulations to better understand the physical mechanisms occurring during the avalanche process.

ACKNOWLEDGMENT

The authors would like to thank: the French National Research Agency (ANR-18-CE24-0010) for the PhD Grant of Dylan Issartel, the Nano2022 research program for the PhD Grant of Shaochen Gao, and also CMP (Grenoble) for its IC prototyping service.

REFERENCES

[1] M.-J. Lee and E. Charbon, "Progress in single-photon avalanche diode image sensors in standard CMOS: From two-dimensional monolithic to three-dimensional-stacked technology," Japanese Journal of Applied Physics, vol. 57, no.10, pp. 1002A3-1/6, Sept. 2018).

[2] R. K. Henderson, N. Johnston, F. Mattioli Della Rocca, H. Chen, D. Day-Uei Li, G. Hungerford, R. Hirsch, D. McLoskey, P. Yip, D. J. S. Birch, "A 192x128 Time Correlated SPAD Image Sensor in 40-nm CMOS Technology," IEEE Journal of Solid-State Circuits, 2019.

[3] F.Acerbi, Z. Bisadi, G. Fontana, N. Zorzi, C. Piemonte, and L. Pavesi, "A Robust Quantum Random Number Generator Based on an Integrated Emitter-Photodetector Structure," in IEEE Journal of Selected Topics in Quantum Electronics, vol. 24, no. 6, pp. 1-7, Nov.-Dec. 2018 , Art no. 6101107

[4] M. Dolatpoor, J.-B. Kammerer, W. Uhring, J.-B. Schell, and F. Calmon, "An Ultrafast Active Quenching for SPAD in CMOS 28nm FDSOI Technology," Conference IEEE Sensors, 2020.

[5] S. Pellegrini, B. Rae, A. Pingault, D. Golanski, S. Jouan, C. Lapeyre, and B. Mamdy, "Industrialised SPAD in 40 nm Technology," International Electron Devices Meeting (IEDM), pp. 16.5.1-16.5.4, 2017.

[6] M. Lee, P. Sun, and E. Charbon, "A first single-photon avalanche diode fabricated in standard SOI CMOS technology with a full characterization of the device," Opt. Express 23, 13200-13209 (2015).

[7] M.-J. Lee, A. R. Ximenes, P. Padmanabhan, T. J. Wang, K. C. Huang, Y. Yamashita, D. N. Yaung, and E. Charbon "A Back-Illuminated 3D-Stacked Single-Photon Avalanche Diode in 45nm CMOS Technology," International Electron Devices Meeting (IEDM), pp. 16.6.1-16.6.4, 2017.

[8] T. Chaves de Albuquerque, D. Issartel, R. Clerc, P. Pittet, R. Cellier, D. Golanski, S. Jouan, A. Cathelin, and F. Calmon, "Indirect Avalanche Event Detection of Single Photon Avalanche Diode Implemented in CMOS FDSOI Technology," Solid-State Electronics, Elsevier, Vol. 163, Jan. 2020, p. 107636.

[9] T. Chaves de Albuquerque, D. Issartel, R. Clerc, P. Pittet, R. Cellier, and F. Calmon, "Lowering the Dark Count Rate of SPAD Implemented in CMOS FDSOI Technology," Conference EUROSOI – ULIS 2019.

[10] D. Issartel, T. Chaves de Albuquerque, R. Clerc, P. Pittet, R. Cellier, D. Golanski, A. Cathelin, and F. Calmon, "SPAD FDSOI cell optimization for lower dark count rate achievement," 2020 Joint International EUROSOI Workshop and International Conference on Ultimate Integration on Silicon (EUROSOI-ULIS), Caen, France, 2020, pp. 1-5.

[11] Sentaurus. Device User Guide, 2018. [Online]. Available: https://www.synopsys.com/silicon/tcad/device-simulation/sentaurus-device.html.

[12] R. Van Overstraeten, H. De Man, "Measurement of the ionization rates in diffused silicon p-n junctions", Solid-State Electronics, 13(5), 583–608, 1970.

[13] Y. Oussaiti, D. Rideau, J-R. Manouvrier, H. Wehbe-Alause, S. Pellegrini, B. Mamdy, and M. Pala "Behavior and Models for quench efficiency in Single-Photon Detection," Single-Photon Workshop, Oct. 2019, Milan, Italy.

First Principles Evaluation of Topologically Protected Edge States in MoS$_2$ 1T$'$ Nanoribbons with Realistic Terminations

Al-Moatasem El-Sayed
Institute for Microelectronics
Technical University of Vienna
Vienna, Austria, and
Nanolayers Research Computing Ltd.
1 Granville Court
London, U.K.
elsayed@iue.tuwien.ac.at

Heribert Seiler
Institute for Microelectronics
Technical University of Vienna
Vienna, Austria
seiler@iue.tuwien.ac.at

Hans Kosina
Institute for Microelectronics
Technical University of Vienna
Vienna, Austria
kosina@iue.tuwien.ac.at

Markus Jech
Institute for Microelectronics
Technical University of Vienna
Vienna, Austria
jech@iue.tuwien.ac.at

Dominic Waldhör
Institute for Microelectronics
Technical University of Vienna
Vienna, Austria
waldhoer@iue.tuwien.ac.at

Viktor Sverdlov
Christian Doppler Laboratory for
Nonvolatile Magnetoresistive Memory and Logic
at the Institute for Microelectronics
Technical University of Vienna
Vienna, Austria
sverdlov@iue.tuwien.ac.at

Abstract—**Exploiting novel materials with advanced properties is necessary to improve device performance and continue with their scaling for high performance applications at reduced power. Among these materials, topological insulators (TIs) present an exciting opportunity where the highly conductive edge states, which are protected against back scattering, can lead to advances in electronic as well as spin controlled devices. Here, we present first principles results that evaluate topologically protected edge states in MoS$_2$ nanoribbons. We vary the width of the nanoribbons and show that they transition to trivial insulators below a critical width. Furthermore, the trivial insulator can be a direct or indirect band gap insulator depending on the width and the edge termination. Our results show that including realistic edge terminations can provide valuable insight, especially for narrow width TIs.**

Index Terms—**Density Functional Theory, Edge States, Topological Insulators, Nanoribbons**

I. INTRODUCTION

Over the past decade, well-known two-dimensional transition metal dichalcogenides (TMDs) – such as MoS$_2$, MoSe$_2$, WS$_2$, and WSe$_2$ – have been shown to exist in topologically non-trivial phases known synonymously as either TIs or quantum spin hall insulators [1]–[5]. This behavior does not occur in their 2H polymorph, which is their ground state, but rather in other high energy polymorphs including the 1T$'$ phase. Their topological nature stems from the underlying symmetries that their Hamiltonians possess, which leads to their electronic structure being classified by a robust $\mathbb{Z}_2$ invariant of 1, whereas that of a trivial insulator is 0 [6]. As a necessary consequence of their non-trivial topology, interfacing them

with a trivial insulator results in chiral, zero gap edge states with outstanding properties including linear dispersion, spin-momentum locking, and resistance to back-scattering [6], [7]. Reliably understanding their behavior is expected to bring about exciting advances in electronic and spintronic devices.

Studies in the literature have used both *ab initio* and semi-empirical methods to describe the electronic structure of the different 1T$'$ TMDs. Various $\mathbf{k} \cdot \mathbf{p}$ and tight-binding model Hamiltonians whose parameters were fitted to the bulk dispersion relations have been used to predict the behavior of the topologically protected edge modes in nanoribbons [1], [3], [4], [8]. At small nanoribbon widths, due to an interaction of the topologically protected states from the opposite edges in a nanoribbon, a small gap in the otherwise gapless linear spectrum of the edge states is predicted to open [8]. This results in a decrease of the ballistic conductance compared to a semi-infinite sample. This gap increases when an electric field is applied perpendicular to the nanoribbon resulting in a further reduction of the conductance. At the same time the separation between the bulk bands decreases with the orthogonal field. The edge-localized modes transform into the bulk-like modes when their dispersion intersect, signifying a transition from a TI to a trivial insulator. As no edge modes protected from back-scattering exist at higher electric fields, the conductance decreases dramatically [9]. Studies using *ab initio* and tight-binding calculations also show that a gap opens in the edge states [4], [5], [10]. However, these studies additionally show that the electronic structure of these narrow nanoribbons are highly dependent on the reconstruction of the edges' atomic

978-1-6654-3746-2/21 $31.00 © 2021 IEEE

structure.

Here, we use density functional theory (DFT) to model monolayer and realistic terminations of narrow nanoribbons of MoS_2 1T'. We focus on nanoribbons cleaved in the y-direction, such as the terminations shown in Fig. 1, in contrast to previous studies in the literature which studied nanoribbons cleaved in the x-direction [5], [10]. We show that at small widths, the nanoribbons no longer have edge states in the band gap. This is in contrast to nanoribbons cleaved in the x-direction which possess gapless edge states even at relatively small widths. Furthermore, we show that depending on the termination, the electronic structure transitions to an indirect band gap.

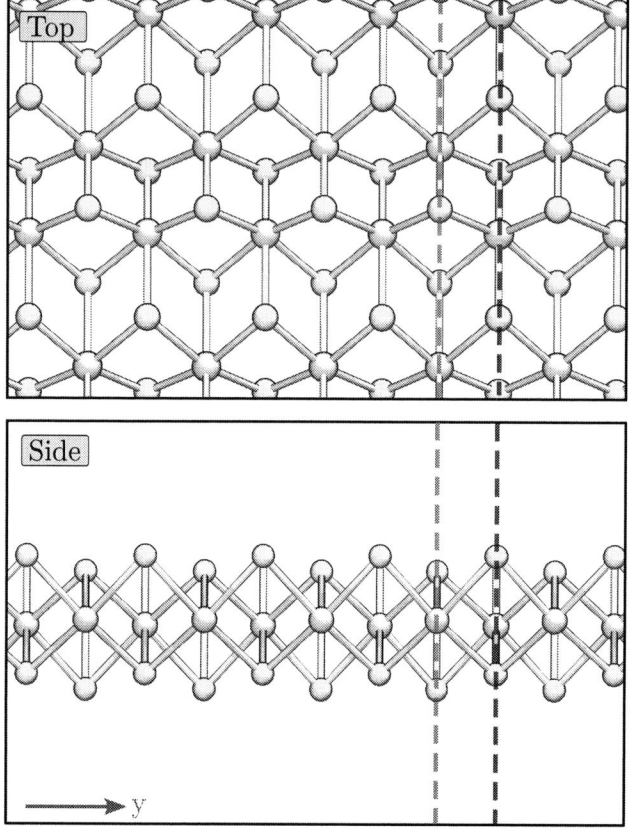

Fig. 1. The MoS_2 1T' structure from the top and the side. The black and red dashed lines indicate the two possible, chemically equivalent, terminations along the y-direction and are made along the same plane for both top and side views. The red arrow in the bottom panel indicates the direction of the y-axis for both the top and side views. The pink spheres are the Mo atoms while the yellow spheres are S atoms.

II. METHODS

We use DFT as implemented in the Vienna Ab-Initio Simulation Package (VASP) to calculate both monolayer and nanoribbon systems of MoS_2 1T'. Exchange and correlation were described using the generalized gradient Perdew-Burke-Ernzerhof (PBE) functional [11]. The projector augmented wave method and its associated pseudopotentials were used in all calculations. The electron density was expanded in a

plane wave basis set whose cutoff was set to 280 eV. The system's k-space was sampled on a Monkhorst pack grid of $8 \times 1 \times 1$ for geometry optimizations and $12 \times 1 \times 1$ for band structure calculations of the nanoribbons. For the bulk monolayer, this changes to $8 \times 8 \times 1$ for geometry optimizations and $12 \times 12 \times 1$ for band structure calculations. Electron densities were converged until the energy change was less than 10^{-4} eV. Geometries were optimized so that the energy changed by less than 10^{-3} eV. The topological edge states of TMDs depend on the band gap opening due to a spin-orbit interaction; therefore, all band structure calculations include the spin-orbit coupling [12].

In this paper, we focus on the 1T' structure of MoS_2. To model the bulk monolayer, a vacuum gap of 2 nm was included in the z-direction and the cell and its atomic coordinates were optimized. The nanoribbon structures were then created by introducing another vacuum gap of 2 nm in the y-direction. Thus, the only remaining periodic dimension is the x-direction. After introduction of the vacuum gap, the atomic structure of the nanoribbons were optimized once more, resulting in significant reorganization of the edges which we refer to as realistic terminations.

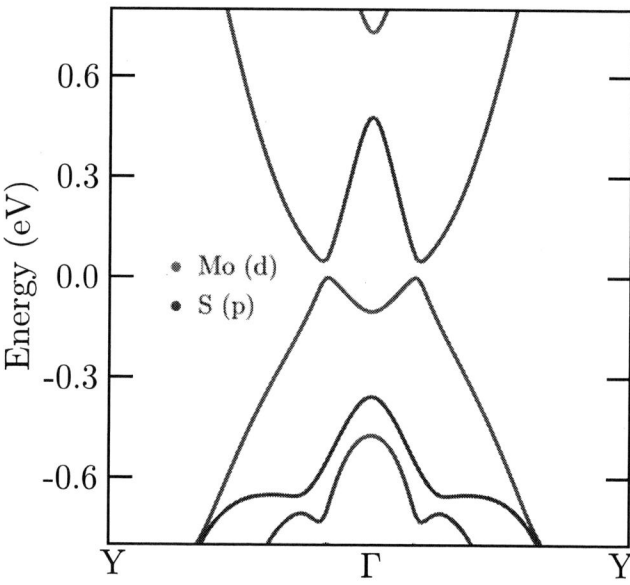

Fig. 2. Band structure of a monolayer of MoS_2 1T'. The band structure is projected onto the Mo 'd' states (red) and S 'p' states (blue).

III. RESULTS

We start by calculating the band structure of a monolayer of MoS_2 1T'. Using a single unit cell containing six atoms, the cell and its atomic coordinates were optimized and the electronic structure was calculated. Fig. 1 shows the optimized atomic structure which is replicated for ease of visualization. The optimized 1T' structure has Mo–S bond lengths of 2.39, 2.48, and 2.51 Å. The band structure of the system can be seen in Fig. 2 and has been projected onto the Mo 'd' and S 'p' orbitals. Around the Fermi level, which is at 0 eV, a

978-1-6654-3746-2/21 $31.00 © 2021 IEEE

band gap of 0.08 eV appears due to the spin-orbit coupling. Without it, the system would be a semi-metal. In addition, an inverted band gap of 0.56 eV appears where the conduction band is mostly made of S 'p' orbitals and the valence band is made of Mo 'd' orbitals. This is in good agreement with what has been reported in the literature and the inverted band gap as well as the spin-orbit band gap are qualitatively what one would expect from a TI [1], [3]. The $\mathbb{Z}_2$ invariant can be numerically calculated as implemented in the Z2pack code and indeed the $\mathbb{Z}_2$ of the MoS$_2$ 1T' was found to be 1 [13].

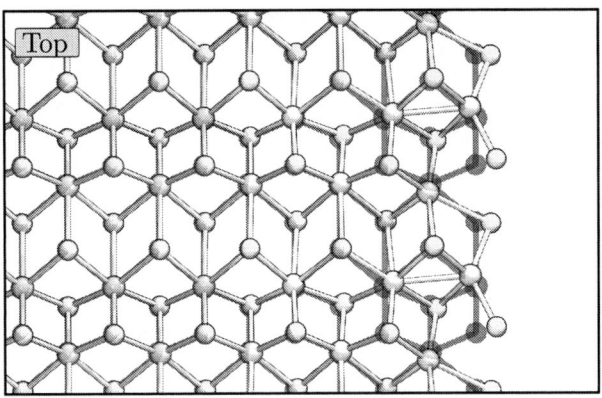

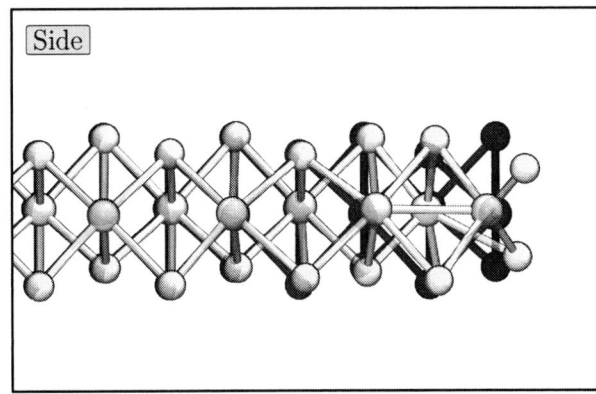

Fig. 3. A nanoribbon of MoS$_2$ 1T' cleaved along the y-direction. The coloured atomic system is the geometry optimized structure, while the black, transparent system is the initial, unrelaxed structure. The color scheme is the same as in Fig. 1

We now turn our attention to nanoribbons of MoS$_2$ 1T' cleaved along the y-direction. Cleaving the MoS$_2$ sheet results in two structural edges. Although they are chemically equivalent, they lead to two possible combinations. When both structural edge types are the same, we refer to this as symmetric, and when they are different they are asymmetric as depicted in Fig. 1. We modeled three different nanoribbon widths: 1.2 nm, 3.8 nm, and 7.9 nm. For all widths, the edges were cleaved and the structure was optimized, leading to reorganization of the edge atoms. The behavior of the relaxation was qualitatively the same regardless of nanotube width and the type of edge termination. The optimization is indicated in Fig. 3, where it is also compared to the initial, unrelaxed structure. The edge can be found on the right side

of the figure, where the strongest atomic displacement occurs. One immediately sees that the displacements only persist over four to five layers of the nanoribbon. The Mo atoms at the edge tend to displace toward the nanoribbon while the S atoms displace away from the nanoribbon. Optimizing the nanoribbon does not introduce a net magnetic moment which would break time reversal symmetry and thus one of the underlying symmetries required for the topologically protected edge states. Conflicting *ab initio* calculations of the magnetic moment do not agree whether nanoribbons cleaved along the x-direction possess a magnetic moment [5], [10].

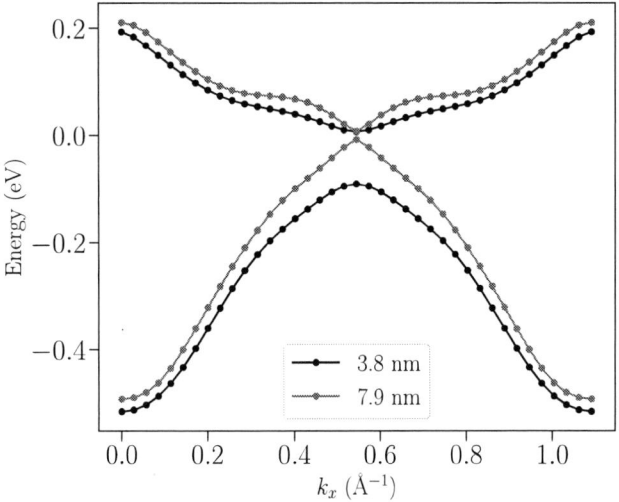

Fig. 4. The highest occupied and lowest occupied bands of two nanoribbon widths of MoS$_2$ 1T'. The narrower, 3.8 nm, nanoribbon's bands are shown in black, while the wider, 7.9 nm, nanoribbon's bands are shown in red. One can see the linear dispersion of the edge states in the wide nanoribbon.

The electronic structure of the nanoribbons was calculated and varies with the width of the nanoribbon. The first thing to note is the appearance of a gap in the topologically protected edge states. This can be seen in Fig. 4, which shows the highest occupied and lowest unoccupied bands of two nanoribbons of different widths. The wider nanoribbon, 7.9 nm, shows linear dispersion of gapless states around the Fermi level at 0 eV, typically associated with the topologically protected edge states. As the width is reduced to 3.8 nm, the states are no longer linearly dispersed and appear to become more parabolic. This behavior was the same regardless whether the edge was symmetrically or asymmetrically terminated. In addition, a band gap of 0.1 eV appears between these edge states. Similar effects have also been seen in MoS$_2$ 1T' nanoribbons that were cleaved in the x-direction [5], [10]. However, in that case, the gap only appeared at widths less than 2 nm and their magnitude was typically much smaller than the gap shown here for nanoribbons cleaved in the y-direction. In addition, *ab initio* calculations showed that of the six possible terminations, only two supported edge states without any other surface states [10]. However, on the y-axis edge, the topological edge states can be seen regardless of the

978-1-6654-3746-2/21 $31.00 © 2021 IEEE

termination.

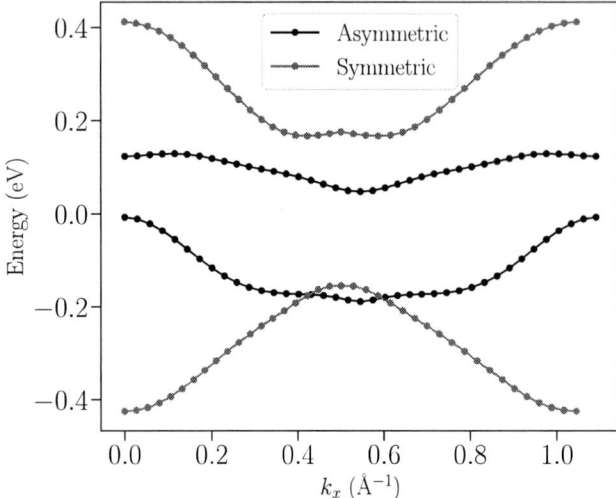

Fig. 5. The highest occupied and lowest occupied bands of two nanoribbon widths of a 1.2 nm strip of MoS_2 $1T'$. The bands of the asymmetric termination are shown in black while those of the symmetric termination are red. The Γ point of the reciprocal cell occurs at a k_x value of 0.54 Å^{-1}.

For narrow nanoribbons, the qualitative behavior changes depending on the termination. Fig. 5 shows the highest occupied and lowest unoccupied bands of a 1.2 nm wide nanoribbon terminated either symmetrically or asymmetrically as depicted in Fig. 1. Due to the narrow width of the nanoribbon and the interaction between the edge states, a large band gap opens up for both terminations. For the symmetric termination, the band gap is 0.3 eV, while it is 0.1 eV for the asymmetric termination. Curiously, for both terminations the qualitative nature of the edge state band gap changes from direct to indirect. For the symmetric nanoribbon, this is a slight shift, so that two minima appear in the conduction band which are shifted by 0.1 Å^{-1} from the valence band maximum. However, for the asymmetric termination, the lowest conduction band minimum is located at the Γ point while the valence band maximum is at the Brillouin zone edge, indicating that band to band transitions in the asymmetric narrow nanoribbons could be suppressed due to the large change in crystal momentum required.

DISCUSSION AND CONCLUSIONS

The highly conductive topologically protected edge states of TIs can have a high impact on electronic devices. However, many fundamental models and theories are based on the properties of the bulk Hamiltonian and ignore the potential changes in the Hamiltonian for downsized systems as well as the effect of edge reorganization. Here, we have calculated the electronic structure of nanoribbons of topologically insulating MoS_2 $1T'$ where the width of the nanoribbon was less than 8 nm. One can see that the edge states begin to interact with each other below 4 nm to open a band gap. This could indicate that some of the fundamental symmetries required for a topologically protected zero energy state have been broken. This effect, although still

disputed, occurs for nanoribbons of MoS_2 $1T'$ cleaved in the x-direction, where the introduction of certain edges introduce a magnetic moment which would break time-reversal symmetry, an essential component of the non-trivial $\mathbb{Z}_2$ invariant [5], [14]. However, our calculations show that edges along the y-direction do not possess a magnetic moment. Furthermore, our results show that the edge termination can have a strong quantitative and qualitative effect on very narrow nanoribbons of MoS_2 $1T'$.

ACKNOWLEDGMENT

The financial support by the Austrian Federal Ministry for Digital and Economic Affairs, the National Foundation for Research, Technology and Development and the Christian Doppler Research Association is gratefully acknowledged. The authors are grateful for the support from the Vienna Scientific Cluster, who provided computer resources on the Austrian high-performance clusters VSC3 and VSC4.

REFERENCES

[1] Q. Xiaofeng, L. Junwei, F. Liang, and L. Ju, "Quantum spin Hall effect in two-dimensional transition metal dichalcogenides," *Science*, vol. 346, no. 6215, pp. 1344–1348, 2014.

[2] T. Olsen, "Designing in-plane heterostructures of quantum spin Hall insulators from first principles: 1T'-MoS2 with adsorbates," *Phys. Rev. B*, vol. 235106, no. 23, pp. 1–9, 2016.

[3] B. Das, D. Sen, and S. Mahapatra, "Tuneable quantum spin Hall states in confined 1T' transition metal dichalcogenides," *Sci. Rep.*, vol. 10, no. 1, pp. 1–13, 2020.

[4] A. Lau, R. Ray, D. Varjas, and A. R. Akhmerov, "Influence of lattice termination on the edge states of the quantum spin Hall insulator monolayer 1T'-WTe2," *Phys. Rev. Mater.*, vol. 3, no. 5, pp. 1–9, 2019.

[5] L. Jelver, D. Stradi, K. Stokbro, T. Olsen, and K. W. Jacobsen, "Spontaneous breaking of time-reversal symmetry at the edges of 1T' monolayer transition metal dichalcogenides," *Phys. Rev. B*, vol. 99, no. 15, p. 155420, 2019. [Online]. Available: https://doi.org/10.1103/PhysRevB.99.155420

[6] M. Z. Hasan and C. L. Kane, "Colloquium: Topological insulators," *Rev. Mod. Phys.*, vol. 82, no. 4, pp. 3045–3067, 2010.

[7] X. L. Qi and S. C. Zhang, "Topological insulators and superconductors," *Rev. Mod. Phys.*, vol. 83, no. 4, 2011.

[8] V. Sverdlov, A. M. B. El-Sayed, H. Kosina, and S. Selberherr, "Conductance in a Nanoribbon of Topologically Insulating MoS2 in the 1T' Phase," *IEEE Trans. Electron Devices*, vol. 67, no. 11, pp. 4687–4690, 2020.

[9] B. Zhou, H. Z. Lu, R. L. Chu, S. Q. Shen, and Q. Niu, "Finite size effects on helical edge states in a quantum spin-hall system," *Phys. Rev. Lett.*, vol. 101, no. 24, pp. 1–4, 2008.

[10] A. Pulkin and O. V. Yazyev, "Controlling the Quantum Spin Hall Edge States in Two-Dimensional Transition Metal Dichalcogenides," *J. Phys. Chem. Lett.*, vol. 11, no. 17, pp. 6964–6969, 2020.

[11] J. P. Perdew, K. Burke, and Y. Wang, "Generalized gradient approximation for the exchange-correlation hole of a many-electron system," *Phys. Rev. B*, vol. 54, no. 23, pp. 16 533–16 539, dec 1996. [Online]. Available: http://www.tandfonline.com/doi/abs/10.1080/13639080.2012.711944 https://link.aps.org/doi/10.1103/PhysRevB.54.16533

[12] E. v. Lenthe, E. J. Baerends, and J. G. Snijders, "Relativistic regular two-component hamiltonians," *J. Chem. Phys.*, vol. 99, no. 6, pp. 4597–4610, 1993. [Online]. Available: https://doi.org/10.1063/1.466059

[13] D. Gresch, G. Autès, O. V. Yazyev, M. Troyer, D. Vanderbilt, B. A. Bernevig, and A. A. Soluyanov, "Z2Pack: Numerical implementation of hybrid Wannier centers for identifying topological materials," *Phys. Rev. B*, vol. 95, no. 7, pp. 1–24, 2017.

[14] K. Chen, J. Deng, X. Ding, J. Sun, S. Yang, and J. Z. Liu, "Ferromagnetism of 1T'-MoS2 Nanoribbons Stabilized by Edge Reconstruction and Its Periodic Variation on Nanoribbons Width," *J. Am. Chem. Soc.*, vol. 140, no. 47, pp. 16 206–16 212, 2018.

978-1-6654-3746-2/21 $31.00 © 2021 IEEE

Impact of the Backgate on the Performance of SOI UTBB nMOSFETs at Cryogenic Temperatures

Yi Han[1,2], Fengben Xi[1,2], Frederic Allibert[3], Ionut Radu[3], Slawomir Prucnal[4], Jin-Hee Bae[1], Susanne Hoffmann-Eifert[5], Joachim Knoch[6], Detlev Grützmacher[1] and Qing-Tai Zhao[1]

[1] *Peter Grünberg Institute (PGI 9) and JARA-Fundamentals of Future Information Technologies, Forschungszentrum Jülich GmbH, 52428 Jülich, Germany*
[2] *Faculty of Mathematics, Computer Science and Natural Sciences, RWTH Aachen University, 52074 Aachen, Germany*
[3] *SOITEC, 38190 Bernin, France*
[4] *Institute of Ion Beam Physics and Materials Research, Helmholtz-Zentrum Dresden-Rossendorf, 01328 Dresden, Germany*
[5] *Peter Grünberg Institut (PGI 10) and JARA-FIT Institut Green IT, Forschungszentrum Jülich GmbH, 52428 Jülich, Germany*
[6] *Institute of Semiconductor Electronics, RWTH Aachen University, 52074 Aachen, Germany*

Abstract—In this paper we present an experimental investigation on SOI UTBB n-MOSFETs at cryogenic temperatures. The device has a silicide source/drain with dopant segregation formed by "Implantation Into Silicide" (IIS) process. The controllability of the back-gate (V_{back}) on the device performance is characterized, showing that V_{back} is essential to tune the threshold voltage V_{th} and improve the subthreshold swing SS, Drain-Induced Barrier Lowering DIBL and mobility at cryogenic temperatures. The cryogenic effect on V_{th}, SS and DIBL with different V_{back} is also studied. Furthermore, using the V_{back} and quantization effect at cryogenic temperature, we can optimize the SS to a lower value, providing a potential way to get ideal value of SS at cryogenic temperature.

Keywords—cryo-CMOS, ultrathin body and BOX fully depleted silicon-on-insulator (UTBB FD-SOI), Cryogenic electronics, quantum computing, back-gate

I. INTRODUCTION

Quantum computing provides a promise for solving intractable exponentially-growing problems by its potentially powerful computing. To achieve this extremely fast computing, an important challenge today is scaling up a larger number of qubits (>1000) [1]. However, until now, typical arrangement of the quantum processor and classical control is separately operating at room and deep-cryogenic temperatures (~mK) [2]. As a result, there are huge numbers of interconnected wires which leads to many problems such as, large temperature gradient, inevitable thermal noise, and high cost. Especially, when increasing the number of qubits from ten to thousands, this wiring requirement becomes an unimaginable disaster. Placing the control and readout circuits in direct vicinity to the qubits is an interesting solution for scalability and fast qubit-information processing [3]. CMOS operating at cryogenic temperature (cryo-CMOS) is a good candidate, which could satisfy stringent requirements on noise, accuracy, and bandwidth [4]. Meanwhile, it can be well compatible with semiconductor qubits, like silicon quantum dot qubits [5], which provides a potential possibility for miniaturized integrated quantum computer.

However, the cryogenic effects of dopant freeze-out, Fermi energy shifting, and band tail effects cause classical CMOS devices, designed for room temperature (RT) application, to experience difficulties in fulfilling the requirements of cryo-CMOS, especially the very low applied voltage (V_{DD}= ~10 mV) requirement by the power limitations of the refrigerators. Such low V_{DD} requires steep subthreshold swing (SS) of <4 mV/dec and controllable threshold voltage (V_{th}) cryo-CMOS [1][4]. Ultra-thin body and ultra-thin buried oxide (UTBB) fully depleted silicon-on-insulator (FDSOI) MOSFETs with low doped channel are promising candidates to fulfill the cryo-CMOS requirements [6]. This technology shows outstanding performances compared with bulk CMOS and Si gate all around nanowires technologies, which eliminates unwanted kink effects, provides low power consumption capabilities, maintains low variability, and offers back-gate control enabling the threshold voltage control at cryogenic temperatures. Moreover, back-gate also offers additional control for electrostatically defined quantum confinement and improves the SS by changing the mean channel position [6][7].

In this work, we designed and fabricated a UTBB FD SOI nMOSFETs for cryo-CMOS as shown in Fig. 1a. The device is characterized at cryogenic temperatures down to 34 K. From the comparison of figure-of merit (FOM) parameters including threshold voltage V_{th}, Drain-Induced Barrier Lowering (DIBL) and the subthreshold swing SS, the impact of the back-gate (V_{back}) on the device performance is systematically investigated at cryogenic temperature.

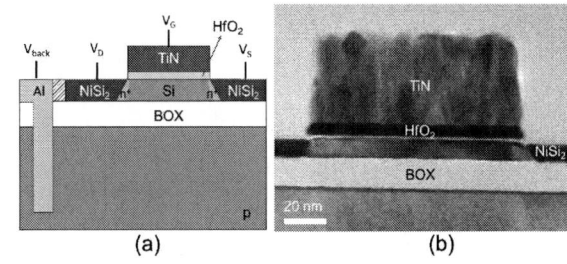

(a)   (b)

Figure 1. The schematic (a) and cross-sectional TEM image (b) of the fabricated device. The initial Si film at source/drain is fully silicided into NiSi$_2$ and well aligned to the gate.

978-1-6654-3746-2/21 $31.00 © 2021 IEEE

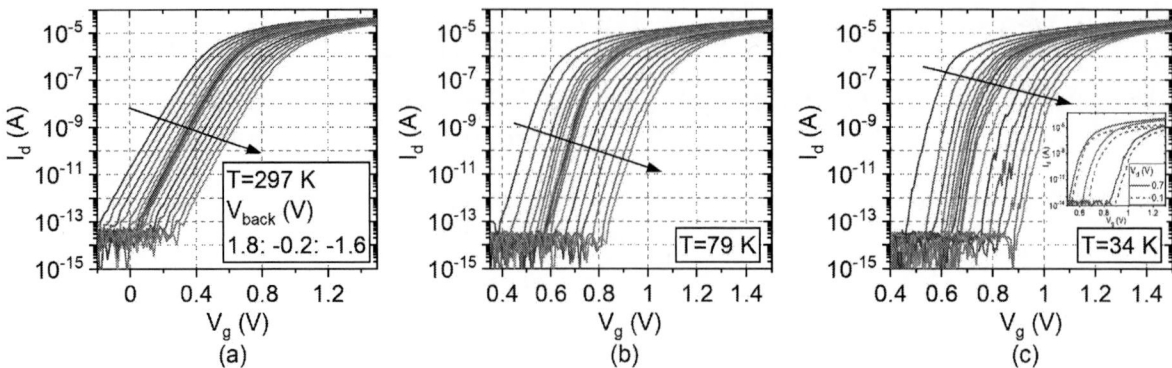

Figure 2. The transfer characteristics obtained with V_{back} = 1.8 to -1.6 V at room temperature (a), 79 K (b) and 34 K (c) for the fabricated 100 nm gate length UTBB n-MOSFET. Curves are close to each other at an intermediate bias range. The inset in (c) shows transfer characteristics with different drain voltage (V_d= 0.7 and 0.1 V) and V_{back} = 1.6, 0, -1.6 V at 34 K to show the DIBL.

II. DEVICE FABRICATION

Transistors were fabricated on UTBB p-type (100) SOI substrates (N_A=1×10^{15} cm^{-3}) with a starting top silicon thickness of 12 nm and a BOX thickness of 15 nm. A gate stack consisting of 5 nm HfO$_2$ and 40 nm TiN deposited by Atomic Layer Deposition (ALD) and sputtering is patterned with e-beam lithography and reactive ion etching (RIE) for different gate lengths from100 nm down to 50 nm. Then an ultra-thin Ni layer Ni (2.4 nm) is sputtered and followed with a rapid thermal annealing (RTP) at 700 °C to form epitaxial NiSi$_2$ layers on source and drain. The unreacted Ni is selectively removed by wet etching with diluted H$_2$SO$_4$. The initial Si film at source and drain is fully silicided into NiSi$_2$ which is aligned to the gate as shown in a cross-sectional transmission electron microscope (TEM) image in Fig. 1b. Perfect NiSi$_2$ source/channel and source/BOX interfaces can be clearly seen. Then, tilted phosphorus implantation at angles of 45° and 135° followed by RTA at 500 °C to drive-out and activate the implanted dopants are performed to form the source/drain. This process is known as "Implantation Into Silicide" (IIS). The source-drain resistance can be effectively reduced with metallic Source/Drain electrodes. It also provides high scalability, compatibility, and low sensitivity to thermal fluctuations due to Schottky barrier device carrier transport dominated by Fowler-Nordheim tunneling at cryogenic temperatures. To prevent surface leakage, a protective SiO$_2$ layer is deposited by plasma enhanced chemical vapor deposition (PECVD). After opening the source, drain and gate contact windows, a 200 nm Al metallization is carried out with a lift-off process, followed by 10 min at 500 °C forming gas anneal (H$_2$/N$_2$: 1/9). The back gate is located close to the channel. No channel doping and no ground plane implant are used for these devices. All devices are characterized from room temperature down to 34 K with a cryogenic probe station. The top gate (V_g) and a back gate on the substrate (V_{back}) are used to control the device.

III. RESULTS AND DISCUSSION

Fig. 2 shows the transfer characteristics of the fabricated 100nm gate length n-MOSFET measured at room temperature, 79 K and 34 K under various V_{back} = 1.8 to -1.6

V. All the transfer curves are shifted gradually from left to the right with decreasing V_{back} at all temperatures, indicating that the threshold voltage V_{th} can be efficiently adjusted by V_{back} at 34 K as shown in Fig. 3a, where V_{th} is extracted by using the maximum transconductance g_m method which allows to remove the temperature variation of the series resistance and the subthreshold swing [8]. An intermediate V_{back} range is obviously found where the V_{th} is less sensitive to the V_{back}, which can be also clearly found and indicated by the plateau of the V_{th} in Fig. 3a. This is due to that a depletion region is formed at the BOX/Si substrate interface [7]. Higher V_{th} (comparing with that at room temperature) leads to that the top gate introduces more electrons in the substrate at cryogenic temperature, impeding the accumulation of holes in the substrate. Therefore, the substrate depletion regime is slightly extended for V_{back} from 1 to 1.4 V with temperature cooling down as shown in Fig. 2 and Fig. 3a. The range of substrate depletion regime at room temperature is indicated by the light yellow in Fig. 3.

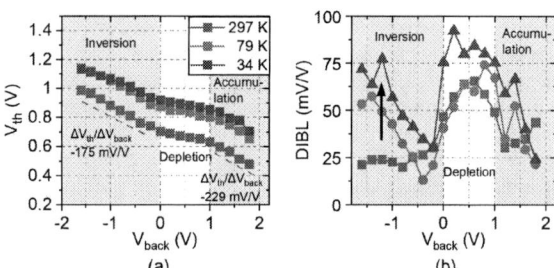

Figure 3. The threshold voltage V_{th} (a), and DIBL (b) as a function of V_{back} at different temperatures. Three regimes (inversion, depletion, and accumulation) of the substrate are indicated by light yellow and gray colors. The arrow indicates the DIBL bumps at inversion substrate regime (negative V_{back}) at cryogenic temperature.

Some humps can be clearly found in the I_d-V_g curves at 34 K (Fig. 2c) with negative V_{back}, which could be related to two quantum dots formed by the Schottky barrier and doping pockets in our dedicated transistors. V_{back} can modify the barrier height of two quantum dots and therefore add or eliminate those humps in transfer curves as shown in Fig. 2c. Meanwhile, by modulating the barrier height, those two coupled quantum dots could work as qubits, which provides a potential for Si quantum devices [5][9]. The body factor

978-1-6654-3746-2/21 $31.00 © 2021 IEEE

$(\Delta V_{th}/\Delta V_{back})$ of V_{th}-V_{back} curves remain unchanged at all temperatures in the substrate inversion and accumulation regimes, demonstrating the Field-assisted ionization is effective at cryogenic temperature. It also lets V_{th} curves maintain similar outline at all temperature as shown in Fig. 3a.

Fig. 3b shows the DIBL which were extracted at a constant drain current (I_d= 100 pA) as a function of V_{back}, as indicated in inset of Fig. 2c. It can be clearly found the DIBL is higher at cryogenic temperature (46 mV/V at room temperature, 75 mV/V at 34 K) at V_{back}= 0 V. In FD-SOI MOS transistors, DIBL mainly depends on three factors: 1) drain-to-channel coupling through Si body; 2) drain-to-channel coupling through the BOX; and 3) drain-to-channel coupling through the substrate [10]. Factors 1 and 2 are mainly affected by the thickness of Si film and BOX respectively, which keeps unchanged at all temperatures. Therefore, in this case, factor 3 dominates especially at cryogenic temperature. At room temperature, the V_{back} deteriorates the DIBL in the substrate depletion regime by depleting the substrate under the BOX, which increases the depth of the space-charge region (t_{sub}) in the substrate and leads to DIBL humps. At substrate inversion/accumulation regime, the mean channel position modified by V_{back} can enhance/reduce the front gate control ability and hence decrease/increase the DIBL. Furthermore, t_{sub} reduces to the thickness of the inversion or accumulation layer in substrate, respectively, and keep constant with V_{back} [11]. At cryogenic temperature, dopant freeze-out effect also depletes the substrate and hence further increases t_{sub}, which aggravates the DIBL especially at substrate depletion regimes. Meanwhile, the extension of the depletion regime also extends the range of DIBL humps at cryogenic temperature. In the substrate inversion regime, bandgap widening and scaling of Fermi-Dirac occupation function reduces overlap of $f(E)$ with the density-of-states (DOS) in the conduction band and hence decreases the density of electros at the BOX/Si substrate interface with the same V_{back}. This weak inversion layer cannot prevent drain-to-channel coupling through the space-charge region. Therefore, t_{sub} still affects the DIBL. The decreasing V_{back} further inverses and depletes the carries in the substrate, and increases t_{sub}, which leads to a larger DIBL in the substrate inversion regime. Quantum humps which is mentioned above in the discussion of the device transfer characteristics may also degrade the DIBL at substrate inversion regime at the same time.

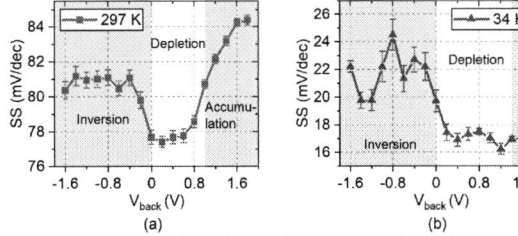

Figure 4. SS as a function of V_{back} for the 100 nm gate length transistor at room temperature (a), and 34 K (b). Three regimes (inversion, depletion, and accumulation) in the substrate are indicated by light yellow and gray colors.

Fig. 4. shows the SS as a function of V_{back} at different temperatures. The error bar of SS is extracted by the linear fit method to show the quantum humps effect. At room temperature, the SS is improved in the substrate depletion regime by modifying the depletion capacitance (C_{dep})

between the channel (SS= $\ln(10)kT/q$ $(1 + C_{it}/C_{ox} + C_{dep}/C_{ox})$). V_{back} depletes carries and forms the capacitance of space-charge region (C_{sub}) in the substrate, which connects with the capacitance of BOX (C_{BOX}) in series and therefore reduces the C_{dep} to C_{dep}=C_{sub}*$C_{BOX}/(C_{sub}+C_{BOX})$. Moreover, by pushing the mean channel position to the surface and thus enhancing the top gate control ability, SS is smaller in the substrate inversion regime than that in the accumulation regime at room temperature. But this enhanced top gate control is weakened at cryogenic temperature, as shown in Fig. 4b. In the substate inversion regime, SS is mainly dominated by the capacitance of interface traps (C_{it}), especially at cryogenic temperature. This is due to that the scaling of Fermi-Dirac occupation function increases the occupied interface trap state density (N_{it}), and therefore increases C_{it}= qN_{it} at cryogenic temperatures. By changing the mean channel position from the back (Si/BOX) to the front (HfO$_2$/Si) interface, a negative V_{back} can increase N_{it} dramatically due to the inferior quality of front interface compared to the back Si/BOX interface. Therefore, it increases the C_{it} and hence degrades the SS in the substrate inversion regime at cryogenic temperatures. On the other hand, a positive V_{back} can reduce the N_{it} by pushing the mean channel position to the back interface, which slightly extends the plateau of SS from the substrate depletion to the accumulation regime, as shown in Fig. 4b. Meanwhile, it can effectively reduce the SS from 20 mV/dec with grounded V_{back} to 16 mV/dec with a positive V_{back} (V_{back}= 1.2 V) at 34 K.

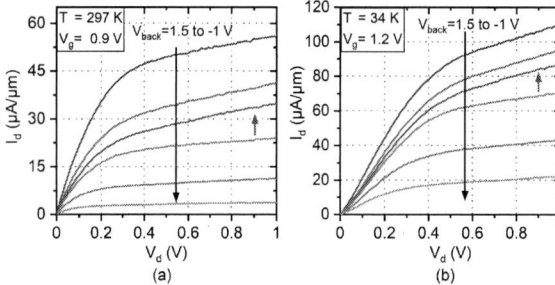

Figure 5. The output characteristics measured at various V_{back} (V_{back}=1.5 to -1 V with a step of 0.5 V indicated by the arrow) and a fixed front gate voltage of V_g= 0.9 V at room temperature (a), and V_g= 1.2 V at 34 K (b).

Fig. 5 shows the output characteristics measured at room temperature and 34 K at various V_{back} with fixed front gate voltage of V_g= 0.9 at 297 K and 1.2 V at 34K, respectively. With the selected V_g, we keep the same overdrive voltage V_{ov}=V_g-V_{th}=0.3 V at V_{back}=0 V at both temperatures of 297 K and 34 K. Comparing Fig. 5a with b, larger currents can be clearly seen at 34 K caused by the mobility enhancement due to the reduced phonon scattering at cryogenic temperature. The saturation current decreases with decreasing V_{back}, because of the V_{th} increasing, especially for high $|V_{back}|$. Although, at V_{back} = 0.5 V, the substrate is still in the depletion regime and the V_{ov} is very close to that at V_{back}=0 V (V_{ov} = 0.3 and 0.32 V for V_{back} = 0 and 0.5 V, respectively) at all temperatures. An obvious current increase is still achieved with V_{back}=0.5 V compared to that at V_{back}= 0 V, as shown by the blue arrow in Fig. 5. This is because that the slight shift of channel centroid to back interface (Si/BOX) can improve the mobility due to the reduction of surface optical phonons and Coulomb scattering at the front interface [12]. Smaller $\Delta I_d/\Delta V_{back}$ is found in the V_{back} range from 1.5

978-1-6654-3746-2/21 $31.00 © 2021 IEEE   59

to 0 V at 34 K compared to the V_{back} range from 1.0 to 0 V at 297 K, which is caused by the extension of V_{th} plateau as mentioned above.

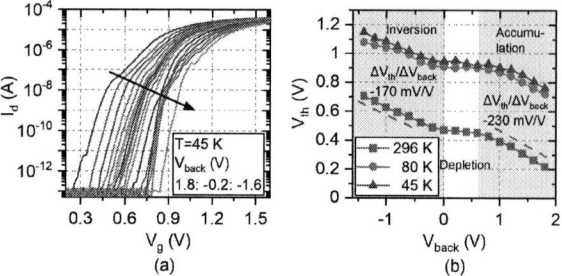

Figure 6. Transfer characteristics (a) and V_{th} (b) measured with various (V_{back} = 1.8 to -1.6 V) for fabricated 50 nm gate length UTBB n-MOSFETs.

The back-gate effect for shorter channel devices is also investigated at cryogenic temperatures. Fig. 6a shows the transfer characteristics measured at 45 K for a 50 nm gate length transistor at various V_{back}. All curves shift gradually with V_{back}, which is similar to that for the 100 nm gate length transistor. More pronounced quantum humps can be found in Fig. 6a. The quantization and V_{back} affect the shape of transfer curves and modify the SS. By balancing both effects, a minimum value of SS=19 mV/dec is achieved with V_{back}= -1.4 V at 45 K. This value is following the kT/q extrapolation after taking the SS of 131 mV/dec at 296 K at V_{back}=0 V. This suggests a back gate provide a potential to break the cryogenic SS saturation caused by the band tail. The threshold voltage V_{th} vs V_{back} is also plotted at various temperatures in Fig. 6b. Obviously, the V_{th} is effectively modified by V_{back}, which is the same as in the case of the 100 nm gate length transistor. Meanwhile, the same body factor value can be also found in Fig. 6b ($\Delta V_{th}/\Delta V_{back}$= -230 and -170 mV/V, respectively, at accumulation and inversion substrate regimes). This shows that the small gate length does not influence the Field-assisted ionization at cryogenic temperature. Furthermore, the substrate depletion regime (the range of V_{back}) does not change with the temperature, which indicates the body-substrate vertical coupling is affected by the Short-Channel Effects (SCE), such as source and drain charge sharing and fringing fields through the BOX. These effects can degenerate the body-substrate vertical coupling and therefore keep the substrate depletion regime constant at all temperature.

IV. CONCLUSION

UTBB FDSOI n-MOSFETs were fabricated and characterized at cryogenic temperatures. The impact of back-gate on the device performance was systematically studied. It shows that V_{back} can effectively adjust V_{th} and improve the SS, DIBL and mobility at cryogenic temperature. Furthermore, by modifying the carrier distribution in the SOI layer, V_{back} can also eliminate the cryogenic effect, like inflection phenomenon, and shows different trends of SS and DIBL at cryogenic temperatures compared with the results at room temperature. The Short-Channel Effects are also investigated with 50 nm gate length transistors, which shows similar V_{th} trend with V_{back}. By optimizing the quantum mechanism effect, a minimum value of SS is achieved with a negative V_{back}, showing a

potential to break the cryogenic SS saturation caused by the band tail for a very steep slope device at cryo-temperatures.

V. ACKNOWLEDGMENT

This work is supported by DFG project "Cryo-CMOS". Support by the Ion Beam Center (IBC) at HZDR and the China Scholarship Council are gratefully acknowledged.

VI. REFERENCE

[1] E. Charbon, "Cryo-CMOS for Quantum Computing," pp. 343–346, 2020.

[2] F. Sebastiano, H. A. R. Homulle, J. P. G. Van Dijk, R. M. Incandela, and B. Patra, "Cryogenic CMOS Interfaces for Quantum Devices," no. August, 2017.

[3] A. Beckers, F. Jazaeri, H. Bohuslavskyi, L. Hutin, S. De Franceschi, and C. Enz, "Design-oriented modeling of 28 nm FDSOI CMOS technology down to 4.2 K for quantum computing," *2018 Jt. Int. EUROSOI Work. Int. Conf. Ultim. Integr. Silicon, EUROSOI-ULIS 2018*, vol. 2018-Janua, pp. 1–4, 2018.

[4] F. Sebastiano *et al.*, "Cryo-CMOS Electronic Control for Scalable Quantum Computing," in *Proceedings of the 54th Annual Design Automation Conference 2017 on - DAC '17*, 2017, pp. 1–6.

[5] S. Bonen *et al.*, "Cryogenic characterization of 22-nm FDSOI CMOS technology for quantum computing ICs," *IEEE Electron Device Lett.*, vol. 40, no. 1, pp. 127–130, 2019.

[6] P. Galy, J. Camirand Lemyre, P. Lemieux, F. Arnaud, D. Drouin, and M. Pioro-Ladriere, "Cryogenic Temperature Characterization of a 28-nm FD-SOI Dedicated Structure for Advanced CMOS and Quantum Technologies Co-Integration," *IEEE J. Electron Devices Soc.*, vol. 6, no. April, pp. 594–600, 2018.

[7] S. Burignat *et al.*, "Solid-State Electronics Substrate impact on threshold voltage and subthreshold slope of sub-32 nm ultra thin SOI MOSFETs with thin buried oxide and undoped channel," *Solid State Electron.*, vol. 54, no. 2, pp. 213–219, 2010.

[8] A. Beckers, F. Jazaeri, A. Grill, S. Narasimhamoorthy, B. Parvais, and C. Enz, "Physical Model of Lowerature to Cryogenic Threshold Voltage in MOSFETs," *IEEE J. Electron Devices Soc.*, vol. 8, no. August, pp. 780–788, 2020.

[9] F. A. Zwanenburg *et al.*, "Silicon quantum electronics," *Rev. Mod. Phys.*, vol. 85, no. 3, pp. 961–1019, 2013.

[10] M. S. Bhoir and N. R. Mohapatra, "Effects of Scaling on Analog FoMs of UTBB FD-SOI MOS Transistors: A Detailed Analysis," *IEEE Trans. Electron Devices*, vol. 67, no. 8, pp. 3035–3041, 2020.

[11] M. K. M. Arshad *et al.*, "Extended MASTAR modeling of DIBL in UTB and UTBB SOI MOSFETs," *IEEE Trans. Electron Devices*, vol. 59, no. 1, pp. 247–251, 2012.

[12] A. Beckers, F. Jazaeri, and C. Enz, "Inflection Phenomenon in Cryogenic MOSFET Behavior," *IEEE Trans. Electron Devices*, vol. 67, no. 3, pp. 1357–1360, 2020.

Constant-current time dependent dielectric breakdown in thick amorphous SiO$_2$ capacitors

Federico Giuliano
ARCES and DEI
University of Bologna
Bologna, Italy
federico.giuliano2@unibo.it

Susanna Reggiani
ARCES and DEI
University of Bologna
Bologna, Italy
susanna.reggiani@unibo.it

Elena Gnani
ARCES and DEI
University of Bologna
Bologna, Italy
elena.gnani@unibo.it

Antonio Gnudi
ARCES and DEI
University of Bologna
Bologna, Italy
antonio.gnudi@unibo.it

Mattia Rossetti
Technology R/D
STMicroelectronics
Agrate Brianza, Italy
mattia.rossetti@st.com

Riccardo Depetro
Technology R/D
STMicroelectronics
Agrate Brianza, Italy
riccardo.depetro@st.com

Giuseppe Croce
Technology R/D
STMicroelectronics
Agrate Brianza, Italy
giuseppe.croce@st.com

Abstract—**Charge transport in thick amorphous silicon dioxide capacitors for integrated galvanic insulators is experimentally investigated and analyzed through numerical simulations carried out with a commercial TCAD tool. The material intrinsic defectivity and the large biases applied to such devices give rise to a leakage current which is responsible of degradation and failure. Hence it is crucial to have a complete understanding of the charge-transport main physical mechanisms in amorphous silicon oxide. For this reason, constant-current time dependent dielectric breakdown measurements have been performed on thick metal-insulator-metal structures and, in order to gain insight on the role of defects on breakdown, numerical simulations have been compared to experiments.**

Index Terms—**Silicon oxide; Insulators; Reliability; TEOS**

I. INTRODUCTION

Metal-insulator-metal (MIM) capacitors embedded in the back-end inter-level dielectric layers have been recently proposed for analog and RF applications [1]–[3]. Silicon dioxide (SiO$_2$) is the main insulator in the electronics industry because of its near-ideal properties; however, ultimate device degradation and failure is still limited by charge buildup in pre-existing defect sites of the oxide layer. Moreover, capacitors for galvanic insulation are thick structures (in the micrometer range) made through several oxidation steps [4].

The tetraethyl orthosilicate (TEOS) is usually adopted as a precursor for the interlayer thick oxides in the plasma-enhanced chemical vapor deposition (PE-CVD). Such technique allows to grow thick SiO$_2$ films in the back-end with good physical properties, but they are known to show a much larger density of preexisting defects leading to higher leakage currents when compared to thermally-grown oxides on silicon [5], [6]. Thus, the internal electric fields can be significantly modified by charge build-up in the bulk of the oxide in addition to the usual charge injection at the contacts, thus further limiting the expected device performance and reliability [7]. For these reasons, a TCAD-based model capable

of correctly handling charge transport and trapping and de-trapping mechanisms in such kind of bulk oxides can be a useful tool for the development and optimization of galvanic insulators in integrated high-voltage systems.

II. TEST STRUCTURES AND EXPERIMENTS

Fig. 1 shows a cross section of the high-voltage MIM capacitor used in this work. The electrodes are in tantalum nitride (TaN). The bottom metal is deposited on silicon and is grounded, while the high voltage is applied to the circular top metal (diameter $d \approx 150~\mu$m). The TEOS material is used as intermetal dielectric. The parallel plate capacitor has a nominal thickness $t_{\mathrm{OX}} = 0.9~\mu$m.

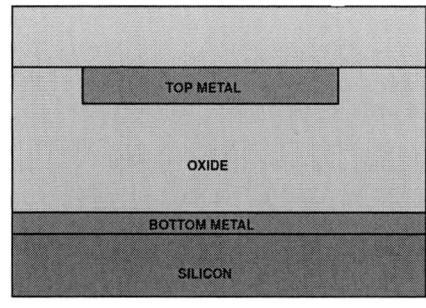

Fig. 1. Schematic view of the TEOS capacitor.

Constant current stresses have been applied to the MIM under study at different temperatures until the breakdown condition was reached measuring the voltage at the top contact. For each temperature, three targets of current density have been used: they are reported in Table I.

Forcing current directly through the capacitor leads to very long voltage rise-times ($dV/dt = I/C$, with I the current flowing through the insulator, C its capacity and V the applied bias). To avoid this issue, the following approach is used:

978-1-6654-3746-2/21 $31.00 © 2021 IEEE

at $t = 0$, a constant voltage is applied to the capacitor, the current is measured and considered as the target current level. In order to avoid current reduction due to charge trapping, a small voltage ramp is applied at fixed time intervals. The new voltage required to force the target current is found and set.

Fig. 2 shows the applied electric field $F_{OX} = |V|/t_{OX}$, with V the voltage applied at the top electrode, as a function of the stress time for each stress and temperature under study.

Table I. Temperature conditions and current targets used in experiments.

	$T = 25\ °C$	$T = 150\ °C$
J_1 (A/cm^2)	$1.4 \cdot 10^{-8}$	$1.6 \cdot 10^{-8}$
J_2 (A/cm^2)	$3.0 \cdot 10^{-8}$	$3.6 \cdot 10^{-8}$
J_3 (A/cm^2)	$6.3 \cdot 10^{-8}$	$7.5 \cdot 10^{-8}$

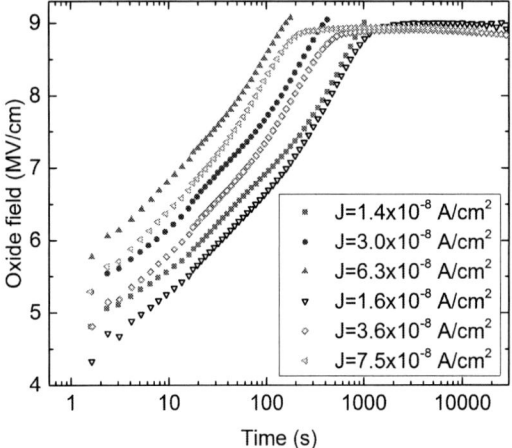

Fig. 2. Constant current measurements performed on the SiO$_2$ TEOS thick capacitor. Two temperature conditions have been tested, namely $T = 25\ °C$ (close symbols) and $T = 150\ °C$ (open symbols).

The application of a constant current stress substantially allows to keep constant the number of carriers injected into the oxide during time. The oxide-field increase is an indication of the charge trapping in the oxide. The voltage increase stops when an oxide field of about 9 MV/cm is reached, and saturates for longer stress times as shown in Fig. 2. The voltage saturation is a clear indication of the onset of avalanche condition leading to breakdown. It should be noted that the blocking field is almost independent of the current forced in the device and temperature, which gives a breakdown field $E_{OX}^{BD} \approx 9$ MV/cm.

III. TCAD MODELING OF THE SiO$_2$

The conduction model of the TEOS SiO$_2$ can be described by using a drift-diffusion (DD) transport model with suitable physical parameters, such as the energy-band structure, the presence of distributed defects in the material band-gap, the

impact-ionization generation and the tunneling injection at the contacts [8]–[10].

As far as the defects are concerned, we mostly based on previous works dealing with thermally-grown SiO$_2$. The experimental data on TEOS oxide structures clearly show the need of a set of almost two traps with different energy levels. Trapping and de-trapping mechanisms have been taken into account by using a first-order detailed balance equation for each trap as available in the TCAD tool [11]; the trap equations are solved consistently along with Poisson and electron and hole transport equations. The trapped charge is explicitly accounted for in the Poisson equation. We have defined only acceptor-type traps for electrons, i.e. defects that are neutral when empty and carry a negative charge when occupied by an electron. A uniform spatial distribution is assumed for all traps. Concerning the energy level of the traps, it should be pointed out that, being SiO$_2$ an amorphous material, energy bands of traps arise instead of discrete trap levels. For this reason, we have defined two uniform distributions of traps in bands of 0.5 eV width with mean energies $E_1 = 6.3$ eV and $E_2 = 6.5$ eV, where the oxide valence band has been taken as the reference level. The energy distribution of traps is represented in Fig. 3, showing also the different trap densities adopted for the bands.

The determination of trap cross-sections requires a special attention to transient responses. In the past years, many authors have reported measurements of electron capture cross sections [12]–[16], with values ranging from 10^{-13} cm^2 to 10^{-18} cm^2. We have used two different cross sections of respectively $\sigma_1 = 1.1 \cdot 10^{-15}$ cm^2 and $\sigma_2 = 9 \cdot 10^{-15}$ cm^2, in fair good agreement with the values reported in [17].

The trap parameters are reported in Table II.

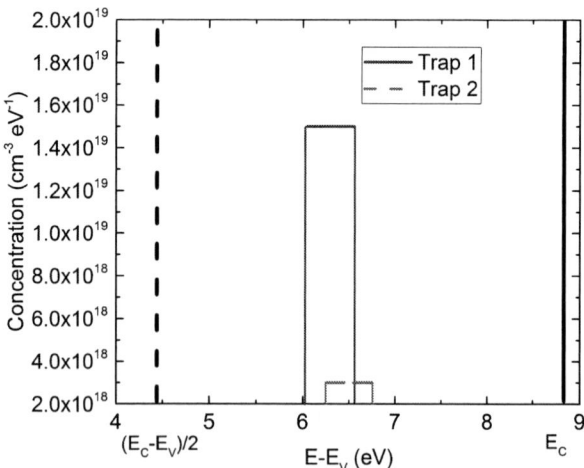

Fig. 3. Energy distribution of traps. The valence band has been taken as the reference level, i.e. $E_V = 0$. The point $(E_C - E_V)/2$ represents the mid-band gap. A band-gap $E_G = 8.9$ eV is assumed for SiO$_2$.

As far as the high-field transport is concerned, the effect of avalanche due to impact ionization cannot be neglected in a complete picture of the relevant physical mechanisms [19].

978-1-6654-3746-2/21 $31.00 © 2021 IEEE

Table II. Trap parameters used for the two trap distributions. Mean energy of the trap level, trap width, electron capture cross section and trap density are reported for each type of trap. The parameter E_T is referred to the top of the valence band, taken as the reference level.

Parameter	Trap 1	Trap 2
E_T (eV)	6.3	6.5
ΔE (eV)	0.5	0.5
σ_e (cm^2)	$1.1 \cdot 10^{-15}$	$9 \cdot 10^{-15}$
N_T (cm^{-3})	$7.5 \cdot 10^{18}$	$1.5 \cdot 10^{18}$

Thus, the impact-ionization generation has been taken into account in our simulation setup using the van Overstraeten-De Man model [10], [20].

IV. SIMULATION RESULTS

The TCAD setup described in the previous Section has been applied to reproduce the experimental data of the constant-current stress. To this purpose, a quasi stationary current ramp has been applied up to the desired current level, and a constant-current stress is directly applied to the simulated device. Fig. 4 shows the TCAD results of the oxide field versus time at $T = 25$ and 150 °C. Experiments are qualitatively reproduced, indicating that the rate at which charge is trapped is in agreement with the experimental data, predicting E_{OX}^{BD} and the corresponding time to breakdown.

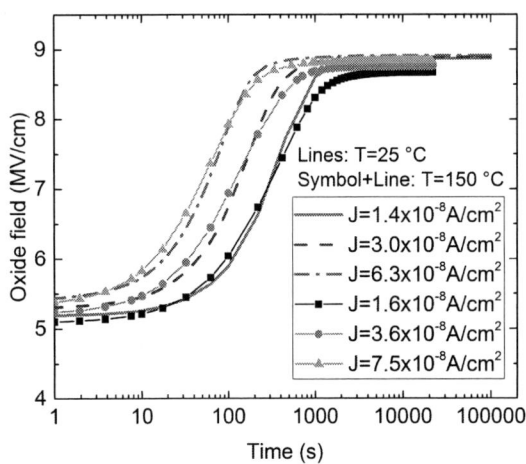

Fig. 4. Simulations of the oxide field plotted as a function of time at different temperatures, namely $T = 25$ °C (lines) and $T = 150$ °C (lines+symbols).

In order to gain insight on the internal phenomena that produce the observed breakdown, the electric field distribution across the device is plotted at $T = 25$ °C, $J = 6.3 \cdot 10^{-8}$ A/cm^2 for three stress times, namely $t_1 = 1$s (at the beginning of the stress), $t_2 = 100$s (during the voltage increase) and $t_3 = 1000$s (after the breakdown condition is reached), as shown in Fig. 5. Charge injection from the top electrode ($x = 0$) leads to a charge trapping dynamics giving rise to a clear increase of the internal field towards the bottom contact.

The maximum field reached near the bottom electrode can be as large as 10.5 MV/cm, comparable with the breakdown strength of bulk SiO$_2$ [18].

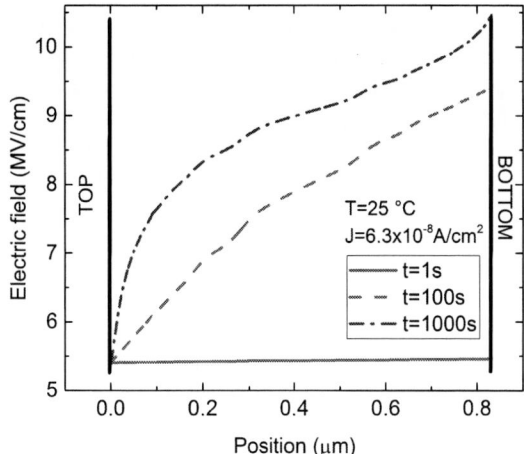

Fig. 5. Electric field distribution across the device at $T = 25$ °C and $J = 6.3 \cdot 10^{-8}$ A/cm^2 for three different stress times.

In order to further assess the role of trapping mechanisms in the observed constant-current TDDB dynamics, the trapped charge across the device has been extracted from simulations at $T = 25$ °C, $J = 6.3 \cdot 10^{-8}$ A/cm^2 at $t_1 = 1$, 100 and 1000 s, as shown in Fig. 6. It can be noted that at short stress times

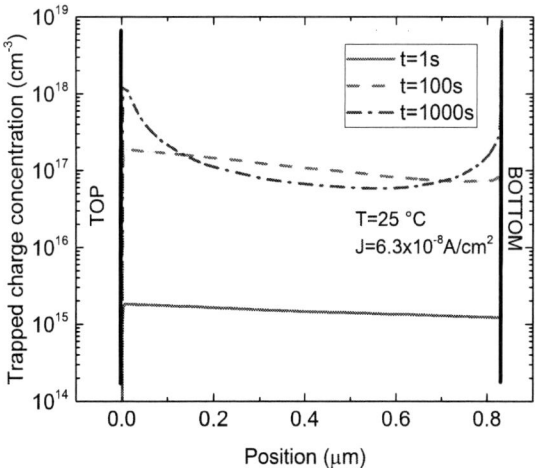

Fig. 6. Total trapped charge across the device at $T = 25$ °C and $J = 6.3 \cdot 10^{-8}$ A/cm^2 for three different stress times.

the trapped charge distribution is substantially flat across the device and simply tends to increase uniformly. Differently, at $t = 1000$ s the trapped charge is greater near both contacts with respect to the central portion of the device. The trapped charge in the proximity of the top contact is mainly due

978-1-6654-3746-2/21 $31.00 © 2021 IEEE

to electron injection from the cathode: an accumulation of charges trapped in the defects is found in the region where the lowest electric field is observed, while a significant trapped-charge emission is expected to play its role at larger electric fields in the middle of the layer. On the other hand, the enhanced trapping near the bottom contact is an effect of avalanche generation which becomes relevant at long stress times due to the even larger field. In fact, the electrons are generated by impact ionization across the device especially near the anode. Those excess electrons can either be emitted from the bottom contact or be trapped, leading to an enhanced trapping concentration in the proximity of the anode.

This is further confirmed by the avalanche generation rate plotted in Fig. 7, where it can be noted that, for long stress times, the generation of electrons gradually increases moving towards the bottom contact. This means that the largest number of generated electrons is close to the anode.

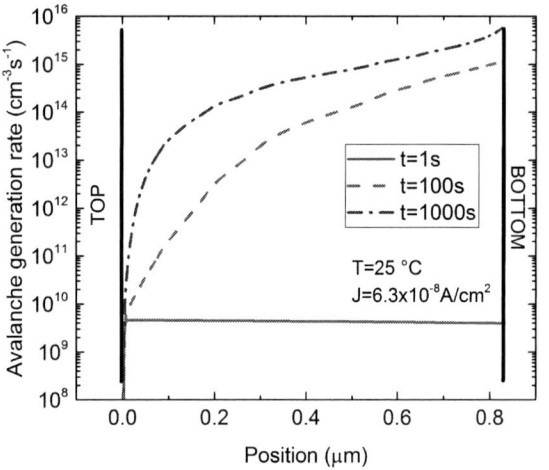

Fig. 7. Avalanche generation rate across the device at $T = 25$ °C and $J = 6.3 \cdot 10^{-8}$ A/cm^2 for three different stress times.

V. CONCLUSIONS

A compact TCAD model has been presented to investigate conduction mechanisms in high-voltage silicon oxide thick TEOS capacitors embedded in the back-end inter-level dielectric layers. The TCAD has been proven to be a useful tool for the study of transport in TEOS oxides, as they tend to show different electrical properties with respect to thermally grown SiO$_2$. The role of traps in the bulk of the oxide has been extensively investigated. From the comparison of numerical simulations to the experiments, we can conclude that impact-ionization is the most relevant mechanism involved in the breakdown of such devices.

REFERENCES

[1] C. C. Hung, A. S. Oates, H. C. Lin, P. Chang, J.L. Wang, C.C. Huang, and Y.W. Yau, *New understanding of Metal-Insulator-Metal (MIM) capacitor degradation behavior*, IEEE 45th Annual International Reliability Physics Symposium, Phoenix, AZ, USA, pp. 630-631, April 15-19, 2007. DOI: 10.1109/RELPHY.2007.369985.

[2] P. Mahalingam, D. Guiling, S. Lee, *Manufacturing challenges and method of fabrication of on-chip capacitive digital isolators*, IEEE International Symposium on Semiconductor Manufacturing, Santa Clara, CA, USA, pp. 1-4, October 15-17, 2007. DOI: 10.1109/ISSM.2007.4446870.

[3] R. Higgins, and J. McPherson, *TDDB Evaluations and Modeling of Very High-Voltage (10 KV) Capacitors*, IEEE 47th Annual International Reliability Physics Symposium, Montreal, QC, Canada, pp. 432-436, April 26-30, 2009. DOI: 10.1109/IRPS.2009.5173292.

[4] S. Shin, Y.P. Chen, W. Ahn, H. Guo, B. Williams, J. West, T. Bonifield, D. Varghese, S. Krishnan, and M. A. Alam *High Voltage Time-Dependent Dielectric Breakdown in Stacked Intermetal Dielectric*, IEEE 56th Annual International Reliability Physics Symposium, Burlingame, CA, USA, pp. P-GD.9-1-P-GD.9-5, March 11-15, 2018. DOI: 10.1109/IRPS.2018.8353669.

[5] W. Wu, S. Rojas, S. Manzini, A. Modelli, D. Re, *Characterization Of SiO2 Films Deposited By Pyrolysis Of Tetraehylorthosilicate (TEOS)* , Journal de Physique Colloques, Vol. 49 (C4), pp.C4-397-C4-400, 1988. DOI: 10.1051/jphyscol:1988483.

[6] M. Sometani, R. Hasunuma, M. Ogino, H. Kuribayashi, Y. Sugahara and K. Yamabe, *Suppression of Leakage Current of Deposited SiO2 with Bandgap Increasing by High Temperature Annealing*, ECS Transactions, Vol. 19 (2), pp. 403-413, 2009. DOI: 10.1149/1.3122105.

[7] E.F. Runnion, S.M. Gladstone, R.S. Scott, D.J. Dumin, L. Lie, and J.C. Mitros, *Thickness Dependence of Stress-Induced Leakage Currents in Silicon Oxide*, IEEE Transactions On Electron Devices, Vol. 44, No. 6, pp. 993-1001, June 1997. DOI: 10.1109/16.585556.

[8] P. C. Arnett, *Transient conduction in insulators at high fields*, Journal of Applied Physics 46, pp. 5236-5243, 1975. DOI: 10.1063/1.321592.

[9] J.F. Verwey, E.A. Amerasekera and J. Bisschop, *The physics of SiO2 layers*, Reports on Progress in Physics, Vol. 53, No. 10, pp. 1297-1331, 1990.

[10] F. Giuliano, S. Reggiani, E. Gnani, A. Gnudi, M. Rossetti, R. Depetro, G. Croce, *Novel TCAD Approach for the Investigation of Charge Transport in Thick Amorphous SiO2 Insulators*, IEEE Transactions on Electron Devices. DOI: 10.1109/TED.2021.3100309

[11] Synopsys Inc., Sentaurus Device User Guide M-2016.12, 2016.

[12] T.H. Ning, *High-field capture of electrons by Coulomb-attractive centers in silicon dioxide*, Journal of Applied Physics, Vol. 47, No. 7, pp. 3203-3208, 1976. DOI: 10.1063/1.323116.

[13] N.S. Saks, and M. G. Ancona, *Determination of Interface Trap Capture Cross Sections Using Three-Level Charge Pumping*, IEEE Electron Device Letters, Vol. 11, No. 8, pp.339-341, Aug. 1990. DOI: 10.1109/55.57927.

[14] D. J. DiMaria, F. J. Feigl, and S. R. Butler, *Trap ionization by electron impact in amorphous SiO2 films*, Applied Physics Letters, Vol. 24, No. 10, pp. 459-461, 1974. DOI: 10.1063/1.1655011.

[15] P. Solomon, *High-field electron trapping in SiO2*, Journal of Applied Physics, Vol. 48, No. 9, pp. 3843-3849, 1977. DOI: 10.1063/1.324253

[16] J. Albohn, W. Füssel, N. D. Sinh, K. Kliefoth and W. Fuhs, *Capture cross sections of defect states at the Si/SiO2 interface*, Journal of Applied Physics, Vol. 88, No. 2, p. 842-849, June 2000. DOI: 10.1063/1.373746.

[17] M.H. Chang, J.F. Zhang, and W.D. Zhang, *Assessment of Capture Cross Sections and Effective Density of Electron Traps Generated in Silicon Dioxides*, IEEE Transactions On Electron Devices, Vol. 53, No. 6, pp. 1347-1354, June 2006. DOI: 10.1109/TED.2006.874155.

[18] D. Arnold, E. Cartier, and D. J. DiMaria, *Theory of high-field electron transport and impact ionization in silicon dioxide*, Physical Review B, Vol. 49, No. 15, pp. 10278-10297, Apr. 1994. DOI: 10.1103/PhysRevB.49.10278.

[19] D. Arnold, E. Cartier, and D. J. DiMaria, *Theory of high-field electron transport and impact ionization in silicon dioxide*, Physical Review B, Vol. 49, No. 15, pp. 10278-10297, Apr. 1994. DOI: 10.1103/PhysRevB.49.10278.

[20] R. van Overstraeten and H. de Man, *Measurement of the Ionization Rates in Diffused Silicon p-n Junctions*, Solid-State Electronics, Vol. 13, No. 1, pp. 583–608, May 1970. DOI: 10.1016/0038-1101(70)90139-5.

TCAD Simulations of High-Aspect-Ratio Nano-biosensor for Label-Free Sensing Application

Rakshita Pritam Singh Dhar[1*], Naveen Kumar[1*], Cristina Medina-Bailon[1],

César Pascual García[3], Vihar Petkov Georigiev[1*]

[1]Device Modelling Group, James Watt School of Engineering, University of Glasgow, UK

[2]Nano-Enabled Medicine and Cosmetics group, Materials Research and Technology Department, Luxembourg Institute of Science and Technology (LIST), Belvaux, Luxembourg

*Correspondence: r.dhar.1@research.gla.ac.uk, Naveen.Kumar@glasgow.ac.uk, Vihar.Georgiev@glasgow.ac.uk

Abstract- **In this paper, we are presenting simulations of junctionless ion-sensitive field-effect transistor (JL-ISFET) as a pH sensor. Our approach is based on a combination of analytical and numerical methods to reveal the impact of the device geometry and structure on its performance. To have a realistic representation of the fabricated device, further simulations are carried which portray the sensing of surface potential by introducing interface trap charges between the oxide layer and electrolyte. Here, we present our initial steps that belong to a more complex and physically more elaborate simulation framework, which will lead to a better device sensing and fabrication choices of more generic biosensors, in a transition from analytical models to numerical simulations to include effects such as surface roughness and defects in the oxide.**

Keywords- ISFET, TCAD simulation, Electrolyte modelling.

I. INTRODUCTION

In the early stages, biosensors based on field-effect transistors (FETs) were fabricated with large planar architectures using heterojunctions which permitted only one-dimensional diffusion of the analytes towards the sensor. However, in the last decade, the progress of the technology and fabrication capabilities in the semiconductor industry allows us to use novel techniques to complete transistors with nanoscale non-planar configuration involving the diffusion of analytes in three-dimensions.

Ion-sensitive field-effect transistors (ISFETs) are devices, initially used as pH sensors, which can also be used to detect different biomolecules using certain modifications. Junction-FET sensors become less reliable with miniaturization due to reduced gate control and complex fabrication process, leading to the development of junctionless transistors. A junctionless (JL) transistor is a device that has uniform doping concentrations throughout the source, channel and drain. Junctionless transistors suffer from a higher threshold voltage that can be tweaked with the use of gate-all-around architecture. In all cases, the limits of detection depend on the diffusion of analytes and the equilibrium of charges at the interface between the liquid and solid phase. The

miniaturization of the devices contributes to a more efficient diffusion of the analytes towards the sensing surface, leading to faster adsorption of the biomarker molecules at the surface of the sensor, while the 3D gating present in nanowires contributes to better transduction [1]. As a result, the sensors can detect lower concentrations in the electrolyte being more sensitive to the changes in the solution. The principle was first applied with silicon nanowires [2], which provided a leap in the limits of detection [3]. However, variability and reliability issues of the sensors have hindered their reach to clinical applications. The variability and the reliability could be improved by using larger devices based on FinFET technology made on silicon-on-insulator (SOI). It is possible to fabricate a high aspect ratio (height>>width of the device) FinFETs, which allows an increase in the total surface area thereby improving the reliability of bio-functionalization and the sensitivity of the sensor [4] [5], while keeping the advantage of 3D gating. Thus, FinFET based JL-ISFETs provides better gate controllability without compromising the ease in fabrication. These kind of ISFETs have great compatibility with the CMOS technology. Nevertheless, there are still challenges to be tackled related to optimisation of the device's architecture and materials chosen [6] [7].

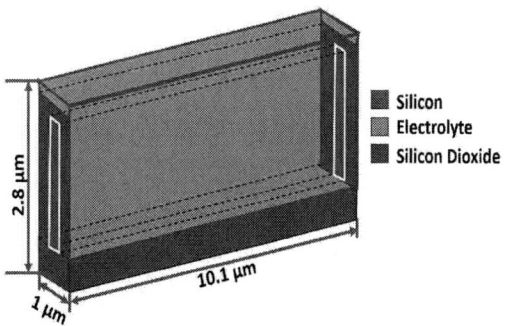

Fig. 1 Schematic of the device implemented in our simulation domain which is based on high-aspect-ratio (height >> width of the device) junctionless ion-sensitive field-effect transistor (JL-ISFET). White rectangular boxes shown on the sides are source/drain electrodes (side contacts).

In this work, we have performed a numerical design-of-experimental (DoE) study to improve the device design based on junctionless ion-sensitive field-effect transistor (JL-ISFET) based on FinFET technology. We have performed device simulations using Synopsys Technology Computer-Aided Design (TCAD) program Sentaurus. Based on the TCAD simulations, current-voltage characteristics (I_D-V_G curve) for p-type FinFET of the device, which are the transfer characteristics of the varying pH response, are obtained explaining the matching behaviour to experimental fabrication data [4]. In our case, we have used the interaction of ions with the oxide surface as an indicator of pH, with the objective in the future, to model other more complex interface functionalization. pH is already meaningful because pH-ISFET technology is widely used.

II. DEVICE ARCHITECTURE AND MATERIAL PROPERTIES

Fig. 1 shows schematics of a high-aspect-ratio JL-ISFET device considered in this work. The pink contour in Fig. 1 is the boundary condition where the gate voltage V_G is applied and the white contours in Fig. 1 correspond to contacts where the drain-source bias, V_{DS} is applied. The implemented geometry and dimensions are based on a fabricated device [4]. The simulation domain has the following dimensions of length, height, and width 10.1µm, 2.8µm and 1µm, respectively. The device has a silicon channel with a width of 200nm and an oxide thickness of 5 nm/10 nm, respectively. In this work, the JL-ISFET is designed to have a high-aspect-ratio meaning the height to width ratio of the fin is very high (h: w ≈10). This ratio allows for a larger total surface area.

Our simulations are based on two types of device configurations. The first one is a JL-ISFET device with a dry gate (a MOSFET device without electrolyte in the gate) and the second one is a JL-ISFET device with an electrolyte. The electrolyte considered in our work is NaCl+H_2O. The channel is p-type doped Si (boron atoms for doping) with a concentration of 10^{17}cm^{-3}. Furthermore, the simulation domain includes an oxide layer which can be either hafnium dioxide or silicon dioxide and the electrolyte layer which surrounds the Si channel on the three sides.

III. RESULTS AND DISCUSSIONS

Based on our simulation, we can calculate device transfer characteristics such as drain current vs. drain voltage (I_D-V_G) for each device. Fig. 2 shows the I_D-V_G curve of the JL-ISFET with a dry gate. Four types of devices are considered having two different oxide thicknesses (t_{ox}) and two oxide materials, respectively. According to MOS equation [8], I_D is directly proportional to permittivity (ε_0) and inversely proportional to t_{ox}. Since HfO$_2$ has a very high dielectric constant (ε~20) in comparison to SiO$_2$ (ε = 1) [9], this leads to high C_{ox} for the I_D at V_G = -1V. Thinner oxide values cause higher values of the drain current, hence the devices with t_{ox}=5 nm have high I_D compared to transistors with t_{ox}=10nm. At V_G = 0 V, there is weak depletion of the hole carriers due to gate metal work-function which leads to less hole current in the channel. For V_G > 0V, the depletion is more, and it will affect the most on the 5nm HfO$_2$ device, leading to the lowest drain current.

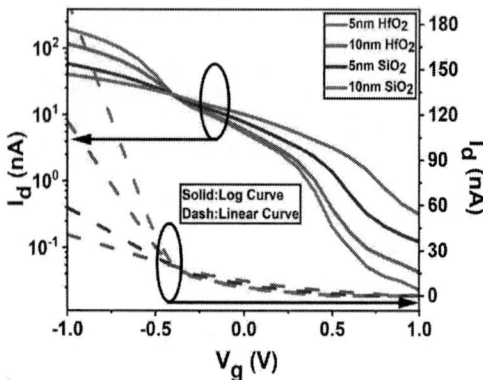

Fig. 2 Current-Voltage (I_D-V_G) characteristics of four different MOSFET devices (without electrolyte in the gate). The device dimensions are the same except the thickness and the type of the gate oxide.

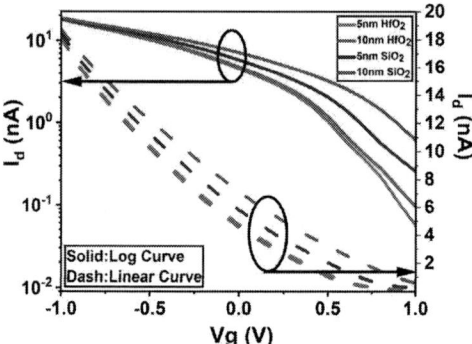

Fig. 3 Current-Voltage (I_D-V_G) characteristics of the different devices with a with an electrolyte [pH @ isoelectric point] above the gate oxide. The device dimensions are the same except the thickness and the type of the gate oxide.

A. Defining Electrolyte in TCAD

As a next step, we have introduced an electrolyte on the top of the gate oxide to simulate the solvent in the biosensor. The electrolyte is modelled as stated in [10]. The electrolyte which is an ionic solution is modelled as a semiconductor material having a dielectric constant of water (ε_r=80) [10]. The electrolyte's charge distribution is represented by the Poisson-Boltzmann equation which is close to the semiconductor equation. By changing the mobility of holes and electrons of the semiconductor, the salt concentration in the solution has been simulated. The electron mobility has a value set to Cl^- ions in water ($6.88*10^{-4} \, cm^2 V^{-1} s^{-1}$) and the hole mobility is set to value of Na^+ ions in water ($4.98*10^{-4} \, cm^2 V^{-1} s^{-1}$). The bandgap in the electrolyte is kept as 1.5eV to represent a solvent [11]. The applied gate bias is kept less than the voltage responsible for the electrolysis of water.

Fig. 3 shows the I_D-V_G curve for all the herein considered devices in which the electrolyte is included considering at equilibrium state (Isoelectric Point), in contact with the gate oxide. In Fig. 3, it is shown that all devices have similar ON-current amongst them [8]. The capacitance of the oxide layer (C_{ox}) varies for different devices and the capacitance of electrolyte)remains the same for all four devices. Our sytem can be described as two series capacitors where one of the capacitors takes into account the charge across the gate oxide and the other capacitor represents the charge accumulation in

978-1-6654-3746-2/21 $31.00 © 2021 IEEE

the electrolyte as shown in the equation below:

$$\frac{1}{C_{eq}} = \frac{1}{C_{ox}} + \frac{1}{C_{electro}} \qquad (1)$$

Where C_{eq}= equivalent capacitance of the device, C_{ox}= Oxide capacitance and $C_{electro}$= Electrolyte capacitance. In such a configuration, the dominant factor, which will determine the device behaviour, is the capacitor with the lowest value. The variation of the gate bias inverts the channel of every device with different oxide material or thickness at the highest negative value. The depletion of the channel is more prominent for the HfO_2 as compared to the SiO_2 for the highest positive gate bias. The devices with a thinner gate oxide and higher dielectric permittivity can comprehend better the ions variation in electrolyte onto the channel.

Fig. 4 2D simulation domain of the adopted ISFET device structure. Si channel is represented with a blue rectangle and has 10nm width. SiO_2 which is marked in brown has 2nm thickness and the electrolyte is depicted in yellow, and it has thickness of 5nm. The red line between the SiO_2 and the electrolyte is the place where the interface trap charges (ITC) are distributed. The pink contours are boundary conditions for the gate voltage (V_G) on both sides of electrolyte and on the top and bottom of Silicon channel are the contacts for V_{DS}.

The next step in our simulation approach is to modify the simulation domain to include pH dependence in the electrolyte section. In order to simulate the effect of the pH on the device performance, we have introduced interface trap charges (ITCs) between the gate oxide and the electrolyte. Moreover, we have modified the simulation domain from 3D to 2D. Moving from 3D to 2D device simulation domain is justified due to the high-aspect-ratio of the devices. As a first approximation, our device can be described as a transistor with gates on both sides and the top gate can be ignored due to significantly smaller surface area in comparison to the side gate area. Fig. 4 reveals the 2D simulation domain that we have adopted for further DoE investigation to save computational time. As a result, the output current will change but the surface potential will be very similar to the full 3D structure.

IV. ANALYTICAL MODEL OF OXIDE AND ELECTROLYTE

When the electrolyte is added on top of the oxide layer, a built-in potential is developed at the interface between oxide and electrolyte called the sensing surface. To represent the

built-in potential at the sensing surface in TCAD simulations, ITCs are introduced (red line on Fig. 4). From a simulation perspective, changes in the values of the ITC density will directly correspond to specific values of pH in the electrolyte. Hence, it is important to be able to evaluate the exact amount of charges on the sensing surface. In order to calculate the ITC density, we have adopted the following analytical model. At 25^0C in pure water $[H^+]=[OH^-]=10^{-7}$mol/L. The Avogadro constant $N_A = 6.022*10^{23}$ mol-1 is converted to 1 mol/L = $6.022*10^{23}$/L = $6.022*10^{20}$cm-3. For electrolyte having pH = 7, the concentration of $[H^+]$ ions = 10^{-7} mol/L which corresponds to p = $10^{-7}*6.022*10^{20}$ cm-3 = $6.022*10^{13}$ cm-3. The hole density of states (N_V) can be calculated using equation below [11]:

$$p = N_v\, e^{-\frac{Ef-Ev}{kT}} \qquad (2)$$

with a result of $2.33*10^{26}$ cm-3. The same calculation can be done for the electron density of states (N_c) using the number of electrons (n) in the following equation [10]:

$$n = N_C\, e^{-\frac{Ec-Ef}{kT}} \qquad (3)$$

with a result of $2.33*10^{26}$ cm-3. The values of the ITC are calculated by representing the built-in potential using the equivalent surface charge density. The surface potential variation with respect to pH is given in equation below [12]:

$$\frac{\partial \Psi_0}{\partial pH} = -2.303\frac{k_B T}{q}\alpha \qquad (4)$$

where Ψ_0 is the surface potential, k_B is the Boltzmann constant, T is the temperature, q is the electron charge, and α is the sensitivity parameter. α is 1 for the ideal device. The value of α is given by the equation below [12]:

$$\alpha = \frac{1}{1 + 2.303\frac{k_B T\, C_{diff}}{q^2 \beta_{diff}}} \qquad (5)$$

where C_{diff} is the diffusion capacitance and β_{diff} is the intrinsic buffering capacitance. C_{diff} depends on electrolyte properties which consider double layer and stern capacitances and β_{diff} depends on the density of surface functionalization [13]. The relation between surface potential (Ψ_0) and the surface charge density (σ_{DL}) is given by equation below [12]:

$$\sigma_{DL} = Q_0\sinh(\frac{q\Psi_0}{2k_B T}) \qquad (6)$$

where $Q_0 = \sqrt{8k_B T\varepsilon_w I_0 N_{avo}}$, I_0 is the ionic concentration, ε_w is permittivity and N_{avo} is Avogadro number. The potential obtained from Eq. 4 is substituted in Eq. 6 to obtain σ_{DL} which is the ITC density (in cm-2) [14]. Based on the analytical model described above, we have calculated the correlation between pH in the electrolyte as a function of the ITC density. The corresponding relationship is presented in Fig. 5 where we can observe the following two main points. First, the curve is highly non-linear, and it has a minimum at around pH=2. The reason for this is that at pH=2, all charges at the oxide-electrolyte interface are canceled (balanced) by the charges in the electrolyte. This point is known in the literature as a point-of-zero-charge (pH$_{PZC}$) or the isoelectric point and for SiO_2, i.e., is equal to pH=2, and the pH$_{PZC}$ is specific for every oxide.

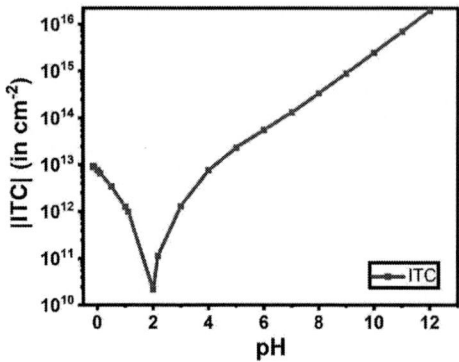

Fig. 5 Relationship of the pH values as a function of the interface trap charge (ITC) density. The data is reported in semi-log scale.

Second, for pH values between 2 and 4, the curve rises steeply for pH values close to the isoelectric point and then increases steadily. For example, at pH=5, the ITC density is $5*10^{13}$ cm^{-2}, while at pH=11 the ITC density is almost 10^{16} cm^{-2}. Hence, ITC density changes around 3 orders of magnitude for 6 values of pH which indeed is a significant change. Based on the results presented in Fig. 5, we have calculated the device behaviour. Fig. 6 shows the I_D-V_G curve of the ISFET varying the pH values from 5 to 11. As the pH increases from 5 to 11, the drain current I_D for a particular value of V_G also increases which is linked to a change of the threshold voltage for each pH.

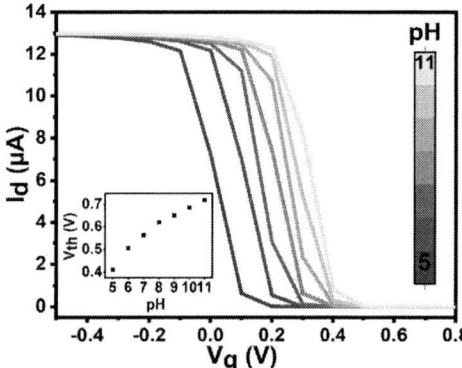

Fig. 6 Current -Voltage (Id-Vg) characteristics as a function of pH for an ISFET device. The pH varies from 5 to 11 which is represented with a blue and yellow line on the graph correspondingly.

Finally, the observed device behaviour in Fig. 6 can be explained in the following way. When the V_g has higher negative values, the I_d is saturated as the channel is inverted and when V_g holds higher positive values, the drain current decreases as the channel gets depleted. In general, higher values of pH mean lower proton concentration (positive charges) in the electrolyte. For e.g., pH = 1 means [H$^+$]= 10^{-1} mol/L, [OH$^-$] =10^{-13} mol/L similarly pH = 10 means [H$^+$] = 10^{-10} mol/L, [OH$^-$] =10^{-4} mol/L. As the V_G lowers to its minimum value (approaching more negative V_G), the drain current increases because the negative potential on the sensing surface layer balances the negative built-up charges and thus attracts more holes from the Si channel, which compensates the width of the depletion region. When pH increases in the electrolyte, the concentration of the protons decreases in the solutions, hence there are fewer positive charges on the surface of the gate oxide which leads to less

repulsion of the holes in the Si channel and as result, the depletion layer in Si decreases leading to more current flow [4]. The increase in pH can also be detected with the linear decrement of the threshold voltage [Fig. 6 inset] (positive values for p-type FET) of the sensing ISFET. A similar methodology can be used to design more accurate and complex biosensors with protein-peptide interactions.

IV. CONCLUSION

In conclusion, we have presented in this work, initial results based on the combination of numerical and analytical models which were developed to simulate high-aspect-ratio ISFET. Our results show that our methodology, which successfully combines numerical TCAD simulations and analytical models, computes the change of the device's current-volage characteristics as a function of pH values in the electrolyte. Moreover, a similar methodology can be used to significantly improve both analytical and numerical models to describe more complex nano-biosensors based on high-aspect-ratio FinFETs.

ACKNOWLEDGMENT

This project has received funding from the European Union's Horizon 2020 research and innovation program under grant agreement No 862539-Electromed-FET OPEN.

REFERENCES

[1] Sheehan, P.E. and L.J. Whitman, *Detection Limits for Nanoscale Biosensors.* Nano Letters, 2005. **5**(4): p. 803-807.

[2] Cui, Y., et al., *Nanowire Nanosensors for Highly Sensitive and Selective Detection of Biological and Chemical Species.* Science, 2001. **293**(5533): p. 1289.

[3] Nair, P.R. and M.A. Alam, *Performance limits of nanobiosensors.* Applied Physics Letters, 2006. **88**.

[4] Rollo, S., et al., *High Aspect Ratio Fin-Ion Sensitive Field Effect Transistor: Compromises toward Better Electrochemical Biosensing.* Nano Letters, 2019. **19**(5): p. 2879-2887.

[5] Rani, D., et al., *On the Use of Scalable NanoISFET Arrays of Silicon with Highly Reproducible Sensor Performance for Biosensor Applications.* ACS Omega, 2016. **1**(1): p. 84-92.

[6] Stern, E., Klemic, J.F., Routenberg, D.A., Wyrembak, P.N., Turner-Evans, D.B., Hamilton, and L. A.D., D.A., Fahmy, T.M., Reed, M.A., *Label-free immunodetection with CMOS-compatible semiconducting nanowires* Nature 445, 2007: p. 519–522.

[7] Gao, A., et al., *Silicon-Nanowire-Based CMOS-Compatible Field-Effect Transistor Nanosensors for Ultrasensitive Electrical Detection of Nucleic Acids.* Nano Letters, 2011. **11**(9): p. 3974-3978.

[8] Van Zeghbroeck, J., *Principles of Semiconductor Devices.* 2011: Bart Van Zeghbroeck.

[9] Robertson, J., *High dielectric constant oxides.* Eur. Phys. J. Appl. Phys., 2004. **28**(3): p. 265-291.

[10] Narang, R., M. Saxena, and M. Gupta, *Analytical Model of pH sensing Characteristics of Junctionless Silicon on Insulator ISFET.* IEEE Transactions on Electron Devices, 2017. **64**(4): p. 1742-1750.

[11] Valiskó, Mónika, Boda, Dezso, *The effect of concentration-and temperature-dependent dielectric constant on the activity coefficient of NaCl electrolyte solutions,* The Journal of Chemical Physics,2014,vol140

[12] van Hal, R.E.G., J.C.T. Eijkel, and P. Bergveld, A general model to describe the electrostatic potential at electrolyte oxide interfaces. Advances in Colloid and Interface Science, 1996. 69(1): p. 31-62.

[13] Medina-Bailon Cristina, Kumar Naveen, Dhar Rakshita P. Todorova, Ilina Lenoble, Damien Georgiev, Vihar P. García, César P., *Comprehensive Analytical Modelling of an absolute pH sensor,* Sensors Journal, 2021, Vol 21, Issue 15.

[14] Alam, M.A., *Principles of Electronic Nanobiosensors.* 2013.

Conductance due to the Edge Modes in Nanoribbons of 2D Materials in a Topological Phase

Viktor Sverdlov
Christian Doppler Laboratory
for Nonvolatile Magnetoresistive
Memory and Logic at the
Insitute for Microelectronics,
Technische Universität Wien
Vienna, Austria
sverdlov@iue.tuwien.ac.at

Heribert Seiler
Institute for Microelectronics
Technische Universität Wien
Vienna, Austria
seiler@iue.tuwien.ac.at

Al-Motasem Bellah El-Sayed
Institute for Microelectronics
Technische Universität Wien
Vienna, Austria and
Nanolayers Research Computing Ltd.
1 Granville Court
London, U.K.
el-sayed@iue.tuwien.ac.at

Hans Kosina
Institute for Microelectronics
Technische Universität Wien
Vienna, Austria
kosina@iue.tuwien.ac.at

Abstract—Employing novel 2D materials with topologically protected current-carrying edge states is promising to boost the on-current in electronic devices. Using nanoribbons is essential to reduce the contribution of the 2D bulk states to the current. Making the nanoribbon widths narrower allows one to put more current-carrying edge states under the gate of a fixed width thus boosting the current. However, the edge states start to interact in narrow nanoribbons. Based on an effective k·p model, we analyze the topologically protected edge states and their conductance for several 2D materials as a function of the normal electric field. We compare the 2D materials MoS₂, MoSe₂, WS₂, and WSe₂ in the topological 1T′ phase and find the largest electric field-induced conductance modulation is in MoS₂ nanoribbons.

Keywords—Topologically Protected Edge States; Topological Insulators; Nanoribbons; k·p Method; Conductance.

I. Introduction

The use of novel materials with advanced properties is mandatory to continue with the device scaling for high performance applications at reduced power. Topological insulators (TIs) belong to a new class of materials possessing highly conductive edge states with a nearly linear dispersion lying in the band gap. These states are topologically protected and therefore immune to backscattering. If the Fermi level lies in the gap, the large on-current is carried by the highly conductive edge states. Applying a gate voltage allows one to move the Fermi level in the conduction or the valence bands. It prompts strong scattering between the edge and the bulk electron or hole states [1]. This leads to a substantial reduction of the current, resulting in an on/off current ratio suitable for device applications [1].

Recently, it was predicted that well-known monolayer-thin two-dimensional (2D) materials with high promise for future microelectronic devices [2] can also be found in a 1T′ TI phase [3]. The band gap within which the edge states exist is opened by the spin-orbit interaction at the intersections (degeneracy points) between the inverted electron and hole bands. The value of the gap can be modulated by an external electric field E_z normal to the 2D sheet. The band gap in the inverted band structure is reduced and can be completely closed upon increasing values of E_z. By further increasing E_z the band gap reopens again; however, the traditional electron and hole band order is restored indicating a topological phase transition from a non-trivial topological to a trivial insulator. In contrast to the TI phase where the highly conductive edge

states carry a large current if the in-plane field is applied, no current-carrying edge states are allowed in the trivial insulating phase. Therefore, there is no current due to the edge states. The electric field induced topological phase transition between the TI and the trivial insulating phases peculiarities of the band structure in topological phases offers an alternative way to modulate the current by the gate voltage and to design the current switches.

To enhance the on-current density due to the edge states it is mandatory to have many edges. One can achieve this by placing several nanoribbons under the gate. The narrower the nanoribbon is, the more nanoribbons one can assemble within a given gate width. At the same time, the contribution due to the 2D bulk states decreases with shrinking width [4].

However, the behaviour of the edge states in a narrow nanoribbon is different from that at the edge of an infinite 2D sheet. Indeed, a small gap in the gapless spectrum of the edge states opens due to an interaction of the topologically protected states from the opposite edges [5,6]. Because of this gap in the dispersion of the edge modes, their conductance is found to be slightly less than the ideal conductance $G_0 = 2e^2/h$.

This gap between the edge states initially increases with increasing normal field E_z until the edge modes' dispersion meets the 2D bulk conduction and valence bands. Their conductance decreases accordingly. It was argued [6] that after the edge modes dispersions as a function of the normal electric field meet the 2D bulk conduction and the valence band, they become indistinguishable from bulk bands. As the fundamental gap in TI decreases with E_z, the gap in the edge states spectrum shrinks and becomes zero, following the bulk dispersion [6]. Therefore, according to [6], the ballistic conductance due to the edge states increases and reaches the ideal value $G_0 = 2e^2/h$ at the critical electric field E_0 when the gap between the electron and hole bands closes. At even larger $E_z > E_0$ the bulk gap opens again signifying a transition to a trivial dielectric. However, the topological edge states are not allowed, and the conductance abruptly drops to zero [6] due to the phase transition from the TI to the trivial insulator state. This discontinuous behaviour of the ballistic conductance is unphysical. Considering a 1T′ MoS₂ nanoribbon as an example, we demonstrated in [7] that the erroneous results [6] were due to the spurious solutions of the dispersion equation. If the spurious solutions are disregarded, a qualitatively different behavior of the ballistic conductance in a 1T′ MoS₂ nanoribbon is obtained [7]. Namely, it was

978-1-6654-3746-2/21 $31.00 © 2021 IEEE

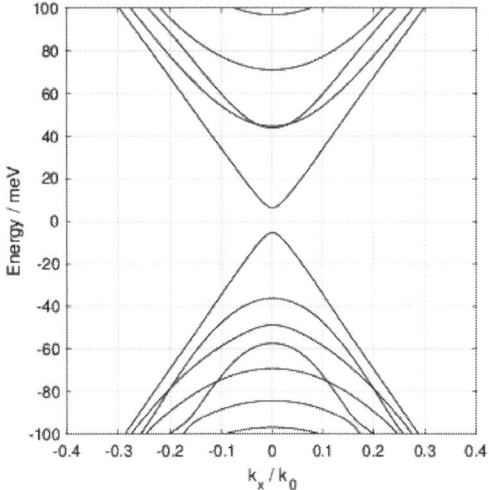

Fig.1 Subband structure in a MoS₂ nanoribbon of width d=20nm at $E_z = 0$. The subbands with the linear dispersion and a small gap at $k_x = 0$ correspond to the edge modes.

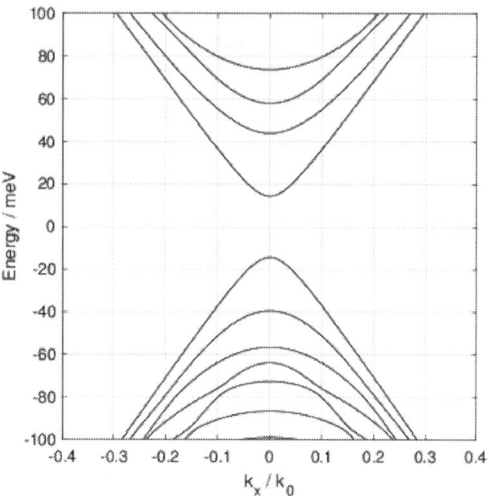

Fig.2 Subband structure in a MoS₂ nanoribbon of width d=20nm at the critical field $E_z = \alpha^{-1}v_2$ at which the spin-orbit gap in the 2D sheet closes. In contrast, the gap in the nanoribbon increases.

demonstrated that the gap between the edge modes always increases with the field E_z and, in contrast to [6], never closes. The increase in the separations between the electron and hole subbands results in a continuous and substantial decrease in the edge channels ballistic conductance.

In this work we evaluate the edge states and their corresponding Landauer conductance in a nanoribbon in the 1T′ TI phase. In addition to MoS₂ we consider potentially relevant 2D materials MoSe₂, WS₂, and WSe₂.

II. METHOD

The subbands in a nanoribbon of a topological 2D material are found by solving the Schrödinger equation with the effective Hamiltonian H [7]:

$$\boldsymbol{H} = \begin{pmatrix} H(\mathbf{k}) & 0 \\ 0 & H^+(-\mathbf{k}) \end{pmatrix}, \quad (1)$$

$$H(\boldsymbol{k}) = \begin{pmatrix} \frac{1}{2} - k_y^2 \frac{m}{m_y^p} - k_x^2 \frac{m}{m_x^p} + U(y) & v_2 k_y - \alpha E_z + i v_1 k_x \\ v_2 k_y - \alpha E_z - i v_1 k_x & \frac{-1}{2} + k_y^2 \frac{m}{m_y^d} + k_x^2 \frac{m}{m_x^d} + U(y) \end{pmatrix} \quad (2)$$

Variable	MoS₂	MoSe₂	WS₂	WSe₂
δ [eV]	0.55	0.76	0.17	0.69
v_1[m/s]	3.38 10⁵	3.42 10⁵	2.93 10⁵	3.54 10⁵
v_2	0.23 10⁵	0.23 10⁵	0.85 10⁵	0.38 10⁵
m_x^p/m_e	0.48	0.28	0.53	0.36
m_y^p/m_e	0.29	0.17	0.28	0.16
m_x^d/m_e	2.32	2.65	3.2	3.28
m_y^d/m_e	0.92	3.14	8.2	8.4
α/e [nm]	0.016	0.027	0.017	0.024
k_0 [nm⁻¹]	1.8	1.9	1.08	1.7

Table I Parameters [6] used in the model. m_e is the electron mass, e is the electron charge.

Here $U(y)$ is the confinement potential. For convenience we introduced the dimensionless units by measuring all energies in units of the separation δ between the electron and hole bulk bands at the Γ-point in the inverted band structure, while the wave vectors $k = (k_x, k_y)$ are in units $k_0 = \frac{2\delta}{\hbar^2} \left(\frac{m_y^d m_y^p}{m_y^d + m_y^p} \right)^{1/2}$, where $m_{y(x)}^{d(p)}$ are the effective masses, $m = \frac{m_y^d m_y^p}{m_y^d + m_y^p}$ and $v_{1(2)}$ are the velocities characterizing the strength of the spin-orbit interaction. The parameters used in simulations as well as the values of k_0 are given in Table I.

The block-diagonal form of (1) simplifies the solution for the subbands as the solution of the lower block is the time-reversal solution [5,7] of the upper block. It is therefore sufficient to solve the Schrödinger equation with (2). If the nanoribbon is cleaved along the OX axis and the confinement potential is approximated by an infinite square well of the

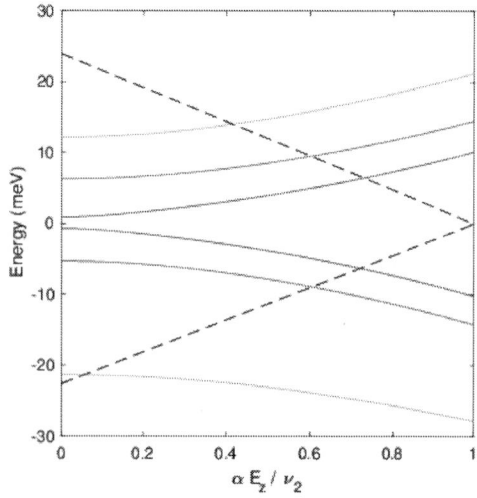

Fig.3 Energies of the electron and hole edge states in a MoS₂ nanoribbon of the width d=30nm (blue), d=20nm (orange) and d=10nm (yellow) as a function of electric field strength. Dashed line: bulk bands extrema.

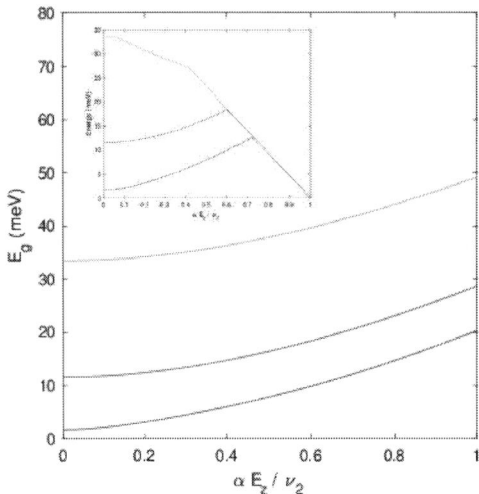

Fig.4 Energy gap in the edge states spectrum in a MoS$_2$ nanoribbon of the width d=30nm (blue), d=20nm (orange) and d=10nm (yellow). Inset: energy separation including the bulk bands extrema shown by dashed line in Fig.3

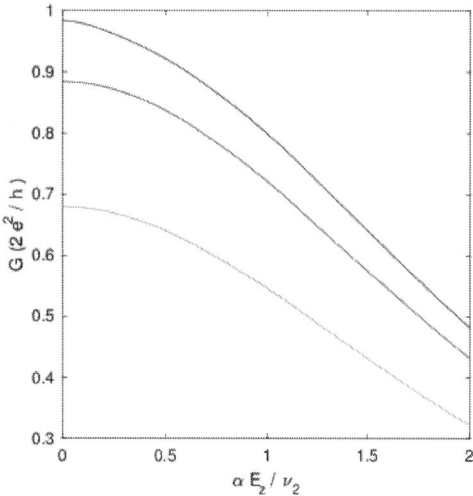

Fig.5 Ballistic conductance due to the lowest electron and topmost hole edge states for a MoS$_2$ nanoribbon of the width d=30nm (blue), d=20nm (orange) and d=10nm (yellow).

width d, the eigenfunction of the upper block in (1) satisfies the boundary conditions

$$\psi_{k_x}(y = \pm d/2) = 0. \tag{3}$$

The wave functions and the edge modes dispersion relations are found numerically using the Newton method [7]. While solving the corresponding dispersion equation, special care must be taken to avoid spurious solutions [7] which lead to erroneous results and unphysical interpretations.

III. RESULTS

Fig.1 shows the subband structure obtained with the Hamiltonian (1,2) and the boundary condition (3) in a 20nm wide 1T' MoS$_2$ cleaved along the OX axis, at $k_x = 0$ and in the absence of the normal electric field ($E_z = 0$). The energies are offset by $\Delta E = \frac{1}{2}\frac{m_y^d - m_y^p}{m_y^d + m_y^p}$ for convenience. Fig.2 demonstrates the subbands at the electric field $E_z = \alpha^{-1}v_2$. This value of the field is critical for the band structure in an infinite sheet of a 2D material. Indeed, at this electric field value the fundamental gap closes. With a further increase of the normal field the gap opens again; however, the material becomes a trivial dielectric, without topologically protected edge states.

In a nanoribbon, however, the subband spectrum remains gapped at the critical electric $E_z = \alpha^{-1}v_2$. Furthermore, the gap between the lowest electron-like and the top-most hole-like subbands seems to increase with the field. To confirm this observation, the dependence of the minimum of the lowest electron-like and the maximum of the topmost hole-like subbands as a function of the normal electric field is shown in Fig.3, for three different nanoribbon widths of 10 nm, 20 nm, and 30 nm. We observe that the gap between the subbands is larger for narrower nanoribbons. At the same time, the gap increases as a function of the field, for all three widths as shown in Fig.4. This contrasts with the dependence of the fundamental band gap in an infinite 2D sheet shown by dashed lines in Fig.3. The 2D bulk band gap decreases with the

normal field and becomes zero as expected at the critical value of $E_z = \alpha^{-1}v_2$ at the electric field.

In a 2D sheet the TI - trivial insulator phase transition happens exactly at the critical value of the normal field. If the behavior of the fundamental gap in the 2D bulk is opposite to that of the subbands in a nanoribbon, the question arises how the phase transition between the TI and a trivial insulator appears in a nanoribbon. It turns out that if the energies of the subbands are within the bulk band gap at small values of E_z, the envelope wave functions are localized at opposite edges of the nanoribbon [7]. When the subbands' dispersions approach the dashed line in Fig.3 corresponding to the 2D bulk bands, the localization becomes weaker and disappears completely at the points of crossing. The transition appears earlier in narrower nanoribbons. With the field further increased, the subbands, however, do not align themselves with the bulk bands, claimed in [6]. Instead, their energies evolve to the bulk conduction and valence bands. Therefore, the nature of the

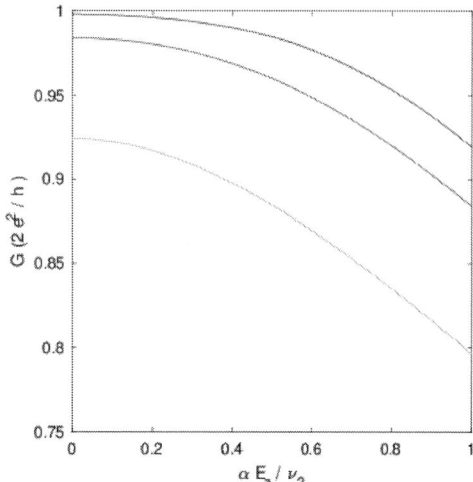

Fig.6 Ballistic conductance due to the lowest electron and topmost hole edge states for a MoSe$_2$ nanoribbon of the width d=30nm (blue), d=20nm (orange) and d=10nm (yellow).

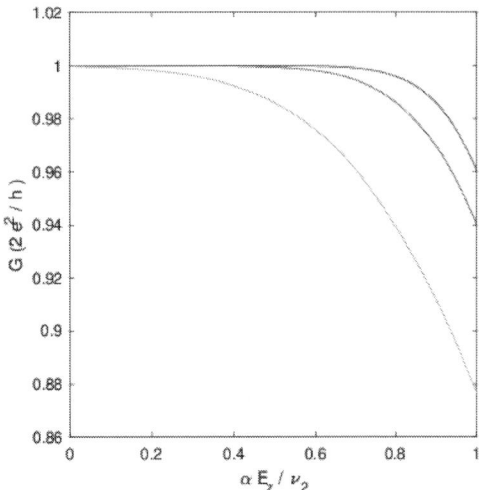

Fig.7 Same as in Fig.6 for WS$_2$ nanoribbons.

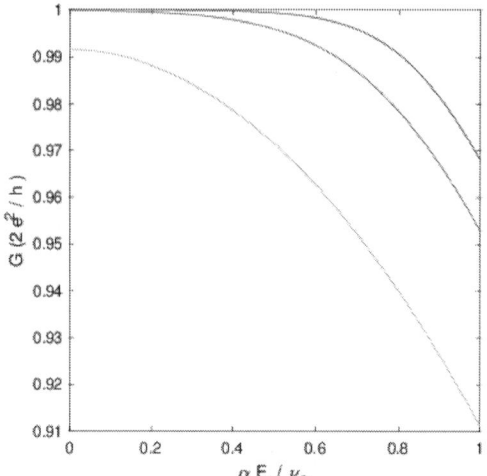

Fig.8 Same as in Fig.6 for WSe$_2$ nanoribbons.

envelope function in the direction normal to the nanoribbon changes from the one localized at the edges to a nonlocalized bulk-like behavior. We can therefore assign a TI to a trivial insulator phase transition in a confined geometry to the change of the wave function behavior appearing precisely at the points of the intersections of the bulk 2D bands with the subbands' dispersions (Fig.3). This transition happens at larger normal fields in broader nanoribbons thus recovering the TI to trivial insulator phase transition in a 2D geometry.

Fig.4 shows that the separation between the lowest electron-like and topmost hole-like subbands, according to Fig.3, increases with E_z and is largest for the 10 nm nanoribbon. If, however, the 2D bulk bands shown in Fig.3 with dashed lines were also included, the separations shown in Fig.4 would look like those in the inset of Fig.4. This is precisely the behavior shown in Fig.5a of the reference [6]. That behavior is not correct, however, as it is caused by spurious solutions of the dispersion equation. As we demonstrated in [7], these spurious solutions coincide with the dispersion of the bulk 2D bands. As they do not depend on the nanoribbon width d, they must be neglected and do not contribute to the ballistic conductance.

Finally, following [6], we evaluate the ballistic conductance due to the lowest electron-like and topmost hole-like subbands as a function of the normal electric field for the nanoribbon made of MoS$_2$, MoSe$_2$, WS$_2$, and WSe$_2$ materials in 1T′ topological phase, with the corresponding parameters listed in Table I. To exploit the Landauer expression for the conductance we need the subbands' dependence on the field shown in Fig.3 but evaluated for all materials. It is assumed that the chemical potential is at zero energy while the temperature is 300 K. Fig.5 shows the ballistic conductance due to the above described subbands. The conductance decreases with the electric field for all nanoribbons' widths including the one of 10 nm. This is in contrast to the behavior predicted in [6] where the conductance of a 10 nm thin nanoribbon is predicted to increase with the normal electric field. As explained, this erroneous behavior of the conductance is due to the spurious solutions of the dispersion equation. The dependence of the corresponding ballistic conductance in MoSe$_2$, WS$_2$, and WSe$_2$ nanoribbons is shown

in Fig.6, Fig.7, and Fig.8, correspondingly. As the separation between the lowest electron-like and topmost hole-like subbands increases with the field, the ballistic conductance decreases. However, the largest conductance modulation is found in MoS$_2$ nanoribbons.

IV. CONCLUSION

A **k·p** method is applied to investigate the topologically protected states at the edges of nanoribbons of several 2D materials as a function of the normal electric field. It is demonstrated that the electric field-induced conductance modulation is largest in 1T′ MoS$_2$ nanoribbons making them more suitable candidates for use in ultra-scaled devices.

ACKNOWLEDGMENT

Financial support by the Austrian Federal Ministry for Digital and Economic Affairs and the National Foundation for Research, Technology and Development and the Christian Doppler Research Association is gratefully acknowledged.

REFERENCES

[1] W.G.Vandenberghe and M. V. Fischetti, "Imperfect two-dimensional topological insulator field-effect transistors", Nature Communications, vol.8, art.14184 (pp. 1-8), 2017. doi: 10.1038/ncomms14184

[2] Yu.Yu. Illarionov, A.G. Banshchikov, D.K. Polyushkin, S. Wachter, T. Knobloch, M. Thesberg, L. Mennel, M. Paur, M. Stöger-Pollach, A. Steiger-Thirsfeld, M.I. Vexler, M. Waltl, N.S. Sokolov, T. Mueller, and T. Grasser, "Ultrathin calcium fluoride insulators for two-dimensional field-effect transistors", Nature Electronics, vol.2, pp.230-235, 2019. doi: 10.1038/s41928-019-0256-8

[3] X. Qian, J. Liu, L. Fu, and Ju Li., "Quantum spin Hall effect in two-dimensional transition metal dichalcogenides", Science, vol. 346, issue 6215, pp.1344-1347, 2014. doi: 10.1126/science.1256815

[4] M.J. Gilbert, "Topological electronics" Communications Physics, vol. 4, art.70 (pp.1-12), 2021. oi: 10.1038/s42005-021-00569-5

[5] B. Zhou, H.-Z. Lu, R.-L. Chu, S.-Q. Shen, and Q. Niu, "Finite size effects on helical edge states in a quantum spin-Hall system" Phys.Rev.Lett., vol.101, art.246807 (pp.1-4), 2008. doi: 10.1103/PhysRevLett.101.246807

[6] B. Das, D. Sen, S. Mahapatra, "Tuneable quantum spin Hall states in confined 1T′ transition metaldichalcogenides," Scientific Reports, vol. 10 (2020), art. 6670. doi: 10.1038/s41598-020-63450-5

[7] V. Sverdlov, A.-M. El-Sayed, H. Seiler, H. Kosina, and S. Selberher. "Subbands in a Nanoribbon of Topologically Insulating MoS$_2$ in the 1T′ Phase", Solid-State Electron., vol. 184, art. 108081 (pp.1 – 9), 2021. doi:10.1016/j.sse.2021.108081

Synaptic transistors based on transparent oxide for neural image recognition

Q N Wang[1-3], C Zhao[1-3,*], W Liu[1,2,*], I Z Mitrovic[2], H van Zalinge[2], Y N Liu[4,5], C Z Zhao[1-3]

[1] School of Advanced Technology, Xi'an Jiaotong-Liverpool University, Suzhou, China.
[2] Department of Electrical Engineering and Electronics, University of Liverpool, Liverpool, UK.
[3] AI University Research Centre (AI-URC), Xi'an Jiaotong-Liverpool University, Suzhou, China
[4] Department of Applied Mathematics, Xi'an Jiaotong-Liverpool University, Suzhou 215123, China
[5] Department of Applied Mathematics, University of Liverpool, Liverpool L69 7ZD, UK.

*E-mail: Chun.Zhao@xjtlu.edu.cn

Wen.Liu@xjtlu.edu.cn

Abstract—Artificial synaptic devices are the critical component for large-scale neuromorphic computing, which surpasses the limitations of von Neumann's structure. Recently the emerging electrolytic gated transistor (EGT) has proven to be a promising neuromorphic application due to the conductance can be updated by the gate voltage stimulation. This paper presents a new low-temperature solution-based oxide thin film transistor, which uses an ion-doped dielectric layer. The suitable ion doping concentration is obtained by the synaptic electrical characteristic. The synaptic transistor also has a low-noise linear conductance update and a relatively high G_{max}/G_{min} ratio.

Keyword—image recognition, solution-processed transistor, synaptic device, MNIST, neural computing.

I. INTRODUCTION

In recent years, computing systems with bionic brain structures have attracted extensive attention [1-3]. The device of simulating biological synaptic behavior and function with artificial synapse as the core unit has been widely studied and has been proposed for multi-level storage and parallel computing [4].

In this paper, the solution-processed metal-oxide synaptic transistors have been proposed with Li^+ ion-doped ZrO_x layers. Several typical synaptic behaviors, including inhibitory/excitatory postsynaptic current (IPSC/EPSC), long-term depression/potentiation (LTD/LTP) The long-term potentiation, and short-term potentiation of biological synapses are successfully simulated by applying the electrical pulse sequence on the gate electrode [3]. The simulated SLP-based ANN consists of 785 presynaptic electrical signals. Then 785 x 10 = 7850 crossbar array simulate biological synaptic weights [4,5].

II. EXPERIMENTAL

Fabrication of artificial synaptic TFT: The ZrO_x-Li precursor solution (1.5M) was prepared by dissolving aluminum nitrate hydrate ($ZrO (NO_3)_2 \cdot xH_2O$) and 0.55M indium nitrate hydrate into deionized water. The InO_x precursor solution (1M) was prepared by dissolving indium nitrate hydrate ($In (NO_3)_3 \cdot xH_2O$) into deionized.
Device Characterization: A semiconductor parameter analyzer

(Agilent B1500) with transistor characterization software under atmospheric conditions was operated to test the electrical properties of the Li doing InO_x/ZrO_x synaptic TFTs. To measure the EPSC/ IPSC current flowing between the S/D electrodes, the 0.1 V steady voltage bias was applied to the postsynaptic terminal.

ANNs simulation: The calculated conductance of synaptic TFT in the crossbar array was applied with the positive synaptic weight value. The measurement of the neurocomputing in ANNs Includes negative values. Subsequently, the synaptic weight ($w = G^+ - G^-$) was expressed as the difference between the state of two synaptic devices (expressed as G^+ and G^-) between each conductance value [6, 7].

III. RESULTS AND DISCUSSION

Biological chemical synapses are the fundamental components of recollection, understanding, and computation that connect the neurons in the human brain [8]. Figure 1a shows the relationship between biological synapses and synaptic devices. Synaptic transmission is a complex transmission process. Figure 1b shows the transfer characteristic curve of Li doped solution-processed synaptic thin film transistor. According to the trend of curve in the figure, the synaptic TFTs has I_{on}/I_{off} ratio (8200) and high mobility which are the most basic characteristic of TFTs. Through in-depth analysis of synaptic performance, it can be found that there is obvious current hysteresis. The action potential produced by the axon end of the presynaptic neuron opens the calcium channel, and calcium diffuses in neurons [9]. The presynaptic corresponds to the gate electrode of the thin film transistor and the postsynaptic corresponds to the drain electrode. EPSC and IPSC are the most basic information flow and processing processes in complex computing [10].

978-1-6654-3746-2/21 $31.00 © 2021 IEEE

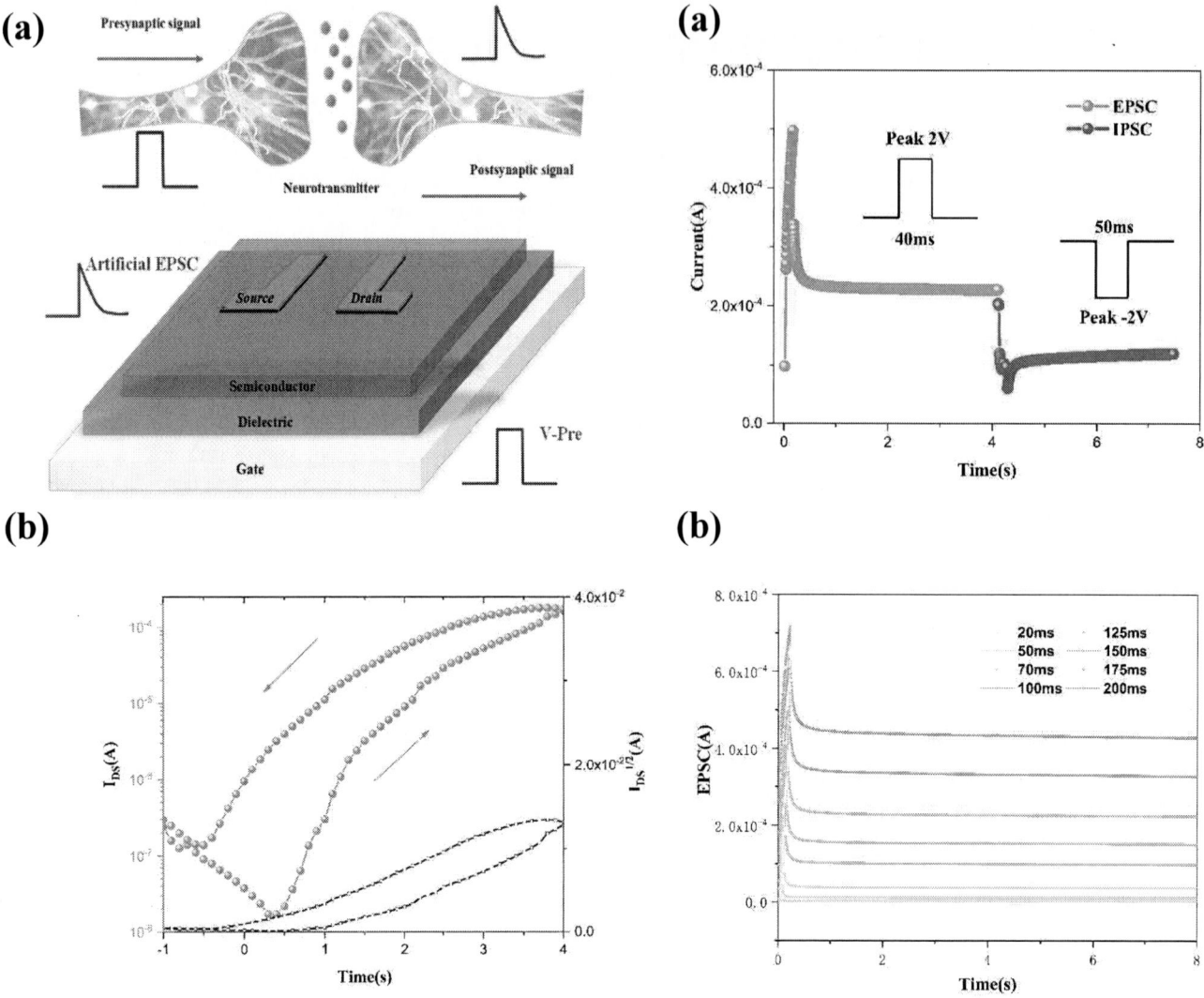

Figure 1. (a) Schematics illustration of the transparent oxide synaptic transistor. (b) Transfer characteristic for Li doped synaptic TFT.

Figure 2. (a) The simulation of EPSC and IPSC behavior. (b) The EPSC triggered by positive pulses (3 V) with eight pulse durations (20 ,50, 70, 100, 125, 150, 175 and 200 ms) at V_{DS}= 1V.

Figure 2 demonstrates the inhibitory/excitatory postsynaptic current behavior. This process achieved by applying a positive voltage (2V 40ms) and a negative voltage (-2V 50ms). The conductivity in the channel rises rapidly to 500μA and then falls back to 210μA. Figure 3 shows the effect of the different number of pulses on LTP / LTE characteristics under different voltage stimulation [11]. The 16 LTP curves can be analyzed that the nonlinearity increases first and then decreases with the increase of voltage. Figure 2b mainly illustrates the effect of electrical pulse width on channel conductivity. With the increase of pulse width, the conductance value after stimulation becomes higher. There is a linear relationship between the electric pulse width and the corresponding conductance. It shows that synaptic weight can be quantified by electrical stimulation. The gate of thin film transistor can be regarded as the pre-synaptic. At the same time, the change of synaptic weight is sensitive to pulse width compared with pulse intensity. This is attributed to the small ion radius and large diffusion coefficient of lithium.

Figure 3a mainly reflects the synaptic device in matrix form. The conductance before electrical stimulation is the original value. In order to reflect the stability and controllability of synaptic weight change. Uniform electrical stimulation is applied to gate of thin film transistors at different positions. After five minutes of electrical stimulation, the synaptic weight will eventually decrease slowly and stabilize in a small range. The main reason for the increase of synaptic weight is the migration of lithium ions from the dielectric layer to the semiconductor layer. The X character composed of eight synaptic thin film transistors of 16 can be clearly distinguished.

Figure 4a present the normalized LTP / LTD curves under different amplitude and pulse width.

978-1-6654-3746-2/21 $31.00 © 2021 IEEE

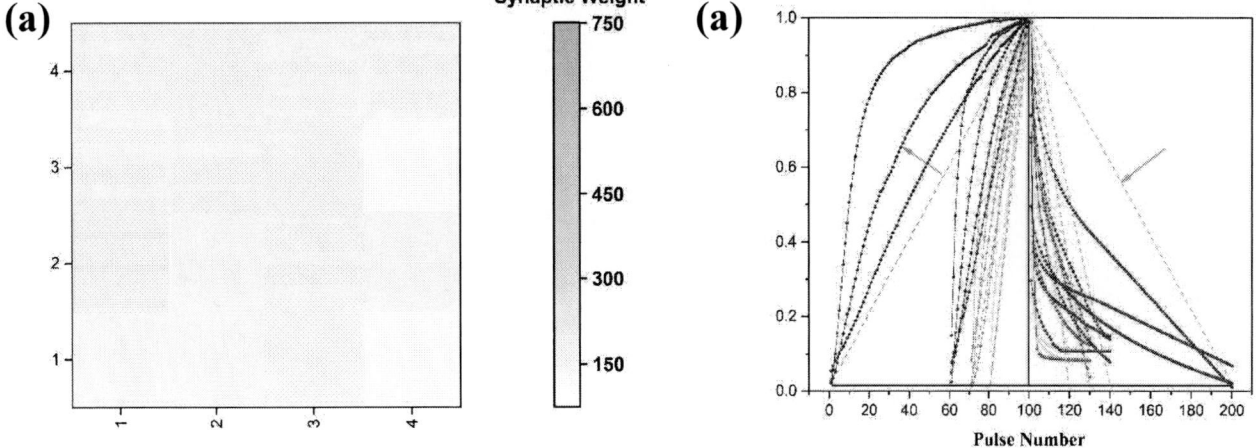

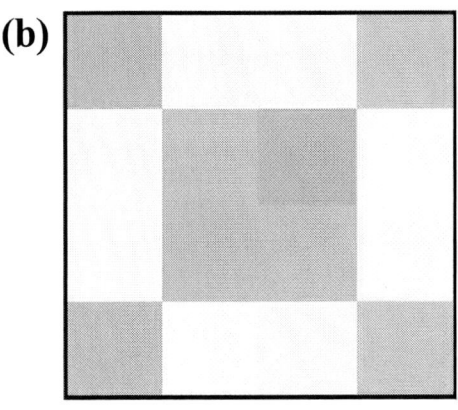

analyze the conductance and nonlinearity that significantly impact learning accuracy based on the LTP/LTD characteristic.

Figure 3. (a) 4◇4 array synaptic device before the volate stimulation.

(b) 4◇4 array synaptic device After the volate stimulation.

Figure 4 shows the effect of the different number of pulses on LTP / LTD characteristics under different voltage stimulation [11]. The 16 LTP curves can be analyzed that the nonlinearity increases first and then decreases with the increase of voltage. The parameter of nonlinearity and G_{max}/G_{min} can be obtained from the curve.

Figure 4. (a) The increasing and decreasing of channel conductance illustrating the long-term potentiation and long-term depression.
(b) The long-term potentiation and Natural forgetting process

IV. CONCLUSION

In conclusion, this work demonstrates the synaptic properties of the solution-processed Li-doped thin-film transistors. The advanced weight update is proposed to improve the recognition rate. EPSC and IPSC are basic information flow and processing processes behind complex calculations. In this work, we

ACKNOWLEDGMENT

This research was funded in part by the Natural Science Foundation of the Jiangsu Higher Education Institutions of China Program (19KJB510059), Natural Science Foundation of Jiangsu Province of China (BK20180242), the Suzhou Science and Technology Development Planning Project: Key Industrial Technology Innovation (SYG201924), University Research Development Fund (RDF-17-01-13), and the Key Program Special Fund in XJTLU (KSF-P-02, KSF-T-03, KSF-A-04, KSF-A-05, KSF-A-07, KSF-A-18).

978-1-6654-3746-2/21 $31.00 © 2021 IEEE

REFERENCES

[1] V. M. Ho, J. A. Lee, and K. C. Martin, "The cell biology of synaptic plasticity," Science, vol. 334, no. 6056, pp. 623-8, Nov 4, 2011.

[2] J. J. Harris, R. Jolivet, and D. Attwell, "Synaptic energy use and supply," Neuron, vol. 75, no. 5, pp. 762-77, Sep 6, 2012.

[3] C. Zhang, W. Tang, L. Zhang, C. Han, and Z. L. Wang, "Contact Electrification Field-Effect Transistor," ACS Nano, vol. 8, no. 8, pp. 8702-8709, 2014/08/26 2014.

[4] Q. Zheng et al., "In Vivo Self-Powered Wireless Cardiac Monitoring via Implantable Triboelectric Nanogenerator," ACS Nano, vol. 10, no. 7, pp. 6510-6518, 2016/07/26 2016.

[5] Y. Dai et al., "A Self-Powered Brain-Linked Vision Electronic-Skin Based on Triboelectric-Photodetecing Pixel-Addressable Matrix for Visual-Image Recognition and Behavior Intervention," Advanced Functional Materials, vol. 28, no. 20, p. 1800275, 2018.

[6] Z. L. Wang, "Triboelectric Nanogenerators as New Energy Technology for Self-Powered Systems and as Active Mechanical and Chemical Sensors," ACS Nano, vol. 7, no. 11, pp. 9533-9557., 2013.

[7] Y. Yang et al., "Liquid-Metal-Based Super-Stretchable and Structure-Designable Triboelectric Nanogenerator for Wearable Electronics," ACS Nano, vol. 12, no. 2, pp. 2027-2034, Feb 27, 2018.

[8] Q. Shen et al., "Self-Powered Vehicle Emission Testing System Based on Coupling of Triboelectric and Chemoresistive Effects," Advanced Functional Materials, vol. 28, no. 10, p. 1703420, 2018.

[9] Z. L. Wang, "On Maxwell's displacement current for energy and sensors: the origin of nanogenerators," Materials Today, vol. 20, no. 2, pp. 74-82, 2017.

[10] Z. L. Wang, "Nanogenerators, self-powered systems, blue energy, piezo tronics and piezo-photo tronics – A recall on the original thoughts for coining these fields," Nano Energy, vol. 54, pp. 477-483, 2018.

[11] X. Xie et al., "Frequency-independent self-powered sensing based on capacitive impedance matching effect of triboelectric nanogenerator," Nano Energy, vol. 65, p. 103984, 2019.

Curvature Based Feature Detection for Hierarchical Grid Refinement in TCAD Topography Simulations

Christoph Lenz*, Alexander Toifl*, Andreas Hössinger†, Josef Weinbub*

*Christian Doppler Laboratory for High Performance TCAD,
Institute for Microelectronics, TU Wien, Vienna, Austria
†Silvaco Europe Ltd, Cambridge, United Kingdom
Email: christoph.lenz@tuwien.ac.at

Abstract—We present a feature detection method for selective grid refinement in hierarchical grids used in process technology computer-aided design topography simulations based on the curvature of the wafer surface. The proposed method enables high-accuracy simulations whilst significantly reducing the run-time, as the grid is only refined in regions with high curvatures. We evaluate our method by simulating selective epitaxial growth of silicon-germanium fins in narrow oxide trenches. The performance and accuracy of the simulation is assessed by comparing the results to experimental data showing good agreement.

Index Terms—TCAD, topography simulation, level-set method, curvature, hierarchical grid, non-planar epitaxial growth

I. Introduction

Many non-planar semiconductor device geometries (e.g., FinFETs) are fabricated by employing strongly anisotropic processing techniques [1]. During selective epitaxial growth (SEG), wafer topographies emerge, which are characterized by crystal facets and thus a combination of high-curvature and essentially flat regions materialize [2]. Local high resolutions of the underlying grids are thus required to accurately resolve high-curvature features and material interfaces during a process simulation, while keeping the overall run-time as low as possible to maximize applicability.

The level-set method is widely used for simulating fabrication processes of semiconductor devices [3], particularly in technology computer-aided design (TCAD) workflows. The wafer surface is described with the zero level-set of a continuous function ϕ which is referred to as the level-set function. The propagation of the wafer surface, i.e., its topographical evolution, is governed by the level-set equation

$$\frac{\partial \phi(x,t)}{\partial t} + V(x)|\nabla \phi(x,t)| = 0\,, \tag{1}$$

which incorporates the velocity field V, allowing to model the growth or etch rates during a process step [3]. The level-set equation is typically solved on a regular grid with resolution Δx using a finite difference scheme [4], [5].

The previously hinted need for local high grid resolutions can be efficiently realized by using hierarchical grids. Hierarchical grids have a *base* grid covering the entire simulation domain, which is complemented by *sub*-grids with higher resolutions, covering areas of interest (e.g., sharp ridges) [6]–[8]. Fig. 1 shows an illustration of a level-set

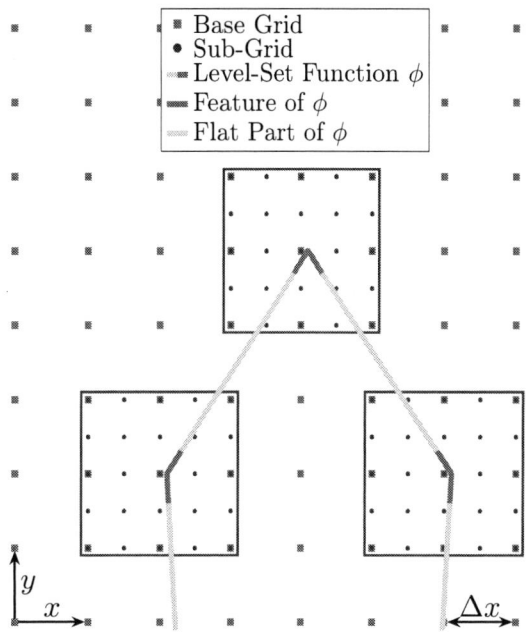

Fig. 1: Illustration of a level-set function ϕ (green/red line segments) with three features (i.e., corners; red line segments) on a hierarchical grid. The base grid has a resolution of Δx, the features of ϕ are covered by sub-grids with a two times higher resolution (blue boxes).

function with three features (i.e., corners) which are resolved with higher spatial accuracy by finer sub-grids. In order to optimize simulation run time, these sub-grids need to be optimally configured and placed.

In this work, we introduce an efficient and automatic detection of level-set features for guiding the sub-grid generation mechanism of hierarchical grid-based process TCAD topography simulations. At the core of the feature detection is the calculation and evaluation of the wafer surface curvature, inspired by other curvature-based applications [9]–[11]. We assess our proposed method based on a representative and cutting-edge process simulation, i.e., selective epitaxy. To that end and to showcase integration into TCAD workflows, our feature detection method has been implemented into Silvaco's *Victory Process* simulator [12].

978-1-6654-3746-2/21 $31.00 © 2021 IEEE

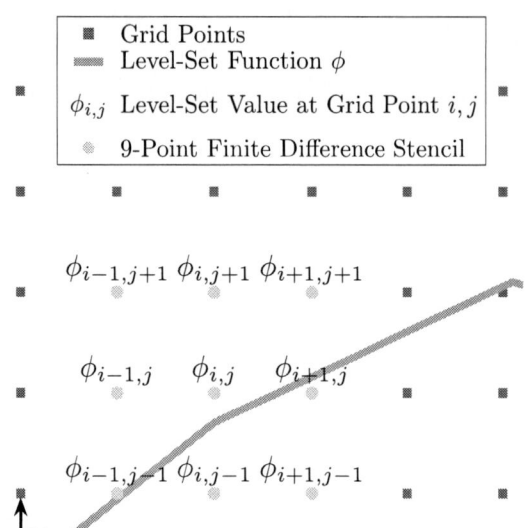

Fig. 2: Illustration of a level-set function and a 9-point stencil which is used to calculate the curvature of the surface at point $\phi_{i,j}$.

The thus augmented simulator is used to selectively grow silicon-germanium (SiGe) fins and to compare the results with recent experiments presented in [2]. Furthermore, we analyze the simulation performance and the accuracy of the results.

II. METHOD

The wafer topography during a fabrication step with pronounced anisotropy is typically characterized by regions of high and low curvature. In the level-set method the local curvature can be calculated directly from the level-set function ϕ

$$\kappa = \frac{\phi_y^2 \phi_{xx} - 2\phi_x \phi_y \phi_{xy} + \phi_x^2 \phi_{yy}}{|\nabla \phi|^3}, \qquad (2)$$

where ϕ_i denotes the partial derivative of ϕ with respect to the coordinate $i \in \{x, y\}$. The partial derivatives are calculated using finite differences:

$$\phi_x \approx \frac{\phi_{i+1,j} - \phi_{i-1,j}}{2\Delta x},$$
$$\phi_{xx} \approx \frac{\phi_{i+1,j} - 2\phi_{i,j} + \phi_{i-1,j}}{\Delta x^2},$$
$$\phi_{xy} \approx \frac{\phi_{i+1,j+1} - \phi_{i-1,j+1} - \phi_{i+1,j-1} + \phi_{i-1,j-1}}{4\Delta x^2}.$$

The calculation of all required finite differences to determine the curvature κ requires a 9-point finite difference stencil around each point of the wafer surface (see Fig. 2).

The curvature is defined on each point of the wafer surface: The absolute value of the curvature $|\kappa|$ lies between 0 and $1/\Delta x$ since the maximal curvature a level-set function can

TABLE I: Simulation parameters employed for the SEG in trench arrays [2]. The number of deposition cycles P_i refers to the number of SEG cycles needed to achieve the topographies in Fig. 3.

Rates [nm/cycle]				Number of deposition cycles for profile P		
R_{100}	R_{110}	R_{311}	R_{111}	P_1	P_2	P_3
13	5	3.1	1.6	5	24	47

describe is bound by the grid resolution [4]. Points with a curvature of $|\kappa| = 0$ describe a flat part of the wafer surface. In contrast, points with a larger value of $|\kappa|$ indicate a feature on the wafer surface. If the absolute curvature of a point exceeds $1/\Delta x$ it indicates that the resolution of the level-set function is not high enough to resolve this feature. Therefore, it is important that such grid points are flagged as features to improve the simulation quality.

Consequently, topography features are detected based on the curvature threshold parameter $0 < C < 1/\Delta x$, which is problem specific. If the curvature of a surface point is larger than C the point is flagged as a feature: The closer C is to 0 the more surface points will be detected as features. Furthermore, in this work, interfaces between stationary material regions are always considered features, which enables a well-resolved level-set description of SiGe material interfaces.

After the feature detection step the flagged surface points are grouped into rectangular *patches*. If patches overlap they are merged together until no more overlaps exist [13]. New sub-grids with, e.g., a four-times smaller Δx (facilitating locally increased resolutions) are then created according to these patches. The level-set values for the sub-grids are calculated with a hierarchical re-distancing step [14].

III. RESULTS

The proposed feature detection method is evaluated by simulating the SEG process presented by Jang *et al.* [2]. There, in an initial dry etching step, $[\bar{1}10]$-aligned SiO$_2$ trenches are formed. A cyclic SEG step follows, which leads to the formation of high-quality $\{311\}$ and $\{111\}$ crystal facets. Each individual cycle is taken as an artificial time step unit during the simulation. As previously indicated, this fabrication process has been simulated with Silvaco's *Victory Process* [12] augmented with the here proposed feature detection method: The method is employed during the SEG step, where we utilize a recently developed numerical stability-enabling level-set method for selective epitaxy [5].

The growth of the SiGe crystal is modeled with a crystal orientation-dependent velocity field V which is constructed from experimentally characterized growth rates [2], as given in Tab. I. The simulation results are in good agreement with the experiment (see Fig. 3).

Tab. II shows the utilized grid resolutions: The results shown in Fig. 3 are based on the *Multi-Grid* parameters, i.e., a base grid (Grid 1) complemented by one additional grid hierarchy level (Grid 2) offering a plethora of sub-grids with higher resolutions.

TABLE II: Grid parameters and run-times (Intel Xeon E5-2680v2) for the entire simulation.

Simulation	Grid 1 Res	Grid 2 Res	Run-Time
Coarse	0.002 µm	-	28 s
Fine	0.0005 µm	-	19 min 54 s
Multi-Grid	0.002 µm	0.0005 µm	13 min 38 s

Fig. 4 compares the results after the final simulation time step (47 SEG cycles) of the simulation using the parameters shown in Tab. II. Fig. 5 shows the error of the *Coarse* and *Multi-Grid* simulation results compared with the *Fine* simulation results: $|(x_1, x_2)| = \sqrt{x_1^2 + x_2^2}$ (L_2-norm). The error is measured from each point of the SiGe surface simulated with either *Coarse* or *Multi-Grid* parameters and its closest point (i.e., the point with the smallest L_2-norm) on the surface simulated with the *Fine* parameters (L_2-error).

The *Coarse* simulation results do not sufficiently match the results using the *Multi-Grid* parameters (Fig. 4), as evident from the mismatch of the peak positions of the SiGe crystal and the maximum of the L_2-error in Fig. 5. Additionally, the maximum L_2-error of *Multi-Grid* is smaller than the minimum L_2-error of *Coarse*, showing the anticipated increase in accuracy of the hierarchical approach.

As to be expected, there is a small L_2-error comparing the *Multi-Grid* results to the the *Fine* results, however, the simulation run with the *Fine* parameters has the disadvantage of significantly increased simulation run-times (see Tab. II). This is due to the higher resolution of the base grid, which

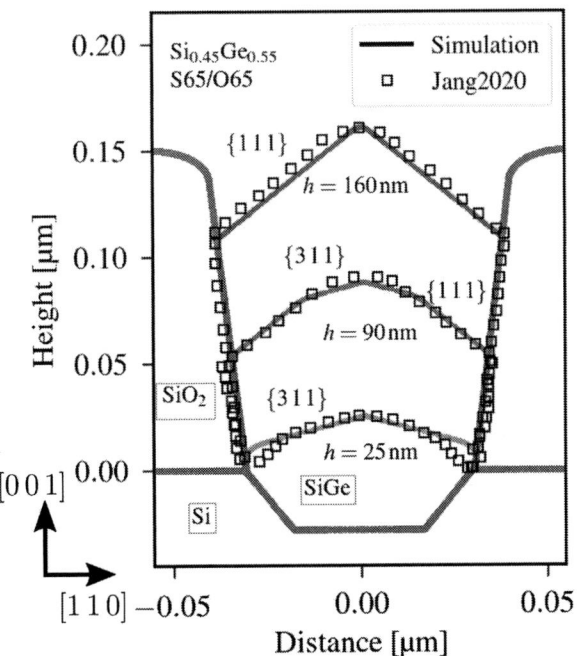

Fig. 3: Simulated surface of the SiGe crystal compared with the experimental results from [2] after 5 (orange), 24 (green), and 47 (red) SEG cycles. The simulation results show good agreement with the experimental data.

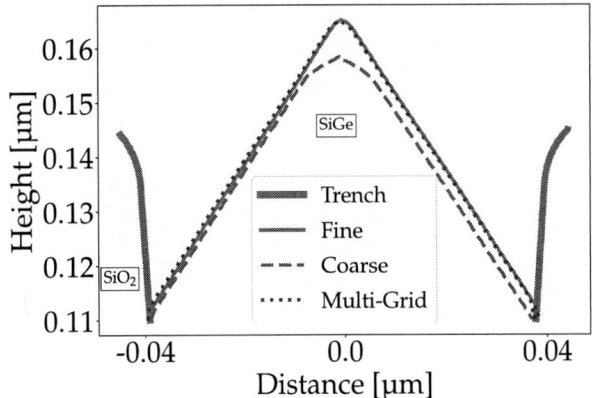

Fig. 4: Surface for the final simulation result of the SEG process after 47 SEG cycles using *Coarse*, *Fine*, and *Multi-Grid* resolutions. The error in the peak of the SiGe crystal using *Coarse* resolution is largest, since the grid resolution is not high enough to properly simulate the SEG process at this feature.

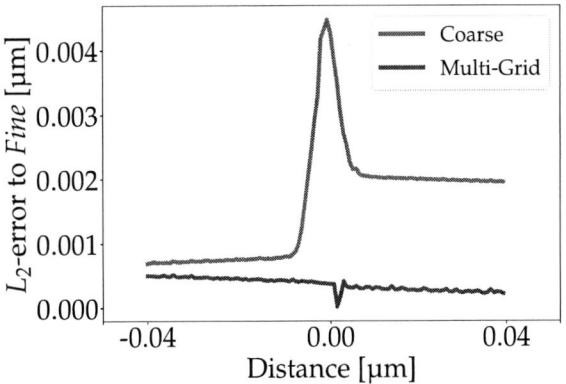

Fig. 5: Smallest L_2-error measured from the surface points of *Multi-Grid* and *Coarse* to the nearest surface point of *Fine*, in the final simulation result of the SEG process (see Fig. 4). The error of the simulation using the *Multi-Grid* parameters is negligible compared to the error when using the *Coarse* parameters.

unnecessarily increases the resolution of many irrelevant flat areas. The *Multi-Grid* result is based on the empirically chosen curvature threshold parameter $C = 0.9$ (and additionally considers the features at the interface of SiO_2 and SiGe, see Section II). To ensure the stability of the solution of (1) the maximal distance the zero level-set can propagate is bound by the Courant-Friedrichs-Lewy (CFL) condition [4]. The CFL condition is determined by the sub-grids with the highest resolution, so the feature detection and following re-distancing step (regridding) do not need to be executed in every time step of the simulation.

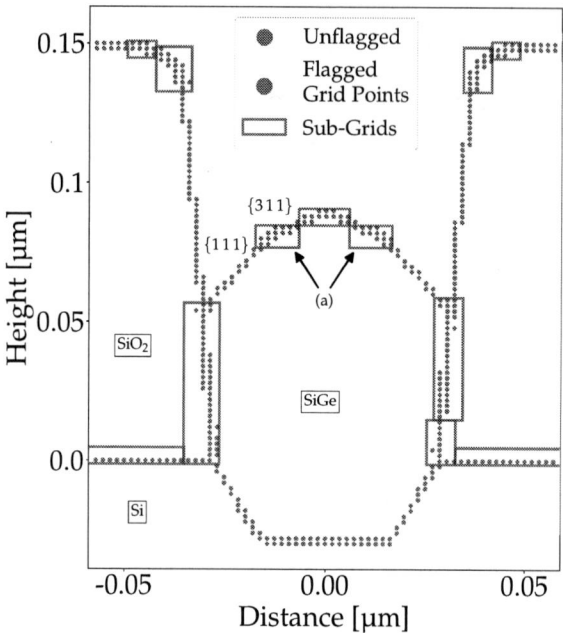

Fig. 6: Grid points near the level-set function for Grid 1 (i.e., base grid) of the *Multi-Grid* simulation after 24 SEG cycles. The flagged grid points (red) and generated sub-grids (blue boxes) for this time step are shown. (a) indicates sub-grids over fine features that develop during the SEG process.

The regridding step only has to be performed when the level-set values, within a sub-grid, required to calculate the finite difference scheme for the propagation of the zero level-set no longer lie within this sub-grid. Thus, to further improve the run-time of the simulation, we empirically determined that it is sufficient to perform the regridding every fourth time step. As discussed before (Fig. 3), *Multi-Grid* enables excellent agreement with the experimental data and a negligible L_2-error (Fig. 5) but with the advantage of a considerable reduced performance penalty as compared to the *Fine* case.

An example of the flagged grid points and thus generated sub-grids after 24 SEG cycles is shown in Fig. 6. This time point was chosen because here, the SEG process develops an additional geometrical feature between the {1 1 1} and {3 1 1} crystal facets, in addition to the peak of the crystal. Our feature detection method detects these fine features and sub-grids are placed accordingly, as indicated by (a) in Fig. 6.

IV. SUMMARY

An efficient and automatic feature detection method for selective grid refinement in hierarchical grids of process TCAD topography simulations has been introduced. The method is based on a feature detection and successive grid refinement strategy which considers the curvature of the level-set function representing a wafer surface. The efficiency

of this method has been demonstrated in a representative simulation of selectively grown epitaxial SiGe fins in oxide trenches. The feature detection, which has been used to optimally create the hierarchical grid, allows to use a low base grid resolution and only use sub-grids with higher resolution where crystal facets emerge during the SEG process and at material interfaces. By using our method the simulation run-time is considerably reduced compared to a simulation using a single high-resolution grid, while maintaining the accuracy of the simulation results.

ACKNOWLEDGMENT

The financial support by the Austrian Federal Ministry for Digital and Economic Affairs, the National Foundation for Research, Technology and Development, and the Christian Doppler Research Association is gratefully acknowledged.

REFERENCES

[1] J. Peng, Y. Qi, H.-C. Lo, P. Zhao, C. Yong, J. Yan, X. Dou, H. Zhan, Y. Shen, S. Regonda, O. Hu, H. Yu, M. Joshi, C. Adams, R. Carter, and S. Samavedam, "Source/Drain eSiGe Engineering for FinFET Technology," *Semiconductor Science and Technology*, vol. 32, no. 9, p. 094004, 2017.

[2] H. Jang, S. Koo, D.-S. Byeon, Y. Choi, and D.-H. Ko, "Facet Evolution of Selectively Grown Epitaxial $Si_{1-x}Ge_x$ Fin Layers in sub-100 nm Trench Arrays," *Journal of Crystal Growth*, vol. 532, p. 125429, 2020.

[3] J. A. Sethian, *Level Set Methods and Fast Marching Methods: Evolving Interfaces in Computational Geometry, Fluid Mechanics, Computer Vision, and Materials Science.* Cambridge University Press, 1999.

[4] S. Osher and R. Fedkiw, *Level Set Methods and Dynamic Implicit Surfaces.* Springer, 2003.

[5] A. Toifl, M. Quell, X. Klemenschits, P. Manstetten, A. Hössinger, S. Selberherr, and J. Weinbub, "The Level-Set Method for Multi-Material Wet Etching and Non-Planar Selective Epitaxy," *IEEE Access*, vol. 8, pp. 115 406–115 422, 2020.

[6] R. A. Trompert and J. G. Verwer, "A Static-Regridding Method for Two-Dimensional Parabolic Partial Differential Equations," *Applied Numerical Mathematics*, vol. 8, no. 1, pp. 65–90, 1991.

[7] S. L. Cornford, D. F. Martin, V. Lee, A. J. Payne, and E. G. Ng, "Adaptive Mesh Refinement Versus Subgrid Friction Interpolation in Simulations of Antarctic Ice Dynamics," *Annals of Glaciology*, vol. 57, no. 73, pp. 1–9, 2016.

[8] F. Löffler, Z. Cao, S. R. Brandt, and Z. Du, "A new Parallelization Scheme for Adaptive Mesh Refinement," *Journal of Computational Science*, vol. 16, pp. 79–88, 2016.

[9] Y. Liu, F. Kong, and F. Yan, "Level Set Based Shape Model for Automatic Linear Feature Extraction from Satellite Imagery," *Sensors and Transducers*, vol. 159, no. 11, pp. 39–45, 2013.

[10] B. Beddad and K. Hachemi, "Brain Tumor Detection by Using a Modified FCM and Level Set Algorithms," in *Proceedings of the International Conference on Control Engineering Information Technology (CEIT)*, 2016, pp. 1–5.

[11] N. Christoff, A. Manolova, L. Jorda, S. Viseur, S. Bouley, and J.-L. Mari, "Level-Set Based Algorithm for Automatic Feature Extraction on 3D Meshes: Application to Crater Detection on Mars," in *Computer Vision and Graphics*, L. J. Chmielewski, R. Kozera, A. Orłowski, K. Wojciechowski, A. M. Bruckstein, and N. Petkov, Eds. Cham: Springer International Publishing, 2018, pp. 103–114.

[12] Silvaco, "Victory Process," 2021. [Online]. Available: www.silvaco.com/tcad/victory-process-3d/

[13] M. Berger and I. Rigoutsos, "An Algorithm for Point Clustering and Grid Generation," *IEEE Transactions on Systems, Man and Cybernetics*, vol. 21, no. 5, pp. 1278–1286, 1991.

[14] M. Quell, G. Diamantopoulos, A. Hössinger, and J. Weinbub, "Shared-Memory Block-Based Fast Marching Method for Hierarchical Meshes," *Journal of Computational and Applied Mathematics*, vol. 392, p. 113488, 2021.

Feature-Scale Modeling of Low-Bias SF$_6$ Plasma Etching of Si

Luiz Felipe Aguinsky*, Georg Wachter[†], Frâncio Rodrigues*, Alexander Scharinger*,
Alexander Toifl*, Michael Trupke[†], Ulrich Schmid[‡] Andreas Hössinger[§], and Josef Weinbub*

*Christian Doppler Laboratory for High Performance TCAD,
Institute for Microelectronics, TU Wien, Gußhausstraße 27-29, 1040 Wien, Austria
[†]Faculty of Physics, University of Vienna, VCQ, Boltzmanngasse 5, 1090, Wien, Austria
[‡]Institute of Sensor and Actuator Systems, TU Wien, Gußhausstraße 27-29, 1040 Wien, Austria
[§]Silvaco Europe Ltd., Compass Point, St Ives, Cambridge, PE27 5JL, United Kingdom
Email: aguinsky@iue.tuwien.ac.at

Abstract—**Low-bias etching of Si using SF$_6$ plasma is a valuable tool in the manufacturing of semiconductor and MEMS devices. This kind of etching has strong isotropic tendencies, since the low voltage bias does not provide enough vertical acceleration and kinetic energy to the ions. This near-isotropy can be difficult to precisely reproduce in a topography simulation, since experimentally realized surfaces cannot be reproduced by a strictly isotropic velocity model. We present a three-dimensional top-down Monte Carlo particle tracing model for calculating the velocity field in a level-set based simulation. We compare it to profilometer measurements of optical cavities, which are of interest to quantum science, fabricated using a two-step SF$_6$ plasma etching process. We contrast our approach to conventional models: a strictly isotropic model and a bottom-up direct flux calculation. We show that our top-down model leads to a more accurate description of the final surface by introducing a sticking probability at the surface and also multiple reflections. We are able to reproduce cavities fabricated from different initial photoresist configurations with a single silicon etch rate ($V_{Si} = 2.15\,\mu m\,min^{-1}$), while the conventional models require a separate V_{Si} for each photoresist geometry. The model successfully reproduces a Si/photoresist selectivity of 10, which, combined with the low calibrated sticking probability ($\beta_{Si} = 7.5\%$), corroborates with F radicals being the main drivers of etching. By exploring the state of the surface after the first etch step, which is not readily available experimentally, we anticipate the phenomena of underetching and photoresist tapering.**

Index Terms—**Plasma etching, process TCAD, topography simulations, ray-tracing, level-set method, SF$_6$ plasma**

I. INTRODUCTION

Plasma etching of silicon (Si) using sulphur hexafluoride (SF$_6$) gases is a standard technology in modern semiconductor fabrication processes [1], such as memory devices, microeletromechanical systems (MEMS), and as a sub-step in the Bosch process [2], [3]. Under low-bias conditions, SF$_6$ plasma etching is known to have a near-isotropic behavior yielding profiles similar, but not identical, to those obtained by wet etching [4]. This is due to the low voltage bias between the plasma and the wafer not accelerating the ions to a large degree. Therefore, the anisotropic component, caused by the kinetic energy of the vertical ions, is minimal [5]. The near-isotropic behavior has proven to be useful in, e.g.,

optical applications, where surface cleanliness requirements favor plasma etching over wet etching [6].

Feature-scale topography simulations are part of process technology computer-aided design (TCAD) workflows which enable, among others, the investigation of etched or deposited materials [7]. While two-dimensional feature-scale modeling of SF$_6$ etching of Si has been reported for anisotropic, high-bias conditions [8], low-bias etching provides a different set of challenges for accurately modeling the topography since the final surfaces are not ideal, that is, they are not equivalent to surfaces etched by a perfectly isotropic process.

Here, we present a three-dimensional feature-scale model tailored to the challenge of the low-bias, near-isotropic regime of SF$_6$ etching of Si. This is achieved using *top-down* Monte Carlo particle tracing [9] including multiple reflections. Our model is contrasted to conventional, strictly isotropic, and bottom-up models [10]. Our top-down model and the conventional models are summarized visually in Fig. 1. The simulated profile is calibrated to an experimentally measured cavity, fabricated with a two-step SF$_6$ low-bias plasma etching process, which is of relevance for the development of optical resonators for quantum science [6]. We then interpret our model with respect to chemical etching mechanisms. Finally, using calibrated simulations, we are able to investigate the state of the surface after the first etch step, a state which is not readily available experimentally, thereby also underlining one of the key advantages of process TCAD simulations.

II. METHOD

In order to simulate the time evolution of an etched surface, we employ the level-set method [11]. The evolving surface is represented as the zero level-set of the signed distance function ϕ. Its propagation is described by the solution of the following level-set equation for ϕ:

$$\frac{\partial \phi(x,t)}{\partial t} + V(x)|\nabla \phi(x,t)| = 0. \qquad (1)$$

The solution of (1) is performed by Silvaco's three-dimensional process TCAD tool *Victory Process* [12]. The modeling of surface reactions and subsequent local etch rates is achieved via the velocity field $V(x)$, as discussed below.

978-1-6654-3746-2/21 $31.00 © 2021 IEEE

a) Strictly isotropic

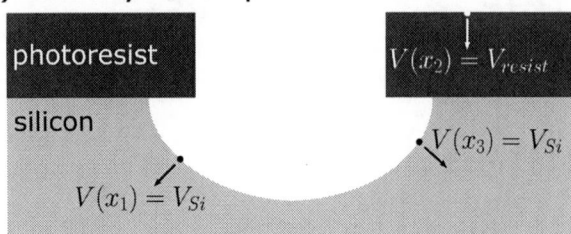

b) Bottom-up

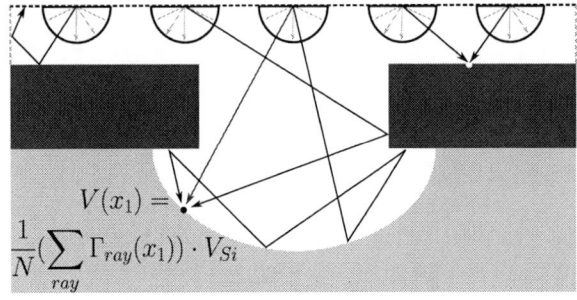

c) Top-down

Fig. 1: Illustration of models for the local surface propagation rate $V(x)$. Our top-down model (c) is contrasted to the conventional strictly isotropic (a) and bottom-up (b) procedures.

As reported previously [1], under low-bias conditions the etching is expected to be near-isotropic. Such etching is mostly performed by highly reactive isotropic F radicals generated in the plasma. They can etch both Si and the photoresist. Due to the low-bias, the directional component caused by the vertically accelerated ions is minimal.

The precise nature of the low-bias behavior requires accurate modelling. In Fig. 1, we present an illustration of three possible approaches to generate $V(x)$ for an isotropic etchant. The straightforward approach, i.e., strictly isotropic, is represented in Fig. 1.a). In this case, the same constant velocity is applied to all exposed surface elements of the same material. This results in a surface equivalent to that processed by ideal isotropic wet etching [13]. Therefore, in the strictly isotropic case the velocity $V_{\text{S-I}}(x)$ is simply a function of the involved material (either Si or the resist):

$$V_{\text{S-I}}(x) = V_{\text{Si/resist}}. \qquad (2)$$

However, since plasma etching is a gas-phase method, the etchant distribution is not always identical across the surface. A more complex model thus requires the construction of an approximation to the local flux of etchant particles $\Gamma(x)$ in order to more accurately model $V(x)$. An elementary way of calculating $\Gamma(x)$, represented in Fig. 1.b), assumes that the incoming particle stream originates isotropically from a source plane. Subsequently, the local visibility of this plane is calculated in a bottom-up fashion for the hemisphere above each surface element [10], leading to a local flux $\Gamma_{vis}(x)$ which is normalized to 1 for a fully-exposed element. The final velocity is the plane-wafer etch rate weighted with $\Gamma_{vis}(x)$. This model allows the capture of some topography-dependent effects, however, it does not take reflections into account. The velocity field in the bottom-up model $V_{\text{B-U}}(x)$ is:

$$V_{\text{B-U}}(x) = \Gamma_{vis}(x) \cdot V_{\text{Si/resist}}. \qquad (3)$$

We propose a physically richer top-down model, as shown in Fig. 1.c), supporting multiple reflections according to the sticking probability β. This is achieved by a Monte Carlo sampling of N particles of a single type, which are generated isotropically in the source plane and carry a flux payload Γ_{ray}. Their trajectories through the domain are computed using a ray-tracing method and reflective boundary conditions. When a simulated particle hits the surface, it is terminated, leaving its payload Γ_{ray} at the surface site. A new reflected particle is generated, following an isotropic reflection distribution and having a new payload Γ_{ref} mediated by $\beta_{\text{Si/resist}}$, i.e.:

$$\Gamma_{ref} = (1 - \beta_{\text{Si/resist}}) \cdot \Gamma_{ray}. \qquad (4)$$

Finally, the local velocity of the surface is calculated from the normalized sum of $\Gamma_{ray}(x)$ for all particles, both generated in the source plane and reflected, and from the plane-wafer etch rate V. This effectively generalizes the two previous models, as the strictly isotropic approach is recovered with the limit $\beta \to 0^+$, and the bottom up, with $\beta \to 1^-$. In summary, the velocity for the top-down model $V_{\text{T-D}}$ is:

$$V_{\text{T-D}}(x) = \frac{1}{N}\left(\sum_{ray} \Gamma_{ray}(x)\right) \cdot V_{\text{Si/resist}}. \qquad (5)$$

The isotropic source and reflection distributions are motivated by the low-bias characteristics of the etching process. That is, the etchants are not accelerated and interact with the surface in a diffuse manner. This is consistent with the expected mechanism of etching: The generation of F radicals in the plasma which chemically etch the surface [14].

Therefore, the free parameters are the silicon plane-wafer etch rates V_{Si} for the first and second steps, and the photoresist etch rate V_{resist}. In addition, the top-down model has as parameters the sticking coefficients β_{Si} and β_{resist}. Although this indicates that the top-down model is successful due to its additional fitting parameters, we will discuss in the next section that the conventional models require different values of V_{Si} for each individual initial photoresist geometry, which is not straightforwardly justifiable from a physical standpoint.

978-1-6654-3746-2/21 $31.00 © 2021 IEEE

Thus, the top-down model involves not only an equivalent number of parameters, but also its values can be physically interpreted and compared to reported results [15].

III. RESULTS

To validate the proposed model, we compare it to three-dimensional profilometer measurements of experimentally fabricated structures [6]. Multiple cavities were etched simultaneously on Si using a two-step SF_6 plasma etching process, each cavity under a different initial photoresist cylindrical opening d. We studied three different cavities with a respective d of $12.4\,\mu m$, $34\,\mu m$, and $52\,\mu m$. The first etch step was performed for 320 seconds and took place having the photoresist present. After photoresist removal using acetone, a second etch step was applied for 48 minutes. We manually calibrate the simulations to the final topography of the fabricated surfaces, i.e., after photoresist removal and the second etch step.

The calibrated parameters for the top-down model are presented in Tab. I. For the strictly isotropic and bottom-up models, the same V_{resist} and second etch step V_{Si} are applied, however, each cavity requires an individually calibrated first step V_{Si}, presented in Tab. II. A cross-section contrasting the simulation approaches to the experiment is shown Fig. 2.

TABLE I: Calibrated parameters for top-down simulation.

Parameter	Calibrated value
First etch step V_{Si}	$2.15\,\mu m\,min^{-1}$
Second etch step V_{Si}	$0.66\,\mu m\,min^{-1}$
V_{resist}	$0.21\,\mu m\,min^{-1}$
β_{Si}	$7.5\,\%$
β_{resist}	$6.1\,\%$

TABLE II: Calibrated first etch step V_{Si} for each photoresist opening d for the strictly isotropic and bottom-up simulations.

Opening d	Strictly isotropic first V_{Si}	Bottom-up first V_{Si}
$12.4\,\mu m$	$1.45\,\mu m\,min^{-1}$	$23.0\,\mu m\,min^{-1}$
$34\,\mu m$	$1.94\,\mu m\,min^{-1}$	$6.0\,\mu m\,min^{-1}$
$52\,\mu m$	$2.09\,\mu m\,min^{-1}$	$3.6\,\mu m\,min^{-1}$

The results show the failure of the bottom-up model, since it cannot correctly capture the curvature, i.e., it underestimates the etch rates at the sidewalls. The strictly isotropic model has a very similar shape to the experiment and to the top-down model, in particular for the cavity with $d = 12.4\,\mu m$. However, since it applies the same rate to all exposed regions, the bottom of the cavity, i.e., the area under the original photoresist opening, remains unrealistically flat. Since the strictly isotropic model is equivalent to having $\beta = 0$, the similarity of the strictly isotropic model to the experimental profile is evidence that a low β is expected, as confirmed in the calibrated values on Tab. I. However, the perfectly flat profiles at the bottom are not observed in the experiment [6] and, additionally, cause the surface to be unsuited for optical applications and incompatible with further numerical investigation [16].

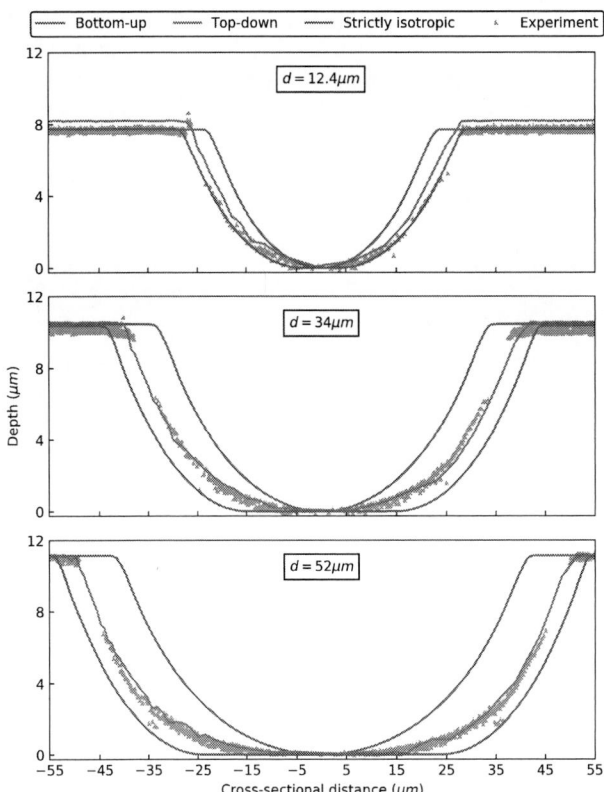

Fig. 2: Cross-sections of the simulated surfaces using the models from Fig. 1 and measurements of experimentally fabricated surfaces with different initial photoresist openings d using a two-step low-bias SF_6 plasma etching process.

In addition to the observed correspondence with the experiment shown in Fig. 2, the top-down model offers a significant advantage. Only a single V_{Si} for the first etch step is required, whereas separate values are required for each cavity for the conventional models. This indicates that the top-down model more accurately captures the real chemical processes involved in low-bias SF_6 etching. The SF_6 plasma is known to be a source of highly reactive F radicals which chemically etch the surface [14]. This is supported by our results, since we can reproduce the topography with a single Monte Carlo particle (representing the F radicals) with a low, but not zero, β which is consistent with reported values [15].

The calibrated parameters in Tab. I show a lower V_{Si} for the second etch step. This is expected due to the effect of reactor loading [1], [17]. In the second etch step there is no photoresist, therefore, there is a larger wafer surface available to consume the reactants. This reduces their local supply, decreasing V_{Si}. The change in reactor loading during the first etch step is neglected, since the increase in exposed Si is small compared to the total wafer area covered by the photoresist. Additionally, the values show a Si/photoresist selectivity of 10, which is consistent with reported results [18].

Using topography simulation, we are able to explore states which are not readily available experimentally. In particular, the profile after the first etch step but before photoresist removal, shown in Fig. 3. For the top-down approach, this figure is obtained with the parameters from Tab. I. The parameters were re-calibrated for the bottom-up and strictly isotropic models in order to achieve the same depth.

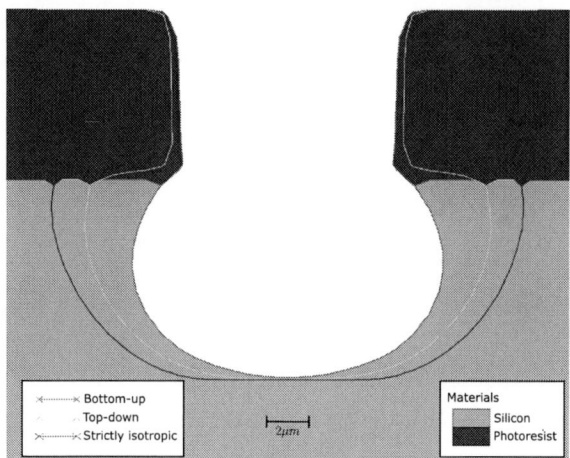

Fig. 3: Simulated etched surfaces for a cavity with photoresist $d = 12.4\,\mu m$ showing underetching and photoresist tapering.

We can see that the conventional bottom-up approach has fundamental limitations. The shape is more bulbous, which leads to the incorrect final curvature seen in Fig. 2. Additionally, the bottom-up model is unable to capture underetching, i.e., the etching of Si directly below the photoresist, which is experimentally a known feature of low-bias SF_6 etching [1]. For the strictly isotropic approach, the flatness of the bottom is even clearer at this step. As discussed previously, this makes the strictly isotropic approach unsuitable. Finally, we would like to highlight that our simulations indicate the presence of photoresist tapering during the etching, which is a phenomenon of interest for further process improvement.

IV. Summary

We have presented a physically rich, calibrated, top-down model to calculating the velocity field $V(x)$ for the surface evolution of low-bias SF_6 plasma etching of Si. We show that the conventional strictly isotropic and bottom-up approaches are insufficient to represent the experimental topography. By introducing multiple reflections and a sticking probability β, the top-down model is able to more accurately represent the final surfaces while simultaneously using the same parameter set for multiple geometries. The success of the top-down approach provides insight into the surface chemistry of SF_6 plasma etching, highlighting the importance of F radicals as the drivers of chemical etching. The calibrated parameter set captures the effect of reactor loading and reports values of β and Si/photoresist selectivity consistent with the literature. Using our calibrated simulations, we are able to explore a state which is not readily accessible experimentally, i.e., after

the first etch step and before photoresist removal, featuring the expected phenomena of underetching and photoresist tapering.

Acknowledgment

The financial support by the Austrian Federal Ministry for Digital and Economic Affairs, the National Foundation for Research, Technology and Development, and the Christian Doppler Research Association is gratefully acknowledged. M. Trupke and G. Wachter gratefully acknowledge support from the EU H2020 framework programme (QuanTELCO, 862721), the FWF (SiC-EiC, I 3167-N27), and the FFG (QSense4Power, 877615).

References

[1] V. M. Donnelly and A. Kornblit, "Plasma Etching: Yesterday, Today, and Tomorrow," *Journal of Vacuum Science & Technology A: Vacuum, Surfaces, and Films*, vol. 31, no. 5, p. 050825, 2013.

[2] C. Waits *et al.*, "Microfabrication of 3D Silicon MEMS Structures using Gray-Scale Lithography and Deep Reactive Ion Etching," *Sensors and Actuators A: Physical*, vol. 119, no. 1, pp. 245–253, 2005.

[3] F. Patocka, C. Schneidhofer, N. Dörr, M. Schneider, and U. Schmid, "Novel Resonant MEMS Sensor for the Detection of Particles with Dielectric Properties in Aged Lubricating Oils," *Sensors and Actuators A: Physical*, vol. 315, p. 112290, 2020.

[4] P. Panduranga, A. Abdou, Z. Ren, R. H. Pedersen, and M. P. Nezhad, "Isotropic Silicon Etch Characteristics in a Purely Inductively Coupled SF_6 Plasma," *Journal of Vacuum Science & Technology B, Nanotechnology and Microelectronics: Materials, Processing, Measurement, and Phenomena*, vol. 37, no. 6, p. 061206, 2019.

[5] M. A. Lieberman and A. J. Lichtenberg, *Principles of Plasma Discharges and Materials Processing*. John Wiley & Sons, 2005.

[6] G. Wachter *et al.*, "Silicon Microcavity Arrays with Open Access and a Finesse of Half a Million," *Light: Science & Applications*, vol. 8, no. 1, pp. 1–7, 2019.

[7] X. Klemenschits, S. Selberherr, and L. Filipovic, "Modeling of Gate Stack Patterning for Advanced Technology Nodes: A Review," *Micromachines*, vol. 9, no. 12, p. 631, 2018.

[8] R. J. Belen, S. Gomez, M. Kiehlbauch, D. Cooperberg, and E. S. Aydil, "Feature-Scale Model of Si Etching in SF_6 Plasma and Comparison with Experiments," *Journal of Vacuum Science & Technology A: Vacuum, Surfaces, and Films*, vol. 23, no. 1, pp. 99–113, 2005.

[9] A. Scharinger, P. Manstetten, A. Hössinger, and J. Weinbub, "Generative Model Based Adaptive Importance Sampling for Flux Calculations in Process TCAD," in *Proceedings of the International Conference on Simulation of Semiconductor Processes and Devices (SISPAD)*, 2020, pp. 39–42.

[10] P. Manstetten, J. Weinbub, A. Hössinger, and S. Selberherr, "Using Temporary Explicit Meshes for Direct Flux Calculation on Implicit Surfaces," *Procedia Computer Science*, vol. 108, pp. 245–254, 2017.

[11] J. A. Sethian, *Level Set Methods and Fast Marching Methods: Evolving Interfaces in Computational Geometry, Fluid Mechanics, Computer Vision, and Materials Science*. Cambridge University Press, 1999.

[12] Silvaco, "Victory Process," 2021. [Online]. Available: www.silvaco.com/tcad/victory-process-3d/

[13] J. Miao, *Silicon Micromachining*. Boston, MA: Springer US, 2008, pp. 1840–1846.

[14] D. L. Flamm, V. M. Donnelly, and J. A. Mucha, "The Reaction of Fluorine Atoms with Silicon," *Journal of Applied Physics*, vol. 52, no. 5, pp. 3633–3639, 1981.

[15] V. M. Donnelly, "Reactions of Fluorine Atoms with Silicon, Revisited, Again," *Journal of Vacuum Science & Technology A: Vacuum, Surfaces, and Films*, vol. 35, no. 5, p. 05C202, 2017.

[16] D. Kleckner, W. T. Irvine, S. S. Oemrawsingh, and D. Bouwmeester, "Diffraction-Limited High-Finesse Optical Cavities," *Physical Review A*, vol. 81, no. 4, p. 043814, 2010.

[17] C. Mogab, "The Loading Effect in Plasma Etching," *Journal of the Electrochemical Society*, vol. 124, no. 8, p. 1262, 1977.

[18] K. R. Williams, K. Gupta, and M. Wasilik, "Etch Rates for Micromachining Processing-Part II," *Journal of Microelectromechanical Systems*, vol. 12, no. 6, pp. 761–778, 2003.

Effect of Temperature on Performance of HZO-Based FD-SOI NCFET

Vullakula Rama Seshu[*], Rameez Raja Shaik[†], K P Pradhan[‡]

[*†‡]Department of ECE, IIITDM Kancheepuram, Chennai, India, 600127.

email:[*]evd17i025@iiitdm.ac.i,[†]rameezraj@ieee.org, [‡]k.p.pradhan@ieee.org

Abstract—In this study, the temperature effect on Zirconium doped HfO$_2$ (HZO) based Metal-Ferroelectric-Metal-Insulator-Semiconductor (MFMIS) type Fully depleted Silicon on Insulator (FDSOI) negative capacitance field effect transistor (NCFET) has been investigated in ATLAS TCAD. The MIS device simulation has been performed in TCAD and its gate charge (Q$_G$) has been extracted. The extracted charge is further utilized in 1D Landau-Khalatnikov (LK) model to find the voltage across ferro-electric (V$_{FE}$), that subsequently used to add the effect of HZO analytically to the device. Performance parameters like minimum sub-threshold swing (min. SS), off current (I$_{off}$) and transconductance (g$_m$) with variation of temperature (280 K - 340 K with step 20 K) have been predicted. The performance variation of device with temperature has been explained using 1D LK theory.

Keywords—FDSOI, g$_m$, HZO, LK model, MFMIS, NCFET, SS

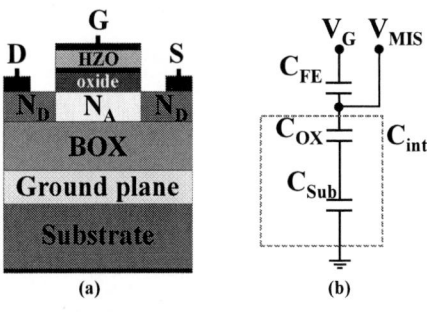

Fig. 1. (a) Cross sectional view of proposed structred, (b) Equivalent capacintance schematic

I. Introduction

WITH the development of technologies like big data and cloud computing, the requirement of very high speed computing hardware is substantial [1]. This type of computing need to process huge number of instructions at the same time leading to more switching activity. Which further lead to very high power dissipation mostly in the form of heat. It is well known that with reduction in supply voltage (V$_{DD}$), power dissipation or heat dissipated comes down significantly at the cost of short channel effects (SCE). To overcome this, field effect transistors (FETs) with steep sub-threshold swing (SS) are essential. SS is limited by 60 mV/dec. at room temperature (T) which is termed as Boltzmann tyranny [2]. FETs with sub 60 mV/dec. can be used to achieve lower power dissipation and also enabling scaling down of V$_{DD}$.

Impact ionization FET (IIFET) and Tunnel FETs (TFET) exhibit steeper SS overcoming Boltzmann tyranny but these pose the drawbacks of reduction in on current (I$_{on}$) and requirement of high V$_{DD}$, respectively. Negative capacitance FET (NCFET) is the other candidate for steeper SS, superior control of SCEs and lower supply voltage. These FETs are achieved by adding ferro-electric material in the gate stack, which amplifies the voltage applied at the gate [3]. The transport mechanism in the channel remains same as that of conventional FETs. NCFET is a promising candidate for achieving ultra low power devices [3], [2].

MFMIS and Metal - Ferroelectric - Insulator - Semiconductor (MFIS) are two prominent and widely researched architectures for NCFET. MFMIS exhibits better performance compared to MFIS for high remnent polarization (P$_r$) ferro-electric materials [6]. The discovery of doped HfO$_2$ has enabled to achieve the NC effect with thinner ferro-electric layer [4]. In this study authors have considered Zirconium doped HfO$_2$ as ferro-electric material due to its advantages of high P$_r$, low annealing temperatures compared Si and Al doped HfO$_2$ and higher endurance towards temperature [5]. High P$_r$ of HZO and MFMIS architecture lead to optimum performance of FDSOI NCFET proposed.

FDSOI with its advantages of reduced SCEs, very thin channel, reduced parasitics and better performance over conventional devices becomes a suitable device structure for low power applications. It is also compatible with current CMOS technology [7], [8], [9]. Despite the implementations of novel devices and structures for energy-efficient performance, temperature remains one of the inevitable factors which affects the performance of the FETs. Depending on the application and run time, the operating temperatures of the device may vary. This effects the performance of the device as well as the properties of ferroelectric [10]. So it is crucial to know the trend of device performance with temperature. Keeping in view of the practical scenario, the authors have performed temperature analysis on the proposed device.

978-1-6654-3746-2/21 $31.00 © 2021 IEEE

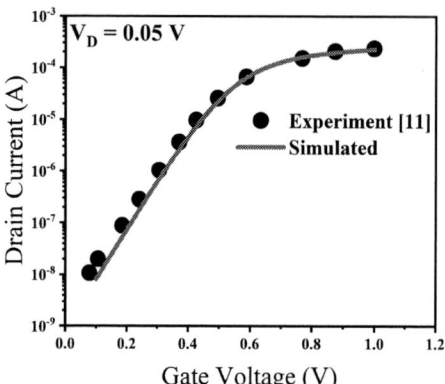

Fig. 2. Calibration with measured data of 28 nm FDSOI n-type MOSFET consists of high-k dielectric as gate stack on top of SiON interfacial layer with gate oxide thickness of 1.2 nm, silicon body of 7 nm without any intentional channel doping, BOX thickness of 25 nm and the ground-plane is under the BOX at T = 300K from [11]

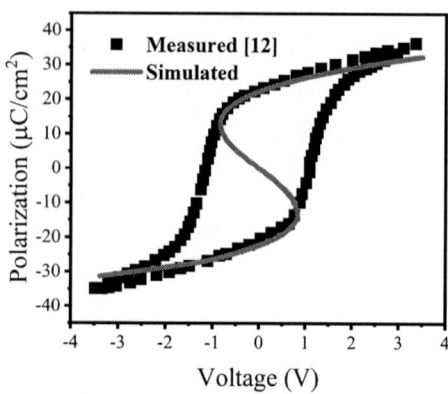

Fig. 3. Calibration Polarization vs Voltage across ferroelectric S-curve derived from LK model with experimental data of HZO based metal-ferro-metal capacitor with $t_{FE} = 10$ nm from [12]

II. SIMULATION METHODOLOGY AND VALIDATION

The cross sectional view of the device considered for this work and its equivalent capacitance circuit are depicted in Fig. 1 (a) and (b) respectively. The simulation methodologies are as follows:

Silvaco Atlas TCAD has been used to design the underlying MIS FDSOI and perform the simulations. The geometric parameters of the device are as follows: gate length (L_G) = 22nm, source/drain doping (N_D) = 10^{20} cm^{-3}, channel doping $N_A = 10^{15}$ cm^{-3} and gate oxide thickness (t_{OX} = 1nm), channel (t_{Si} = 6nm), substrate (t_{SUB} = 100nm). HfO$_2$ which is a high-k dielectric material and HZO have been considered as oxide layer and ferro-electric material, respectively.

Initially, all the models are set to match the simulation and experimental data from [11]. Fig. 2 shows a good agreement between the simulated and experimental data. The models used for this work are field-dependent mobility, concentration dependent mobility, concentration dependent Shockley-Read-Hall and Auger. The above mentioned models will capture the effect of field and concentration on mobility and also recombination mechanisms. And here temperature parameter has been included to account for the temperature effect. The gate charge (Q_G) extracted from the MIS simulation have been used to predict the voltage across the ferro-electric (V_{FE}) using 1D- LK equation.

$$V_{FE} = t_{FE}(2\alpha Q_G + 4\beta Q_G^3) \quad (1)$$

The LK parameters (α & β) are tuned and set to match the P-V_{FE} with the experimental data from [12]. Fig. 3 exhibit good fitting of the P-V_{FE} S-curve from simulation with experimental hysteresis. The fitting is achieved with a proper tuning of the material parameters as $\alpha = -4.89 * 10^{10} V cm/C$, $\beta = 4.9 * 10^{19} V cm^5/C^3$

and $\gamma = 0 V cm^5/C^5$ for T= 300K. The temperature dependency of α with temperature in LK equation has been considered to account the temperature effect on ferro-electric behaviour [10].

$$\alpha = \alpha_0(T - T_C) \quad (2)$$

$$C_{FE} = \frac{\partial Q_{FE}}{\partial V_{FE}} = \frac{1}{t_{FE} * (2\alpha_0(T - T_C) + 12\beta Q_{FE}^2)} \quad (3)$$

Here T represents the operating temperature of the device, T$_C$ is the Curie temperature of HZO and C$_{FE}$ is the capacitance of HZO. Beyond T$_C$ = 723.15K ferro-electric property of HZO degrades significantly [5]. Fig. 1 (b) shows that the oxide capacitance (C$_{OX}$) and C$_{FE}$ are in series which can be used to infer that the charge in HZO (Q$_{FE}$) is same as Q$_G$. The obtained V_{FE} has been analytically added to V_{MIS} to obtain total gate voltage (V_G) using Kirchoff's Voltage law as per the schematic diagram shown in Fig. 1 (b).

$$V_G = V_{MIS} + V_{FE} \quad (4)$$

III. RESULTS AND DISCUSSIONS

A. Effect of ferro-electric thickness

In this subsection the influence of ferro-electric thickness (t_{FE}) on the performance of the FDSOI NCFET at temperature (T)= 300K and drain voltage (V$_D$ = 0.5V) have been discussed. It can be noted that with increment in t_{FE} the ferro-electric capacitance ($|C_{FE}|$) decreases. Which exhibits better matching with total internal capacitance (C_{int}) leading to more voltage amplification and improvement in device performance parameters (SS, I_{off}, g_m) [2].

Fig. 4 shows the transfer characteristic of the device for various t_{FE}. It can be observed that with increment in t_{FE} the I_{on}/I_{off} ratio improves. The decrement in $|C_{FE}|$ with increment in t_{FE} leads to a point when $|C_{FE}| < C_{int}$. This leads to unstability and manifests as hysteresis

978-1-6654-3746-2/21 $31.00 © 2021 IEEE

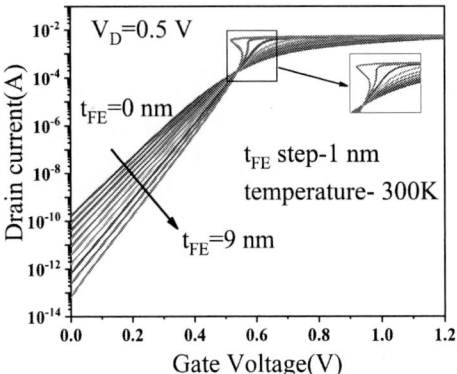

Fig. 4. Transfer characteristics with variation in HZO thickness at $V_D = 0.5V$, T $= 300K$ for proposed FDSOI NCFET

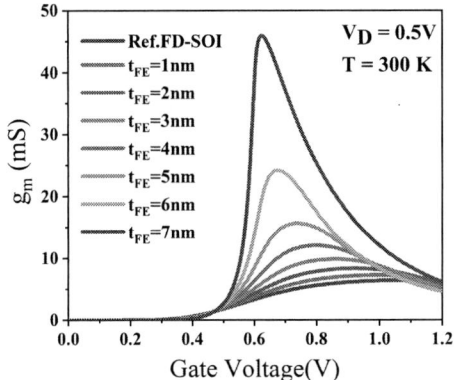

Fig. 5. Transconductance with variation in HZO thickness at $V_D = 0.5V$, T$= 300K$ for proposed FDSOI NCFET

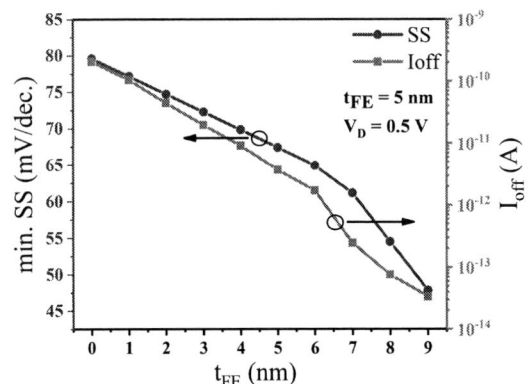

Fig. 6. min. SS and I_{off} with variation in HZO thickness at $V_D = 0.5V$, T$= 300K$ for proposed FDSOI NCFET

in the transfer characteristics [2]. It can be observed that hysteresis has started from $t_{FE} = 8$ nm which is the critical thickness for this device [13].

Fig. 5 depicts the trend in g_m with respect to V_G for t_{FE}

less than the critical thickness to keep the device operating as linear device. An upward shift in the peak has been exhibited with increment in t_{FE} due to better capacitance matching. This implies boost in g_m with increment in t_{FE}.

In Fig. 6 the trend of min. SS and I_{off} with respect to t_{FE} have been plotted. The results are inline with previous observations made at the beginning of this subsection. Increment in t_{FE} helps to achieve lower sub-threshold swing and I_{off} as the negative capacitance effect becomes more pronounced.

B. Effect of temperature

In this section the variation in the performance of the device for different temperatures (280 K, 300 K, 320 K, 340 K) has been exhibited and discussed. From the conclusions of the previous section to avoid hysteresis t_{FE} and V_D are considered to be 5 nm and 0.5 V respectively for this section. It is to be noted that with increment in temperature $|C_{FE}|$ increases by virtue of the (2) and (3). This leads to greater mismatch between C_{int} and $|C_{FE}|$ which leads to degradation in the device performance.

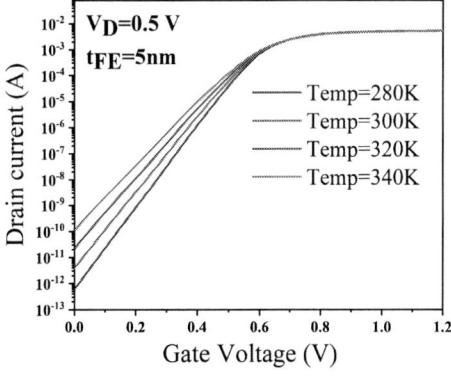

Fig. 7. Transfer characteristics with variation in temperature at $V_D = 0.5V$, $t_{FE} = 5nm$ for proposed FDSOI NCFET

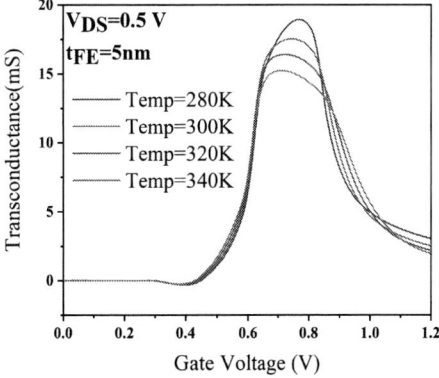

Fig. 8. Transconductance with variation in temperature at $V_D = 0.5V$, $t_{FE} = 5nm$ for proposed FDSOI NCFET

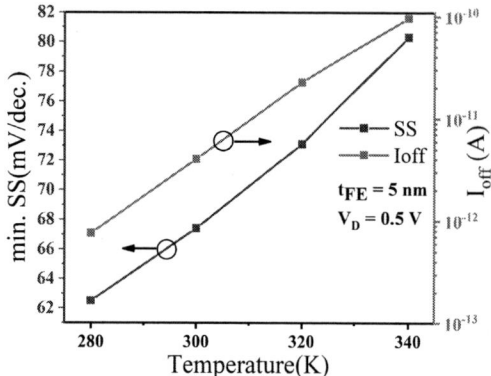

Fig. 9. min. SS and I_{off} with variation in temperature at $V_D = 0.5V$, $t_{FE} = 5nm$ for proposed FDSOI NCFET

It is essential to observe how transfer characteristics of the device are effected by temperature, Fig. 7 shows the transfer characteristic of the device for different temperatures. It can be observed that increase in temperature there is decrement in the I_{on}/I_{off} ratio. Fig. 8 depicts the trend in g_m with V_G for different temperatures. Decrement in peak can be observed indicating degradation in performance with temperature [10].

The trend in min. SS and I_{off} with temperature have been plotted in Fig. 9. It can be observed that with increase in temperature min. SS and I_{off} increase. The results are inline with the argument that with greater mismatch between C_{int} and $|C_{FE}|$ lesser amplification of gate voltage which further to lesser surface potential and manifests as degradation in device performance.

IV. CONCLUSION

In summary, investigation on effect of ferro-electric thickness and temperature on performance of HZO based FDSOI NCFET has been presented. It is observed that by incorporating higher ferro-electric thickness there is a significant improvement in SS, g_m, I_{on}/I_{off} compared to reference FDSOI device. Hysteresis behaviour is observed from the critical thickness of 8 nm. The performed simulations predict that, there is an enhancement in performance below T = 300 K and degradation for higher temperatures compared to T = 300 K. This work provides the understandings of device performance under temperature variation.

ACKNOWLEDGMENT

Thanks to Dr. S R Routray, SRM Institute of Science and Technology, for providing the simulation tool.

REFERENCES

[1] S. Salahuddin, K. Ni, and S. Datta, "The era of hyper-scaling in electronics," *Nature Electronics*, vol. 1, no. 8, pp. 442–450, 2018.

[2] S. Salahuddin and S. Datta, "Can the subthreshold swing in a classical fet be lowered below 60 mv/decade?" in *2008 IEEE International Electron Devices Meeting*. IEEE, 2008, pp. 1–4.

[3] S. Salahuddin and S. Datta, "Use of negative capacitance to provide voltage amplification for low power nanoscale devices," *Nano Letters*, vol. 8, no. 2, pp. 405–410, 2008, pMID: 18052402. [Online]. Available: https://doi.org/10.1021/nl071804g

[4] J. Müller, T. S. Böscke, U. Schröder, S. Mueller, D. Bräuhaus, U. Böttger, L. Frey, and T. Mikolajick, "Ferroelectricity in simple binary zro2 and hfo2," *Nano Letters*, vol. 12, no. 8, pp. 4318–4323, 2012, pMID: 22812909. [Online]. Available: https://doi.org/10.1021/nl302049k

[5] D. Wang, J. Wang, Q. Li, W. He, M. Guo, A. Zhang, Z. Fan, D. Chen, M. Qin, M. Zeng *et al.*, "Stable ferroelectric properties of hf0. 5zr0. 5o2 thin films within a broad working temperature range," *Japanese Journal of Applied Physics*, vol. 58, no. 9, p. 090910, 2019.

[6] G. Pahwa, T. Dutta, A. Agarwal, and Y. S. Chauhan, "Physical insights on negative capacitance transistors in nonhysteresis and hysteresis regimes: Mfmis versus mfis structures," *IEEE Transactions on Electron Devices*, vol. 65, no. 3, pp. 867–873, 2018.

[7] S. A. Vitale, P. W. Wyatt, N. Checka, J. Kedzierski, and C. L. Keast, "Fdsoi process technology for subthreshold-operation ultralow-power electronics," *Proceedings of the IEEE*, vol. 98, no. 2, pp. 333–342, 2010.

[8] Z. Krivokapic, W. Maszara, and M.-R. Lin, "Manufacturability of 20-nm ultrathin body fully depleted soi devices with fusi metal gates," *IEEE Transactions on Semiconductor Manufacturing*, vol. 18, no. 1, pp. 5–12, 2005.

[9] R. Muralidhar, R. H. Dennard, T. Ando, I. Lauer, and T. Hook, "Advanced fdsoi device design: The u-channel device for 7 nm node and beyond," *IEEE Journal of the Electron Devices Society*, vol. 6, pp. 551–556, 2018.

[10] C. Wang, J. Wu, H. Yu, G. Han, X. Miao, and X. Wang, "Effects of temperature on the performance of hf$_{0.5}$zr$_{0.5}$o$_2$-based negative capacitance fets," *IEEE Electron Device Letters*, vol. 41, no. 11, pp. 1625–1628, 2020.

[11] B. K. Esfeh, N. Planes, M. Haond, J.-P. Raskin, D. Flandre, and V. Kilchytska, "28 nm fdsoi analog and rf figures of merit at n2 cryogenic temperatures," *Solid-State Electronics*, vol. 159, pp. 77–82, 2019. doi: doi.org/10.1016/j.sse.2019.03.039.

[12] K. Jang, N. Ueyama, M. Kobayashi, and T. Hiramoto, "Experimental observation and simulation model for transient characteristics of negative-capacitance in ferroelectric hfzro 2 capacitor," *IEEE Journal of the Electron Devices Society*, vol. 6, pp. 346–353, 2018.

[13] B. Awadhiya, P. N. Kondekar, and A. D. Meshram, "Effect of ferroelectric thickness variation in undoped hfo 2-based negative-capacitance field-effect transistor," *Journal of Electronic Materials*, vol. 48, no. 10, pp. 6762–6770, 2019.

On the Breakdown Voltage Temperature Dependence of High-Voltage Power Diode Passivated with Diamond-Like Carbon

Luigi Balestra[a], Susanna Reggiani[a],
Antonio Gnudi[a], Elena Gnani[a]
[a]ARCES Research Center and DEI, University of Bologna
University of Bologna
Bologna, Italy
luigi.balestra5@unibo.it

Jagoda Dobrzyńska[b], Jan Vobecký[b,c]
[b]*Hitachi-ABB Power Grids*
CH-5600 Lenzburg, Switzerland
[c]*Microelectronics Department*
Czech Technical University in Prague
CZ-166 27 Prague 6, Czech Republic

Abstract—**Diamond-Like Carbon (DLC) is well established material for the passivation of high voltage negative beveled power diode. In our previous works, the conduction mechanism of the DLC has been carefully described through the characterization and the physical modeling of Metal-Insulator-Semiconductor (MIS) structures. In addition, the effects on the breakdown voltage and leakage current have been clarified comparing the available experiments with numerical simulations. However, the role played by the DLC on the breakdown voltage temperature dependence is still lacking. In this work, it has been investigated assuming a release of the trapped charges with increasing temperatures.**

Index Terms—**Diamond-Like Carbon, Power diode, Junction Termination, Breakdown Voltage**

I. INTRODUCTION

The ability to predict the breakdown voltage (BV) of negative beveled large-area power diodes up to the maximal allowed junction temperature is fundamental to guarantee high blocking capability. Silicon power diodes usually show a strong increase of the breakdown voltage with temperature which follows from the reduction of the impact ionization generation coefficients [1] [2]. However, the role played by the material deposited on the bevel termination is fundamental to obtain an accurate estimation of the BV. If the DLC is used as semi-insulating passivation material the BV shows a weaker temperature dependence when compared with respect to the silicon bulk material. The I-V curves of DLC passivated power diodes under reverse bias condition are reported in Figs. 2, 3 and 4. In our previous works [3] it has been demonstrated that BV is affected almost exclusively by the amount of fixed charge available in the DLC layer originating from the depletion condition experienced by the doped DLC. In amorphous materials, such as DLC, the dominant charge transport mechanism is the trap-to-trap hopping as confirmed

This work was supported by Hitachi-ABB Power Grids, Switzerland.

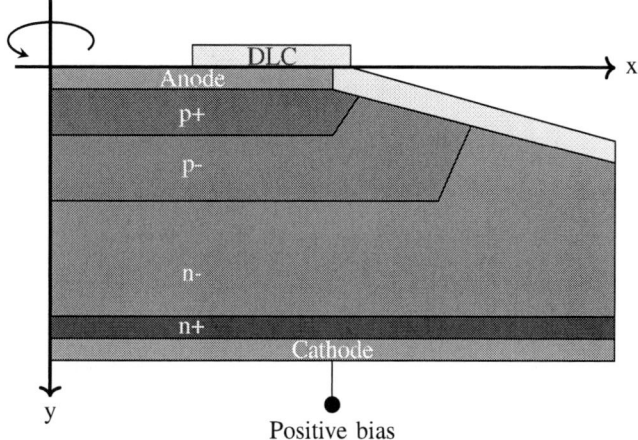

Fig. 1. Schematic representation of a power diode with negative bevel termination and DLC as passivation layer. Structure not in scale

by the Poole-Frenkel-like conductivity observed in MIS structures [3] [4]. It can be expressed as [5]:

$$\sigma \propto \exp\left\{-\frac{q}{k_B T}\left[\Phi - \sqrt{\left(\frac{qE}{\pi \varepsilon_r \varepsilon_0}\right)}\right]\right\} \tag{1}$$

With Φ the barrier height, k_B the Boltzmann constant, T the absolute temperature, q the elementary charge, E the electric field, ε_0 and ε_r the vacuum and the relative dielectric constant respectively. At high temperature trapped charges acquire enough energy to overcome the barrier and, as a consequence, the amount of fixed charge available to modify the depletion region width along the bevel is drastically reduced. This leads to a small increase of the BV with temperature. Assuming a temperature dependent doping concentration, it is possible to emulate the charge de-trapping inside the DLC passivation layer.

978-1-6654-3746-2/21 $31.00 © 2021 IEEE

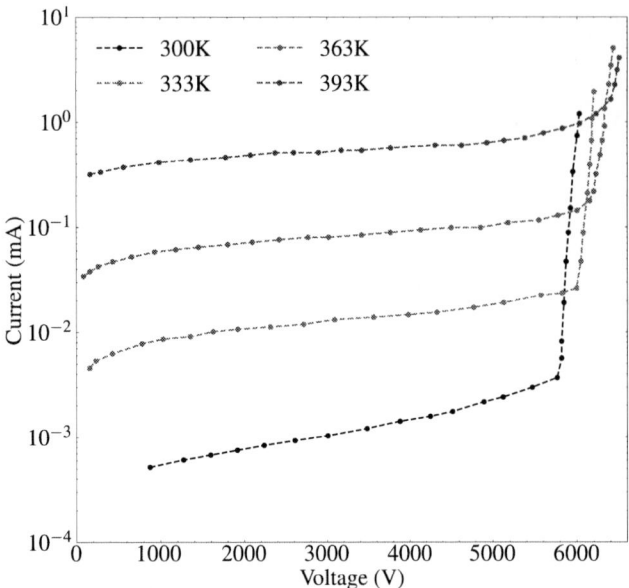

Fig. 2. IV curves of undoped DLC passivated power diode under reverse bias condition for different ambient temperature

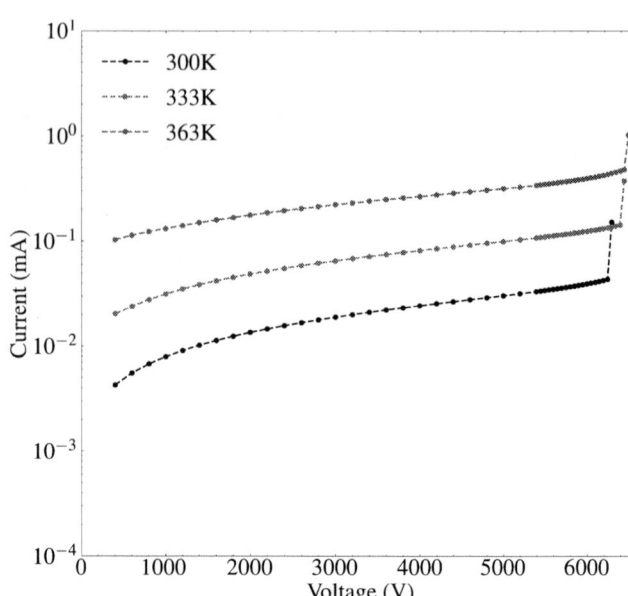

Fig. 4. IV curves of NDLC passivated power diode under reverse bias condition for different ambient temperature

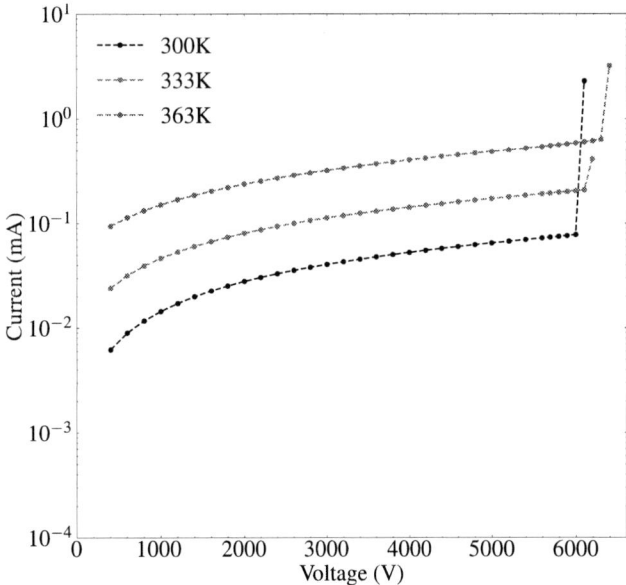

Fig. 3. IV curves of NDLC passivated power diode under reverse bias condition for different ambient temperature

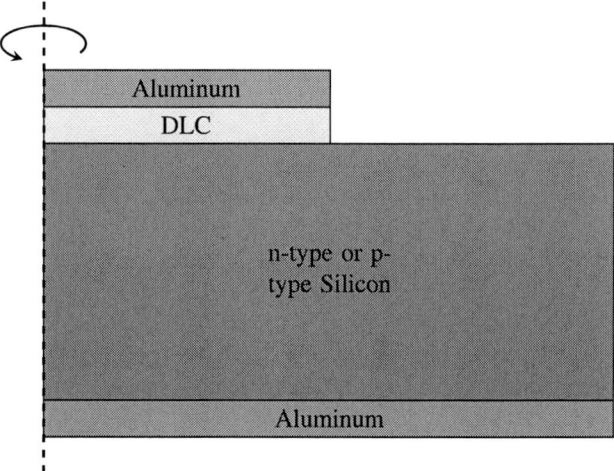

Fig. 5. Schematic view of the cross-section of a Metal-DLC-Si device. The structure is not in scale. TCAD simulations were carried out with a cylindrical symmetry in order to predict the current spreading in the Si substrate.

II. SIMULATIONS RESULTS AND COMPARISON WITH EXPERIMENTS

The device under test is a 4.5kV negatively-beveled high voltage diode with a diameter of 90mm as schematically represented in Fig.1. The DLC has been used as passivation layer directly in contact with the junction termination. The 2D radial cross-section has been simulated with the TCAD tool assuming cylindrical coordinates [7]. The DLC is modeled by using the drift-diffusion (DD) transport equation with Gaussian density of states [8], a Poole-Frenkel-like hopping mobility and the first-order Debye equation of the ferroelectric model giving the polarization effect. The silicon impact ionization generation has been accounted for by using the Van Overstraeten model with default parameters.

Metal-Insulator-Semiconductor (MIS) structures have been used to identify the main charge transpost features in the DLC layers. The schematic view of the test structures is reported in Fig. 5. The DLC is deposited on a lightly doped n-type or p-type silicon substrate. Top and bottom metal

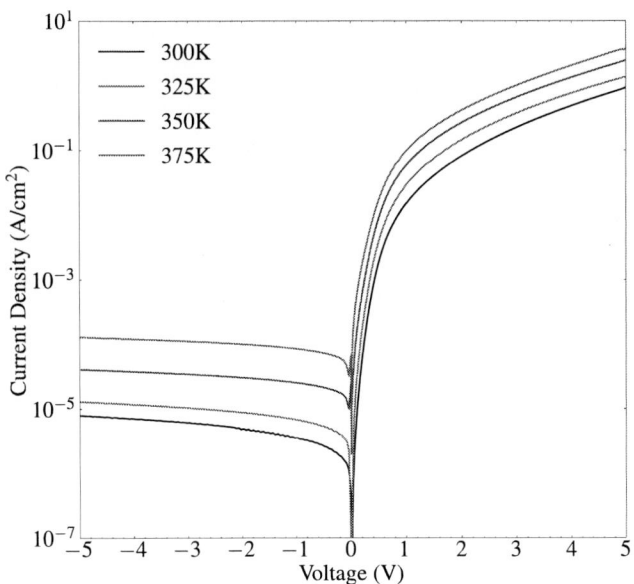

Fig. 6. Current density of MIS structure with n-type Si substrate under forward bias condition extracted at ±1.25V for different temperatures and different DLC recipes. Symbols: experiments. Dashed lines: exponential fitting

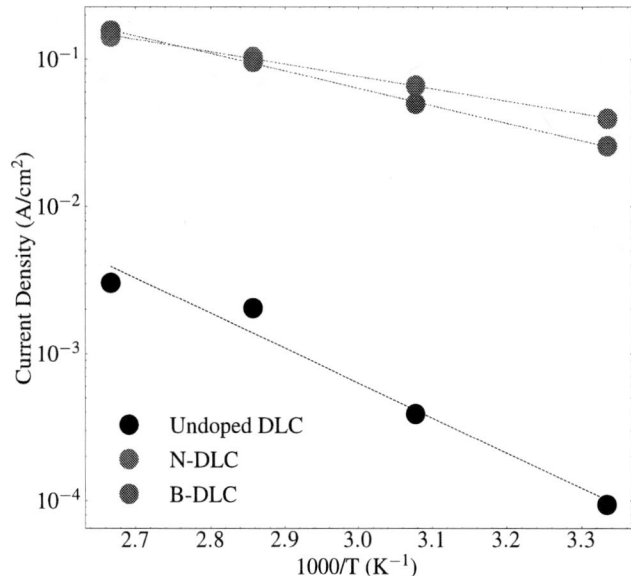

Fig. 7. Current density of MIS structure with n-type Si substrate under forward bias condition extracted at +1.25V for different temperatures and different DLC recipes. Symbols: experiments. Dashed lines: exponential fitting

contacts have been realized with Aluminum. Different recipes of the passivation layer have been tested, namely, undoped DLC, Nitrogen-doped DLC (N-DLC) and the Boron-doped DLC (B-DLC). [3] [4]. The JV curves have been obtained by measuring the current density as a function of the top electrode bias for different ambient temperatures [3]. As an example, the JV curves for the MIS structure with n-type Silicon and B-DLC are shown in Fig. 6. The forward-bias current densities extracted at ±1.25V for the n-type and p-type silicon substrates, respectively, are reported in Figs. 7 and 8. All the different DLC recipes show an Arrhenius-like temperature dependence. Such value of the applied voltage is well above the onset of the space-charge limited current but it is low enough to neglect the energy barrier lowering due to the local electric field in the Poole-Frenkel conductivity equation. The corresponding activation energy reported in Table I are in perfect agreement with those used in the simulator setup [3]. It can be clearly observed that the undoped DLC shows a higher activation energy when compared with the doped cases (N-DLC, B-DLC). This suggests the presence of shallower trap levels in the doped DLC, which according with Eq.1, results in a lower temperature dependence of the breakdown voltage since the energy barrier Φ is reduced.

In Figs. 9 and 10 the measured BV as function of the ambient temperature is reported. The undoped DLC behaves as a lightly doped p-type semiconductor [9] [10] and provides a small amount of negative trapped charge which results in a slightly increase of the BV at low temperatures when compared with simulations with ideal SiO_2. At higher temperatures, for doped DLCs, the TCAD predictions show an overestimation of BV. The experimental reduction can be explained assuming that above 300K traps tend to empty [6].

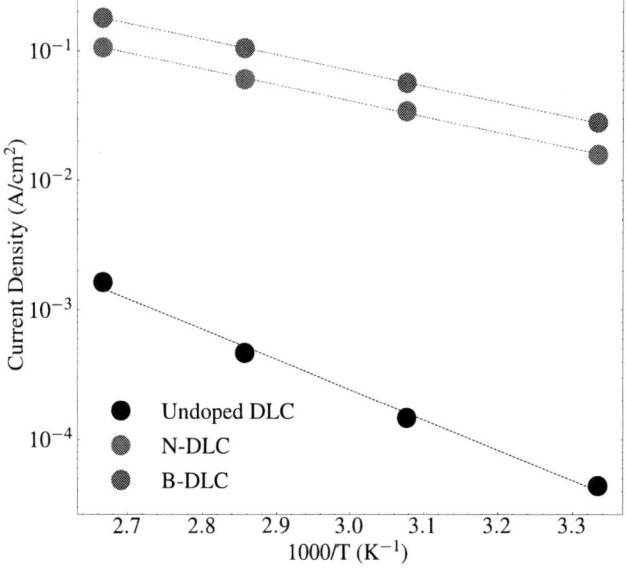

Fig. 8. Current density of MIS structure with p-type Si substrate under forward bias condition extracted at -1.25V for different temperatures and different DLC recipes. Symbols: experiments. Dashed lines: exponential fitting

In order to reproduce this feature through TCAD simulations, it has been assumed, differently to our previous model [3], that the doping concentration is a decreasing function of temperature following an exponential curve as shown in Fig. 11.

III. CONCLUSION

In this work, the breakdown voltage temperature dependence of DLC passivated power diodes has been measured

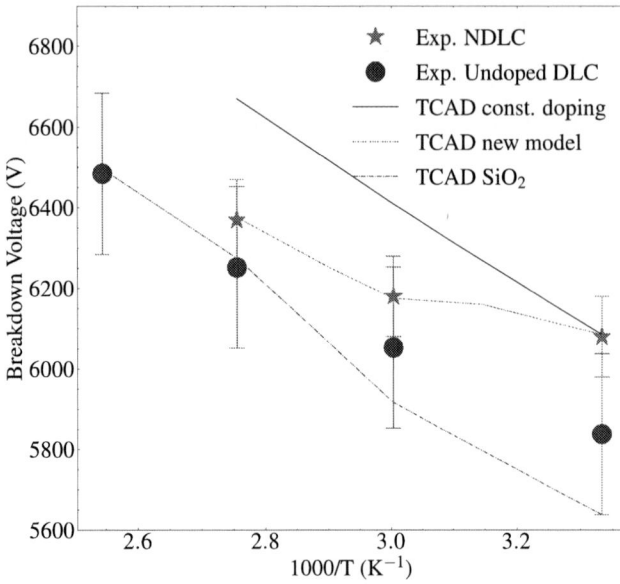

Fig. 9. Breakdown voltage of negative beveled power diode with NDLC as passivation. Error bars refer to room temperature measurements. Symbols: experiments. lines: TCAD simulations

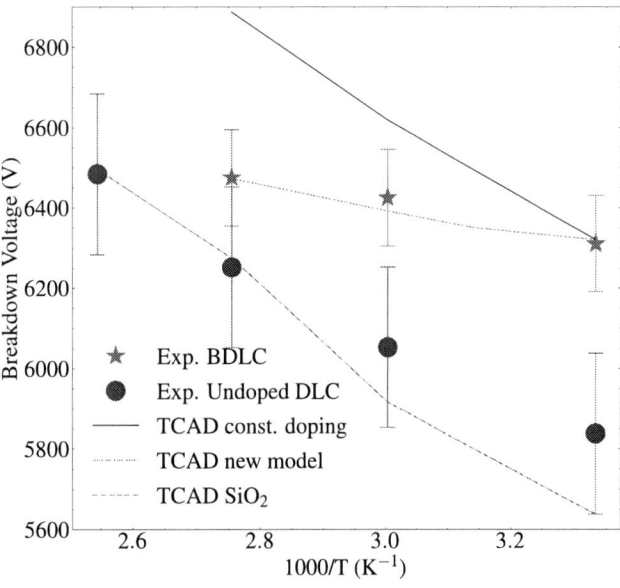

Fig. 10. Breakdown voltage of negative beveled power diode with BDLC as passivation. Error bars refer to room temperature measurements. Symbols: experiments. Lines: TCAD simulations

and compared with TCAD simulations. Following the Poole-Frenkel model which predicts a reduction of the trapped charge as the temperature increases, experiments have been explained assuming a temperature dependent equivalent doping concentration.

REFERENCES

[1] Ershov, M., and V. Ryzhii. "Temperature dependence of the electron impact ionization coefficient in silicon." Semiconductor science and technology 10.2 (1995): 138.

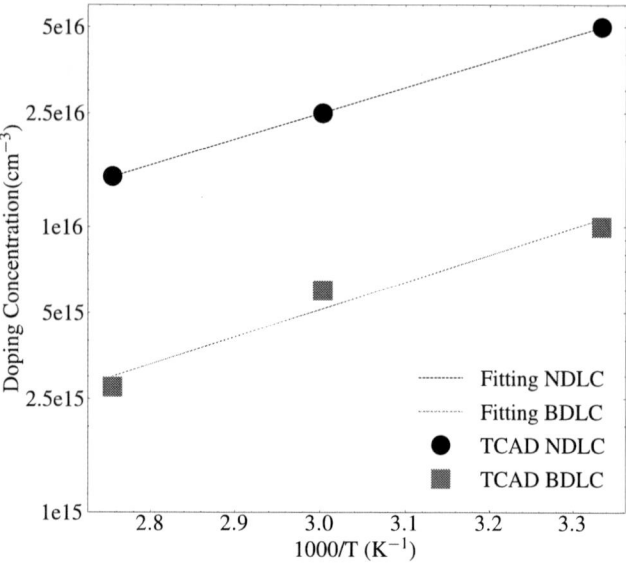

Fig. 11. DLC Doping concentration as function of temperature

TABLE I
POOLE-FRENKEL ACTIVATION ENERGY

Silicon Substrate	Undoped DLC	NDLC	BDLC
n-type	0.473	0.166	0.234
p-type	0.464	0.244	0.240

[2] Reggiani, S., et al. "Investigation about the high-temperature impact-ionization coefficient in silicon." Proceedings of the 30th European Solid-State Circuits Conference (IEEE Cat. No. 04EX850). IEEE, 2004.

[3] L. Balestra, S. Reggiani, A. Gnudi, E. Gnani, J. Dobrzynska and J. Vobecký, "Influence of the DLC Passivation Conductivity on the Performance of Silicon High-Power Diodes Over an Extended Temperature Range," in IEEE Journal of the Electron Devices Society, vol. 9, pp. 431-440, 2021, doi: 10.1109/JEDS.2021.3073232.

[4] S. Reggiani et al., "TCAD-based investigation on transport properties of Diamond-like carbon coatings for HV-ICs," 2016 IEEE International Electron Devices Meeting (IEDM), San Francisco, CA, 2016, pp. 36.7.1-36.7.4, doi: 10.1109/IEDM.2016.7838557

[5] Herbert Schroeder, "Poole-Frenkel-effect as dominating current mechanism in thin oxide films—An illusion?!", Journal of Applied Physics 117, 215103 (2015) https://doi.org/10.1063/1.4921949

[6] S. Kundoo and S. Kar, "Nitrogen and Boron Doped Diamond Like Carbon Thin Films Synthesis by Electrodeposition from Organic Liquids and Their Characterization," Advances in Materials Physics and Chemistry, Vol. 3 No. 1, 2013, pp. 25-32. doi: 10.4236/ampc.2013.31005.

[7] Synopsys Inc., Sentaurus Device User Guide M-2016.12, 2016.

[8] Paasch, Gernot, and Susanne Scheinert. "Charge carrier density of organics with Gaussian density of states: analytical approximation for the Gauss–Fermi integral." Journal of Applied Physics 107.10 (2010): 104501, https://doi.org/10.1063/1.3374475

[9] Robertson, John. "Amorphous carbon." Current Opinion in Solid State and Materials Science 1.4 (1996): 557-561.

[10] Robertson, John. "Diamond-like amorphous carbon." Materials science and engineering: R: Reports 37.4-6 (2002): 129-281.

In-depth Cryogenic Characterization of 22 nm FDSOI Technology for Quantum Computation

Hung-Chi Han*, Farzan Jazaeri*, Antonio D'Amico*, Zhixing Zhao[†],
Steffen Lehmann[†], Claudia Kretzschmar[†], Edoardo Charbon*, and Christian Enz*
*Ecole Polytechnique Fédérale de Lausanne (EPFL), Switzerland
[†]GlobalFoundries, Germany
Email: hung.han@epfl.ch

Abstract—In this paper, the influence of temperature and back-gate bias is experimentally investigated on 22 nm FDSOI CMOS process. Cryogenic DC characterization was carried out under various back-gate voltages, V_{back}, from 2.95 K back to 300 K. An abrupt drop-off in drain current due to intersubband scattering is experienced in the transfer characteristic with a certain V_{back}. Moreover, resonant and source-to-drain tunneling transports are observed in devices with minimal channel length at cryogenic temperatures. The threshold voltage, V_T, and free carrier mobility, μ_{eff}, and their dependence on the back-gate voltage over a wide range of temperatures are extracted and discussed in detail. This work aims at investigating the impact of back gate potential on V_T and carrier transport at cryogenic temperatures, further paving the way towards up-scaling of quantum computers.

Index Terms—FDSOI, Cryogenic CMOS, Quantum Computing

I. INTRODUCTION

Silicon spin qubits have recently gained attention towards the up-scaling of quantum computers. They have remarkably proven their strength to achieve the so-called quantum integrated circuit by leveraging the well-established Complementary Metal-Oxide-Semiconductor (CMOS) technology [1, 2]. This is where the Fully-Depleted Silicon-On-Insulator (FDSOI) CMOS technology with electrostatically confined quantum dots holds the promise to reach the full monolithic integrated quantum processor [3]. Nevertheless, the lack of a cryogenic physics-based compact model of FDSOI is still a challenge to enable efficient circuit design at deep cryogenic temperatures. In light of recent efforts regarding the cryogenic DC characterization of a FDSOI technology [4]–[6]; the first study of such technology down to 4.3 K was reported in [4], further down to 20 mK in [5], and a design-oriented model was proposed in [6]. Moreover, [3, 7] revealed the carrier transport in an ultra-thin channel with respect to the intersubband scattering in a long device, and the resonant tunneling in a short channel, respectively. However, the effect of sweeping the back-gate voltage over a wide voltage range in short devices has not been reported yet. Additionally, the impact of the back-gate voltage on the free carrier transport at deep cryogenic temperatures is not well physically explained. Therefore, this paper aims at addressing the less-studied effects

This work was supported in part by the EU H2020 RIA project SEQUENCE under Grant No. 871764.

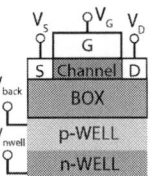

Fig. 1: The 2-D cross-section of 22 nm FDSOI technology.

TABLE I: Measured devices (22 nm FDSOI process [8]).

Type	W/L
nMOS	wide / long
pMOS	wide / long
nMOS	wide / short
pMOS	wide / short

of the back-gate voltage on the DC performance for transistors of an up-to-date 22 nm FDSOI technology at deep cryogenic temperatures.

II. CRYOGENIC DC MEASUREMENTS

Table. I lists the measured devices with different types and geometries (nMOS/pMOS with long/short channels), fabricated on a 22 nm FDSOI process [8]. We assume the short channel effects are negligible in the long devices. On the contrary, short devices have the minimum length of such technology. Devices were probed on a LakeShore cryogenic probe station, CRX-4K, and characterized by the Keysight B1500A from ultra-low temperature, i.e., 2.95 K, back to room temperature, i.e., 300 K. The intermediate temperatures were taken at 20, 36, 77, 150, 210 K. A two-dimensional cross-section of a device is illustrated in Fig. 1. The p-WELL bias is represented by V_{back}. The p-WELL and n-WELL are reverse biased, and the device could operate either in forward back bias (FBB) or in reverse back bias (RBB). The FBB stands for the negative V_{back} for pMOS and positive V_{back} for nMOS, and vice versa for RBB. Hence, it allows a freedom to modulate V_T.

The transfer characteristics of devices in a linear mode, $|V_{DS}| = 10$ mV, and $V_{back} = 0$ V are plotted in Fig. 2 over a wide range of temperatures. Nevertheless, the room-temperature data of long-channel nMOS (Fig. 2(a)) and short-channel pMOS (Fig. 2(d)) are missing due to electrostatic discharge during the abrupt warming up process. Nevertheless, Fig. 2 demonstrates the typical cryogenic behavior of CMOS technologies, i.e., subthreshold swing reduction, increasing V_T, and μ_{eff} enhancement, which provides a promising solution for low-power and high-performance modern electronics for quantum computations.

978-1-6654-3746-2/21 $31.00 © 2021 IEEE

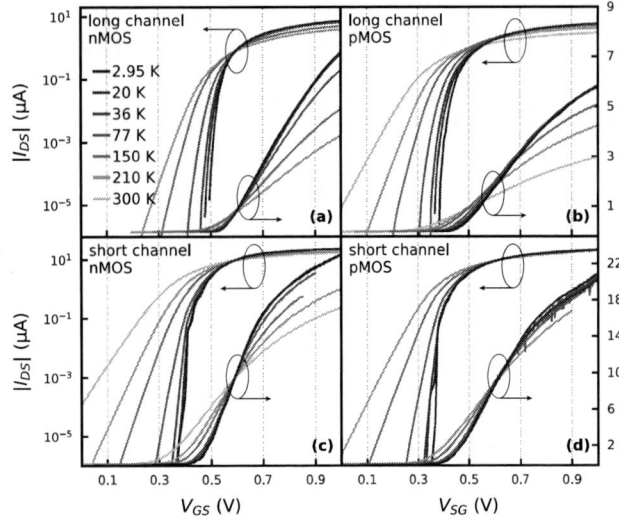

Fig. 2: Transfer characteristics of devices measured from 300 K and down to 2.95 K with $|V_{DS}| = 10\,\text{mV}$ and $V_{back} = 0\,\text{V}$, (a,b) correspond to the long-channel nMOS and pMOS, (c,d) are with respect to short-channel nMOS and pMOS.

The output characteristics in strong inversion ($|V_{GS}| = 0.9\,\text{V}$ for long devices and $|V_{GS}| = 0.7\,\text{V}$ for short devices) and $V_{back} = 0\,\text{V}$ are plotted in Fig. 3 over a wide range of temperatures. No kink effect is observed in this voltage range. The μ_{eff} was extracted from the linear and strong inversion regimes using g_{DS} function proposed by Jazaeri et al. [6, 9, 10]. This approach relies on the drift-diffusion transport model without any pre-assumption on the gate-voltage-dependent mobility. The results are detailed in Sec. IV.

III. RESULT AND DISCUSSION

A. Intersubband scattering

As demonstrated in Fig. 4(a,b), I_D-V_G of long-channel devices at 2.95 K is modulated by V_{back}, which ranges from -4 to 4 V with an interval of 2 V. The V_T is increased by RBB and decreased by FBB. It is worth noting that a sudden drop-off in drain current is experienced for long-channel nMOS with $V_{back} = 4\,\text{V}$ at around $400\,\text{mV}$ above V_T, which is 0.22 V (see Fig. 4(a)), leading to a negative $\delta I_{DS}/\delta V_{GS}$. The phenomenon is attributed to the intersubband scattering for two-dimensional charge gas density [7], where the scattering rate follows such step-like density of states [11]. Hence, the drop of the current is due to the onset of the higher-order subbands. The relaxation time [11],

$$\frac{1}{\tau} = \sum_{p'} S(p, p') \left[1 - f(p') \right] \approx \sum_{p'} S(p, p'), \qquad (1)$$

is introduced to further explain why the intersubband scattering appears for this device at cryogenic temperatures and for some values of the back-gate voltage. The term of S in Eq. (1) is the scattering rate from the initial momentum p to the final

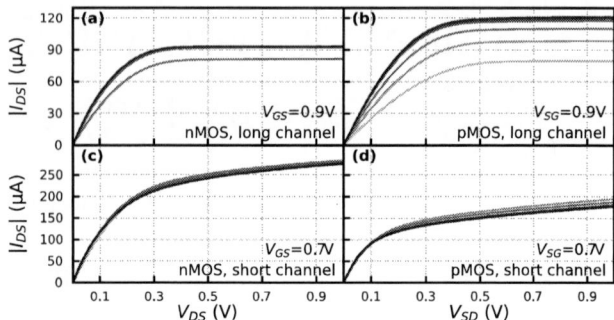

Fig. 3: Output characteristics of devices measured from 2.95 K back to 300 K with $V_{back} = 0\,\text{V}$ (the legend follows the one in Fig. 2).

momentum p' (Note that $f(p')$ is the occupancy rate of state at p' described by the Fermi-Dirac distribution, which is often equal to zero due to cryogenic temperatures). The scattering rate could be elaborated by *Fermi's Golden Rule*, the relaxation time, only considering the intersubband scattering between the first two subbands at cryogenic temperatures, is expressed by [11]

$$\frac{1}{\tau} = \sum_{p'} \frac{2\pi}{\hbar} |H_{p',p}|^2 \delta(E_2(p') - E_1(p) - \Delta E) \qquad (2)$$

with reduced Plank constant $\hbar$, matrix element $|H_{p',p}|$, and δ-function for energy conservation, in which, E_1 and E_2 stand for the eigenenergy of 1st and 2nd subbands, respectively, and ΔE is the energy change due to the scattering event. When $k_B T$ (Boltzmann constant k_B and temperature T) is large enough to cross numerous subbands, the abrupt drop-off in free carrier mobility is smoothed out. Additionally, subbands are split out in the strong vertical field due to potential confinement, where the δ-function in Eq. (2) is difficult to be satisfied at cryogenic temperatures because larger ΔE is required. Thus, the significant impact of intersubband scattering is only experienced for FBB at cryogenic temperature, but not for RBB conditions. Whereas, intersubband scattering is not experienced for pMOS device in the measured range since the smaller effective mass for holes leads to a wider gap between subbands. Moreover, intersubband scattering experienced in pMOS devices has not been reported yet. To observe the intersubband scattering for pMOS devices, a stronger FBB should be required.

B. Free Carrier Tunneling Transports

Fig. 4(c-f) emphasize the V_{back}-dependence of the measured subthreshold I_D-V_G at 2.95 K in linear and saturation regimes. The long-channel devices show the consistency in subthreshold regime between low and high horizontal fields (see Fig. 4(c,d)). However, the oscillatory drain current is pronounced for short-channel devices, especially in linear operation mode (see dashed lines in Fig. 4(e,f)). This is due to the resonant tunneling current, where free carriers tunnel the barrier via discrete energy levels of ionized quantum dots, and current flows in parallel to the drift-diffusion component (or even

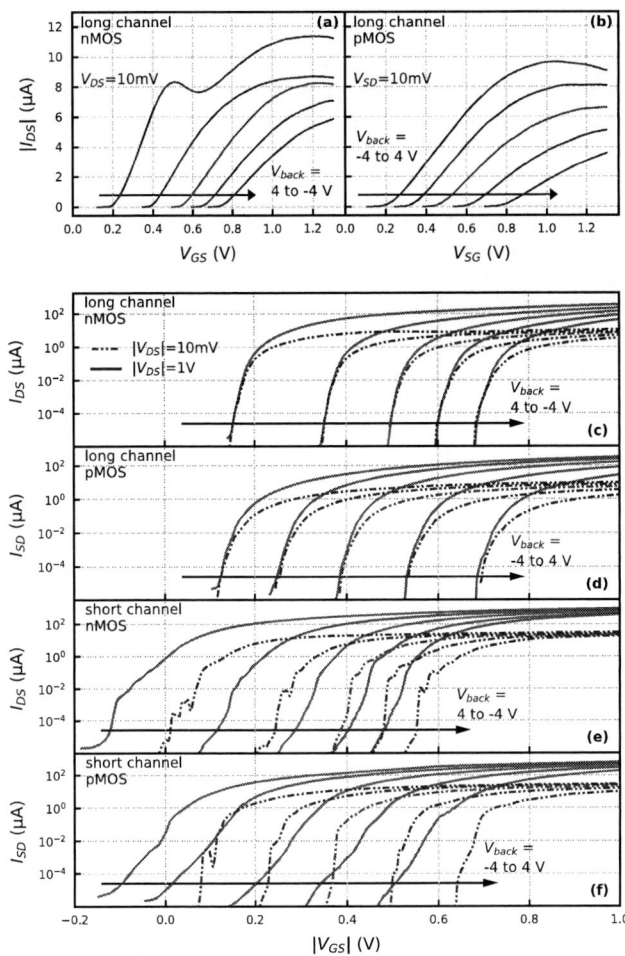

Fig. 4: Transfer characteristics of long/short nMOS/pMOS with V_{back} sweeping from -4 to 4 V at 2.95 K. The I_D-V_G curves with $V_{back} = 0$ V are highlighted by lighter blue. (a, b) I_D-V_G in linear scale for long-channel nMOS and pMOS devices, (c-f) logarithmic-scale I_D-V_G measured with $|V_{DS}| = 10$ mV (broken lines) and 1 V (solid lines) for long/short nMOS/pMOS, respectively.

the ballistic transport) [12]. Moreover, the resonant tunneling manifests differently with respect to V_{back}. It implies that the eigenvalue of a quantum dot is influenced by V_{back} as well. As evidenced by the solid lines in Fig. 4(e,f), the oscillatory behavior in I_D-V_G becomes less dominant in the saturation regime, which is due to the drain-induced barrier lowering (DIBL) and source-to-drain tunneling. The prior allows more free carriers following the drift-diffusion transport, and the latter is an additional current component that degrades the subthreshold performance owing to the drain-induced barrier thinning [13].

C. Summary of Oscillatory Drain Current

Two free carrier transport mechanisms, i.e., intersubband scattering and resonant tunneling, have been discussed previously. Although they both manifest as an oscillatory drain current in the transfer characteristic, they are due to different

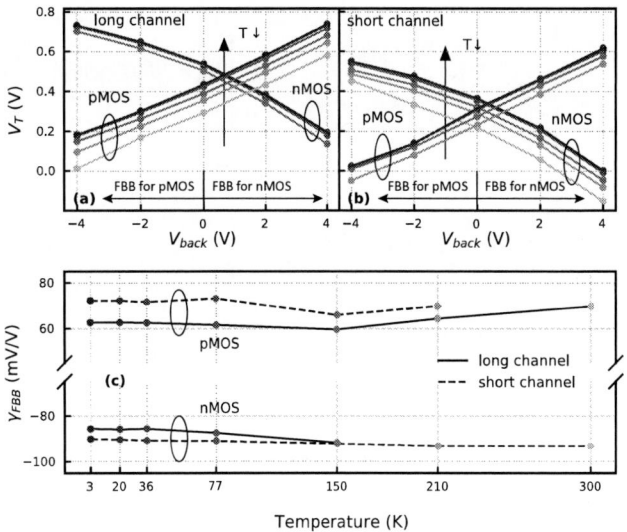

Fig. 5: (a,b) V_{back}-dependent threshold voltage at different temperatures, which is extracted from I_D-V_G with $|V_{DS}| = 1$ V (the legend of curves follows the one in Fig. 2), (c) back-body coefficient of FBB (γ_{FBB}) versus temperature.

	Intersubband scattering	Resonant tunneling
at cryogenic temperature	Yes	Yes
negative $\delta I_D / \delta V_G$	Yes	Yes
in subthreshold regime	No	Yes
above threshold	Yes	No
requires quantum confinement	Yes	No
experiences only in short channel	No	Yes

TABLE II: Summary of the properties of intersubband scattering and resonant tunneling, which experience in this study.

reasons. Hence, Table. II compares these two mechanisms in different conditions in terms of channel length, inversion status, device configuration, and etc.

IV. BACK-GATE EFFECTS ON V_T AND μ_{eff}

The extracted V_T from I_D-V_G in saturation regime is plotted versus V_{back} in Fig. 5(a,b) for different temperatures. With the fixed V_{back} at 0 V, the increase in V_T at 2.95 K with respect to that at room temperature is around 150 mV for long-channel pMOS and short-channel nMOS, which is due to the shift of the quasi-Fermi level [14]. Although the comprehensive threshold voltage model for FDSOI technology, including the back-gate and low-temperature influence, has not been established yet, the consistency of the correlation between V_T and V_{back} at various temperatures implies that the front-to-back gate coupling does not lead to significant temperature dependence. Therefore, taking advantage of back-gate configuration, the increase in V_T could be compensated by FBB. Fig. 5(c) further elaborates the back-body coefficient for FBB, γ_{FBB}, with respect to different temperatures, describing how much V_T is modulated by 1 V of V_{back} while in FBB. The result suggests that room-temperature γ_{FBB} could sufficiently estimate the compensation for V_T at cryogenic temperatures.

978-1-6654-3746-2/21 $31.00 © 2021 IEEE

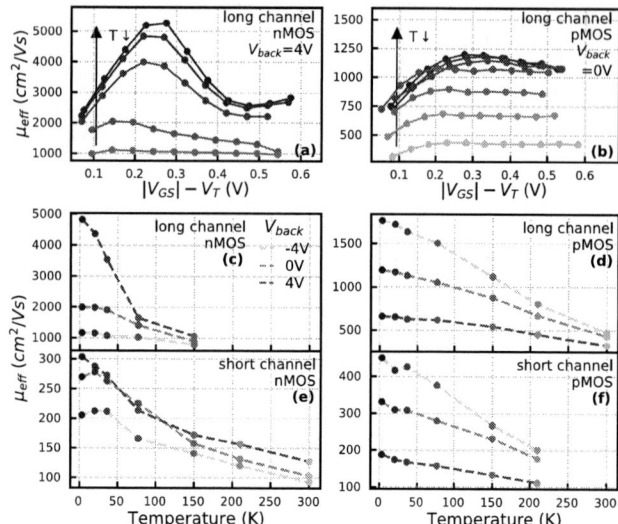

Fig. 6: Analysis of low field μ_{eff} with respect to various voltage conditions and temperatures, where μ_{eff} is extracted from I_D-V_D based on the method proposed in [9]. Low field μ_{eff} versus overdrive voltage at various temperatures for (a) long-channel nMOS at $V_{back} = 4\,\mathrm{V}$ and (b) long-channel pMOS at $V_{back} = 0\,\mathrm{V}$, (c-f) low field μ_{eff} as a function of temperature at $|V_{GS}| - V_T \approx 0.3\,\mathrm{V}$ with respect to various V_{back} (the legend of (a,b) follows to the one in Fig. 2).

The μ_{eff} is plotted in Fig. 6(a,b) versus the overdrive voltage, V_{ov}, for long-channel devices, where nMOS and pMOS are biased with $V_{back} = 4\,\mathrm{V}$ and $0\,\mathrm{V}$, respectively. At the same V_{ov}, it is clear that the μ_{eff} at cryogenic temperatures is enhanced in comparison to that at room temperature due to the reduced phonon scattering. The μ_{eff} trends in Fig. 6(b) show the universal mobility behavior [15]. For μ_{eff} at cryogenic temperatures, i.e., $T < 77\,\mathrm{K}$, decreases along the V_{ov} suggests that the surface roughness scattering has a relatively large impact on μ_{eff} due to less-dominant phonon scattering. Whereas, in Fig. 6(a), μ_{eff} at $T < 77\,\mathrm{K}$ dramatically drops off at $V_{ov} \approx 0.4\,\mathrm{V}$. It further verifies the substantial influence of intersubband scattering by I_D-V_D characterization, which has been already seen in Fig. 4(a).

Nevertheless, Fig. 6(c-f) demonstrate μ_{eff} as the function of temperature with respect to various V_{back}, while at $V_{ov} \approx 0.3\,\mathrm{V}$. The μ_{eff} overall shows the slight difference between each V_{back} at room temperature but diverges at cryogenic temperatures. It can be inferred from the vertical position of the mobile charge. In comparison to the back-gate inversion, i.e., strong FBB, mobility enhancement of the front-gate inversion is much limited by the remote Coulomb scattering from ionized charges in gate due to the ultra-thin gate dielectric [16]. Besides, it should be noted that mobility is enhanced less efficiently by the temperature and back gate in short-channel devices because of the neutral defects [17].

V. Conclusion

An experimental study is presented for a 22 nm FDSOI technology from 2.95 K back to 300 K for different device types and channel lengths. The transfer and output characteristics were accurately measured to investigate the influence of V_{back} on the DC performance in terms of I_D-V_G, V_T, and μ_{eff}. Quantum transports are pronounced in short-channel devices at cryogenic temperatures, i.e., resonant tunneling and source-to-drain tunneling. Additionally, the intersubband scattering significantly impacts the behavior of long-channel nMOS in strong FBB at cryogenic temperatures. The V_T versus a wide range of V_{back} is presented for various temperatures, which further suggests that the γ_{FBB} does not show the notable dependence on temperature. Furthermore, μ_{eff} with respect to temperature for different V_{back} is presented and reveals that μ_{eff} of front-gate inversion is enhanced less efficiently by temperature due to the remote Coulomb scattering.

References

[1] R. Maurand, X. Jehl et al., "A CMOS silicon spin qubit," Nat. Commun., vol. 7, no. 1, pp. 1–6, Nov. 2016.

[2] F. Jazaeri, A. Beckers et al., "A review on quantum computing: From qubits to front-end electronics and cryogenic mosfet physics," in 2019 MIXDES - 26th International Conference "Mixed Design of Integrated Circuits and Systems", 2019, pp. 15–25.

[3] S. Bonen, U. Alakusu et al., "Cryogenic characterization of 22-nm FDSOI CMOS technology for quantum computing ICs," IEEE EDS, vol. 40, no. 1, pp. 127–130, Jan. 2019.

[4] H. Bohuslavskyi, S. Barraud et al., "28nm Fully-depleted SOI technology: Cryogenic control electronics for quantum computing," in 2017 Silicon Nanoelectronics Workshop (SNW), 2017, pp. 143–144.

[5] P. Galy, J. Camirand Lemyre et al., "Cryogenic Temperature Characterization of a 28-nm FD-SOI Dedicated Structure for Advanced CMOS and Quantum Technologies Co-Integration," IEEE Journal of the Electron Devices Society, vol. 6, pp. 594–600, 2018.

[6] A. Beckers, F. Jazaeri et al., "Characterization and modeling of 28-nm FDSOI CMOS technology down to cryogenic temperatures," Solid. State. Electron., vol. 159, no. 688539, pp. 106–115, 2019.

[7] M. Cassé, B. Cardoso Paz et al., "Evidence of 2D intersubband scattering in thin film fully depleted silicon-on-insulator transistors operating at 4.2 K," APL, vol. 116, no. 24, 2020.

[8] R. Carter, J. Mazurier et al., "22nm FDSOI technology for emerging mobile, Internet-of-Things, and RF applications," in IEDM, Dec 2016, pp. 2.2.1–2.2.4.

[9] F. Jazaeri, A. Pezzotta, and C. Enz, "Free Carrier Mobility Extraction in FETs," IEEE TED, vol. 64, no. 12, pp. 5279–5283, Dec 2017.

[10] F. Jazaeri and J.-M. Sallese, Modeling Nanowire and Double-Gate Junctionless Field-Effect Transistors. Cambridge University Press, 2018.

[11] M. Lundstrom, Fundamentals of Carrier Transport. Cambridge University Press, 2010.

[12] R. Wacquez, M. Vinet et al., "Single dopant impact on electrical characteristics of soi NMOSFETs with effective length down to 10nm," in VLSIT, Jun. 2010, pp. 193–194.

[13] J. Wang and M. Lundstrom, "Does source-to-drain tunneling limit the ultimate scaling of MOSFETs?" in IEDM, Dec. 2002, pp. 707–710.

[14] A. Beckers, F. Jazaeri et al., "Physical model of low-temperature to cryogenic threshold voltage in mosfets," IEEE Journal of the Electron Devices Society, vol. 8, pp. 780–788, 2020.

[15] S. Takagi, A. Toriumi et al., "On the universality of inversion layer mobility in si mosfet's: Part i-effects of substrate impurity concentration," IEEE Transactions on Electron Devices, vol. 41, no. 12, pp. 2357–2362, 1994.

[16] J. Koga, T. Ishihara, and S. Takagi, "Effect of gate impurity concentration on inversion-layer mobility in mosfets with ultrathin gate oxide layer," IEEE Electron Device Letters, vol. 24, no. 5, pp. 354–356, 2003.

[17] M. Shin, M. Shi et al., "Low temperature characterization of mobility in 14nm fd-soi cmos devices under interface coupling conditions," Solid-State Electronics, vol. 108, pp. 30–35, 2015, selected papers from the 15th Ultimate Integration on Silicon (ULIS) conference.

978-1-6654-3746-2/21 $31.00 © 2021 IEEE

Impact of different types of planar defects on current transport in Indium Phosphide (InP)

Christian Dam Vedel[1,2,*], Enrico Brugnolotto[1,3], Søren Smidstrup[2], Vihar P. Georgiev[1]

Device Modelling Group, School of Engineering, University of Glasgow, Glasgow G12 8QQ, United Kingdom[1]

Synopsys Denmark ApS, Fruebjergvej 3, 2100 Copenhagen, Denmark[2]

IBM Research Europe-Zurich, 8803 Rüschlikon, Switzerland[3]

Correspondence author: christianvedel@hotmail.com[*]

Abstract—**In this paper we show first-principles simulation results of the three most commonly occurring types of planar defects in Indium Phosphide (InP), which are Rotational Twin Planes (RTPs) and two types of Stacking Faults (SFs). We have found that only the two less common of these defects, the extrinsic and intrinsic SFs, have an impact on the current flow in the semiconductor. These two types of defects cause an increase in the resistivity of the semiconductor and a remarkable decrease in currents for low voltages. The most commonly occurring defect type, RTPs, were revealed to have little to no effect on the electrical properties of the semiconductor.**

Index Terms—**Density Function Theory (DFT), First-principles simulation, Defects, III-V semiconductor, Indium Phosphide (InP)**

I. Introduction

III-V semiconductors, such as Indium Phosphide (InP) and its alloys, are widely researched due to their unique electrical and optical properties, such as high carrier mobilities [1], [2], direct and tunable band-gaps [3], [4] and low exciton binding energies [5], [6]. These properties can be exploited to improve on existing devices such as MOSFETs [7] and PIN photodiodes [8] or to pioneer novel new devices such as lasing microdisks [9] and topological photonics [10].

When growing III-V semiconductors with conventional growth methodologies, a large number of planar defects such as Stacking Faults (SFs) and Rotational Twin-Planes (RTPs) occur. Growth processes which mitigate and control the formation of these defects have been developed [11]–[14], as it has been experimentally validated that these defects degrade device performance. Specifically it was shown that RTPs reduce mobility, carrier lifetime and quantum efficiency in III-V nanowires [15]–[17]. Theoretical investigations have shown that twin-planes acts as scattering centres in Silicon [18], but also that in low densities, they have no effect on the current in InP nanowires [19]. The work reported here is an attempt to shine some more light on this complicated issue, using the state-of-the-art simulation software Synopsys QuantumATK [20]. Our aim is to calculate the resistance induced by the defects, as well as their impact on the current-voltage (I-V) characteristics of bulk InP for device usages.

This project has received funding from the European Union's Horizon 2020 research and innovation program under grant agreement No 860095 MSCA-ITN-EID DESIGN-EID.

II. Simulation Methodology

All calculations in this paper were performed using the atomistic first-principles simulation methodology called Density Functional Theory (DFT), wherein the electron density is treated as the fundamental variable [21]–[23]. Atoms are treated explicitly through basis sets consisting of Linear Combination of Atomic Orbitals (LCAO) and their corresponding pseudopotential. In this work the "High accuracy" version of the PseudoDojo set [24], as implemented in QuantumATK, were used. Exchange and correlation effects were approximated at the level of a Generalized Gradient Approximation (GGA), especially made for solids by Perdew, Burke and Emzerhof (PBES) [25]. To simulate current transport, DFT was combined with Non-Equilibrium Green Functions (NEGF), to accurately describe electrodes at finite bias [26]. The Brillouin zone were sampled with a k-point density of $300\,\text{Å}$ in the transport direction and $8\,\text{Å}$ in the transverse directions. The real space density mesh cutoff used were $85\,\text{Ha}$.

To construct the systems of interest, first bulk InP in the Zinc-Blende (ZB) phase were relaxed, until the forces between the atoms were no larger than $0.05\,\text{eV/Å}$. The resulting lattice constant of $5.890\,\text{Å}$ agrees well with the experimental value of $5.869\,\text{Å}$. The crystal were then cleaved along the $[1\,1\,1]$-direction, in which the crystal has an ABC stacking sequence of polarised layers, each layer having both a plane of Indium and Phosphor atoms. For the semi-infinite electrodes, three layers were used in order to repeat the stacking sequence. The finite central region were built with 38 layers, corresponding to $149\,\text{Å}$, to allow for any induced potential to be screened in the electrodes. The simple unit cell were $4.165\,\text{Å}$ wide in the transverse directions, and periodic boundary conditions were imposed to reproduce an infinite bulk. An n-type intrinsic background doping, corresponding to $1 \times 10^{17}\,\text{cm}^{-3}$, were added as compensation charges to all atoms in the system.

Six systems were inspected in total, a pristine InP system, as reference, and five defect systems with the defects located in the middle of the central regions. Both types of SF defects were considered, the extrinsic SF, where a layer is missing in the stacking sequence (ABC_BCA), and the intrinsic SF, where a layer is added (ABCBABC). Three RTP defect systems were also considered, one with a single RTP, one with two RTPs seperated by 10 layers and finally two RTPs back-to-back.

In literature the two terms RTPs or simply "twins" and SFs are often used interchangeably and referring to different kind of defects, sometimes even in different crystallographic phases. This necessitates a brief explanation of the terms used in the current work. In this work, a SF is when a layer is either removed or added to the [1 1 1] ZB stacking sequence, but the sequence is otherwise uninterrupted and the added layer is identical to other layers in the crystal, (see Fig. 1b & 1c). RTPs on the other hand is a type of twin-plane wherein the crystal is rotated 60° around the [1 1 1]-direction at the plane. The resulting crystal appears as a mirror-image with the atomic species swapped, (see Fig. 1a).

In a crystal where one species is much larger than the other, such as InP, two RTPs back-to-back is indistinguishable from an intrinsic SF, which explains the confusion of the terms in the literature. The stacking sequence of a ZB crystal in any ⟨1 1 1⟩-direction can be investigated by HR-STEM images taken from the ⟨1 1 0⟩-directions. Two such examples can be seen in Fig. 2, where in 2a the defects are clearly RTPs, since the stacking changes direction after the defect, but in 2b it is not clear whether the defect is an intrinsic SF or two RTPs.

III. RESULTS AND DISCUSSION

To investigate the defects' perturbation of the system, the Hartree difference potential was projected onto the transport direction of the systems. The Hartree difference potential is the electrostatic potential of the electron density, calculated from the Poisson equation, except that the compensation charge density, from the added doping, is subtracted. There was a notable difference between the SFs and the RTPs potential, but only small variations between the different RTP systems, thus only one of each type is shown in Fig. 3. The rapid oscillations in the potential is the inter-atomic potential between subsequent atomic layers. To better visualise the perturbation of the potential, the envelope were also plotted. The perturbation from a single RTP, seen in Fig. 3a, is insubstantial compared to the periodic inter-atomic potential. In contrast the potential barrier caused by the extrinsic SF, seen in Fig. 3b, is large enough to severely hinder carrier transport.

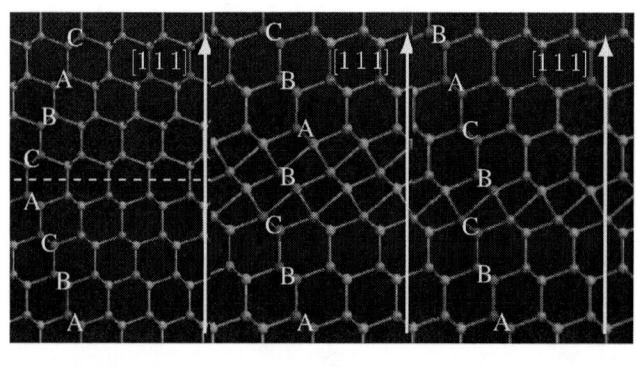

(a) A single RTP   (b) Extrinsic SF   (c) Intrinsic SF

Fig. 1. Three types of planar defects in a ZB InP crystal.

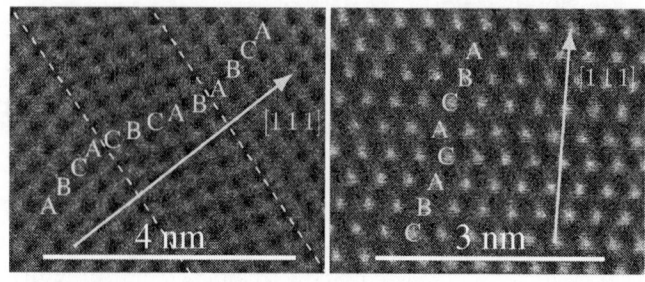

(a) Two RTPs separated by three layers   (b) An intrinsic SF or two RTPs

Fig. 2. HR-STEM images of InP in the zinc-blende phase, seen from the ⟨1 1 0⟩-direction, which allows vision of the stacking in the ⟨1 1 1⟩-direction.

To quantify how much the potential barriers, caused by the defects, hinder current transport, a contact resistance was calculated.

$$R_{contact} = R_{cent} - R_{elec} = \frac{1}{G_{cent}} - \frac{1}{G_{elec}} \quad (1)$$

Where in (1) $R_{cent(elec)}$ and $G_{cent(elec)}$ is the resistance and conductance of the central (electrode) region respectively. The conductance is calculated in the usual fashion from the transmission spectrum:

$$G_{cent} = \int T_{cent}(E) \left(-\frac{\partial f(E)}{\partial E} \right) dE \quad (2)$$

where T_{cent} is the transmission spectrum and $f(E)$ is the Fermi-Dirac distribution. The contact resistance (1) was evaluated for all systems for various Fermi level shifts (i.e. doping

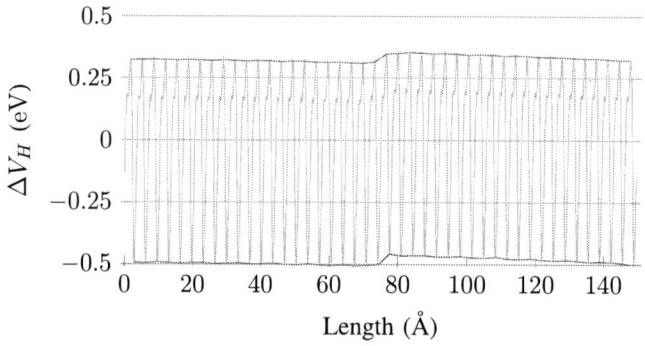

(a) A single twin-plane

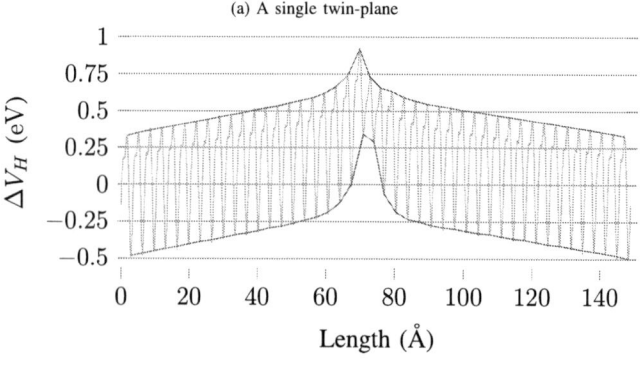

(b) Extrinsic SF

Fig. 3. Hartree difference potential, ΔV_H, along the transport direction. The potentials' envelopes are highlighted in solid red.

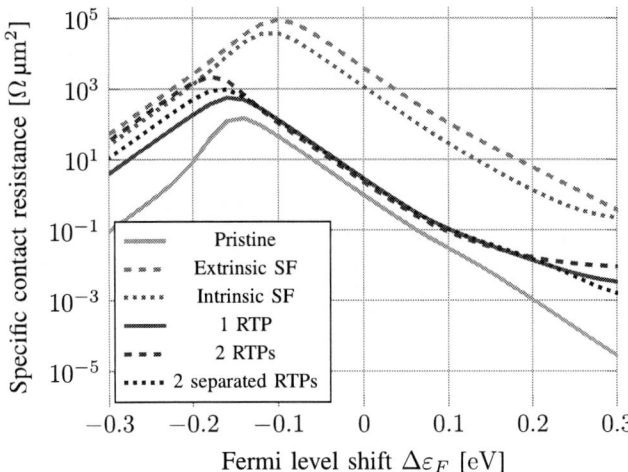

Fig. 4. Specific contact resistance of the five defects and a pristine reference system as a function of the Fermi level shift. $\Delta\varepsilon_F = 0\,\text{eV}$ corresponds to the n-type intrinsic background doping of $1 \times 10^{17}\,\text{cm}^{-3}$ and $\Delta\varepsilon_F = 0.17(-0.24)\,\text{eV}$ corresponds to a n(p)-type doping of $6 \times 10^{19}\,\text{cm}^{-3}$.

levels), the results are shown in Fig. 4. To interpret the shown results, it is worth mentioning that a high p-type doping of $6 \times 10^{19}\,\text{cm}^{-3}$ corresponds to a Fermi shift of $-0.24\,\text{eV}$ whereas a similar high n-type doping corresponds to a Fermi shift of $0.17\,\text{eV}$. We find an increase in resistance across all defect systems as expected. Due to the large potential barrier from the SFs, the specific contact resistance is increased by four orders of magnitude for these systems, for intermediate doping levels, with the extrinsic SF having a four times higher resistance compared to the intrinsic SF. For the RTP systems on the other hand, the resistance is only increased three times compared to the pristine system for intermediate doping levels. Interestingly there is no further increase in the resistance from the additional RTP at these doping levels. At high doping levels the resistance increase from the RTPs rises, and especially for p-type doping we start to see a large effect in comparison to the pristine system. Notably at high p-type doping there is also a difference between the RTP systems, with the RTP system with two RTPs back-to-back having the larger increase in resistance.

To further investigate how much the defects perturb the current flow, I-V curves for the systems were calculated for a bias-range of $0\,\text{V}$ to $1.3\,\text{V}$, these can be seen in Fig. 6. Despite the triple resistance of the systems with RTPs as compared to the pristine InP, we do not see any noticeable difference in the current of these systems. All these systems shows app. constant current values until the breakdown voltage at app. $0.6\,\text{V}$, after which they increase linearly on a log scale. The breakdown voltage at $0.6\,\text{V}$ corresponds to the band gap achieved with the used simulation parameters.

It is well known that DFT is notorious for underestimating band gaps, but otherwise yields good results in agreement with experiments. While there does exist methods to achieve a better band gap, these methods are more computationally expensive and does not improve on any achieved results that are not directly related to the band gap. As this work is an investigation of electrical properties, these methods were not used and the value of the breakdown voltage should not be considered seriously. The current in the SF systems are strongly reduced, by several orders of magnitude even, for low voltages. As the voltage is increased, the potential barriers are slowly lowered compared to the conduction states, and the current approaches that of the pristine and RTP systems. The barrier still blocks out a few conduction states and as such the current does not actually reach the pristine and RTP systems. The intrinsic SFs current approaches the pristine system quicker than the extrinsic SF, which fits with it having a lower barrier to be overcome. Interestingly once it has been overcome, the intrinsic SFs barrier actually blocks out more conduction states than the extrinsic SF. Unfortunately despite great efforts and much time spent on it, it was not possible to converge the I-V curves for the SFs above app. $0.3\,\text{V}$. We think that this is due to the induced potential barrier, shown in Fig. 3b, not being completely screened out in the electrodes. To visualise the conduction states and the SFs influence on these, the projected Local Density Of States (LDOS), projected along the transport direction, is plotted for all systems at zero bias, these are shown in Fig. 5. The conduction and valence band edges, marked by the yellow lines, are estimated automatically, and due to the slightly coarse sampling of the LDOS, artefacts appear as small sharp peaks that should be disregarded.

When comparing the LDOS of the pristine system in Fig. 5a with the RTP systems in Fig. 5d, 5e and 5f, we see that the RTPs barely disturbs the states of the systems. There is a slight relocation of high- and low-energy states and a few of the states in the conduction (valence) band disperses (congregates) near the RTP, causing a small area at the RTP (next to the RTP) having less states for transport, which explains the three times increase in the resistance.

The SFs seen in Fig. 5c and 5b on the other hand shows a huge difference in the position of states. For the intrinsic SF, some of the conduction and valence band states are shifted upwards in energy, with the largest shift occurring at the SF. The upwards shift of valence band states effectively reduces the band gap, suggesting that breakdown voltage would have occurred at lower voltages than for the pristine system, had the I-V curve been converged to completion. The upward shift of the conduction band states, lowers the current for low voltages and explains the large increase in resistance calculated earlier. Interestingly we here see a big difference between the extrinsic and intrinsic SF, namely that for the extrinsic SF, several trap levels occur in the band gap. In fact the amount of trap levels and the gap between them almost makes the system metallic at the SF. The conduction band states are also shifted more in energy and further away from the SF as compared to the intrinsic SF, which explains the higher resistance of this SF.

To estimate the likelihood of the formation of the planar defects considered in this paper, we look at their formation energies relative to the pristine InP system shown in Table I. The SFs, which have large detrimental effect on the current transport of the semiconductor, luckily also has relatively high

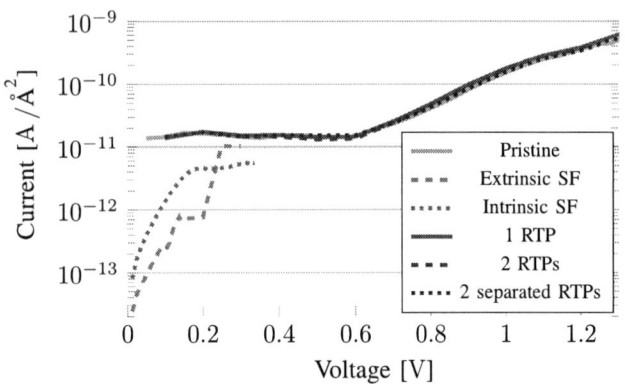

Fig. 6. I-V characteristic of the five defects and a pristine reference system.

formation energies, meaning that it is very unlikely that they will ever form. Especially the extrinsic SF, which was the more disruptive of the two, has a very high formation energy. In contrast the formation energy of the single RTP is so low, that it is very likely that it will form. In fact there seems to be a negative formation energy for the systems with two RTPs, suggesting that the true ground state of InP could be a twinning super-lattice. More investigation than was done in this paper, needs to be done for that result to be conclusive though. At the very least we can say that RTPs will be very hard to avoid forming, luckily as was shown in this paper, RTPs have almost no influence on current transport, at least at low densities. Yet more research needs to be done, to investigate the effect of

high densities of RTPs on current transport, to conclude that RTPs are generally unimportant for electrical properties.

IV. CONCLUSION

In conclusion we have found that low densities of rotational twin-planes, have little to no effect on the current flow in undoped or lowly doped InP. Stacking faults of both the extrinsic and intrinsic kind, have a huge detrimental effect on current flow, but could perhaps be utilised in optical applications for their band gap narrowing effects. Their high formation energies mean that they will not occur naturally and great effort would be needed to fabricate them and as such optical applications seems unlikely.

V. ACKNOWLEDGMENT

We thank the operations team of the Binning and Rohrer Nanotechnology Center for their support. Results were obtained using the ARCHIE-WeSt High Performance Computer (www.archie-west.ac.uk) based at the University of Strathclyde.

TABLE I
FORMATION ENERGIES

System	Intrinsic SF	Extrinsic SF	1 RTP	2 RTPs	2 separated RTPs
Energy [eV]	0.5449	1.1820	0.0037	-0.2443	-0.2454

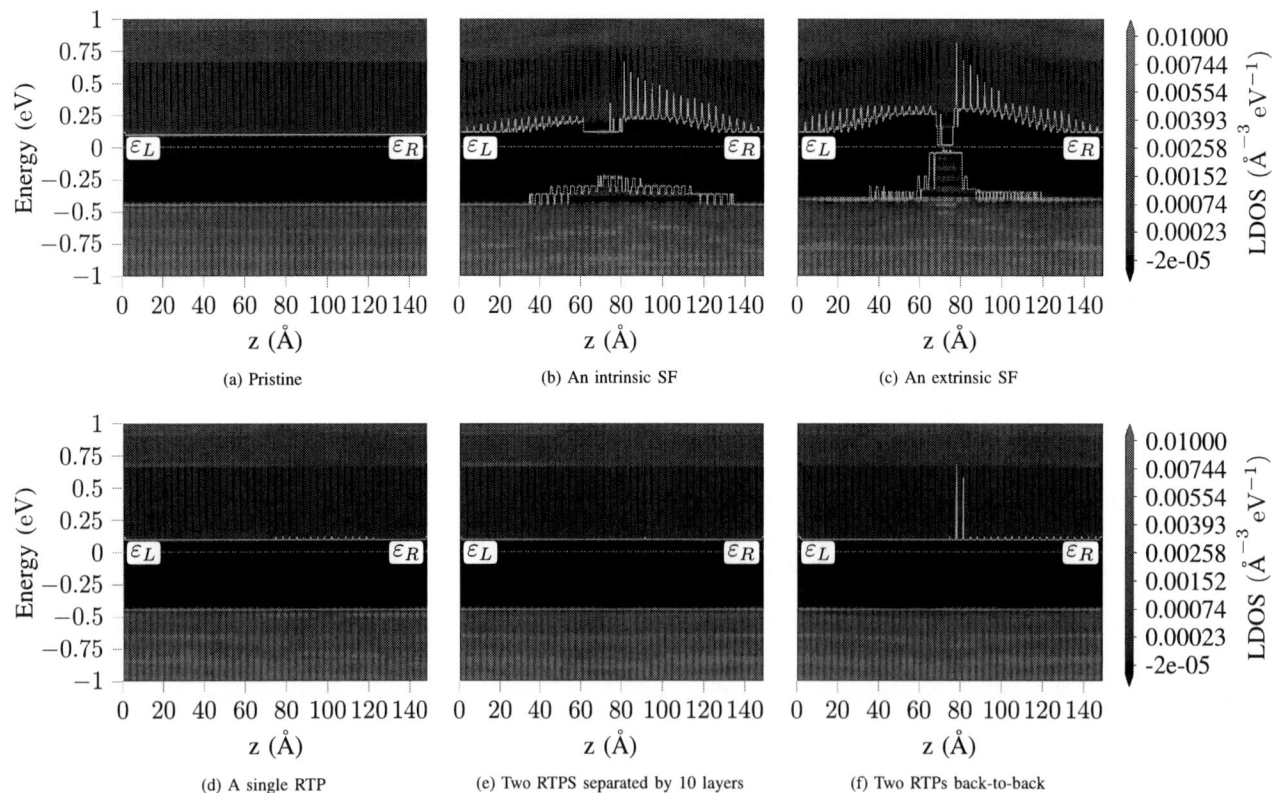

(a) Pristine   (b) An intrinsic SF   (c) An extrinsic SF

(d) A single RTP   (e) Two RTPS separated by 10 layers   (f) Two RTPs back-to-back

Fig. 5. Projected LDOS of calculated systems at zero bias. Conduction band minimum and valence band maximum is indicated by yellow lines.

REFERENCES

[1] T. Skotnicki and F. Boeuf, "How can high mobility channel materials boost or degrade performance in advanced cmos," in *2010 Symposium on VLSI Technology*, 2010, pp. 153–154.

[2] P. Ong and L. Teugels, "5 - cmp processing of high mobility channel materials: Alternatives to si," in *Advances in Chemical Mechanical Planarization (CMP)*, S. Babu, Ed. Woodhead Publishing, 2016, pp. 119–135. [Online]. Available: https://www.sciencedirect.com/science/article/pii/B978008100165300005X

[3] I. Vurgaftman and J. R. Meyer, "Band parameters for nitrogen-containing semiconductors," *Journal of Applied Physics*, vol. 94, no. 6, pp. 3675–3696, 2003. [Online]. Available: https://doi.org/10.1063/1.1600519

[4] A. De and C. E. Pryor, "Predicted band structures of iii-v semiconductors in the wurtzite phase," *Phys. Rev. B*, vol. 81, p. 155210, Apr 2010. [Online]. Available: https://link.aps.org/doi/10.1103/PhysRevB.81.155210

[5] F. W. Wise, "Lead salt quantum dots: the limit of strong quantum confinement," *Accounts of Chemical Research*, vol. 33, no. 11, pp. 773–780, 2000, pMID: 11087314. [Online]. Available: https://doi.org/10.1021/ar970220q

[6] P. Reiss, M. Carrière, C. Lincheneau, L. Vaure, and S. Tamang, "Synthesis of semiconductor nanocrystals, focusing on nontoxic and earth-abundant materials," *Chemical Reviews*, vol. 116, no. 18, pp. 10 731–10 819, 2016, pMID: 27391095. [Online]. Available: https://doi.org/10.1021/acs.chemrev.6b00116

[7] C. Convertino, C. Zota, H. Schmid, D. Caimi, M. Sousa, K. Moselund, and L. Czornomaz, "Ingaas finfets directly integrated on silicon by selective growth in oxide cavities," *Materials*, vol. 12, no. 1, 2019. [Online]. Available: https://www.mdpi.com/1996-1944/12/1/87

[8] C. M. Oliver, K. E. Moselund, and V. P. Georgiev, "Evaluation of material profiles for iii-v nanowire photodetectors," in *21st International conference on Numerical Simulation of Optoelectronic Devices*, 2021, in press.

[9] P. Staudinger, S. Mauthe, N. V. Triviño, S. Reidt, K. E. Moselund, and H. Schmid, "Wurtzite InP microdisks: from epitaxy to room-temperature lasing," *Nanotechnology*, vol. 32, no. 7, p. 075605, nov 2020. [Online]. Available: https://doi.org/10.1088/1361-6528/abbb4e

[10] H. Zhao, P. Miao, M. H. Teimourpour, S. Malzard, R. El-Ganainy, H. Schomerus, and L. Feng, "Topological hybrid silicon microlasers," *Nature Communications*, vol. 9, no. 1, Mar 2018. [Online]. Available: http://dx.doi.org/10.1038/s41467-018-03434-2

[11] H. J. Joyce, Q. Gao, H. H. Tan, C. Jagadish, Y. Kim, X. Zhang, Y. Guo, and J. Zou, "Twin-free uniform epitaxial gaas nanowires grown by a two-temperature process," *Nano Letters*, vol. 7, no. 4, pp. 921–926, 2007, pMID: 17335270. [Online]. Available: https://doi.org/10.1021/nl062755v

[12] P. Staudinger, S. Mauthe, K. E. Moselund, and H. Schmid, "Concurrent zinc-blende and wurtzite film formation by selection of confined growth planes," *Nano Letters*, vol. 18, no. 12, pp. 7856–7862, 2018, pMID: 30427685. [Online]. Available: https://doi.org/10.1021/acs.nanolett.8b03632

[13] U. Krishnamachari, M. Borgstrom, B. J. Ohlsson, N. Panev, L. Samuelson, W. Seifert, M. W. Larsson, and L. R. Wallenberg, "Defect-free inp nanowires grown in [001] direction on inp (001)," *Applied Physics Letters*, vol. 85, no. 11, pp. 2077–2079, 2004. [Online]. Available: https://doi.org/10.1063/1.1784548

[14] R. E. Algra, M. A. Verheijen, M. T. Borgström, L.-F. Feiner, G. Immink, W. J. P. van Enckevort, E. Vlieg, and E. P. A. M. Bakkers, "Twinning superlattices in indium phosphide nanowires," *Nature*, vol. 456, pp. 369–372, 2008. [Online]. Available: https://doi.org/10.1038/nature07570

[15] P. Parkinson, H. J. Joyce, Q. Gao, H. H. Tan, X. Zhang, J. Zou, C. Jagadish, L. M. Herz, and M. B. Johnston, "Carrier lifetime and mobility enhancement in nearly defect-free core-shell nanowires measured using time-resolved terahertz spectroscopy," *Nano Letters*, vol. 9, no. 9, pp. 3349–3353, 2009, pMID: 19736975. [Online]. Available: https://doi.org/10.1021/nl9016336

[16] S. Perera, M. A. Fickenscher, H. E. Jackson, L. M. Smith, J. M. Yarrison-Rice, H. J. Joyce, Q. Gao, H. H. Tan, C. Jagadish, X. Zhang, and J. Zou, "Nearly intrinsic exciton lifetimes in single twin-free gaas/algaas core-shell nanowire heterostructures," *Applied Physics Letters*, vol. 93, no. 5, p. 053110, 2008. [Online]. Available: https://doi.org/10.1063/1.2967877

[17] R. L. Woo, R. Xiao, Y. Kobayashi, L. Gao, N. Goel, M. K. Hudait, T. E. Mallouk, and R. F. Hicks, "Effect of twinning on the photoluminescence and photoelectrochemical properties of indium phosphide nanowires grown on silicon (111)," *Nano Letters*, vol. 8, no. 12, pp. 4664–4669, 2008, pMID: 18983127. [Online]. Available: https://doi.org/10.1021/nl802433u

[18] M. D. Stiles and D. R. Hamann, "Electron transmission through silicon stacking faults," *Phys. Rev. B*, vol. 41, pp. 5280–5282, Mar 1990. [Online]. Available: https://link.aps.org/doi/10.1103/PhysRevB.41.5280

[19] C. Thelander, P. Caroff, S. Plissard, A. W. Dey, and K. A. Dick, "Effects of crystal phase mixing on the electrical properties of inas nanowires," *Nano Letters*, vol. 11, no. 6, pp. 2424–2429, 2011, pMID: 21528899. [Online]. Available: https://doi.org/10.1021/nl2008339

[20] S. Smidstrup, T. Markussen, P. Vancraeyveld, J. Wellendorff, J. Schneider, T. Gunst, B. Verstichel, D. Stradi, P. A. Khomyakov, U. G. Vej-Hansen, M.-E. Lee, S. T. Chill, F. Rasmussen, G. Penazzi, F. Corsetti, A. Ojanperä, K. Jensen, M. L. N. Palsgaard, U. Martinez, A. Blom, M. Brandbyge, and K. Stokbro, "Quantumatk: an integrated platform of electronic and atomic-scale modelling tools," *Journal of Physics: Condensed Matter*, vol. 32, no. 1, 2020. [Online]. Available: https://iopscience.iop.org/article/10.1088/1361-648X/ab4007

[21] P. Hohenberg and W. Kohn, "Inhomogeneous electron gas," *Phys. Rev.*, vol. 136, pp. B864–B871, Nov 1964. [Online]. Available: https://link.aps.org/doi/10.1103/PhysRev.136.B864

[22] W. Kohn and L. J. Sham, "Self-consistent equations including exchange and correlation effects," *Phys. Rev.*, vol. 140, pp. A1133–A1138, Nov 1965. [Online]. Available: https://link.aps.org/doi/10.1103/PhysRev.140.A1133

[23] S. Smidstrup, D. Stradi, J. Wellendorff, P. A. Khomyakov, U. G. Vej-Hansen, M.-E. Lee, T. Ghosh, E. Jónsson, H. Jónsson, and K. Stokbro, "First-principles green's-function method for surface calculations: A pseudopotential localized basis set approach," *Phys. Rev. B*, vol. 96, p. 195309, Nov 2017. [Online]. Available: https://link.aps.org/doi/10.1103/PhysRevB.96.195309

[24] M. Schlipf and F. Gygi, "Optimization algorithm for the generation of oncv pseudopotentials," *Computer Physics Communications*, vol. 196, pp. 36–44, 2015. [Online]. Available: https://www.sciencedirect.com/science/article/pii/S0010465515001897

[25] J. P. Perdew, A. Ruzsinszky, G. I. Csonka, O. A. Vydrov, G. E. Scuseria, L. A. Constantin, X. Zhou, and K. Burke, "Restoring the density-gradient expansion for exchange in solids and surfaces," *Phys. Rev. Lett.*, vol. 100, p. 136406, Apr 2008. [Online]. Available: https://link.aps.org/doi/10.1103/PhysRevLett.100.136406

[26] M. Brandbyge, J.-L. Mozos, P. Ordejón, J. Taylor, and K. Stokbro, "Density-functional method for nonequilibrium electron transport," *Phys. Rev. B*, vol. 65, p. 165401, Mar 2002. [Online]. Available: https://link.aps.org/doi/10.1103/PhysRevB.65.165401

Temperature Increase in MRAM at Writing: A Finite Element Approach

Tomáš Hadámek
*Christian Doppler
Laboratory for Nonvolatile
Magnetoresistive Memory
and Logic at the Institute
for Microelectronics
TU Wien*
Vienna, Austria
hadamek@iue.tuwien.ac.at

Mario Bendra
*Christian Doppler
Laboratory for Nonvolatile
Magnetoresistive Memory
and Logic at the Institute
for Microelectronics
TU Wien*
Vienna, Austria
bendra@iue.tuwien.ac.at

Simone Fiorentini
*Christian Doppler
Laboratory for Nonvolatile
Magnetoresistive Memory
and Logic at the Institute
for Microelectronics
TU Wien*
Vienna, Austria
fiorentini@iue.tuwien.ac.at

Johannes Ender
*Christian Doppler
Laboratory for Nonvolatile
Magnetoresistive Memory
and Logic at the Institute
for Microelectronics
TU Wien*
Vienna, Austria
ender@iue.tuwien.ac.at

Roberto L. de Orio
*Institute for
Microelectronics
TU Wien*
Vienna, Austria
orio@iue.tuwien.ac.at

Wolfgang Goes
Silvaco Europe
Cambridgeshire, United
Kingdom
wolfgang.goes@silvaco.com

Siegfried Selberherr
*Institute for
Microelectronics
TU Wien*
Vienna, Austria
selberherr@TUWien.ac.at

Viktor Sverdlov
*Christian Doppler
Laboratory for Nonvolatile
Magnetoresistive Memory
and Logic at the Institute
for Microelectronics
TU Wien*
Vienna, Austria
sverdlov@iue.tuwien.ac.at

Abstract — **The writing process in spin transfer torque magnetoresistive random access memories is facilitated by elevated temperatures. In this work we investigate the temperature in the free layer (FL) during switching. With our fully three-dimensional (3D) finite element method simulation approach, we numerically solve the heat transport equation coupled to the electron, spin, and magnetization dynamics and demonstrate that the FL temperature is highly inhomogeneous due to non-uniform magnetization of the FL during switching. While the average temperature in the FL can be obtained based on an average current density and an averaged potential drop across the tunnel barrier in a one-dimensional model, a fully 3D model is required to evaluate the large local temperature variations.**

Keywords — MRAM, magnetic tunnel junction, current-induced heating, heating asymmetry, temperature variations

I. Introduction

The ongoing miniaturization of semiconductor components has pushed the chip technology to its limits due to rapidly increasing leakage currents. Moreover, in the commonly employed von Neumann architecture, where the processing unit is separated from the memory, significant energy losses due to the data transfer back and forth, the so called von Neumann bottleneck, is present. The magnetoresistive random access memory (MRAM) is a promising emerging candidate to overcome these issues. MRAM is complementary metal-oxide semiconductor (CMOS) compatible [1-3] and has zero stand-by power consumption as it is intrinsically nonvolatile. It can also be integrated into logic [4] and offers a wide temperature operation range [5].

A magnetic tunnel junction (MTJ), the basic building block of MRAM cells, consists of 3 layers: a pinned ferromagnetic layer, an oxide barrier, and a free ferromagnetic layer. In spin-transfer torque MRAM (STT-MRAM) [6],

relatively high current densities through the structure are required to switch the magnetization of the FL. This results in an increased temperature of the MRAM cell and mediates the switching of the FL magnetization [7,8]. On the other hand, the increased FL temperature caused by self-heating can result in an information loss as it compromises the thermal stability [8]. To preserve the data, the temperature must be rapidly relaxed after writing. In [9], the heating asymmetry in the MTJ was observed for reversed current direction. This asymmetry was further numerically studied in [10], showing a non-linear increase of the saturation temperature with increasing heating power.

In this work we investigate the inhomogeneity of the temperature in the FL at switching. This inhomogeneity can be expected due to the fast magnetization dynamics of the MRAM cell, which results in a significant inhomogeneity of the current density across the FL plane caused by non-uniform magnetization [11,12]. Moreover, during switching, the current densities change significantly due to the rapid change of the magnetization direction. Therefore, to model the temperature in MRAM, the heat transport equation has to be coupled to current, magnetization, and spin dynamics in a fully three-dimensional (3D) model.

II. Method

A. Temperature Modelling

To describe the dynamics of the temperature T in the structure at time t and position r, the heat flow equation is used.

$$c_v\rho \frac{\partial T(r,t)}{\partial t} - \nabla \cdot [\kappa \cdot \nabla T(r,t)] = q(r,t) \qquad (1)$$

c_v, ρ, and κ stand for the heat capacity, mass density, and heat conductivity of the material, respectively. $q(r,t)$ is the heat source term. In MTJs, two main heat sources can be identified.

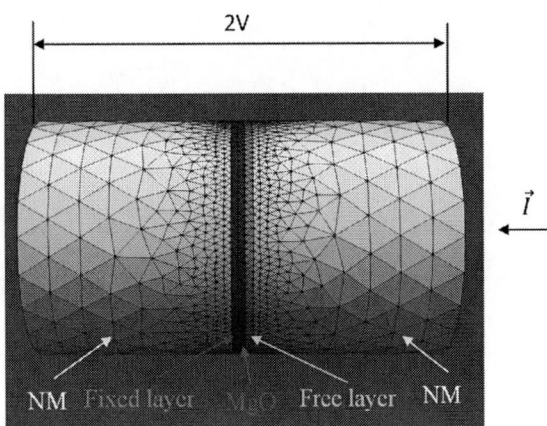

Fig. 1. The *simulated structure consisting of CoFeB(1 nm)/MgO(1 nm)/CoFeB(1.2 nm) MTJ connected to normal metal contacts (30 nm). The diameter is 40 nm. The ends are kept at a constant temperature. A bias of 2 V is applied across the structure.*

The first heat source is attributed to the Joule heat in the ferromagnetic layers and the metal contacts. It can be expressed as: $q_r(r) = j^2(r)\rho_E$, where ρ_E is the material resistivity and j is the current density.

The second heat source is associated with electrons tunneling through the insulating barrier [7,8]. When a potential difference is applied across the MTJ, the electron tunneling from the low potential side (source) arrives at the high potential side (receiver) as a hot electron. Its energy is above the receiver Fermi energy and must be dissipated through various scattering processes. Similarly, the hole left behind the electron at the source must be filled by an electron from a higher energy level, the excess energy of which is therefore released. Setting the x-axis to coincide with the main axis along the structure, the hot electron/hole heat source is described by

$$q_t(r) = \left(1 \pm \alpha(\Delta U)\right)\frac{j_x(y,z)\Delta U(y,z)}{2\lambda}\exp\left(-\frac{|x - x_{F/P}|}{\lambda}\right), (2)$$

where j_x and $\Delta U(y,z)$ stand for the x-component of the current density and the potential drop at the position (y,z) along FL. The x-coordinate of the position of the free/pinned-barrier interface is denoted by $x_{F/P}$ and λ is a characteristic length at which hot electrons/holes lose their energy. The heat production imbalance between the receiver and the source side is accounted for by the asymmetry coefficient $\alpha(\Delta U)$ [10]. In this work α is set to zero – no asymmetry in the receiver and source sides is considered – in order to fully investigate the details of inhomogeneous temperature development due to the non-uniform current density. The system relevant parameters are listed in Table I and Table II.

B. Magnetization Dynamic

To describe the magnetization dynamics in the micromagnetic model the Landau-Lifshitz-Gilbert equation is used.

$$\frac{\partial m}{\partial t} = -\gamma\mu_0\, m \times H_{eff} + \alpha m \times \frac{\partial m}{\partial t} + \frac{1}{M_S}T_S \quad (3)$$

m is the normalized magnetization, $\gamma\mu_0$ is the scaled gyromagnetic ratio, α is the Gilbert damping, and M_S stands for the saturation magnetization. The effective filed H_{eff} consists of several components, with the demagnetization field H_{Demag} and the anisotropy field H_{Aniso} contributing the most. In the following, H_{Demag} is determined by an optimized hybrid FEM-BEM approach [13] and H_{Aniso} is calculated with the parameters listed in Table I. T_S stands for the STT and couples the spin with magnetization dynamics. To determine T_S, the spin accumulation in the structure is computed [11,14]. The current densities and potentials were determined by solving the coupled spin, charge, and magnetization dynamics in an MTJ [12]. The described approach was implemented in a fully 3D finite element method solver based on an open source library MFEM [15]. For the time integration, an implicit Euler method was used.

III. RESULTS

In Fig. 1, the simulated structure is shown. The MTJ consisting of CoFeB(1 nm)/MgO(1 nm)/CoFeB(1.2 nm) is connected to non-magnetic metal (NM) contacts (30 nm). The outer ends of the contacts are kept at constant temperature. For the parallel to anti-parallel (P-AP) switching and anti-parallel to parallel (AP-P) switching positive and negative voltages are applied to the structure, respectively. When the voltage is applied, the electric current starts to flow through the structure and the STT acts on the free layer magnetization. Eventually, the FL magnetization is flipped.

TABLE I. SIMUALTION PARAMETERS - MAGNETIZATION AND SPIN

Parameter	Value
Gilbert damping, α	0.02
Gyromagnetic ratio, γ	$1.76 \cdot 10^{11}$ rad s^{-1} T^{-1}
Vacuum permeability, μ_0	$4\pi \cdot 10^{-7}$ H m^{-1}
Saturation magnetization, M_S	$1.2 \cdot 10^6$ A m^{-1}
Exchange constant, A	$1 \cdot 10^{-11}$ J m^{-1}
Anisotropy constant, K	$0.9 \cdot 10^6$ J m^{-3}
Current spin polarization, β_σ	0.7
Diffusion spin polarization, β_D	1.0
Electron diffusion coefficient, D_e	$1 \cdot 10^{-4}$ m^2/s
Spin-flip length, λ_{sf}	10 nm
Spin dephasing length, λ_φ	5 nm
Exchange length, λ_J	0.5 nm
Tunnel megnetoresistance ratio (TMR)	200%

TABLE II. SIMUALTION PARAMETERS - HEAT

	MgO	CoFeB	NM
ρ [kgm^{-3}]	3600	8200	8050
c_v [J K^{-1}kg^{-1}]	735	440	500
κ [W K^{-1}m^{-1}]	0.38	83	43
ρ_e [Ω m]	-	2×10^{-5}	2×10^{-5}
λ [nm]	-	1	1

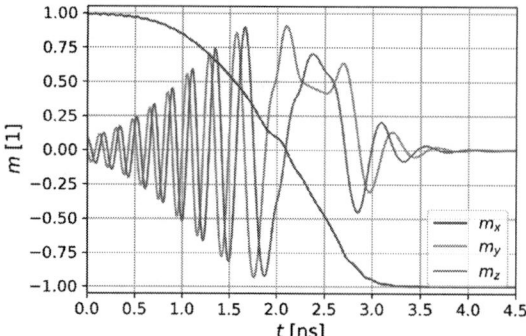

Fig. 2. Averages of normalized free-layer magnetization in x-, y- and z-direction during switching from parallel to anti-parallel. The initial magnetization was tilted by 5° in the z-direction from the x-direction to eliminate the slow incubation phase. The parallel to anti-parallel switching shows an initial oscillating behaviour.

Fig. 2 and Fig. 3 show the P-AP and AP-P switching, respectively. In both simulations, three different phases can be identified: Initial magnetization oscillation in the y-z plane, faster m_x component change, and a final magnetization oscillation in the y-z plane. While the fast m_x component changes are comparable in both P-AP and AP-P switching, the oscillation phases vary significantly. In the P-AP switching, the initial oscillation phase is much longer than in the AP-P switching. This is caused by an FL stabilization due to H_{Demag}, which favors the parallel orientation of the magnetic layers. Similarly, the end oscillation phase is much longer for the AP-P switching. The initial FL magnetization is tilted by 5° from the easy axis along the structure to eliminate the slow incubation phase.

Fig. 4 and Fig. 5 show the temperature development in the FL during the P-AP and AP-P switching, respectively. The average temperature T_{avg} (solid green, coincides with the dotted orange), the minimum temperature T_{min} (in dot-dashed gray) and the maximum temperatures T_{max} (in dashed black) at the FL obtained from a fully 3D model are shown. The average temperature T_{avg-1D} calculated with an average current density and an average potential drop across the barrier is also shown (dotted orange) and coincides with T_{avg}

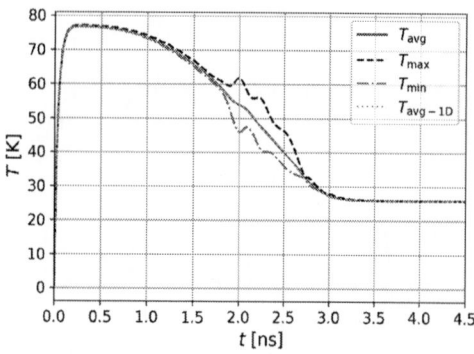

Fig. 4. Temperature in the free layer during parallel to anti-parallel switching. An average temperature T_{avg} (solid green), maximum temperature (dashed black), minimum temperature (dot-dashed grey) and T_{avg-1D} (dotted orange, coincides with green) was calculated using averages of current densities and potential drop.

from the 3D simulations. At the beginning, a fast heating is observed after which the structure reaches thermal saturation for a specific magnetization arrangement in about 150 ps. When the saturation is achieved, the temperature in the FL changes slowly synchronously with the magnetization change. At the beginning, the temperature profile of the FL is almost homogeneous and T_{avg}, T_{min}, and T_{max} do nearly coincide. However, during the switching, an inhomogeneous temperature profile develops and T_{min} and T_{max} differ significantly. The difference is caused by the current density variations across the FL due to the non-uniform magnetization distribution shown in Fig. 6. The left plot depicts the temperature profile in the FL after 1 ns during AP-P switching. The low (high) temperature regions visible at the bottom left (top right) clearly correlate with the low (high) value regions of m_x shown in the right plot. A temperature difference of about 14 K for the applied voltage of 2 V is observed.

The ratio of the maximum temperature difference ΔT_{max} to T_{avg} in the FL is displayed in Fig. 7 and Fig. 8. In both, the P-AP and AP-P, switching simulations, this ratio increases slowly at the beginning. It then shows a sudden and considerable relative increase exceeding 30 %. After about

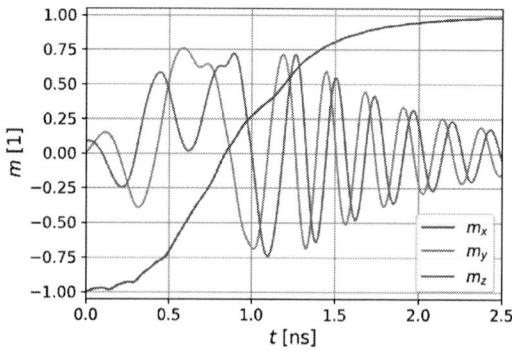

Fig. 3. Averages of normalized free-layer magnetization in x-, y- and z-direction during switching from anti-parallel to parallel. The initial magnetization was tilted by 5° in the z-direction from the x-direction to eliminate the slow incubation phase. A strong oscillating behavior is visible at the final switching phase.

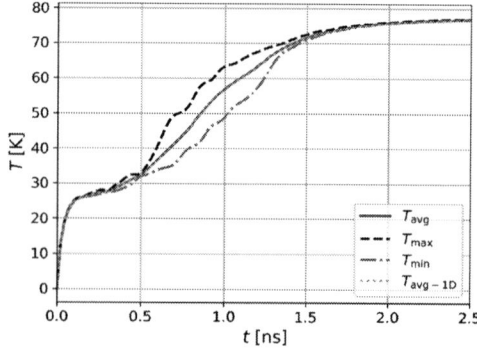

Fig. 5. Temperature in the free layer during anti-parallel to parallel swtiching. An average temperature T_{avg} (solid green), maximum temperature (dashed black), minimum temperature (dot-dashed grey) and T_{avg-1D} (dotted orange, coincides with green) was calculated using averages of current densities and potential drop.

978-1-6654-3746-2/21 $31.00 © 2021 IEEE

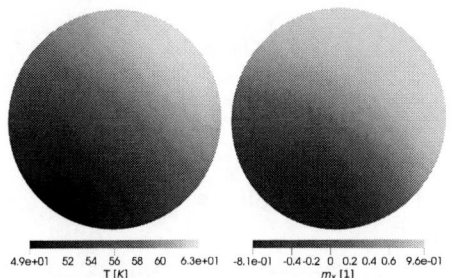

Fig. 6. Temperature profile (left) and m_x (right) in the FL at 1 ns during AP-P switching. The minimum/maximum temperature coincides with the minimum/maximum of m_x.

0.75 ns, this ratio drops again. When compared to Fig. 2 and Fig. 3, the sudden increase can easily be associated with the region of fast m_x component change, when magnetization oscillations in the y-z plane are not present.

IV. CONCLUSION

We have studied the FL temperature development in STT-MRAM utilizing our fully 3D finite element simulation approach coupling heat, charge, and spin transport with the magnetization dynamics. Strong temperature inhomogeneity across the FL was predicted rooting in an inhomogeneous current density distribution attributed to non-uniform magnetization of the FL. The ratio of the maximum temperature difference across the barrier to the average temperature of the FL exceeded 30 %, when the FL magnetization rapidly switches. While the average temperature in the FL can be obtained from an average current density and an average potential drop across the barrier, a fully 3D simulation is required to model the temperature variations in the FL.

ACKNOWLEDGMENT

Financial support by the Austrian Federal Ministry for Digital and Economic Affairs, the National Foundation for Research, Technology and Development, and the Christian Doppler Research Association is gratefully acknowledged.

REFERENCES

[1] H. Ohno, "A hybrid CMOS/magnetic tunnel junction approach for non-volatile integrated circuits," 2009 Symposium on VLSI Technology, pp.122–123, August 2009

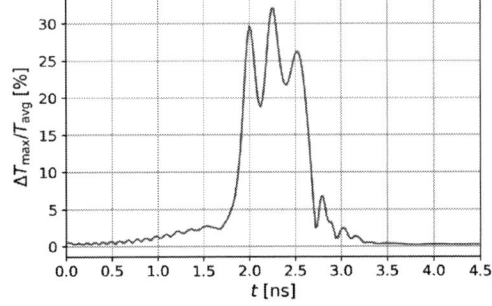

Fig. 7. Ratio of the maximum temperature difference ΔT_{max} of the free layer to the average free layer temperature T_{avg} during parallel to anti-parallel switching. ΔT_{max} reaches 30 % of the average temperature, when m_y and m_z stop to oscillate.

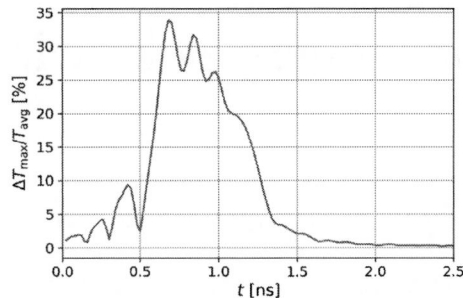

Fig. 8. Ratio of the maximum temperature difference ΔT_{SAT} of the free layer to the average free layer temperature T_{avg} during anti-paralell to parallel switching. ΔT_{max} reaches 30 % of the average temperature, when m_y and m_z stop to oscillate.

[2] K. L. Wang and P. K. Amiri, "Nonvolatile spintronics: perspectives on instant-on nonvolatile nanoelectronic systems," Spin, vol. 2, no. 2, art. no. 1250009, November 2012

[3] V. K. Joshi, P. Barla, S. Bhat and B. K. Kaushik, "From MTJ device to hybrid CMOS/MTJ circuits: A review," IEEE Access, vol. 8, pp.194105-194146, October 2020

[4] D. Suzuki, M. Natsui, A. Mochizuki, S. Miura, H. Honjo *et al.*, "Fabrication of a 3000-6-input-LUTs embedded and block-bevel power-gated non-volatile FPGA chip using p-MTJ-based logic-in-memory structure," *2015 Symposium on VLSI Circuits (VLSI Circuits)*, pp. C172–C173, June 2015.

[5] L. Thomas, G. Jan, J. Zhu, H. Liu, Y.-J. Lee, *et al.*, "Perpendicular spin transfer torque magnetic random access memories with high spin torque efficiency and thermal stability for embedded applications (invited)," Journal of Applied Physics, vol. 115, no. 17, art. no. 172615, April 2014.

[6] M. Cubukcu, O. Boulle, M. Drouard, K. Garello, C. O. Avci *et al.*, "Spin-orbit torque magnetization switching of a three-terminal perpendicular magnetic tunnel junction," Applied Physics Letters, vol. 104, no. 4, art. no. 042406, January 2014.

[7] I. L. Prejbeanu, M Kerekes, R C Sousa, H Sibuet, O Redon *et al.*, "Thermally assisted MRAM," Journal of Physics: Condensed Matter, vol. 19, no. 16, art. no. 165218, April 2007.

[8] T. Taniguchi and H. Imamura, "Thermally assisted spin transfer torque switching in synthetic free layers," Physical Review B, vol. 83, no. 5, art. no. 054432, February 2011.

[9] E. Gapihan, J. Hérault, R. C. Sousa, Y. Dahmane, B. Dieny *et al.*, "Heating asymmetry induced by tunneling current flow in magnetic tunnel junctions," Applied Physics Letters, vol. 100, no. 20, art. no. 202410, May 2012.

[10] T. Hadámek, S. Selberherr, W. Goes, V. Sverdlov, "Heating asymmetry in magnetoresistive random access memories," Proceedings of the World Multi-Conference on Systemics, Cybernetics and Informatics (WMSCI), in press, July 2021

[11] S. Fiorentini, J.Ender, S.Selberherr, R.L.de Orio, W.Goes and V.Sverdlov, "Coupled spin and charge drift-diffusion approach applied to magnetic tunnel junctions,", Solid-State Electronics, vol. 186, art. no. 108103, June 2021

[12] S. Fiorentini, J. Ender, M. Mohamedou, R. Orio, S. Selberherr *et al.*, "Computation of torques in magnetic tunnel junctions through spin and charge transport modeling", *2020* International Conference on Simulation of Semiconductor Processes and Devices (SISPAD), pp. 209-212, October 2020.

[13] D. R. Fredkin, T. R. Koehler, "Hybrid method for computing demagnetizing fields, " IEEE Transactions on Magnetics, vol. 26, no. 2, pp. 415-417, March 1990.

[14] C. Abert, M. Ruggeri, F. Bruckner, C. Vogler, G. Hrkac *et al.*, "A three-dimensional spin-diffusion model for micromagnetics, " Scientific Reports, vol. 5, art. no. 14855, October 2015.

[15] R. Anderson, J. Andrej, A. Barker, J. Bramwell, J.-S. Camiera *et al.*, "MFEM: A modular finite element library," Computers & Mathematics with Applications, vol. 81, pp.42-74, January 2021

978-1-6654-3746-2/21 $31.00 © 2021 IEEE

Junctionless Nanowire Transistors Based Wilson Current Mirror Configuration

André B. Shibutani[1], Michelly de Souza[1], Member, IEEE, Renan Trevisoli[2], Member, IEEE and
Rodrigo T. Doria[1], Member, IEEE

[1]Centro Universitário FEI, Electrical Engineering Department – São Bernardo do Campo, Brazil
[2]Universidade Federal do ABC, UFABC – Santo André, Brazil
e-mail: abshibutani@fei.edu.br

Abstract— **In this paper, a Wilson current mirror based on junctionless nanowire transistors (JNTs) is evaluated for the first time. Considering that the Wilson current mirror exhibits an enhanced output resistance with respect to the common source configuration, this study is focused on verifying the mirroring accuracy of different transistor dimensions. Also, the work examines the impact of the transistor feedback circuit on the output resistance and the current transfer ratio. The current mirror has been evaluated through numerical simulations, which were calibrated to experimental data of single devices.**

Keywords- JNT; Wilson Current Mirror; Current transfer ratio.

I. INTRODUCTION

Considering the complex fabrication of small transistors with junctions and Short Channel Effects (SCEs) related to the drain control over the channel charges with the transistor reduction, the Junctionless Nanowire Transistor raised as a solution to minimize the drain and source effect over the channel with the reduction of the transistor dimensions [1]. Since the device presents a constant doping profile along all the silicon active layer, the device seems to have a better electrical characteristic with respect to the MOSFET with junctions, reflecting on a better subthreshold slope and a low drain induced barrier lowering (DIBL) [2].

Regarding the analog properties, the literature shows that the junctionless device has a low output conductance and a high output gain, considering the basic circuit topologies such as the source-follower and the common-source circuit configurations [3][4]. The latter indicates that the building block is suitable as a current source since it presents a high output resistance.

The literature also points out a JNT current mirror functionality, in which the current transfer ratio depends not only on the change of the channel width to length ratio between input and output devices, i.e. $(W/L)_2/(W/L)_1$, but

This work was supported by São Paulo Research Foundation (FAPESP) grant #2019/15500-5, National Council for Scientific and Technological Development (CNPq) grant #303938/2020-0.

also on the transistor bias [5], which is attributed to the JNT operation mode. Different from conventional inversion mode devices, JNTs operate in partial depletion/ accumulation regimes [6]. Additionally, when junctionless transistors of different widths are applied in input and output, the output current does not vary with the same ratio of W as in inversion mode devices [5]. Also, narrower devices present better mirroring precision due to the better electrostatic control of the triple gate electrodes over the channel charges [7].

Since the cascode arrangement exhibits a higher output resistance, different current mirror configurations were designed considering the cascode arrangement, aiming to improve the current mirror as a current source building block [8]. One of those current mirrors is the Wilson configuration depicted in Figure 1, which is studied in this work. In Figure 1, the association of JN3 and JN4 provides a negative feedback for the current mirror formed by JN1 and JN2, compensating the effect of the Early voltage in the output.

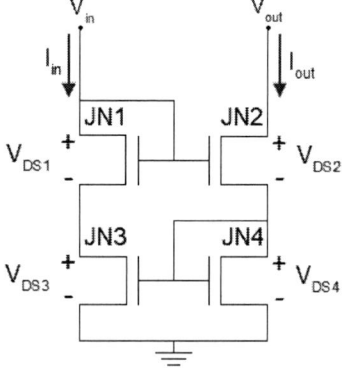

Figure 1. JNT Wilson current mirror configuration.

II. DEVICE CHARACTERISTICS AND CALIBRATION OF SIMULATIONS

The JNT devices used in this work present a silicon active layer with channel length of 100 nm, silicon layer thickness (t_{si}) of 10 nm, arsenic doping concentration (N_D) of 1×10^{19} cm^{-3} and silicon width varying from 20 to 440 nm. The buried oxide layer is 145 nm thick, and the gate stack presents an effective oxide thickness of 1.5 nm. Additionally, the devices evaluated were fabricated in CEA-Leti in accordance with the fabrication process described in [9]. Since there are no experimental samples of Wilson current mirror, the curves I_D vs. V_{GS} were measured individually from different transistors of the same fabrication process, with the premise to calibrate each transistor through simulations.

After that, all the simulations were performed through Synopsys Sentaurus TCAD [10] taking into account the effects of electron-holes mobility degradation, high-field saturation, band gap narrowing, and carriers' generation and recombination. Figure 2 shows the measured and calibrated drain current (I_{D1}) as a function of the gate-to-source voltage (V_{GS}) curves for a JNT with a width of 40 nm biased at several drain voltages (V_{DS}).

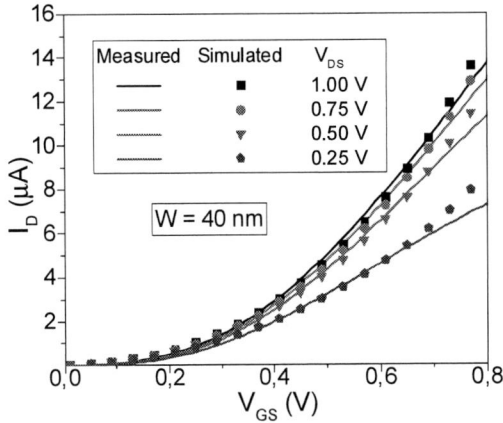

Figure 2. Measured and simulated curves of I_D as a function of V_{GS} of a transistor with W=40 nm biased at different V_{DS}.

III. WILSON CURRENT MIRROR SIMULATION AND DISCUSSIONS

The analysis of the Wilson current mirror has been performed considering that all transistors that compose the structure present the same physical characteristics, i.e., W and L. Figures 3 and 4 illustrate the mirroring precision (I_{out}/I_{in}) as a function V_{out} and I_{in}, respectively, for transistors' width varying from 10 to 440 nm. It is worth to mention that the curves for the transistor of $W = 10$ nm were extrapolated through calibration data. Figure 3 presents the structures biased at a fixed $V_{in} = 1.2$ V and the latter at a fixed $V_{out} = 1.2$ V. As it is shown in both figures, the mirroring precision seems to have a current transfer ratio closer to unity for small transistor widths in a wide range of operation. This behavior is justified by the better gate control over the channel charges as the transistor is scaled down. Complementary, the negative feedback of the configuration provided by transistors JN3 and JN4 acting as resistors seems to minimize the variation of I_{out} with V_{out} for small transistors, which ensures that a constant I_{out} will be

delivered independent of variations in the circuit load. On the other hand, as the transistor width increases, Figure 4 shows that the current transfer ratio seems to reach values close to unity for higher values of I_{in}, which makes those transistors appealing to supply load circuits that require high currents, using smaller bias voltages. However, for the Wilson configuration compounded by larger transistors, the devices are never in saturation simultaneously. Also, larger transistors that have a drain and gate short-circuited are always near to the threshold between triode and saturation regimes as their threshold voltages are close to zero ($V_{TH} = -0.08$ for $W = 290$ nm and $V_{TH} = -0.09$ for $W = 440$ nm).

Besides that, Figure 3 also reflects the width dependence on the threshold voltage [11]. As the narrower devices present larger capacitive coupling, the threshold voltages become larger. Consequently, the overdrive voltage ($V_{GT} = V_{GS} - V_{TH}$) is lower for narrow devices, which makes the transistors of the configuration reach the saturation for lower V_{out}. In contrast, the stability observed for larger transistors with the increase in V_{out} is a consequence of JN2 reaching saturation alone.

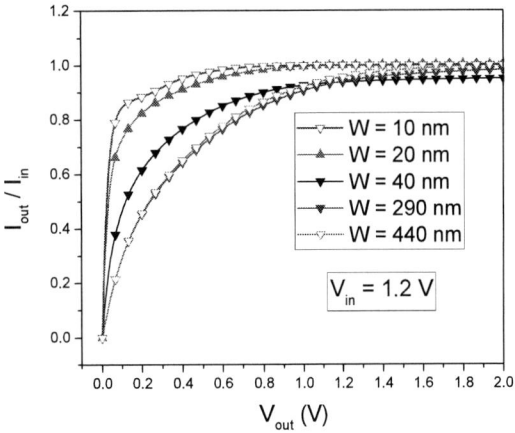

Figure 3. The mirroring precision of a Wilson configuration as a function of the output voltage.

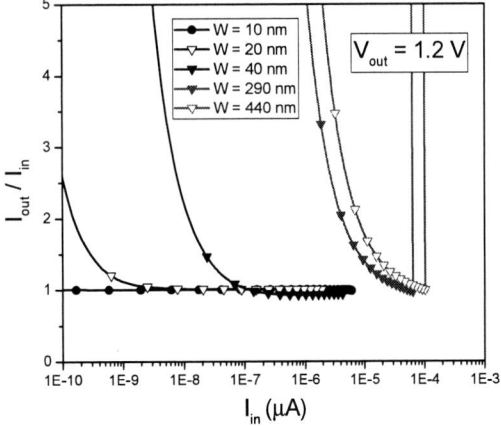

Figure 4. Mirroring precision of a Wilson configuration as a function of I_{in}.

The larger capacitive coupling advantages can also be extended to asymmetrical current mirror configurations. Figure 5 depicts the mirroring precision (I_{out}/I_{in}) as a function of I_{in}, for transistor widths of JN2 and JN4 varying together from 20 to 440 nm, while the transistor widths of the transistors that compound the input stage are kept constant at a fixed width equal to 10 nm. As can be seen in Figure 5, the association between small transistors seems to reach a mirroring precision closer to inversion mode devices, considering the mirroring precision is approximately 2, when I_{in} is around 3 µA for W_{out} = 20 nm. Thus, the better mirroring precision in small transistors can also be explained by the better capacitive coupling.

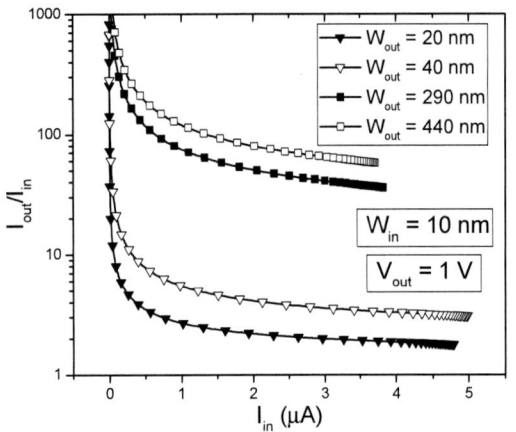

Figure 5. Iout/Iin as a function of Iin for Win = 10nm.

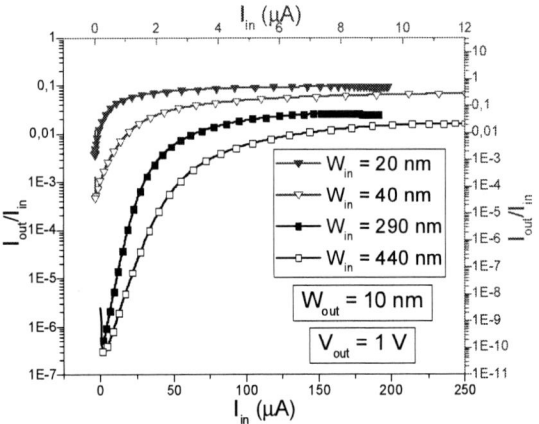

Figure 6. Iout/Iin as a function of Iin for Wout = 10nm.

Figure 6 shows the mirroring precision (I_{out}/I_{in}) of asymmetrical Wilson current mirror configurations as a function of I_{in}, for W_{out} = 10 nm, while W_{in} varies between 20 to 440 nm. As one can note, the asymmetrical configurations between small transistors are easier to estimate than larger transistors as stated before. Thus, the results point out that the smaller transistors are more adequate to work as current mirrors once the mirroring

precision is closer to $(W/L)_2/(W/L)_1$ in symmetrical and asymmetrical configurations besides the transistors of the association that reaches saturation for higher input bias simultaneously.

In order to examine the performance of the Wilson current mirror as a current source, in Figure 7 the output resistance (R_{out}) is presented as a function of V_{out} for a fixed V_{in} = 1.2 V and transistors' width varying from 10 to 440 nm. Figure 7 also shows R_{out} as a function of V_{in} for a fixed V_{out} = 1.2 V for the same transistor widths. As one can note, the increase of R_{out} for small values of V_{out} or V_{in} for small transistor widths ($W \leq 40$ nm) is the result of the transistor JN2 or JN4 operating in the subthreshold regime, where the transistors exhibit a high resistance to the current conduction (off state). However, as the wider transistors ($W \geq 290$ nm) have a negative threshold voltage, the devices are always operating at on-state for positive voltages, making this resistance increase not remarkably pronounced.

Moreover, for a fixed V_{in} = 1.2 V, the output resistance of associations compounded by small transistors increases smoothly for higher values of V_{out}, when all the devices get deeper in saturation. Thus, the growth in R_{out} for those transistors is a result of the channel length modulation in the saturation region as the output current increases with V_{out}. Also, the Figure 7 shows that the output resistances of the blue and black curves for the same width do not cross each other at the point $V_{in} = V_{out}$ = 1.2 V because the feedback mechanism works differently for the input and output variations. Additionally, even though, the circuit bias can improve R_{out}, the output resistance is considered high for any circuit bias. Such resistance is higher than the one obtained for a JNT based common-source current mirror [5]. In order to compare the resistance of those configurations, the Figure 8 was depicted. In that figure, it can be seen the output resistance as a function of the transistor widths for $V_{in} = V_{out}$ =1.2 V. As one can note, the difference in resistance increases with the transistor width reduction, reaching a difference of a decade for a transistor width equal to 10 nm.

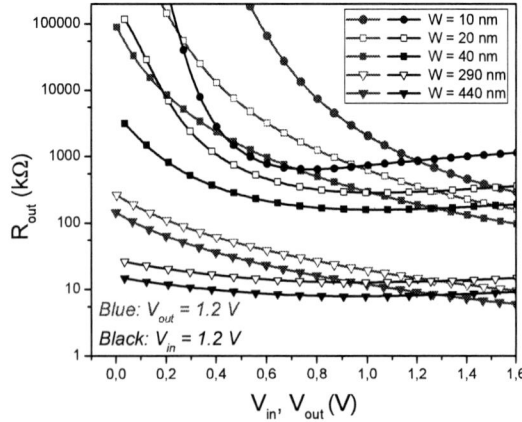

Figure 7. Rout as a function of Vin and Vout.

Considering the larger resistance showed by the Wilson configuration current mirror with respect to a common-source one, Figure 9 illustrates I_{out}/I_{in} of both current mirror configurations as a function of I_{in} for different fixed V_{out}. As one can note, the transfer current ratio is closer to unity in the Wilson current mirror of small dimensions. Additionally, for larger JNTs the Wilson arrangement also improves the current transfer ratio, getting closer to the unity with respect to the common-source-configuration, an enhancement obtained by the transistor JN2 reaching the saturation for higher input current bias. Besides that, as it can be seen in Figure 9, the Wilson current tends to the unity for higher currents.

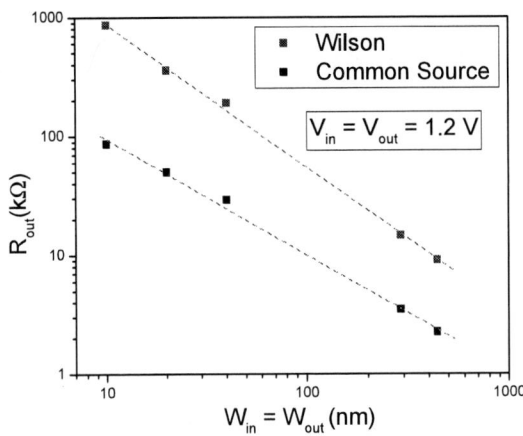

Figure 8. R_{out} of Wilson and common-source configurations vs. $W_{in} = W_{out}$.

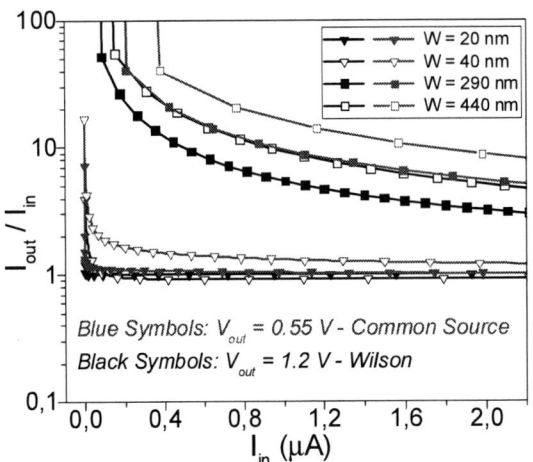

Figure 9. I_{out}/I_{in} of Wilson and common-source configurations vs. I_{in}.

IV. CONCLUSIONS

This work has presented an evaluation of the performance of junctionless transistor-based Wilson current mirrors. It was shown that the negative feedback in such structures minimizes the output current variation and increases the range, over which the output current remains around the

unity. Smaller transistors operate at lower currents, which cannot be done with larger transistors. Oppositely, the larger transistors ($W \geq 290$) can lead larger currents, yet the transistors of the Wilson association are never in saturation simultaneously. Also, the larger transistors present a poorer mirroring precision with respect to the small ones due to its lower capacitive coupling. Besides that, the Wilson output cascode configuration composed of junctionless devices offers a high output resistance, which can reach higher magnitudes than offered by a common-source current mirror topology. A symmetrical configuration compounded by transistor widths equal to 10 nm presents an output resistance a decade higher than in a common-source topology. On the other hand, a Wilson association compounded of transistor width equal to 440 nm presents an output resistance difference of about 7 kΩ. Thus, the Wilson current mirror exhibits a better mirroring precision and a better output resistance with respect to the common-source current mirror counterpart.

ACKOWLEDGEMENTS

The authors thank Sylvain Barraud and CEA-Leti for supplying the experimental devices.

REFERENCES

[1] J. P. Colinge et al., "Nanowire transistors without junctions," Nat. Nanotechnol., vol. 5, no. 3, pp. 225–229, 2010.

[2] C. W. Lee et al., "Performance estimation of junctionless multigate transistors," Solid. State. Electron., vol. 54, no. 2, pp. 97–103, 2010.

[3] R. T. Doria et al., "Junctionless multiple-gate transistors for analog applications," IEEE Trans. Electron Devices, vol. 58, no. 8, pp. 2511–2519, 2011.

[4] M. de Souza, R. T. Doria, R. Trevisoli, S. Barraud, and M. A. Pavanello, "On the Application of Junctionless Nanowire Transistors in Basic Analog Building Blocks," IEEE Trans. Nanotechnol., vol. 20, pp. 234–242, 2021.

[5] A. B. Shibutani, M. de Souza, R. Trevisoli, R.T. Doria, "Junctionless Nanowire Transistors Based Common-Source Current Mirror", accepted to be presented in SBMicro 2021, p. 1-4, 2021.

[6] J. P. Colinge et al., "Junctionless Nanowire Transistor (JNT): Properties and design guidelines," Solid. State. Electron., vol. 65–66, no. 1, pp. 33–37, 2011.

[7] J. P. Colinge and A. Chandrakasan, FinFETs and other multi-gate transistors. 2008.

[8] J. Ramirez-Angulo, R. G. Carvajal, and A. Torralba, "Low Supply Voltage High-Performance CMOS Current Mirror With Low Input and Output Voltage Requirements," IEEE Trans. Circuits Syst. II Express Briefs, vol. 51, no. 3, pp. 124–129, 2004.

[9] S. Barraud et al., "Scaling of trigate junctionless nanowire MOSFET with gate length down to 13 nm," IEEE Electron Device Lett., vol. 33, no. 9, pp. 1225–1227, 2012.

[10] Sentaurus, "Sentaurus sDevice 2015", Simulation, no. June, p. 2015, 2009.

[11] R. Trevisoli, R. T. Doria, M. de Souza, M. A. Pavanello, "Threshold Voltage in Junctionless Nanowire Transistors", Semicond. Sci. Technol., vol. 26, p 105009, 2011.

Improving the Photon Detection Probability of SPAD implemented in FD-SOI CMOS Technology with light-trapping concept

S. Gao
*Univ Lyon, INSA Lyon, CNRS, INL,
UMR5270*
Villeurbanne, France
shaochen.gao@insa-lyon.fr

D. Issartel
*Univ Lyon, INSA Lyon, CNRS, INL,
UMR5270*
Villeurbanne, France
dylan.issartel@insa-lyon.fr

R. Orobtchouk
*Univ Lyon, INSA Lyon, CNRS, INL,
UMR5270*
Villeurbanne, France
regis.orobtchouk@insa-lyon.fr

F. Mandorlo
*Univ Lyon, INSA Lyon, CNRS, INL,
UMR5270*
Villeurbanne, France
fabien.mandorlo@insa-lyon.fr

D. Golanski
STMicroelectronics
Crolles, France
dominik.golanski@st.com

A. Cathelin
STMicroelectronics
Crolles, France
andreia.cathelin@st.com

F. Calmon
*Univ Lyon, INSA Lyon, CNRS, INL,
UMR5270*
Villeurbanne, France
francis.calmon@insa-lyon.fr

Abstract—This article proposes a 3D electro-optical simulation method to estimate the Photon Detection Probability (PDP) of Single-Photon Avalanche Diodes (SPAD) implemented in 28nm Fully Depleted Silicon-On-Insulator (FD-SOI) CMOS technology. In order to improve the PDP of SPAD implemented in FD-SOI CMOS technology, a light-trapping approach is studied, thanks to the patterning of Shallow Trench Insolation (STI) layer and the patterning of Silicon substrate, respectively in the case of Front Side Illumination (FSI) and Back Side Illumination (BSI). An average gain of 50% for wavelengths between 400nm-1000nm and of 200% for wavelengths between 800-1000nm respectively in the case of FSI and BSI are achieved. Based on this study, IC fabrication including several designs of SPAD implemented in FD-SOI CMOS technology with different pattern sizes is launched for future electro-optical characterization.

Keywords—SPAD; 28nm FD-SOI CMOS; Photon Detection Probability PDP; TCAD Simulation; light-trapping

I. INTRODUCTION

Single-Photon Avalanches Diodes (SPAD), also known as Geiger-mode Avalanches Diodes are detectors of choice for many applications requiring low light condition and high temporal resolution [1]. SPAD devices are widely studied for the last few years and their sub-nanosecond response time and high light sensitivity have been proved [2]. Recently, SPAD devices have been successfully implemented in 28 nm Fully Depleted Silicon-on-Insulator (FD-SOI) CMOS technology from STMicroelectronics, allowing intrinsic 3D pixel [3] and indirect sensing of SPAD avalanche event [4]. In this technology, SPAD devices are located below the buried oxide (BOX) layer, on top of which the electronic readout circuits are inserted in the thin silicon layer. SPAD architecture is thus intrinsically 3D stacked and provides much higher fill-factor, which is one of the key figures of merit (FOM) of SPAD devices.

Another key FOM of SPAD devices is the Photon Detection Probability (PDP), which is defined as the probability of an absorbed incident photon to generate an avalanche event and represents the single-photon sensitivity of SPAD devices normalized by the active area. Several factors can greatly impact the PDP : wavelength, applied voltage, photogeneration rate, diode architecture with junction depth and doping levels, etc... To enhance the PDP, the approach of spreading the multiplication region [5] has been proposed with a complete revision of fabrication process and frequently compromised SPAD relevant performances. Other approaches of modifying the SPAD internal structure for instance by extending horizontally and vertically the photo-sensitive region (charge-focusing) [6] are also proposed. The PDP can also be improved by optical approaches with nanostructuration of the SPAD devices, which are inspired by the photon absorption management in thin solar cells and recently applied to SPAD devices [7-9]. In this study, an efficient 3D electro-optical simulation method is proposed in order to estimate the PDP . A light-trapping approach to improve the PDP of SPAD devices implemented in 28nm FD-SOI CMOS technology, using the proposed simulation flow, is proposed and demonstrated through a complete study both in the case of Front Side Illumination (FSI) and Back Side Illumination (BSI).

In section II, the PDP estimation method is described and the electrical properties of SPAD structure are also considered, while in section III, the simulation methodology with light-trapping approach is presented. Simulation results for both FSI and BSI cases are reported and discussed in section IV.

II. PHOTON DETECTION PROBABILITY SIMULATION METHOD

A. PDP simulation flow

In order to estimate the PDP of SPAD devices, we integrated the product of the Avalanche Triggering Probability (ATP) and the electron-hole photogeneration rate (G) over the active volume (V) of the SPAD. Then we divided by the incident photon flux $\Phi_{Photons}$ (number per second) at the considered wavelength λ:

$$PDP(\lambda) = \frac{1}{\Phi_{Photons}(\lambda)} \iiint ATP(x,y,z) * G(x,y,z,\lambda)dV \quad (1)$$

978-1-6654-3746-2/21 $31.00 © 2021 IEEE

As Eq. 1 suggested, the calculation of *PDP* included electrical and optical parts. The electrical and optical simulations were respectively performed with *Synopsys Sentaurus* and *Lumerical FDTD* in order to obtain data for both the electrical part (*ATP*, electric field *E*, carrier mobility µ) and the optical part (photogeneration rate *G*). An external *Matlab* routine was developed to combine and to postprocessed electrical and optical data, as shown in Fig. 1.

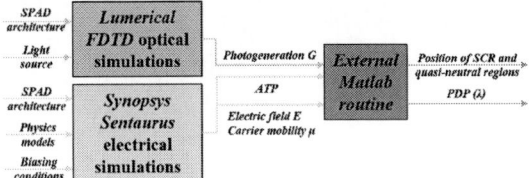

Figure 1. Block diagram of Photon Detection Probability estimation

For the electrical part, the simulations under the environment of *Synopsys Sentaurus* were 2D cylindrical symmetry. Whereas the optical simulations in *Lumerical FDTD* were 3D. The developed *Matlab* routine was able to extrapolate the 2D electrical data according to the cylindrical symmetry and to create 3D matrix of electrical data, in order to realise the integration of *ATP* and *G* 3D matrices and thus to calculate the *PDP*.

B. Electrical behavior of the investigated SPAD strucutre

In our consideration, the SPAD structure was composed of three regions: one Space Charge Region (SCR, also known as multiplication region) and two quasi-neutral regions, one of which was p-type doped and the other n-type doped. The limits between each region were defined by investigating the electric field distribution across the structure and the ratio between drift and diffusion velocities of minority carriers V_{drift}/V_{diff} in each quasi-neutral region, to evaluate the contribution of each region to the *PDP*. In Fig .2 and Fig. 3, the electrical field distribution and the velocities' ratio V_{drift}/V_{diff} distribution respectively at the p-side and the n-side limits of SCR are represented (electrical behavior at an excess voltage V_{ex} of 1.5V, which equals to 15% of the breakdown voltage V_{bd}).

We considered that when the direction of electric field was favorable for electrons, the p-type quasi-neutral region began (upper limit). The end of p-type quasi-neutral region (beginning of SCR) was given by the position where the electric field magnitude was below the threshold of -10^{-5}V/cm. The end of SCR (beginning of n-type quasi-neutral region) was the position where the electrical field magnitude was

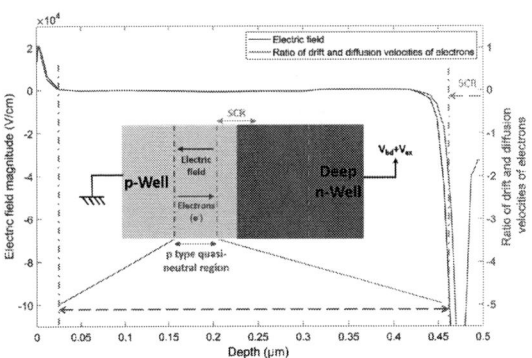

Figure 2. Electric field distribution and velocities' ratio ditribution V_{drift}/V_{diff} of electrons at the p-side limit of SCR

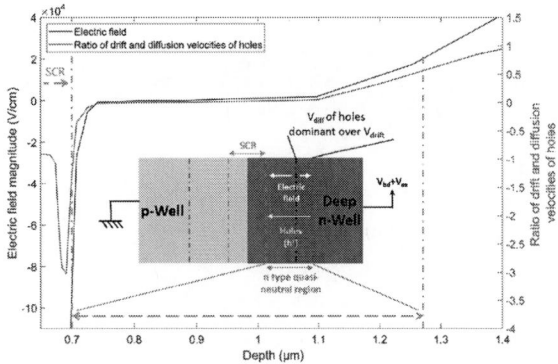

Figure 3. Electric field distribution and velocities ratio ditribution V_{drift}/V_{diff} of holes at the n-side limit of SCR

above the threshold. The end of the n-type quasi-neutral region was determined by the velocities' ratio V_{drift}/V_{diff}. In fact, after the position where the direction of electric field was no longer favorable for holes, the diffusion velocity of holes was dominant over the drift velocity of holes to a certain depth. We considered that when the ratio V_{drift}/V_{diff} was surpassed 0.5, the n-type quasi-neutral region ended.

III. ELECTRO-OPTICAL SIMULATION METHODOLOGY

Fig 4. shows the cross section of the SPAD implemented in 28nm FD-SOI CMOS technology. Due to design rules, small blocks of Shallow Trench Isolation (STI) exist in the active region above the p-Well and serve as patterns in the case of FSI. For BSI, patterning of silicon substrate with circular holes could be realised by using the technique of holographic insolation. The dimensions, the period of patterns and their positions need to be properly chosen in order to realize a photonic crystal layer providing light-trapping effects to locate the maximum of interferences in the SCR.

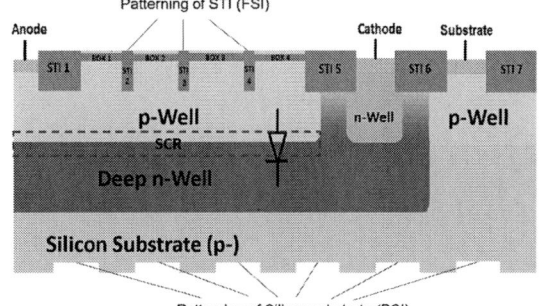

Figure 4. Schematic of SPAD implemented in 28nm FD-SOI CMOS technology (Scales are not respected and Back-End-of-Line (BEOL) layers are not represented)

The SPAD structure could be simplified from the fully optical point of view, as shown in Fig. 5. The SiO_2 layer and silicon substrate are both considered semi-infinite.

As aforementioned, the diffraction grating with proper dimensions could allow maximum of light intensity in the SCR and thus improve the *PDP*. Here the diffraction grating play two roles. Firstly, the diffraction grating serves as an antireflection coating allowing to minimize the reflection coefficient R_1 at interface SiO_2/diffracting grating in the case of FSI and the reflection coefficient R_4 at interface air/diffraction grating in the case of BSI. Both reflection coefficients are function of the optical refractive index of

978-1-6654-3746-2/21 $31.00 © 2021 IEEE

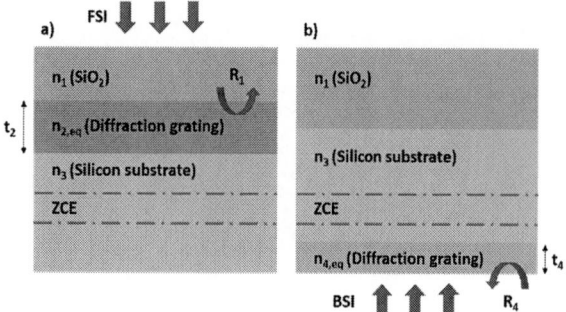

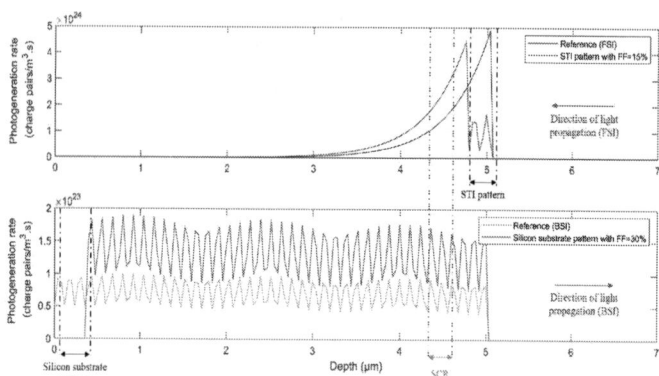

Figure 5. Simplified equivalent optical schematic of the SPAD structure in the case of FSI (a) and of BSI (b)

diffraction grating and adjacent layers and also of the thickness of the diffraction grating. Eq 2. and Eq. 3 show the equivalent optical refractive index $n_{2,eq}$ and $n_{4,eq}$ respectively in the case of FSI and BSI, which follow an effective medium law [10]:

$$n_{2,eq}^2 = n_{SiO_2}^2 * (1 - FF) + n_{silicon}^2 * FF \qquad (2)$$

$$n_{4,eq}^2 = n_{air}^2 * (1 - FF) + n_{silicon}^2 * FF \qquad (3)$$

The equivalent refractive indexes are function of the Filling Factor (FF) which is the ratio between the silicon area and the total area. The second role of diffraction grating is to create interferences and to locate the maximums of light intensity in the SCR where the ATP is the highest. The positions of the maximums of light intensity are function of the pattern period. Appropriate pattern periods could increase the photogeneration rate in the SCR and improve the PDP.

Our strategy of the PDP optimization is to first simulate the reflection behavior with small pattern period (less than 100nm) in order to obtain the suitable FF. The second step is to find the optimal pattern period which allows the maximums of interferences in the SCR.

IV. SIMULATION RESULTS

A. Technological constrains in FSI

In the case of FSI, several constrains exist due to the design rules and fabrication process. Firstly, the thickness and the depth of the STI blocks are determined by the technological process. Moreover, the design rules restrict the width, length of STI blocks and distance between them. Secondly, in the standard process, the Back-End-of-Line (BEOL) layers exist above the SPAD structure in order to insure the electrical connection, the passivation and the encapsulation. The BEOL layers are not optimized for light transmission and the characterization of their optical properties are not fully realised. Nonetheless, the optimisation strategy could still be applied.

B. Optimization of FF

The first step consisted of studying the antireflective effect of the diffraction grating. A pattern period of 10nm and several values of FF were applied to both cases of FSI and BSI. Fig. 6 illustrates the distribution of photogeneration rate of one STI pattern with FF=15% (FSI) and of one silicon substrate pattern with FF=30% (BSI). These profiles were extracted near the centre of one pattern. The incident light power was set to 1W/m² and the thickness of STI pattern and silicon substrate pattern were set to 330nm and 370nm respectively.

Figure 6. Photogeneration rate of one pattern for a reference SPAD and for a patterned SPAD respectively at wavelength of 500nm and 900nm for FSI (top) and BSI (bottom)

As showed in Fig. 6, the diffracting gratings in both cases of FSI and BSI reduced significantly the reflection. Relative enhancement of photogeneration rate up to 60% for FSI and up to 90% for BSI were obtained.

C. Optimization of pattern period

In the second step, the value of pattern period was varied so as to locate the maximums of interferences in the SCR. Several values of pattern period for each FF were simulated.

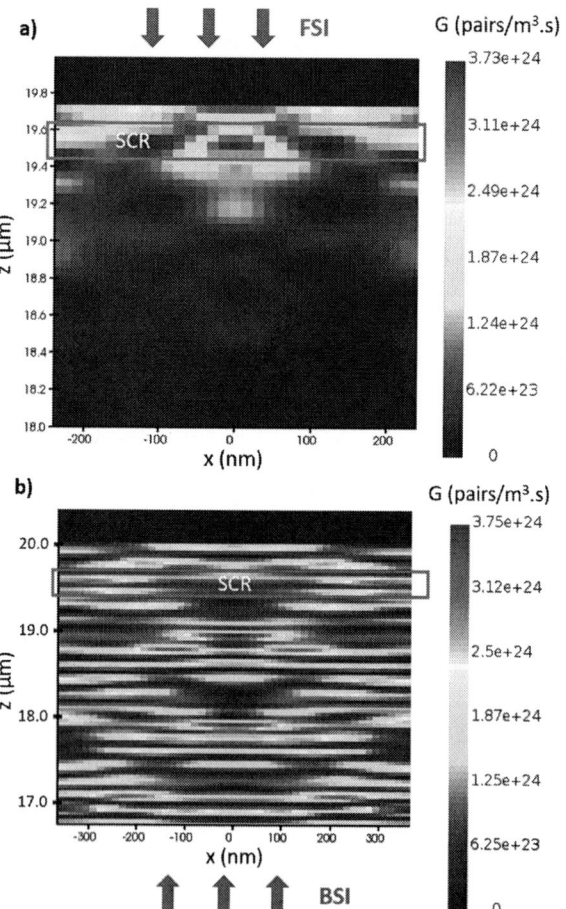

Figure 7. Cartography of photogeneration rate of patterned SPAD with FF=15% and period=0.48µm at λ=500nm for FSI (a) and with FF=30% and period=0.73µm at λ=900nm for BSI(b)

Fig. 7 shows respectively an example of cartography of photogeneration rate with a pattern period of 0.48µm and an FF of 15% for FSI and with a pattern period of 0.73µm and an FF of 30% for BSI. The thickness of STI pattern and silicon substrate pattern were 330nm and 370nm. Fig. 7 illustrates the existence of maximums of interferences in the SCR, showing clearly the diffracting effect of the pattern.

D. PDP calculation

In order to observe the *PDP* improvement with the diffracting pattern for other wavelengths, simulations with a larger range of wavelengths of 400-1000nm for FSI and of 800-1000nm for BSI were performed with the same values of FF and pattern period. Fig. 8 and Fig. 9 show respectively the *PDP* distribution at V_{ex} of 0.6V (6% of V_{bd}) and 1.5V (15% of V_{bd}) for both FSI (FF=15% and period=0.48µm) and BSI (FF=30% and period=0.73µm). The second value of V_{ex} allows observe greater difference between reference SPAD and patterned SPAD, since higher values of V_{ex} lead to greater values of *ATP*.

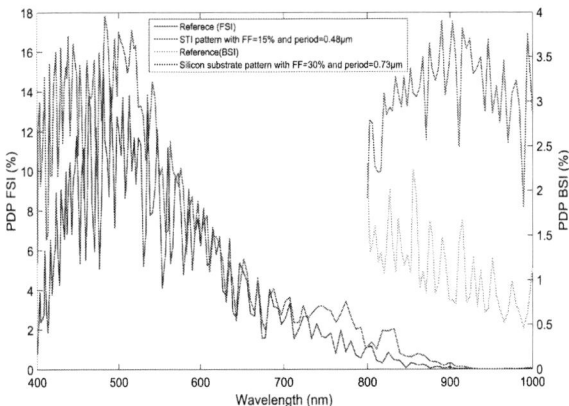

Figure 8. PDP (λ) for reference SPAD and for patterned SPAD at V_{ex}=0.6V (FSI: FF=15% and pattern period=0.48µm; BSI: FF=30% and pattern period=0.73µm

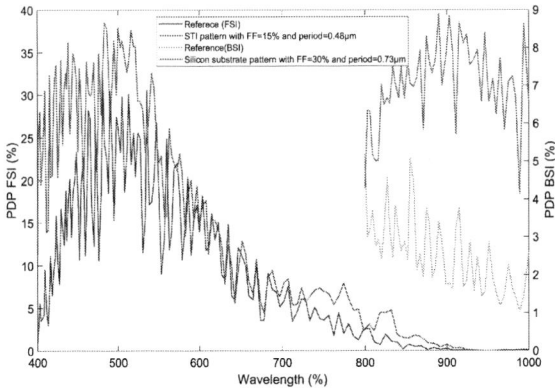

Figure 9. PDP (λ) for reference SPAD and for patterned SPAD at V_{ex}=1.5V (FSI: FF=15% and pattern period=0.48µm; BSI: FF=30% and pattern period=0.73µm

In the case of FSI, the curves show a significant relative *PDP* increase for the short- and long-range wavelengths. The relative gain could respectively reach over 100% and 300%. For the mid-range wavelengths, the relative gain was smaller (over 25%) but with higher absolute values of *PDP*. An average gain of over 100% was obtained for the whole range of wavelengths. In the case of BSI, the difference between reference SPAD and patterned SPAD was more significant

and an average gain of over 200% was achieved for wavelengths of 800nm and 1000nm. Thanks to these simulations results, several pattern dimensions for IC fabrication and future characterization were submitted.

V. CONCLUSION

In this article, a 3D electro-optical simulation method for SPAD Photon Detection Probability (*PDP*) calculation is presented. Then the light-trapping approach thanks to the patterning of STI layer (Front Side Illumination) and of silicon substrate layer (Back Side Illumination) is proposed to improve the *PDP* of a SPAD fabricated in 28nm FD-SOI CMOS technology. An average gain of over 50% (FSI) for wavelengths between 400-1000nm and of 200% (BSI) for wavelengths between 800-1000nm are achieved. These encouraging results have permitted IC fabrication including several designs of FD-SOI CMOS SPADs with different STI pattern sizes for electro-optical characterization for FSI. Future work will address the characterization of samples with/without patterns.

ACKNOWLEDGMENT

The authors would like to thank the Nano2022 research program for the PhD Grand of Shaochen Gao, the French national research agency ANR (ANR-18-CE24-0010) for the PhD Grand of Dylan Issartel and CMP (Grenoble) for IC prototyping services.

REFERENCES

[1] D. Bronzi, F. Villa, S. Tisa, A. Tosi and F. Zappa, "SPAD figures of merit for photon-counting, photon-timing, and imaging applications: A review", IEEE Sensors J., vol. 16, no. 1, pp. 3-12, 2016.

[2] M-J Lee and E. Charbon, "Progress in single-photon avalanche diode image sensors in standard CMOS: From two-dimensional monolithic to three-dimensional-stacked technology", Jpn. J. Appl. Phys., vol. 57, no. 1002A3, 2018.

[3] T. Chaves de Albuquerque, F. Calmon, R. Clerc, P. Pittet, T. Benhammou, D. Golanski, S. Jouan, D. Rideau and A. Cathelin, "Integration of SPAD in 28nm FDSOI CMOS technology", 48th European Solid-State Device Research Conference (ESSDERC), pp. 82-85, 2018.

[4] T. Chaves de Albuquerque, D. Issartel, R. Clerc, P. Pittet, R. Cellier, W. Uhring, A. Cathelin and F. Calmon, "Body-biasing considerations with SPAD FDSOI: advantages and drawbacks", 49th European Solid-State Device Research Conference (ESSDERC), pp. 210-213, 2019.

[5] A. Gulinati, I. Rech, F. Panzeri, C. Cammi, P. Maccagnani, M. Ghioni and S. Cova, "New silicon SPAD technology for enhanced red-sensitivity, high-resolution timing and system integration", J. Mod. Opt., vol. 59, no. 17, pp. 1489-1499, 2012.

[6] K. Morimoto, "Megapixel SPAD Cameras and Time-Resolved Applications", PhD thesis, Advanced Quantum Architecture Laboratory, Ecole polytechnique fédérale de Lausanne, Switzerland. http://www.kunter-fonds.ethz.ch/APP_Themes/default/datalinks/Morimoto_PhD_Thesis_%202020.pdf

[7] J. Ma, M. Zhou, Z. Yu, X. Jiang, Y. Huo, K. Zang, J. Zhang, J.S Harris, G. Jin, Q. Zhang and J. Pan, "Simulation of a high-efficiency and low-jitter nanostructured silicon single-photon avalanche diode", Optica, vol. 2, issue 11, pp. 974-979, 2015.

[8] K. Zang, X. Jiang, Y. Huo, X. Ding, M. Morea, X. Chen, C. Lu, L. Ma, M. Zhou, Z. Xia, Z. Yu, T.I. Kamins . Q. Zhang and J.S. Harris, "Silicon single-photon avalanche diodes with nano-strucutured light trapping", Nat. Commun., vol. 8, no. 628, 2017.

[9] L. Frey, M. Marty, S. André and N. Moussy, "Enhancing near-infrared photodetection efficiency in SPAD with silicon surface nanostructuration", J. Electron Devices Soc., vol. 6, pp. 392-395, 2018.

[10] F. Mandorlo, M. Amara, H.S. Nguyen, A. Charlty-Meano, A. Belarouci and R. Orobtchouk, "Color management of semi-transparent nano-patterned surfaces", Opt. Eng., vol. 60(5), p. 055101, 2021

Characterization and Lambert – W Function based modeling of FDSOI five-gate qubit MOS devices down to cryogenic temperatures

E. Catapano[1,2], A. Aprà[3], M. Cassé[2], F. Gaillard[2], S. de Franceschi[3], T. Meunier[4], M. Vinet[2] and G. Ghibaudo[1]

1) IMEP-LAHC, Univ. Grenoble Alpes, Minatec, 38016 Grenoble, France. 2) CEA-LETI, Univ. Grenoble-Alpes, Minatec, 38054 Grenoble, France. 3) CEA-IRIG, Univ. Grenoble-Alpes, Minatec, 38042 Grenoble, France.

4) CNRS, Institut Néel, Univ. Grenoble Alpes, 38042 Grenoble, France

Abstract — FD-SOI five-gate (5G) qubit MOS devices are electrically characterized in linear regime down to deep cryogenic temperatures. The Lambert-W function is successfully used for the modelling of such 5G MOS devices from subthreshold regime to strong inversion. Its applicability is demonstrated down to 20 K. The 5G device is modeled as a series of five independent transistors: the "active" one, that directly controls the current, and the "external" ones, that act as access resistances. The Lambert-W function enables to accurately determine the inversion charge and the active channel resistance from weak to strong inversion. This approach allows reconstructing the drain current characteristic avoiding the evaluation of the mobility attenuation factors. The main device parameters are extracted versus temperature. Finally, the role of different scattering mechanisms has been investigated, underlying the impact of neutral defects for the gates in proximity of source and drain.

Keywords – FDSOI, five-gate qubit device, characterization, Lambert function, modeling, parameter extraction.

I. Introduction

Quantum computing is an appealing technology for many fields, since it promises a boost in computational performances making it suitable to deal with exponentially-growing problems. In the last twenty years, many efforts have been devoted in developing solid-state qubits, and more recently, the fabrication of spin qubits based on Silicon-On-Insulator (SOI) CMOS platform has been demonstrated [1] [2] [3]. Their interest mainly relies on the possibility of converting the expertise in transistor manufacturing into large-scale qubit technology. Moreover, silicon quantum bits pave the way to the possibility to integrate the control CMOS – based electronics on the quantum processor itself [4]. Indeed, one of the main issues of nowadays quantum computers is the limitation in the number of wiring connections between the room temperature control electronics and the quantum chip. Since future quantum computers, in order to accomplish high demanding tasks, will require several hundreds or even millions of qubits, their co – integration with the classical electronics seems to be the only viable solution. Furthermore, for qubit mass production [5] [6], fast screening and selection of functional devices is mandatory, and electrical parameters extraction methods effective in a wide range of temperatures must be developed accordingly. Indeed, industrial approach to qubit manufacturing requires not only the technology to be mature and reliable, but also easy-to-use device compact models for device rapid characterization.

In this paper, we characterize a p-type FD-SOI five-gate qubit MOS transistor biased in linear regime down to very low temperatures. Particular attention is devoted to subthreshold regime at deep cryogenic temperature, where carriers are

known to become degenerate, and Boltzmann statistics is not reliable anymore. Afterwards, we propose a model based on the Lambert – W function, whose validity is demonstrated, for the first time, down to deep cryogenic temperatures, fitting both the drain current and the $Y - function$ and providing the dependence of the main figures of merit with temperature. Moreover, the values of parameters such extracted are compared with those obtained using standard methods. Finally, transport scattering mechanisms are studied, in order to investigate the mobility behavior as function of the distance from source and drain.

II. Device and Experimental Details

The p-type five-gate device (5G) presented in this work is sketched in Fig. 1. It has been fabricated starting from

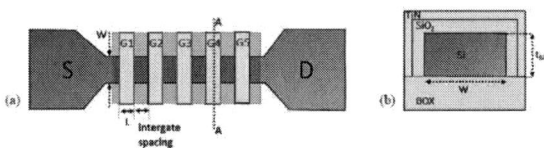

Figure 1. (a) 5G device top-view schematic. (b) AA cross-section view

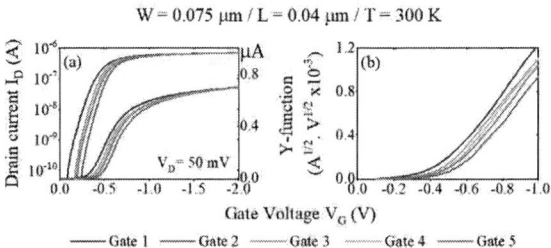

Figure 2. (a) $I_D(V_G)$ and (b) $Y(V_G)$ for every gate at T = 300K.

CEA-LETI FD-SOI NanoWire (NW) process flow [1] [3]. NWs are fabricated from $300\,mm$ SOI wafer, with $145\,nm$ thick buried oxide. The silicon film thickness is $t_{Si} = 17.9\,nm$, defining the NW height, and covered by a $6\,nm$ thick SiO_2 oxide and by a $5\,nm$ TiN metal gate (Fig.1). Si_3N_4 spacers, whose length is nominally equal to the gate one, i.e. $L_G = 40nm$, separate the gates. The channel width is $W_G = 75nm$. The device has been tested in a dilution fridge from room temperature down to 20K. Static measurements of the drain current in linear regime ($|V_{DS}| = 50\,mV$) were performed sweeping the voltage on one gate (*active gate*), while keeping the other gates (*external gates*) at a fixed

978-1-6654-3746-2/21 $31.00 © 2021 IEEE

potential, namely $V_{G,ext} = -2V$. In such a way, the current was controlled by the active gate.

III. EXPERIMENTAL RESULTS

A. Drain current transfer characteristics

Drain current $I_D(V_G)$ transfer characteristics for the five gates at room temperature are shown in Fig. 2(a). From $I_D(V_G)$ curves, it is clear that each gate exhibits a different threshold voltage V_{th} and a different low field mobility μ_0. Furthermore, access resistance strongly affects the behavior of the device at high gate voltages. It is worth to point out that here the access resistance comprehends both the source/drain resistances and the external gate channels. In Fig.2.b, we plot the corresponding $Y-function$, given by $Y(V_G) = \frac{I_D}{\sqrt{g_m}} = \beta^{\frac{1}{2}}(V_G - V_{th})$, where $g_m = \frac{dI_D}{dV_G}$ is the transconductance, $\beta = \frac{W_G}{L_G}C_{ox}\mu_0 V_D$ is the gain current factor and C_{ox} is the gate oxide capacitance. $Y-function$ is known to eliminate series resistance influence [7]. Both $I_D(V_G)$ and $Y(V_G)$ for different temperatures from 300K to 20K are presented in Fig. 3. Note the strong increase in the subthreshold slope as the temperature decreases (see also Fig. 4), as well as the increase of both the

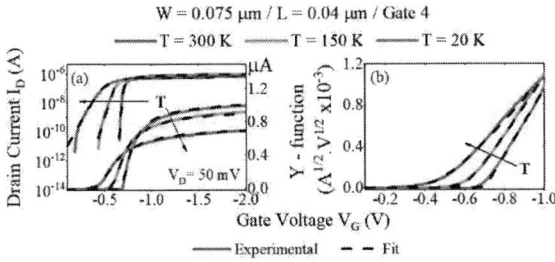

Figure 3. Experimental (straight lines) and fitted (dashed lines) (a) $I_D(V_G)$ and (b) $Y(V_G)$ curves for different temperature

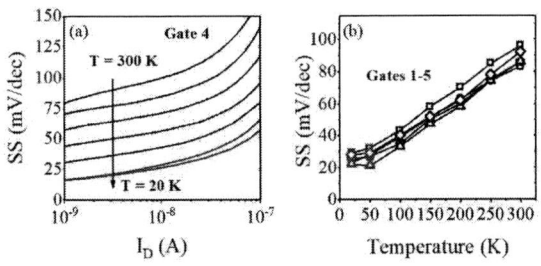

Figure 4. (a) $SS(I_D)$ for different temperatures. Below $T = 100K$ it starts to saturate. (b) $SS(T)$ for all the five gates.

threshold voltage and the drain current in strong inversion. The latter is due to the increase of the mobility μ_0, consequence of the phonon number reduction at cryogenic temperatures, as confirmed by the $Y-function$ slope, steeper at lower temperatures. Concerning the subthreshold regime, the subthreshold swing $SS = \frac{dV_G}{dlog(I_D)}$ follows the Boltzmann limit, i.e. it shows a linear behavior with temperature, down to $T = 70 - 80K$, before saturating to a value close to $20\ mV/dec$ (Fig. 4) [8] [9]. The trend is the same independently of the gate, and it is consistent with what is observed in standard CMOS technology. The saturation of SS at deep cryogenic temperature is due to the presence of exponential band tails in the density of states below band edge [8] [9]. Indeed, at low temperature the Fermi level falls into the band tail making the carriers degenerate. In this case,

Boltzmann statistics is not valid anymore, and the metallic statistics prevails. The exponential band tails in MOSFET are likely related to potential-fluctuations-induced disorder [10] [11].

B. Lambert function based modeling

Lambert – W function allows to model the channel inversion charge $Q_i(V_G)$ using as fitting parameters both the subthreshold ideality factor n and the threshold voltage V_{th}, and, by turn, the drain current characteristic $I_D(V_G)$, for a given mobility μ_0 [12]. Here, we developed the model introduced in [13], assuming that the 5G device can be modeled in linear operation with five independent gate-controlled resistances placed in series, whose values depend on V_G (active channel) and $V_{G,ext}$ (access channels). Each resistance is characterized by different intrinsic parameters, i.e. V_{th} , n and μ_0 . The active channel resistance, corresponding to gate i, is given by:

$$R_{Ch,i}(V_G,\mu_{0,i},V_{th,i},n_i) = \frac{1}{\frac{W}{L_G}Q_i(V_G,V_{th,i},n_i)\mu_{0,i}} \quad (1)$$

with $Q_i(V_G,V_{th,i},n_i) = n_i.C_{ox}.\frac{k_B T}{q}.LW\left(e^{\frac{V_G-V_{th,i}}{n_i.kT/q}}\right)$,

where k_B is the Boltzmann constant, T is the temperature in kelvin and q is the electron charge. The access resistance is computed as:

$$R_{Acc}(V_{G,ext}) = R_{ch,j}(V_{G,ext},\mu_{0,j},V_{th,j},n_j) +$$
$$R_{ch,k}(V_{G,ext},\mu_{0,k},V_{th,k},n_k) +$$
$$R_{ch,l}(V_{G,ext},\mu_{0,l},V_{th,l},n_l) +$$
$$R_{ch,m}(V_{G,ext},\mu_{0,m},V_{th,m},n_m) + R_{Series} \quad (2)$$

where $R_{ch,i}, R_{ch,j}, R_{ch,k}, R_{ch,l}$ are the channel resistances of gates $i,j,k\ and\ l$, and R_{Series} is an additional fitting parameter to take into account the contribution of source and drain access. Since the number of fitting parameters is quite high, each gate has been considered individually, allowing the extraction of the intrinsic parameters from $Y(V_G)$ and the subthreshold current. After, the access resistance has been automatically computed according to Eq. (2), using R_{Series} as input parameter to fit $I_D(V_G)$ in strong inversion. It is worth noting that no degradation mobility laws as function of V_G have been introduced to account for current saturation at high voltages, actually dominated by access resistance effect. The fittings for both $I_D(V_G)$ and $Y(V_G)$ characteristics and for temperatures varying from 300K to 20K are shown in Fig. 3. The model successfully fits the drain current from subthreshold to strong inversion regime, allowing the extraction of the main device parameters (Fig. 5.(a)-(d)), whose values and trends are in good agreement with those extracted using classical methods. Indeed, both V_{th} and μ_0 have been extracted using the $Y-function$, while R_{Acc} has been derived from the expression of the first order attenuation factor $\theta_1 = \theta_{10} + R_{Acc}G_m$ [7], where θ_{10} is the intrinsic mobility reduction factor and $G_m = \frac{W}{L}C_{ox}\mu_0$. θ_{10} has been neglected, since in such short channels the access resistance effect prevails. Finally, n has been extracted from the subthreshold swing according to $SS =$

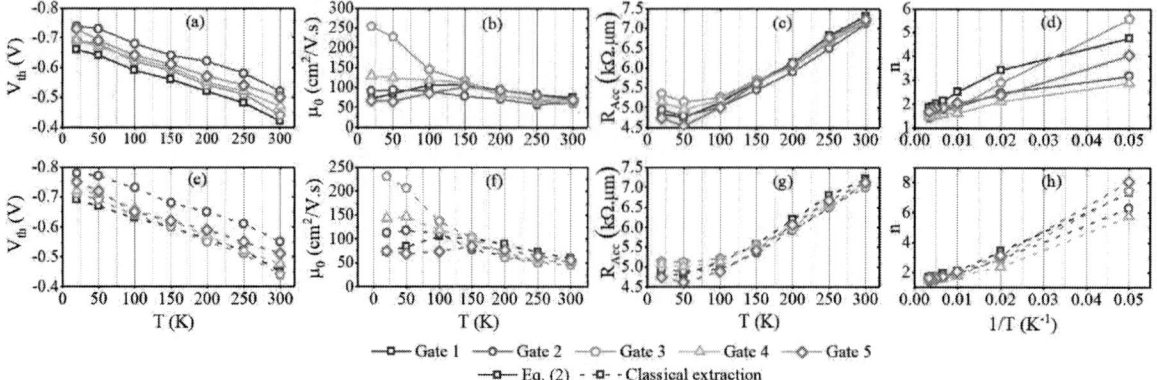

Figure 5. Parameters extracted for the every gate from room T down to 20K using eq.(2) (a)-(d) and classical extraction methods (e)-(h).

$n\left(\frac{k_B T}{q}\right) ln10$. It is important to underline that this expression of the subthreshold slope relies on the Boltzmann statistics; therefore, it is valid as far as the semiconductor is non-degenerate, i.e. down to $T \approx 70 - 80K$ in the present case. At lower temperatures, when SS in Fig. 4(b) saturates, the values of the ideality factor are not physically meaningful anymore [9]. n increases to keep SS constant while temperature lowers down. In the Lambert – based model, that relies on Boltzmann statistics as well, n is used as fitting parameter, and the comparison with the classical expression previously introduced has the sole scope of validating the consistency of the compact model with respect to the existing extraction methods. It must be noted that the discrepancy between Boltzmann and Fermi – Dirac statistics is mostly relevant in subthreshold region, because in strong inversion the Fermi level goes over the valence (conduction) band edge and $Q_i(V_G) \approx C_{Ox}(V_G - V_{th})$.

The threshold voltage decreases quasi linearly with temperature (Fig. 5(a),(e)), independently on the gate, whereas the ideality factor n varies nearly as 1/T below 70-80K since SS is plateauing (Fig. 5(d),(h)). The low field mobility (Fig. 5(b),(f)), instead, shows different trends for different gates, revealing that the central device (gate 3) has a better mobility at low temperature. Exploiting the Mathiessen's rule it is possible to have the quantitative analysis of mobility behavior [14] [15]:

$$\frac{1}{\mu_0} = \frac{T}{300 \cdot \mu_{ph}} + \frac{300}{T \cdot \mu_C} + \frac{1}{\mu_{nd}} \quad (3)$$

where μ_{ph}, μ_C and μ_{nd} are the phonon, Coulomb and neutral defects scattering mechanisms, respectively. Eq. (3) is used to fit the experimental data (Fig. 6(a)), allowing the extraction of the different scattering mechanisms for every gate. It is clear from Fig. 6(b) that neutral defects are the limiting mechanism for gates closer to source and drain, whereas they are less impacting the central device. Indeed, the latter shows a mobility increase as the temperature is lowered down, coherently with a transport dominated by phonon scattering. These results are consistent with recent studies of mobility and transport in FD-SOI nano-devices, where it is shown that the scattering processes in highly scaled devices are led by source and drain neutral defects [14] [15].

Finally R_{Acc}, computed using Eq. (2), also shows a linear decrease with T down to $100K$, before saturating for temperatures below $100K$. In Fig. 7 are plotted both R_{Series} and $R_\beta = R_{Acc} - R_{Series}$. The latter decreases down to $T = 100K$, then it increments again for almost all the gates. This is coherent with the definition given in Eq. (1), since R_{ch} is

inversely proportional to the low field mobility, that has been shown to be not very sensitive to temperature. The behavior of R_{Acc} is therefore driven by R_{Series}, that shows a metallic-like trend versus temperature.

It should be pointed out that, except for R_{Acc}, the values of the parameters extracted at room temperature are in line with standard FD-SOI MOSFET technology [12].

The highest value of both R_{Acc} and R_β below $T = 100K$ is found for gate 3, consistently with the mobility trend. Indeed, the resistance viewed from the other four gates is smaller because it includes the contribution of the central gate, which shows the highest mobility. For the same reason, the access resistance of gate 3 is the highest.

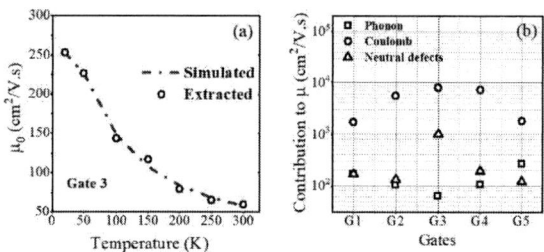

Figure 6. (a) Experimental (symbols) and simulated (dashed line) low field mobility as function of temperature. (b) Contribution of different scattering mechanisms to the total mobility

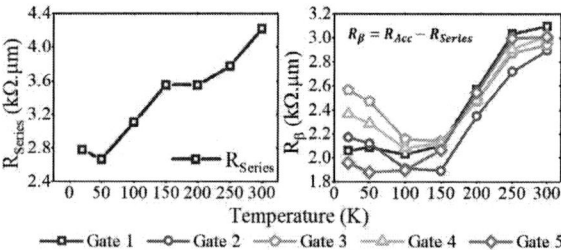

Figure 5 Series resistance R_{Series} (a) and R_β (b) down to 20K

It is worth to highlight that, with the exception of the central gate, the drain current increase at deep cryogenic temperature cannot be justified neither with an increased mobility nor a decreased R_β. Actually, it is mainly related to a reduction of R_{Series}.

IV. CONCLUSIONS

In this paper, we reported the functionality of a five-gate MOS based qubit device, biased in linear region, down to deep cryogenic temperature. Particular emphasis has been devoted to the subthreshold regime, where the subthreshold slope shown the well-known linear temperature dependence, before saturating below $70 - 80\ K$. Furthermore, a compact model based on the Lambert $-$ W function has been developed for both the drain current and the $Y - function$. Its validity has been demonstrated from subthreshold regime to strong inversion and from room down to deep cryogenic temperatures. Exploiting this model, the main device figures of merit have been extracted. Their values are in agreement with those obtained using classical extraction methods, and are in line with the state-of-the-art FD-SOI technology. Finally, the transport scattering processes have been investigated for different gates as function of temperature, highlighting different mechanisms depending on the proximity to source and drain. Consistently with previous studies, neutral defects scattering turned out to be the limiting mechanism in nano-scale devices transport.

In conclusion, we proposed a simple compact model for multi-gate MOS based qubit devices, valid in a wide range of temperatures, that could easily be extended to non-linear operation region and advantageously exploited as a characterization tool for the future qubit mass production.

ACKNOWLEDGMENT

The author are grateful to Labex MINOS of French ANR ANR-10-LABX-55-01, the ERC Synergy QuCube (Grant No. 810504), and EU H2020 RIA project SEQUENCE (Grant No. 871764).

REFERENCES

[1] L. Hutin et al., "Si CMOS platform for quantum information processing," in 2016 IEEE Symposium on VLSI Technology, Honolulu, HI, USA, Jun. 2016, pp. 1–2. doi: 10.1109/VLSIT.2016.7573380.

[2] R. Maurand et al., "A CMOS silicon spin qubit," Nat. Commun., vol. 7, no. 1, p. 13575, Dec. 2016, doi: 10.1038/ncomms13575.

[3] S. De Franceschi et al., "SOI technology for quantum information processing," in 2016 IEEE International Electron Devices Meeting (IEDM), San Francisco, CA, USA, Dec. 2016, p. 13.4.1-13.4.4. doi: 10.1109/IEDM.2016.7838409.

[4] X. Xue et al., "CMOS-based cryogenic control of silicon quantum circuits," Nature, vol. 593, no. 7858, pp. 205–210, May 2021, doi: 10.1038/s41586-021-03469-4.

[5] R. Li et al., "A flexible 300 mm integrated Si MOS platform for electron- and hole-spin qubits exploration," in 2020 IEEE International Electron Devices Meeting (IEDM), San Francisco, CA, USA, Dec. 2020, p. 38.3.1-38.3.4. doi: 10.1109/IEDM13553.2020.9371956.

[6] R. Pillarisetty et al., "High Volume Electrical Characterization of Semiconductor Qubits," in 2019 IEEE International Electron Devices Meeting (IEDM), San Francisco, CA, USA, Dec. 2019, p. 31.5.1-31.5.4. doi: 10.1109/IEDM19573.2019.8993587.

[7] G. Ghibaudo, "New method for the extraction of MOSFET parameters," Electron. Lett., vol. 24, no. 9, p. 543, 1988, doi: 10.1049/el:19880369.

[8] G. Ghibaudo, M. Aouad, M. Casse, S. Martinie, T. Poiroux, and F. Balestra, "On the modelling of temperature dependence of subthreshold swing in MOSFETs down to cryogenic temperature," Solid-State Electron., vol. 170, p. 107820, Aug. 2020, doi: 10.1016/j.sse.2020.107820.

[9] A. Beckers, F. Jazaeri, and C. Enz, "Theoretical Limit of Low Temperature Subthreshold Swing in Field-Effect Transistors," IEEE Electron Device Lett., vol. 41, no. 2, pp. 276–279, Feb. 2020, doi: 10.1109/LED.2019.2963379.

[10] E. Arnold, "Disorder-induced carrier localization in silicon surface inversion layers," Appl. Phys. Lett., vol. 25, no. 12, pp. 705–707, Dec. 1974, doi: 10.1063/1.1655369.

[11] N. F. Mott, M. Pepper, S. Pollitt, R. H. Wallis, and C. J. Adkins, "The Anderson transition," Proc. R. Soc. A, vol. 345, pp. 169–205, doi: https://doi.org/10.1098/rspa.1975.0131.

[12] T. A. Karatsori et al., "Full gate voltage range Lambert-function based methodology for FDSOI MOSFET parameter extraction," Solid-State Electron., vol. 111, pp. 123–128, Sep. 2015, doi: 10.1016/j.sse.2015.06.002.

[13] E. Catapano et al., "Statistical and Electrical Modeling of FDSOI Four-Gate Qubit MOS Devices at Room Temperature," IEEE J. Electron Devices Soc., vol. 9, pp. 582–590, 2021, doi: 10.1109/JEDS.2021.3082201.

[14] M. Shin et al., "In depth characterization of electron transport in 14nm FD-SOI CMOS devices," Solid-State Electron., vol. 112, pp. 13–18, Oct. 2015, doi: 10.1016/j.sse.2015.02.012.

[15] M. Shin et al., "Low temperature characterization of mobility in 14nm FD-SOI CMOS devices under interface coupling conditions," Solid-State Electron., vol. 108, pp. 30–35, Jun. 2015, doi: 10.1016/j.sse.2014.12.013.

Device simulations of ion-sensitive FETs with arbitrary surface chemical reactions

Leandro Julian Mele*, Pierpaolo Palestri*‡ and Luca Selmi†

*DPIA, University of Udine, 33100, Udine, Italy.

†DIEF, University of Modena and Reggio Emilia, 44100, Modena, Italy.

‡Corresponding author. Email: pierpaolo.palestri@uniud.it

Abstract—In this work, we exploit the general-purpose solver COMSOL, equipped with electrolyte and semiconductor physics modules, to implement a versatile model of potentiometric chemical sensors including arbitrarily complex surface reactions at the oxide/electrolyte interface with examples on 2D device-level simulations of an ISFET. Firstly, Multiphysics simulations of V_{TH} sensitivity to pH sensing are compared with analyses based on semiconductor TCAD. Then, more complex Na^+ sensing experiments are examined and numerical simulations are compared against 1D electrochemical models.

Index Terms—ion-sensitive FETs, TCAD, device simulations, surface chemical reactions, potentiometric sensors

I. Introduction

Ion-sensitive FETs (ISFETs) have widespread application as on-chip integrated biosensors, offering advantages in terms of miniaturization, low-cost and CMOS compatibility [1]. In ISFETs, the transduction process takes place at the interface between a (functionalized) solid surface and an electrolyte containing ions/analytes. Many ISFET-based sensors have been studied for a variety of analytes [2]; in a few of them the fabrication of the sensing surface occurs in the back-end of line of an unaffected CMOS process flow suited to integrate readout and digitalization functionalities [2], [3].

To predict and engineer the sensor response at the device level, one should combine into a comprehensive model the electrochemical processes taking place in the electrolyte and at the interfaces (i.e. surface chemical reactions), as well as their coupling with the underlying FET device. Such a combination, however, is not always straightforward with commercial TCAD for semiconductor device modeling. Recently, a few solutions to this issue have been proposed based on, e.g. Sentaurus™ [4], where the electrolyte is modeled as a "generic" semiconductor (with appropriate gap, density of states and mobility to describe cations and anions as electrons and holes) and surface reactions with protons are introduced via physical model interfaces (PMIs) [5], [6]. Clearly, such a workaround is possible only for electrolytes where the ionic concentration is dominated by 1:1 dissolved salts (e.g., Na^+ and Cl^- dissolved in water) [7]. An alternative is offered by general-purpose multiphysics solvers such as COMSOL Multiphysics® [8], where semiconductor modeling modules

We acknowledge funding from the European Unions Horizon 2020 research and innovation programme under grant agreement No 862882 (IN-FET project) via the IUNET Consortium.

can be combined with electrolyte modules to simulate multi-ion solutions.

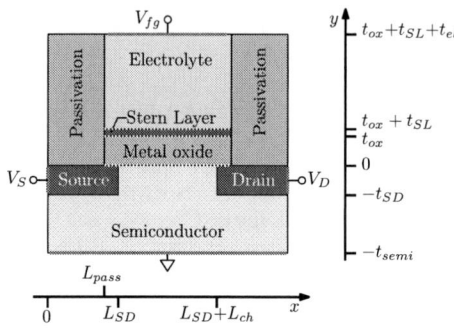

Fig. 1. Template ISFET structure (not to scale) used for the simulations in Sentaurus TCAD and COMSOL. Parameter values are reported in Table I. Electrical contacts are highlighted in magenta whereas dotted lines indicate the boundaries where ad-hoc equations for boundary conditions are used to replace simulation domains and to include surface reactions.

In this paper we investigate the pros' and cons' of TCAD and Multiphysics approaches for the simulation of pH ISFETs by simulating the template device in Fig. 1 and validating the COMSOL implementation of a multiphysics model with a Sentaurus model empowered with the workaround presented in [6] to handle pH sensitive surfaces. Moreover, we demonstrate the versatility of multiphysics tools in capturing arbitrary sets of surface reactions, by implementing in COMSOL the recently proposed model [9].

II. Description of the template ISFET

We consider the 2D structure shown in Fig. 1 for the TCAD and COMSOL models, with the physical and geometrical parameters reported in Table I. The bulk is p-type semiconductor ($N_A = 10^{15}$ cm^{-3}) whereas the source and drain have $N_D = 10^{20}$ cm^{-3}. The device width is $W = 1$ μm.

TABLE I
PHYSICAL AND GEOMETRICAL PARAMETERS USED IN THIS WORK

Param.	Value	Units	Param.	Value	Units
t_{SL}	1	[nm]	t_{semi}	0.5	[μm]
t_{el}	0.5	[μm]	L_{SD}	200	[nm]
t_{ox}	3	[nm]	L_{pass}	180	[nm]
t_{SD}	45	[nm]	L_{ch}	200	[nm]
ε_{ox}	3.9	-	ε_{el}	78.5	-
ε_{Si}	11.7	-	ε_{Stern}	22.58	-

978-1-6654-3746-2/21 $31.00 © 2021 IEEE

The ion-sensitivity properties of ISFETs stem from surface chemical reactions taking place at the oxide/electrolyte interface. In order to take into account steric limitations due to finite size of ions, models should include the Stern Layer (i.e. a very thin insulating layer) between the oxide and the electrolyte (see Fig. 1) [6], [10]. In this work, the metal oxide and Stern Layer are explicitly simulated in Sentaurus, whereas in COMSOL these regions are not simulated explicitly but instead described using boundary equations; in particular for the gate oxide, the built-in function 'Thin Insulating Gate Boundary' is used to set the gate oxide parameters whereas the equation of an ideal parallel plates capacitor is employed for the Stern Layer. In this way, the otherwise fine mesh grids required in these layers are avoided without loss of accuracy. Clearly, this simplification neglects the presence of traps in the oxide layer.

Surface chemical reactions are implemented at the oxide/Stern Layer interface (dotted line in Fig. 1). In Sentaurus TCAD we used PMIs as explained in [6], whereas in COMSOL surface reactions have been implemented using a boundary equation for the surface charge that depends on the considered surface chemistry. The sensitivity of the ISFET is extracted as the threshold voltage shift, ΔV_{TH}, at $I_{DS}=0.3$ μA, upon changes of the analyte concentration.

III. IMPLEMENTATION IN COMSOL OF THE MODEL FOR ARBITRARY SURFACE REACTIONS

The interactions between ions in the electrolyte and solid surfaces, requires models with different level of complexity depending on the number of participating reactions and ions. The simplest case is represented by the ion-to-ligand surface adsorption, usually studied with the Langmuir isotherm [11]. pH-sensitive ISFETs, instead, are described by two reactions for the double protonation of negatively charged oxygen terminations of metal oxides [12]. A three-reaction model was also proposed to take into account chloride ions adsorption at the doubly protonated sites [13]. In the literature, these models have been presented case-by-case; the complexity of the underlying math increases rapidly with the number of reactions. Recently, we developed a general and systematic methodology [9] to model an arbitrary number of surface chemical reactions involving different types of binding sites. The model uses graphs to represent the changes of the site "states" upon binding/unbinding of ionic species. In fact, any surface reaction can be decomposed in the consecutive binding of ions, each one leading to a different state of the binding site/ligand complex, as shown in Fig. 2.

Therefore, for a set of N possible states there are $N-1$ reactions, i.e., $N-1$ arrows linking the nodes in Fig. 2. Each arrow has a coefficient given by the ratio between the concentration of the ion/analyte $[I^{z_i}]$ and the corresponding dissociation constant K_I. This coefficient sets the relationship between the occupation probabilities of two adjacent states of the ligand/binding site:

$$f_{i+1} = \frac{[I^{z_i}]}{K_I} f_i. \tag{1}$$

Since there are $N-1$ reactions, the system is completed using the normalization condition $\sum_1^N f_i = 1$. The final equation can be written in matrix form as:

$$\boldsymbol{M} \cdot \boldsymbol{f} = \begin{pmatrix} 0 \cdots\cdots 0 & 1 \end{pmatrix}^T, \tag{2}$$

where $\boldsymbol{f} = \begin{pmatrix} f_1 \cdots\cdots f_N \end{pmatrix}^T$ is a N-elements column vector and the $N{\times}N$ matrix $\boldsymbol{M}$ contains the electrochemical reaction parameters [9]:

$$\boldsymbol{M} = \begin{pmatrix} & \vdots & & \vdots & & & \vdots & & & \vdots \\ 0 & \cdots & 0 & \frac{[I^{z_I}]}{K_I} & 0 & \cdots & 0 & -1 & 0 & \cdots & 0 \\ & \vdots & & \vdots & & & \vdots & & & \vdots \\ & \vdots & & \vdots & & & \vdots & & & \vdots \\ 1 & \cdots & \cdots & \cdots & \cdots & \cdots & \cdots & & & 1 \end{pmatrix}$$

By solving Eq. 2, one finds $\boldsymbol{f} = \boldsymbol{M}^{-1} \cdot \begin{pmatrix} 0 \cdots\cdots 0 & 1 \end{pmatrix}^T$.

From the state probabilities f_i, one can finally compute the surface charge density at the sensing surface in equilibrium as the sum of charged states extended to a number of sites per unit of area, N_S, that is

$$Q_S = qN_S \sum_{i=1}^N z_i f_i, \tag{3}$$

where q denotes the absolute value of the elementary charge. Notice that Eq. 3 identifies the surface reactions of a single type of site. When multiple site types coexist on the same sensing surface, the total surface charge density, $Q_{S,tot}$, is given by the sum over each single type of site, j:

$$Q_{S,tot} = q \sum_{j=1}^M N_{S_j} \sum_{i=1}^N z_{j,i} f_{j,i}. \tag{4}$$

In this work, Eq. 2 is solved in symbolic form, where the concentration of ionic species are expressed using Boltzmann statistics and thus dependent only on the electrostatic potential at the oxide/Stern Layer interface. The resulting $\boldsymbol{f}$ is then inserted into Eq. 3 or 4 and plugged in COMSOL, that uses it as a charge boundary when solving the device electrostatics.

IV. MODEL COMPARISON FOR PH SENSING

pH-sensitive ISFETs are the most studied FET-based potentiometric sensors thanks to the high sensitivity to the pH of the solution featured by most common metal oxides. The surface chemical reactions at each binding site are well explained in the literature by the site-binding (SB) model [10], [12] and

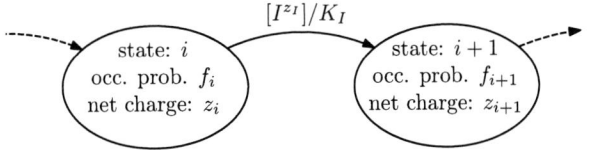

Fig. 2. Graph representing the steady-state relationship between two arbitrary states of a ligand/binding site: from state i to $i+1$ with associated occupation probabilities, f, and net signed numbers of elementary charges, z, upon adsorption of the ion I with valence z_I (i.e. giving $z_{i+1} = z_i + z_I$). K_I denotes the dissociation constant of the reaction.

978-1-6654-3746-2/21 $31.00 © 2021 IEEE

consist of a two-reaction model that describes the interactions of metal hydroxyl groups, MOH, with H^+ ions in the solution.

The SB model is a special case of the model presented in Section III. Its graph representation has three states as reported in Fig. 3.a. The equilibrium dissociation constants

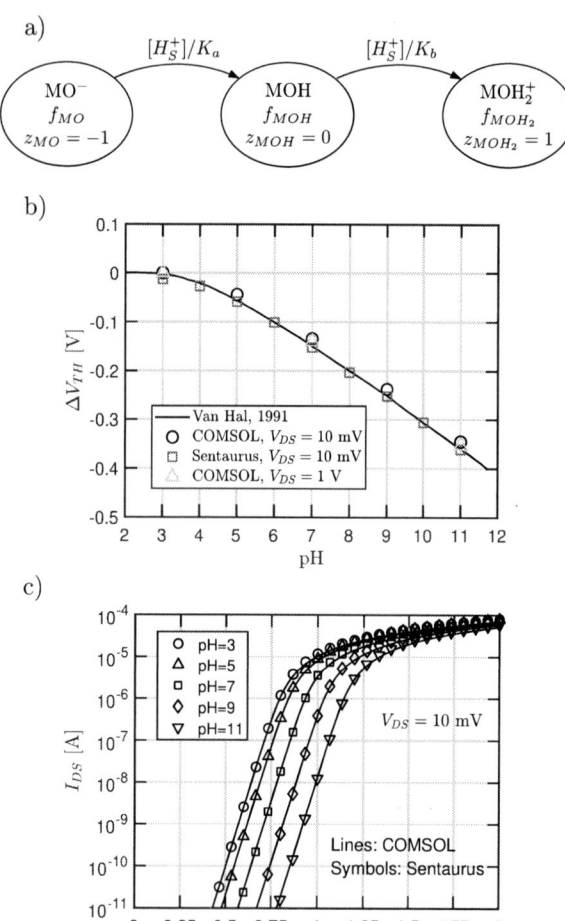

Fig. 3. a) Graph of the SB model following our framework (Fig. 2) for arbitrary surface reactions [9]. b) Simulation of the ISFET threshold voltage shift employing the SB model at the Stern Layer/electrolyte interface of Fig. 1, using Sentaurus (red squares), COMSOL (black circles) and the model from [10]. c) comparison of the full I_{DS}-V_{fg} curves obtained using the two simulation platforms for different pH values. In all cases the electrolyte ionic strength is set to 100 mM and the reaction parameters of the SB model are from SiO$_2$, that is, K_a=10^{-6} M, K_b=10^2 M and N_S=5·10^{18} m^{-2} [10].

of the two reactions relate to the dimensionless fractional occupation functions as:

$$K_a = \frac{f_{MO^-}[H_S^+]}{f_{MOH}}, \quad K_b = \frac{f_{MOH}[H_S^+]}{f_{MOH_2^+}}, \quad (5)$$

where $[H_S^+]$ is the proton concentration at the metal oxide/electrolyte surface, that is calculated by the solvers assuming Boltzmann statistics [9]. Since Sentaurus only considers 1:1 ions, an effective electrolyte taking into account the ionic strength must be set to correctly reproduce the electrostatics in the liquid part of the device and $[H_S^+]$ is computed from the

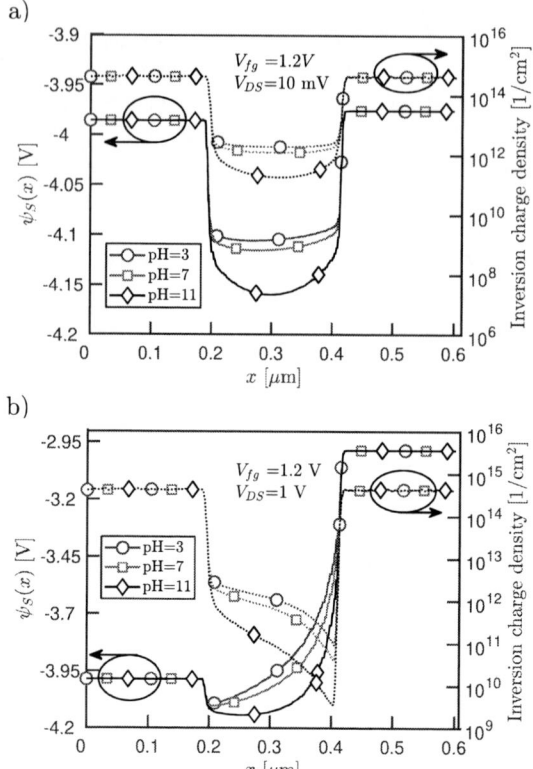

Fig. 4. Surface channel potential ψ_S (left axis) and inversion charge density (right axis) along the x (horizontal) direction for three representative pH values and, $V_{DS} = 10$ mV (a) and $V_{DS} = 1$ V (b). The results are obtained with COMSOL, simulating the structure of Fig. 1 featuring the SB model at the electrolyte/oxide interface.

pH in the bulk looking at the potential difference between bulk and surface [6], [7]. In COMSOL, instead, the ions H^+, OH^-, Na^+ and Cl^- are explicitly simulated. Figure 3.b, compares the simulated ΔV_{TH} vs pH obtained using COMSOL (circles) and Sentaurus (squares) for the structure depicted in Fig. 1 with device parameters as in Table I and an electrolyte with background ionic concentration of $[N_a^+] = [Cl^-] = 100$ mM. The two solvers are in very good agreement, and both reproduce well the model results in the literature [10] (black line). We also see that the V_{DS} value has a minor effect on the ΔV_{TH} vs the pH as expected for a well-tempered FET. The comparison of the I_{DS}-V_{fg} trans-characteristics is reported in Fig. 3.c for a few pH values and proves the quantitative agreement between the two solvers. Further analysis is provided in Fig. 4, which reports the results obtained with COMSOL for the electrostatic potential, ψ_S, and the inversion charge density, along the x direction at the channel surface of the ISFET as a function of the electrolyte pH and for $V_{DS} = 10$ mV (a) and $V_{DS} = 1$ V (b). We see that charge and potential are not uniform along the channel, even at low V_{DS}, due to the influence of source and drain. However, 1D models give a fairly accurate estimate of the V_{TH} shift, but 2D ones are needed to model the device as a whole and capture the coupling of 2D short channel electrostatics with the surface charge build-up mechanism.

978-1-6654-3746-2/21 $31.00 © 2021 IEEE

V. Application to Complex Sensing Scenarios

As an example of model use and validation against experimental data and independent models, we consider the experiment in [14], where a differential setup consisting of an active and a control ISFET (see Fig. 5.a) was used to sense the concentration of sodium ions, Na^+. The control ISFET features a gold layer sensitive to pH and Cl^- (i.e. modelled with a modified SB [13]) whereas the active ISFET employed a gold layer functionalized with Na^+-sensitive ligands. Therefore, $Q_{S,tot}$ in Eq. 4 is the sum of two contributions stemming from a four-states and two-state reaction graphs, respectively (see Fig. 5.a).

a)

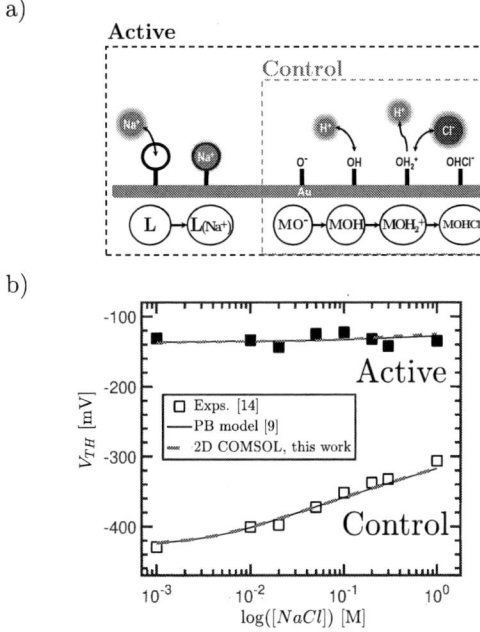

b)

Fig. 5. (a) Sketch of the active and control device surfaces for the differential setup used in [14]. (b) Comparison of the V_{TH} vs the concentration of NaCl in the solution between the experimental data reported in [14] (symbols), the PB model in [9] (black solid line) and the COMSOL implementation of this work (red dashed line). A rigid vertical shift has been applied to the different model results, for the sake of easier comparison.

Figure 5.b shows that experimental data from [14] (symbols) and COMSOL simulations (dashed lines) provide the same ΔV_{TH}, despite in our simulations we use the template ISFET in Fig. 1 that has not the same geometry and only shares with the one in [14] the sensing surface. This underlines that the static response (i.e. the ΔV_{TH}) of these potentiometric electrochemical sensors is marginally affected by the architecture of the underlying FET. This conclusion is corroborated by the good agreement with the steady-state Poisson-Boltzmann (PB) model (solid lines in Fig. 5.b) developed in [9], not including the semiconductor part of the device. Hence, the good agreement between the one-dimensional PB model and 2D simulations of the complete device in COMSOL highlights the capability of the former model to provide accurate predictions with considerably less effort, as far as geometrical factors, FET operation conditions and other effects can be neglected.

VI. Conclusions

We have reported a Multiphysics framework implementation of the model in [9] that overcomes the limitations of commercial TCAD, allowing for a more complete treatment of the multi-ion nature of the electrolyte and for complex, coupled chemical reactions at the electrolyte/oxide interface. On the other hand, TCAD is more efficient in simulating the semiconductor part of the ISFET especially if using comprehensive sets of S.O.A. microscopic physical models. Our COMSOL simulations suggests that 1D, electrolyte-only models, (e.g., as those in [9]) may be sufficient to interpret experimental data of large ISFETs free of significant short channel effects or at low V_{DS}. Device simulations with TCAD or COMSOL including the semiconductor part, however, can turn useful when studying the impact of technology options on the ISFET response.

References

[1] N. Moser, T. S. Lande et al., "ISFETs in CMOS and Emergent Trends in Instrumentation: A Review," *IEEE Sensors Journal*, vol. 16, no. 17, pp. 6496–6514, 2016. DOI: 10.1109/JSEN.2016.2585920

[2] M. Kaisti, "Detection principles of biological and chemical FET sensors," *Biosensors and Bioelectronics*, vol. 98, pp. 437–448, 2017. DOI: 10.1016/j.bios.2017.07.010

[3] F. Bellando, L. J. Mele et al., "Sensitivity, Noise and Resolution in a BEOL-Modified Foundry-Made ISFET with Miniaturized Reference Electrode for Wearable Point-of-Care Applications," *Sensors*, vol. 21, no. 5, 2021. DOI: 10.3390/s21051779

[4] *Synopsys Sentaurus Device simulator user manual*, Synopsis Inc., Mountain View, CA, USA, 2016.

[5] I.-Y. Chung, H. Jang et al., "Simulation study on discrete charge effects of SiNW biosensors according to bound target position using a 3D TCAD simulator," *Nanotechnology*, vol. 23, no. 6, p. 065202, 2012. DOI: 10.1088/0957-4484/23/6/065202

[6] A. Bandiziol, P. Palestri et al., "A TCAD-Based Methodology to Model the Site-Binding Charge at ISFET/Electrolyte Interfaces," *IEEE Transactions on Electron Devices*, vol. 62, no. 10, pp. 3379–3386, Oct 2015. DOI: 10.1109/TED.2015.2464251

[7] F. Pittino, P. Palestri et al., "Models for the use of commercial TCAD in the analysis of silicon-based integrated biosensors," *Solid-State Electronics*, vol. 98, pp. 63–69, 2014. DOI: 10.1016/j.sse.2014.04.011

[8] *COMSOL Multiphysics user manual*, COMSOL Inc., Sweden, 2021.

[9] L. J. Mele, P. Palestri, and L. Selmi, "General Approach to Model the Surface Charge Induced by Multiple Surface Chemical Reactions in Potentiometric FET Sensors," *IEEE Transactions on Electron Devices*, vol. 67, no. 3, pp. 1149–1156, March 2020. DOI: 10.1109/TED.2020.2964062

[10] R. Van Hal, J. C. Eijkel, and P. Bergveld, "A general model to describe the electrostatic potential at electrolyte oxide interfaces," *Advances in colloid and Interface Science*, vol. 69, no. 1-3, pp. 31–62, 1996. DOI: 10.1016/S0001-8686(96)00307-7

[11] I. Langmuir, "The adsorption of gases on plane surfaces of glass, mica and platinum," *Journal of the American Chemical society*, vol. 40, no. 9, pp. 1361–1403, 1918. DOI: 10.1021/ja02242a004

[12] D. E. Yates, S. Levine, and T. W. Healy, "Site-binding model of the electrical double layer at the oxide/water interface," *J. Chem. Soc., Faraday Trans. 1: Phys. Chem. Condensed Phases*, vol. 70, pp. 1807–1818, 1974. DOI: 10.1039/F19747001807

[13] A. Tarasov, M. Wipf et al., "Understanding the electrolyte background for biochemical sensing with ion-sensitive field-effect transistors," *ACS nano*, vol. 6, no. 10, pp. 9291–9298, 2012. DOI: 10.1021/nn303795r

[14] M. Wipf, R. L. Stoop et al., "Selective Sodium Sensing with Gold-Coated Silicon Nanowire Field-Effect Transistors in a Differential Setup," *ACS Nano*, vol. 7, no. 7, pp. 5978–5983, 2013. DOI: 10.1021/nn401678u

978-1-6654-3746-2/21 $31.00 © 2021 IEEE

Random Telegraph Noise real time testing based on downsampling for mass data extraction

Maximilian Juettner
HTW Dresden
Dresden, Germany
Maximilian.juettner@htw-dresden.de

Michael Otto
GLOBALFOUNDRIES Fab1 LLC &
Co. KG
Dresden, Germany
Michael.Otto2@globalfoundries.com

Jan Hoentschel
GLOBALFOUNDRIES Fab1 LLC &
Co. KG
Dresden, Germany
Jan.Hoentschel@Globalfoundries.com

Abstract—To meet the rising demand for Random Telegraph Noise (RTN) analysis, a new method to quickly separate RTN impacted devices from devices that show no RTN has been proposed and validated via existing analysis methods.

Keywords— Random Telegraph Noise; parameter extraction; CMOS; characterization

I. INTRODUCTION

With the Internet of Things (IoT), applications are more connected and more integrated with their environment than ever before. This is only possible via a large amount of sensors that connect to analog/mixed-signal (AMS) circuits and numerous radio frequency (RF) interfaces to allow for connectivity. Both AMS and RF applications are significantly more sensitive towards noise then digital only applications.

RTN as one of the mayor contributors to noise in modern semiconductor devices shows a huge variation even on a device-to-device basis [1]. Therefore, many devices must be investigated to model the statistical impact of RTN on a given device type. In [2] the RTN measurement of a high number of devices has been demonstrated and further enhanced in recent years. Therefore, an efficient algorithm for the RTN evaluation is needed.

This paper first gives an overview of the measurement setup which allows the measurement of the many devices needed for a sophisticated RTN analysis. Afterwards the algorithm used for quickly categorizing the signals into RTN-affected and none RTN-affected is introduced. Lastly the algorithm is evaluated by comparing the detection against a variation of the already evaluated WTLP extraction [3],[4].

II. MEASUREMENT PARAMETERS AND SYSTEM

A. Measurement Setup and Conditions

The system is based on a National instruments (NI) PXI test system equipped with source measurement units (SMU) that allow for sample rates of up to 1.8 MS/s. It is coupled with a fully automated prober for 24/7 operation and can be controlled via a custom test program, which allows reconfiguration of the full test flow through a single centralized control file.

This solution is scalable to >100 SMU channels in a multi chassis configuration, while maintaining trigger capabilities across all channels and chassis. A typical setup is for example sampling for 10s @ 100kHz for > 20 devices in parallel. In another configuration the devices under test (DUTs) are placed in a 4-bit multiplexer structure with each address

probing 6 DUTs in parallel while allowing individual biasing for all devices. This system allows for a throughput of over 120k DUT's per wafer in around 2 hours, thus providing a sufficient amount of data to achieve the statistical significance needed for RTN analysis. A more in-depth view on the structure can be found in [5].

III. RTN CATEGORIZATION

A. Downsampling

The sampling frequency f_s is directly correlated to the maximum observable frequency bandwidth B for a given signal, as dictated by the Nyquist-Shannon sampling theorem [6]:

$$B < \frac{f_s}{2} \tag{1}$$

The measurement frequency of a SMU can be adjusted by the aperture time, which determines how long a signal is averaged by the analog digital converter (ADC) before calculating the final value for a sample. If the aperture time is longer than either the capture time τ_c or the emission time τ_e the RTN event will be averaged to an in between value of trapped and not-trapped state.

The software solution is based on averaging n samples of the raw data together, thus emulating a longer aperture time of the measurement card. This method does not need extra test time or tester resources. The only prerequisite is that the complete signal was sampled and recorded. In Fig. 1 the impact of a downsampling factor of 2, that has been used in this paper, has been demonstrated for a signal with and without RTN.

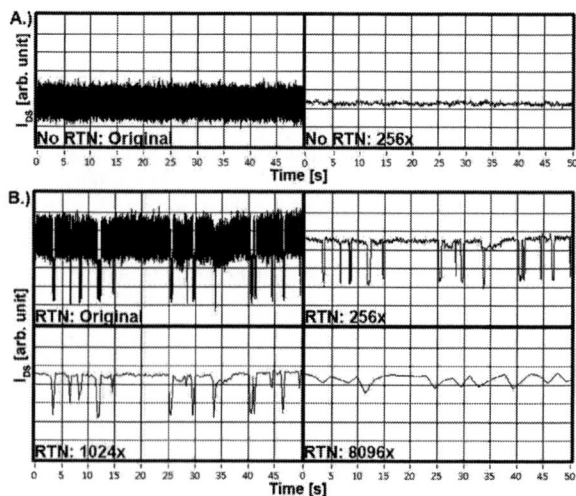

Fig. 1 Downsampling A.) without RTN and B.) with RTN

B. Algorithm for extraction

If the signal shows no RTN it still contains other noise sources (i.e. thermal noise, flicker noise). The outliers get averaged first and with every downsampling step i each remaining sample gets closer to the mean.

If the signal contains RTN, the other noise sources behave in the same way as without RTN. Due to the RTN events having discrete levels that affect multiple samples, they do not converge towards the mean immediately. Instead, they average out at the value for i at which the new aperture time becomes bigger than either τ_c or τ_e. This causes a huge shift in different metrics like RMS, Max-Min and various percentile ranges for I_{DS}.

The shift is best observed by looking at the normalized derivative. In this paper we choose P99.9-P0.1 which has proven to be a good compromise between robustness and noise detection. The slope of the derivative can then be used to assess whether a signal is affected by RTN or not. For a signal without RTN the normalized derivative is steeply declining with only small local maxima occurring for bigger i, caused by other noise phenomena like flicker noise. If the signal contains RTN, the derivative shows a more significant peak even for larger values of i.

Each downsampling step i can also be associated with a corresponding maximum observable frequency Bandwidth based on a variation of (1):

$$B_i < \frac{f_s}{2^{i+1}} \tag{2}$$

This allows for a rough estimate of the time constant that gets averaged out first in the signal. The derivative contains one downsample step less, because it calculates the change from one step to another. Thus, for the derivative the lower bandwidth is used. Examples for the downsampling, as well as the normalized derivative for a none RTN and a RTN-impacted signal based on P99.9-P0.1 are shown in Fig. 2.

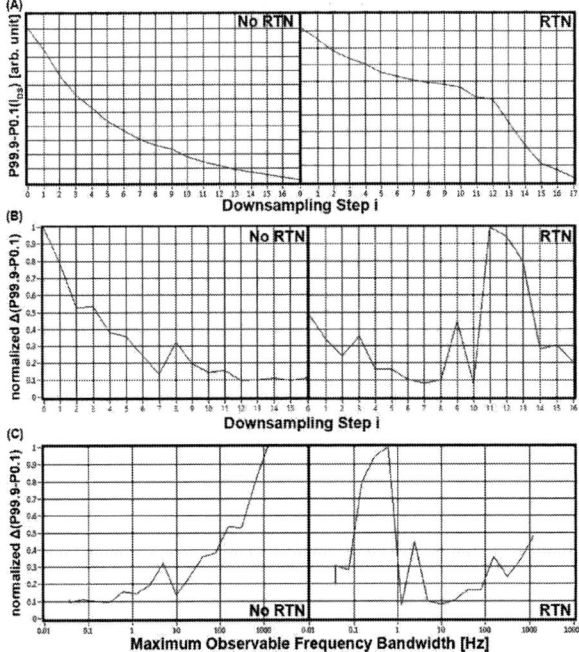

Fig. 2 (A) Downsampled P99.9-P0.1 range; (B) Normalized derivative of downsampled P99.9-P0.1 range; (C) Normalized derivative over observable bandwidth for signals without RTN (left) and with RTN (right)

For a real time based system, the computational load is a critical factor, thus the algorithm needs to be lightweight, while still having a high detection rate. Therefore, the algorithm needs to automatically terminate execution once an RTN event has been found. Three different tests to determine a significant peak have been implemented in this work. An overview of the algorithm is outlined in Fig. 3.

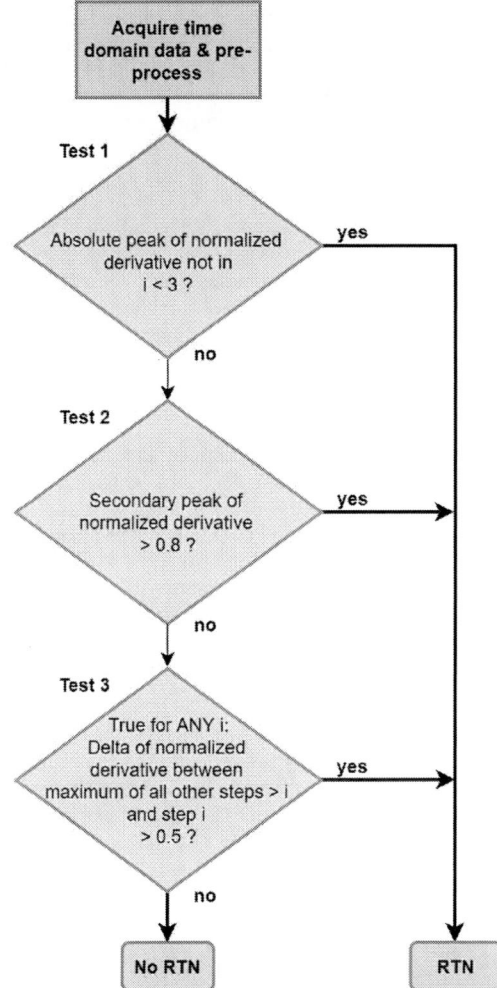

Fig. 3 Flow chart for RTN-test algorithm with example parameters

The unsteadiness caused by outliers is the strongest for the first i steps. Therefore, if the maximum of the normalized derivative is inside those values it cannot be used as an indicator for RTN. If the normalized derivative contains any significant peak outside the first i steps, the signal is very likely to contain RTN. In this paper i < 3 has been chosen (Test 1 in Fig. 3).

If the maximum is inside the first i values, the algorithm searches for a secondary peak that has an absolute value greater than an adjustable parameter, here set to 0.8 (Test 2 in Fig. 3). This value can be set lower for signals with clear RTN and needs to be set higher for signals where other noise sources are dominating.

The last test of the algorithm takes the relative position of a peak into consideration. For bigger values of i, the overall change of the normalized derivative gets lower and lower, therefore even a smaller peak can indicate RTN. To account for this the algorithm iterates over i from left to right and splits the curve into two parts for every downsample step. It than calculates the delta between the current value of the curve at i and the maximum of the remaining right part for each iteration. If the delta is bigger than an adjustable value, here 0.5 was chosen, for any iteration n, the signal likely contains RTN (Test 3 in Fig. 3).

There are multiple ways to improve the detection rate of the algorithm. E.g., measuring at multiple frequencies reduces the uncertainty for values at the corners of the detection range, splitting the measurement in multiple chunks and running the algorithm in parallel increases precision at the cost of performance and using multiple different percentile ranges increases robustness

C. RTN-Simulation

Due to the varying nature of RTN in both time constants and amplitude it is not possible to calibrate the extraction algorithm based on measurement data. Therefore, a RTN simulation was developed that can generate known RTN signals. Besides the standard parameters like bias point, standard deviation, sample frequency, number of RTN traps and number of samples it allows for a multitude of other adjustments.

The RTN parameters are defined as a percentage of the maximum signal amplitude and the range of their time constants as well as the event-to-event-variation for both parameters can be defined individually. It is recommended to set the minimum simulated time constants to match the value of test 1, because faster RTN events will get ignored by the algorithm to increase reliability. For the same reason the maximum capture/emission time needs to be set below the number of generated samples to guarantee that the generated time domain data shows RTN.

Furthermore, the simulation incorporates an adjustable additive white gaussian noise (AWGN) and flicker noise component that can be freely changed to account for different technologies and structures. Different exemplary time domain signals and their corresponding normalized derivatives with and without flicker noise and RTN are shown in Fig.4.

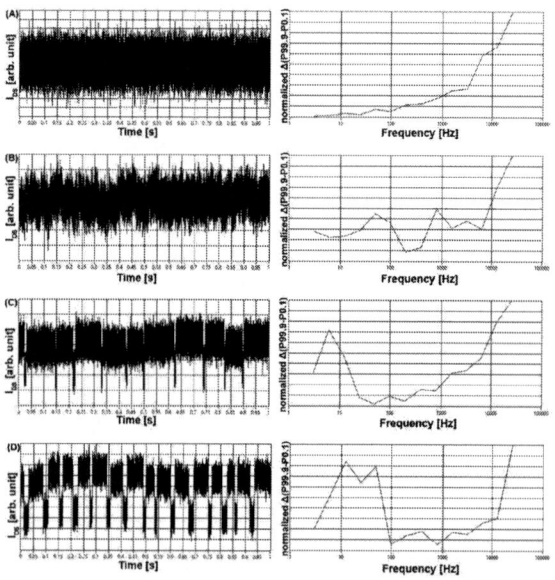

Fig. 4 Signal (A) without RTN/flicker noise (B) without RTN and with flicker noise (C) with RTN and without flicker noise (D) with RTN/flicker noise

The impact of flicker noise on the readout is especially important due to it having a higher sample to sample correlation than AWGN, as can be derived by the Wiener–Khinchin theorem, which links auto covariance and therefore the autocorrelation and noise spectrum [7].

Thus, it can be the cause of more drastic shifts for bigger i than AWGN. Therefore, the flicker noise component of the signal needs to be adjusted in a way that matches the power spectral density of the simulated signal with that of the measurement data. Once the RTN-simulation is calibrated to the measurement data it can be used to fine tune the extraction algorithm.

Out of the three tests in Fig. 3 for RTN the one with the most significant impact is Test 3, because it covers all the cases missed by the less computational expensive peak detections. The dependency of the detection rate on the delta between local minima and local maxima is shown in Fig. 5.

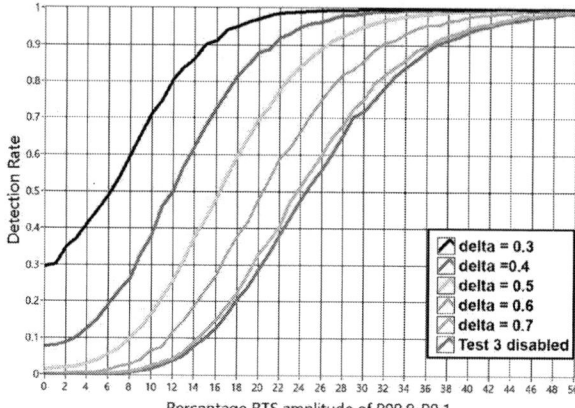

Fig. 5 Influence of delta between local maxima and previously occurred minima (Test 3 disabled -> result of Test 3 = no RTN)

The detection rate gets higher the bigger the RTN jump is compared to P99.9-P0.1, nearing 100% when the RTN-amplitude is about half as big as the overall noise impact on the signal when test 3 is disabled (output always no RTN). The smaller the required delta, the more RTN is detected even for small amplitudes. This however comes with the drawback of signals without RTN events being detected as having RTN (Percentage RTN amplitude = 0). Thus, the tradeoff needs to be considered. For the given measurement data, a delta of 0.5 has been chosen.

Another way to increase the detection rate is by splitting the time domain signal into multiple parts and analyze them individually. There are 2 considerations that need to be taken into account when splitting the curve.

One is if the number of samples is increased according to the number of curves the original data is split into (blue curve in Fig. 6). This will result in an increase of measurement time, but otherwise there will be a loss of information, due to less downsampling steps being available than before, this can result in some of the RTN not being detected. The other one is how the decision RTN/no RTN is reached: Is 1 detection enough (green curve) or is a unanimous decision (red curve) required. This affects the trade-off between detection rate and misdetection significantly

The results of the simulation for different implementations of a split curve are shown in Fig. 6.

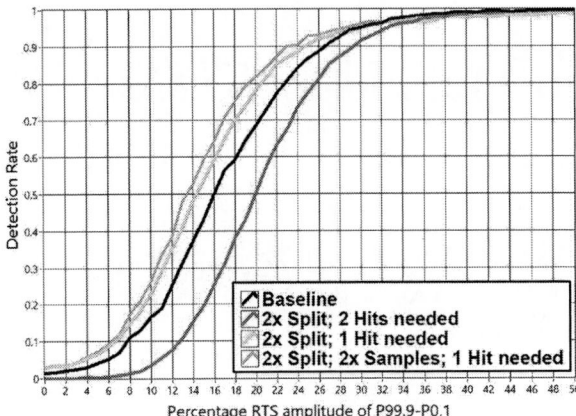

Fig. 6: Different variations of curve splitting with different decision criteria

D. Validation

For validation purposes an extended version of the weighted time lag plot (WTLP) method is used. It introduces a Monte-Carlo-Simulation to fit gaussian curves into the compressed diagonal and can detect RTN-events hidden due to superposition not showing a local maximum [1]. This extraction method, while precise, proves to be too slow to keep up with the ever-increasing number of devices measured.

The downsampling algorithm results matched with the WTLP method on ~88.5% of the cases for the sample data set. Therefore, the algorithm is suitable to be used for either real time analysis and classification during the measurement or as a pre-filter for other algorithms.

IV. SUMMARY

A new method for quickly categorizing devices into RTN-affected and none-RTN-affected has been proposed. Due to the comparatively low computational load the algorithm can be used for split-analysis of large data sets while maintaining a good detection rate for RTN effects. This in turn enables a significant time reduction for more in-depth analysis methods. Furthermore, the algorithm can be used for on-chip RTN analysis.

V. ACKNOWLEDGMENT

This project was funded as Important Project of Common European Interest (IPCEI) by the Ministry of Economics and Industry of Germany.

REFERENCES

[1] T. Nagumo, K. Takeuchi, S. Yowogawa, K. Imai and Y. Hayashi, "New analysis Methods for Comprehensive Understanding of Random Telegraph Noise" in IEDM Tech. Dig., pp. 32.1.1-32.1.4, 2009.

[2] M. Otto, M. Juettner, J. Hoentschel," Mass data analysis of Random Telegraph Noise in 22nm FDSOI back biased transistors", EUROSOI-ULIS 2017: abstracts session 10, 2017, on pp. 5-6.

[3] M. Maestro, J. Diaz, M.B. Gonzalez, J. Martin-Martinez, R. Rodriguez, M. Nafria, et. al., "A new high resolution random telegraph noise (RTN) characterization method for resistive RAM ", EUROSOI-ULIS 2015: 2015 Joint International EUROSOI Workshop and International Conference on Ultimate Integration on Silicon, 2015.

[4] T. Obara, A.Teramoto, A. Yonezawa, R. Kuroda, S. Sugawa, T. Ohmi,"Analyzing Correlation between Multiple Traps in RTN Characteristics", 2014 IEEE International Reliability Physics Symposium, 2014.

[5] A. Jayakumar; N. Chan; L. Pirro; O. Zimmerhackl; M. Otto; T. Kleissner; J. Hoentschel ,"Area-Efficient and Bias-Flexible Inline Monitoring Structure for Fast Characterization of RTN and Transistor Local Mismatch in Advanced Technologies", 2020 IEEE 33rd International Conference on Microelectronic Test Structures (ICMTS), 2020.

[6] C. E. Shannon, "Communication in the Presence of Noise", Proceedings of the IRE, 1949.

[7] F. Sischka, "Semiconductor Noise in the Time and Frequency Domain"; Technische Universität Dresden, 2014.

Charge Pumping-Based Method for Traps Density Extraction in Junctionless Transistors

E. T. Fonte[1], R. Trevisoli[2] and R. T. Doria[1]

[1]Electrical Engineering Department, Centro Universitário FEI, São Bernardo do Campo, Brazil
[2]CECS, Universidade Federal do ABC, Santo André, Brazil
e-mail: ewertonteixeira@fei.edu.br

Abstract—**A study of Junctionless Transistors (JNTs) is presented in this work, with emphasis on verifying the extraction of the interface traps density using the charge pumping method. To the best of our knowledge, this is the first work to use this method in JNTs. The method was applied to both simulated and experimental data and has shown satisfactory results.**

Keywords-SOI Technology; Junctionless nanowire transistors; Interface traps; Charge pumping.

I. INTRODUCTION

Junctionless transistors (JNTs) have been proposed as an alternative for ultimate technologies since the absence of junctions can enable the fabrication of extremely short devices [1]. These devices present constant and heavy doping concentration from source to drain and work similarly to accumulation mode transistors. At low absolute gate bias (V_{GS}), the silicon layer is completely depleted. In a nMOS transistor, the increase of V_{GS} reduces the depletion depth and a neutral region is formed deep into the silicon layer, which enables the formation of a bulk current. The neutral region increases with the increment of V_{GS} and, when the device reaches the flatband, the entire silicon layer is neutral. For even larger V_{GS}, an accumulation current is observed.[1]

The peculiar operation mode of JNTs makes the interface traps impact in the device behavior different from the expected in inversion mode transistor, since the surface potential in JNTs varies with V_{GS} even in the on-state condition. Thus, it is important to determine the active interface trap density (N_{it}) in such devices. In this work, the applicability of a modified version of the charge pumping method has been proposed [2]. Initially, two-dimensional simulations were performed aiming at the extraction of N_{it}, as shown in section II. After that, the method applicability has been verified in experimental devices, whose results will be discussed throughout section III of this work. Section IV points out the main conclusions obtained.

This work was supported by São Paulo Research Foundation (FAPESP) grant #2019/15500-5, National Council for Scientific and Technological Development (CNPq) grant #303938/2020-0 and financed in part by the Coordenação de Aperfeiçoamento de Pessoal de Nível Superior - Brasil (CAPES) - Finance Code 001.

II. MODIFIED CHARGE PUMPING METHOD

As the charge pumping method described in [2] cannot be directly applied to JNTs, a modification has been proposed. A voltage pulse was applied to the gate of the device and the current transient (I_D) was measured at the source/drain, which were short-circuited and grounded. The pulse applied to the gate presents amplitude of 1.2 V with a reference value of 50 mV, where the JNT is in the off-state condition. The pulse amplitude defines the energy range of the interface traps that are activated.

Initially, the proposed method has been validated to 2D numerical simulations performed in Sentaurus TCAD [3], which considered drift-diffusion transport mechanism and models accounting for bandgap narrowing, mobility dependence on lateral electrical field and carriers' generation/recombination. The physical characteristics of the simulated devices have been based on the experimental ones fabricated according to the process described in [4] and present channel length (L) of 50 nm, silicon layer thickness (t_{Si}) of 10 nm, thickness of the gate oxide layer (t_{OX}) equals to 1.3 nm, source/drain and channel doping concentration (N_d) of 1×10^{19} cm^{-3}. The simulations were performed at 300 K. The trap profile has initially been considered as uniform along the middle of the bandgap and the total trap density (N_0) is given in cm^{-2}eV^{-1}. Although this is not the expected profile in experimental devices, such profile simplifies the theoretical calculation of N_{it} in order to validate the method.

In Figure 1, the simulated I_D vs. time characteristics for different trap densities are shown. When the input signal is changed from 50 mV to 1.2 V, a peak current is observed independently on the trap density. This current is due to the population of the channel region as well occupation of activated traps. All the current curves present another peak just after the transition of the input signal from 1.2 V to 50 mV, which can be associated to the depopulation of the channel. In the sequence, the current presents an exponential decay, denominated as

978-1-6654-3746-2/21 $31.00 © 2021 IEEE

pumping current, which can be observed in the interval from 600 ns to 10 µs. This current is associated to the detrapping of the carriers. As one can note, for larger trap densities, a longer trap decay is observed, indicating that the decay time could be used for the N_{it} extraction.

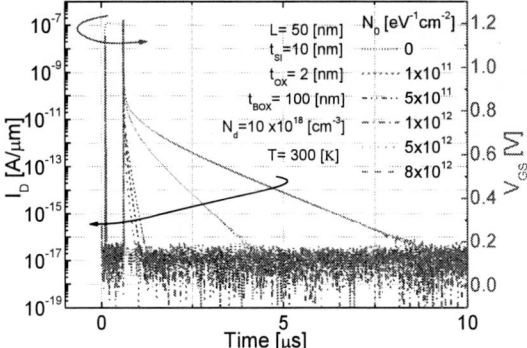

Figure 1. I_D vs. time curve for devices with different trap densities.

Aiming to verify the dependence of the drain current variation with the trap density, I_D was obtained at 1 µs (along the exponential decay) and plotted against N_0 as shown in figure 2 [5]. As one can note, the results indicate that the application of the adapted charge pumping in JNTs could indicate the magnitude of active traps density in experimental devices.

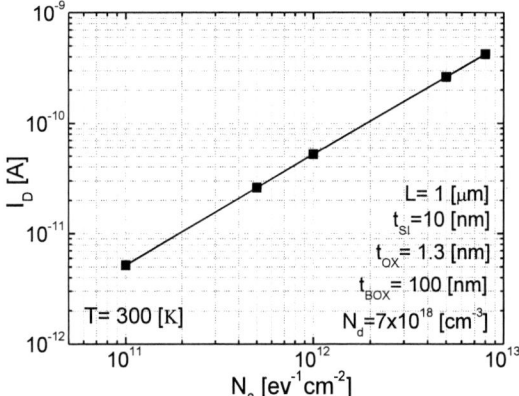

Figure 2. Drain current at 1 µm as a function of the trap density in simulated devices [5].

Thus, N_{it} could be extracted by the integral of the pumping current as a function of time. When divided by the elementary charge (q) of the electron and the area of the transistor gate ($W.L$, where W is the channel width and L is its length), the trap density per unit area is obtained, as shown in (1) [6].

$$N_{it} = \frac{1}{q.L.W} \int_{t_i}^{t_2} I_D(t) dt \qquad (1)$$

where t_1 and t_2 represent the time interval of pumping current exponential decay.

The trap density extracted by the proposed method has been applied to the simulated curves and the results are shown in the central column of Table I. It is worth to mention that the total trap density (N_0) considered in the simulations (first column of the table) is given in cm^{-2}eV^{-1} whereas the extracted one is given in cm^{-2}.

Table I. Comparison between the extracted by method and simulated N_{it}.

Trap Density (N_0) [cm^{-2}eV^{-1}]	Extracted Interface Traps (N_{it}) [cm^{-2}]	Simulated Interface Traps (N_{it}) [cm^{-2}]
1.00x10^{11}	9.88x10^{10}	3.48x10^{10}
5.00x10^{11}	2.53x10^{11}	1.80x10^{11}
1.00x10^{12}	3.28x10^{11}	3.78x10^{11}
5.00x10^{12}	1.23x10^{12}	2.40x10^{12}
8.00x10^{12}	2.24x10^{12}	4.30x10^{12}

As the traps are distributed along the bandgap, only traps whose activation energy are in a certain interval of the bandgap are activated. This interval is given by the surface potential variation between low and high states of V_{GS}. In the simulations, only traps in an interval of 0.5 eV within the middle of the bandgap were active.

In order to determine the quantity of trapped charges in the simulations in cm^{-2}, the trap density per eV (N_{it}^*) was plotted as a function of the energy both at the beginning and the end of the pumping current decay. The curve obtained at the beginning of the decay showed that practically all the traps were occupied. However, not all of them were vacated at the end of the decay, as can be seen in figure 3.

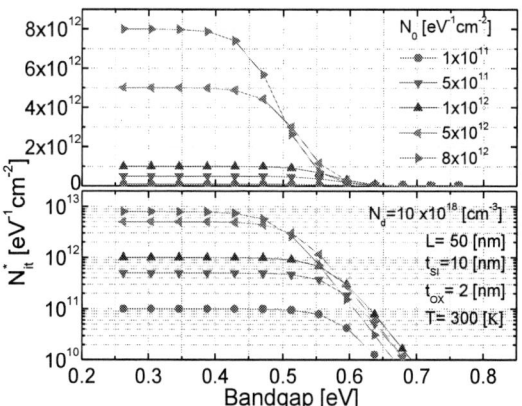

Figure 3. N_{it}^* versus energy for simulated curves at the end of the pumping current (t=10µs).

To calculate the total activated trap density (cm^{-2}) in the V_{GS} range, the integral of N_{it}^* was obtained as a function of energy at the end of the decay and this value was subtracted to the N_{it} in the beginning of the decay. The values for N_{it} obtained from the simulations are also presented in Table I, where it can be seen that both extracted and simulated values are in accordance.

Aiming to extend the results to a trap profile closer to the ones observed in experimental devices, the proposed method was applied to a simulated device in which an exponential profile with a maximum trap density of $N_0 = 1 \times 10^{14}$ cm^{-2}eV^{-1} close to the conduction band of the bandgap was considered. Similarly to the previous analysis, the present one evaluated the pupping current observed after that a pulse of 1.2 V was applied to the gate of the device. The applied pulse and the pumping current are shown in figure 4.

The pumping current decays in the interval from 600 ns to 610 ns approximately. This current is associated to the detrapping of the carriers similarly to the results shown in figure 1. The proposed method has been applied to the simulated curves and resulted in a N_{it} of 2.4×10^{11} cm^{-2}.

uniform trap density along the middle of the bandgap, in the exponential profile a significant part of the interface traps were not occupied at the beginning of the decay

The density of trapped charges at the beginning and at the end of the current decay were equal to 3.57×10^{11} and 1.37×10^{11} cm^{-2}, respectively. The difference between both values results in an effective trap density of $N_{it} = 2.2 \times 10^{11}$ cm^{-2}. The concordance between N_{it} calculated through the method and the simulated one demonstrate the effectiveness of the proposed method for an exponential trap profile.

III. EXPERIMENTAL RESULTS

The proposed method was also applied to an experimental device with similar physical characteristics to the simulated ones, considering a pulse with the same magnitude. The measured device is triple gate SOI junctionless and was fabricated in Leti, France, according to the process described in [4]. The device presents channel doping concentration of 5×10^{18} cm^{-3}, channel length of 400 nn and channel width (W) of 10 μm. The transistor presents a silicon layer thickness (t_{Si}) of 10 nm, an EOT of about 1.55 nm and a buried oxide thickness of 145 nm. The pumping current obtained against time is shown in the figure 6.

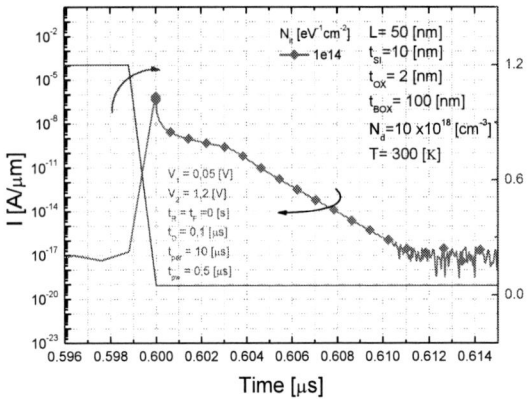

Figure 4. I_D vs. time curve for devices with an exponential trap density along the bandgap with $N_0 = 1 \times 10^{14}$ cm^{-2}eV^{-1}.

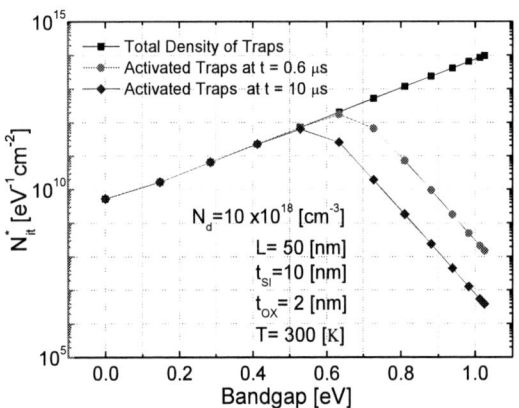

Figure 5. N_{it} versus energy for simulated curves at the beginning and the end of the pumping current as well as the total density of traps.

In order to determine the amount of trapped charges that were occupied at the beginning and at the end of the pumping current decay in the simulations, N_{it}^* was plotted in figure 5 as a function of the energy at $t = 0.6$ μs and 10 μs as well as the total trap density. Differently from the

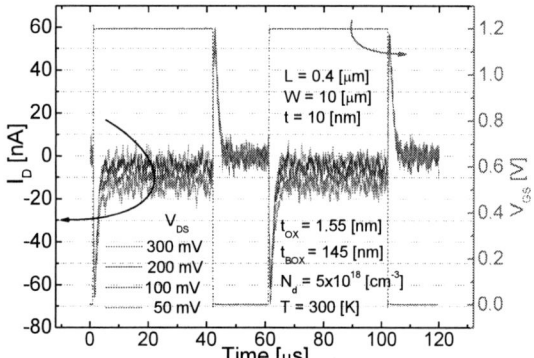

Figure 6. Pumping current extracted from experimental measurements, with different voltage drain bias drain and source values.

In figure 6, one can observe a different behavior in relation to the simulated results when the gate bias varies from 0 V to 1.2V. In this case, a negative peak of pumping current is observed and then the absolute current decreases exponentially. This negative current may be related to the movement of holes inside the silicon layer, which was not taken into account in the simulations, whose main objective was to verify the method accuracy. Anyway, this current can indicate the presence of traps not only in the upper part of the prohibited band, but also in the lower one, indicating that experimental devices present a U-shape trap density along bandgap [7].

In figure 6, it can also be observed that after the negative peak, the absolute current reduces as the holes closer to the valence band are detrapped. However, the overall current stabilizes in the order of -20 nA. It is estimated that this current is related to a leakage occurring through the gate oxide (I_G), which receives a 1.2 V bias, which is close to the limit supported by the technology [4, 8,9]. In order to validate this assumption, the I_G vs V_{GS} curve was extracted for this device and is shown in figure 7. According to the figure, for V_{GS} = 1.2 V, it is obtained that I_G = 18 nA. This value is very close when compared to the negative current in the experimental device at the application of the charge pumping method adapted to the junctionless transistor, which demonstrates the source of such current.

In order to verify the effect of the leakage current in the applicability of the proposed method, the pumping current was obtained for different drain biases (V_{DS}) between 50 and 300 mV and the method was applied to all the curves. The value of N_{it} extracted by the method resulted in approximately 9 x 10^{12} cm^{-2}, for all the cases, which is very similar to the results from [7] ($1x10^{13}$ cm^{-2}) and [10] ($4.75x10^{12}$ cm^{-2}) , which have been obtained for similar devices, through differnet methods. It is woth to mention that in [7], the trap density of junctionless transistors was estimated through the body factor variation with substrate biasing of the devices whereas in [10] it was obtained thorugh the low frequency noise analyisis.

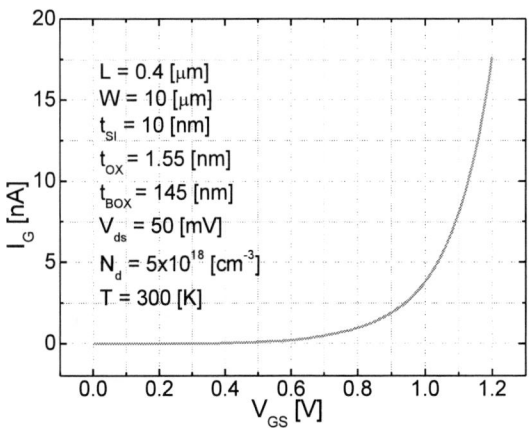

Figure 7. I_G vs V_{GS} curve to extract the current value when V_{GS} = 1.2 V

I. CONCLUSIONS

This work has proposed an adaptation of the charge pumping method to evaluate the quality of the interfaces in junctionless transistors. The applicability of the proposal has been verified through 2D numerical simulations and measurements. Analyzing all the simulations it was possible to conclude that charge pumping method works in junctionless nanowire transistors and can be used as a method to indicate the interface traps density, since the exponential decay of the pumping current along time has demonstrated to be directly correlated to the trap density in such devices. The method was successfully applied to experimental devices and showed that the evaluated junctionless present N_{it} of about 9 x 10^{12} cm^{-2}, which is very close to the results from previous works from literature.

ACKNOWLEDGMENT

The authors thank CEA-Leti and Sylvain Barraud for supplying the experimental devices.

REFERENCES

[1] J.P. Colinge, C.-W. Lee, A. Afzalian, N.D. Akhavan, R. Yan, I. Ferain, P. Razavi, B. O'Neill, A. Blake, M. White, A.-M. Kelleher, B. McCarthy, R. Murphy , "Nanowire transistors without junctions", *Nature Nanotechnology*, v. 5, n. 3, pp. 225-229, 2010.

[2] J.S. Brugler, P.G.A. Jespers, "Charge pumping in MOS devices", *IEEE Tranactions on. Electron Devices*, v. 16, p 297, 1969.

[3] Sentaurus Device User's Manual, Synopsys, 2019.

[4] S. Barraud, M. Berthome, R. Coquand, M. Casse, T. Ernst, M.-P. Samson, P. Perreau, K.K. Bourdelle, O. Faynot, T. Poiroux, "Scaling of trigate junctionless nanowire MOSFET with gate length down to 13 nm", *IEEE Electron Device Letters,* vol. 33, pp. 1225-1227, 2012.

[5] E.T. Fonte, R. Trevisoli, R.T. Doria, "Applicability of Charge Pumping Technique for Evaluating the Effect of Interface Traps in Junctionless Nanowire Transistors", in. Proc. *2019 34th Symposium on Microelectronics Technology and Devices (SBMicro)* , p. 1-4, 2019.

[6] G. Barbottin, A. Vapaille, *Instabilities in Silicon Devices: Silicon Passivation and Related Instabilities*, Elsevier, 1999.

[7] R. Trevisoli, R.T. Doria, M. de Souza, M.A. Pavanello, "Extraction of the interface trap density energetic distribution in SOI Junctionless Nanowire Transistors", *Microelectronic Engineering*, v. 147, pp. 23-26, 2015.

[8] D. Bosch, J.P. Colinge, G. Ghibaudo, X. Garros, S. Barraud, J. Lacord, B. Sklenard, L. Brunet, P. Batude, C. Fenouillet-Béranger, J. Cluzel, R. Kies, J. M. Hartmann, C. Vizioz, G. Audoit, F. Balestra, F. Andrieu "All-Operation-Regime Characterization and Modeling of Drain Current Variability in Junctionless and Inversion-Mode FDSOI Transistors", *2020 IEEE Symposium on VLSI Technology*, pp. 1-2, 2020.

[9] D.-Y. Jeon, S.J. Park, M. Mouis, M. Berthomé, S. Barraud, G.-T. Kim, G. Ghibaudo, "Revisited parameter extraction methodology for electrical characterization of junctionless transistors", *Solid-State Electronics,* vol. 90, pp. 86-93, 2013.

[10] R.T. Doria, R. Trevisoli, M. de Souza, M.A. Pavanello, "Low-Frequency Noise and Effective Trap Density of Short Channel P- and N-Types Junctionless Nanowire Transistors", *Solid-State Electronics*, v. 96, pp. 22–26, 2014.

978-1-6654-3746-2/21 $31.00 © 2021 IEEE

Field-Effect Passivation of Lossy Interfaces in High-Resistivity RF Silicon Substrates

Martin Rack, Lucas Nyssens, Massinissa Nabet, Dimitri Lederer, Jean-Pierre Raskin

Université catholique de Louvain, Louvain-la-Neuve, Belgium

Abstract—In this paper, a novel method for increasing effective resistivity in low-doped silicon substrates is presented. The parasitic surface conduction effect is known to be the dominant cause of RF substrate losses in high-resistivity silicon materials, and the proposed approach passivates the Si/SiO$_2$ interface through the field-effect. The technique is applied below CPW lines in the 45RFSOI node from GLOBALFOUNDRIES on alternate high-resistivity silicon substrates. Small-signal measurement results reveal a strong increase in substrate effective resistivity as a function of the applied bias to the passivation structures, and therefore an appreciable decrease in line loss α. Large-signal measurements performed at 900 MHz further reveal a strong increase in substrate linearity, of approximately 20 dB, achieved through the passivation technique. Finally, a TCAD-based simulation approach enables the investigation into alternate biasing-based passivation structures and the achievable RF substrate figures of merit.

Keywords- Substrate losses, substrate linearity, effective resistivity, parasitic surface conduction, interface passivation.

I. INTRODUCTION

Nowadays, advanced silicon CMOS are offering competitive radio-frequency (RF) figures of merit [1]. In RF Integrated circuits electrical signals propagate in a conductive metal layer atop an insulator-semiconductor stack. A coplanar waveguide (CPW) is depicted in Fig. 1a atop such a stack of materials (half of the CPW is depicted). To avoid losses in such lines, and also to mitigate coupling between them, it is important for the underlying substrate to have a high *effective resistivity*. Furthermore, it is required that the substrate be as linear as possible to minimize distortion of the line signals. Low-doped silicon is then favoured with high nominal resistivity (ρ_{nom}) above 1 kΩcm. However, the effective resistivity (ρ_{eff}), which is sensed by the overlying planar circuits, can differ strongly from the nominal resistivity ρ_{nom}. This is usually due to the presence of parasitic fixed oxide charges present at semiconductor interface. Then, a highly conductive channel-like layer is induced beneath the insulator, as depicted in Fig. 1a. These charges are inevitable in the IC fabrication process, and induce free electrons at the interface in very high concentrations, locally lowering the resistivity beneath the circuits by a factor of 10^3 to 10^6. This in turn lowers the sensed effective resistivity by a factor of 10 to 10^4, to values as low as 1 Ωcm. This is referred to as the *parasitic surface conduction (PSC) effect* that renders the use of a high-

resistivity (HR) substrate ineffective for improving the performance of overlying ICs as compared to standard-doped Si (ρ_{nom} ~10 Ωcm) [1]. A breakthrough was made in the early 2000s with the introduction of a thin polysilicon layer rich in defects (traps) beneath the buried oxide (BOX) in SOI technology [2, 3]. This layer effectively mitigates the PSC effect by pinning the Fermi-level near mid-gap at the interface [4] in a highly resistive state, enabling ρ_{eff} of over 1 kΩcm.

In this paper, a novel concept for combatting the PSC effect is presented, which relies on locally depleting the interface, making it highly resistive. This is achieved by biasing dedicated polysilicon or metal shapes implemented below RF passive devices. Thanks to this *field-effect passivation scheme*, significantly improved linearity and reduced substrate losses are demonstrated.

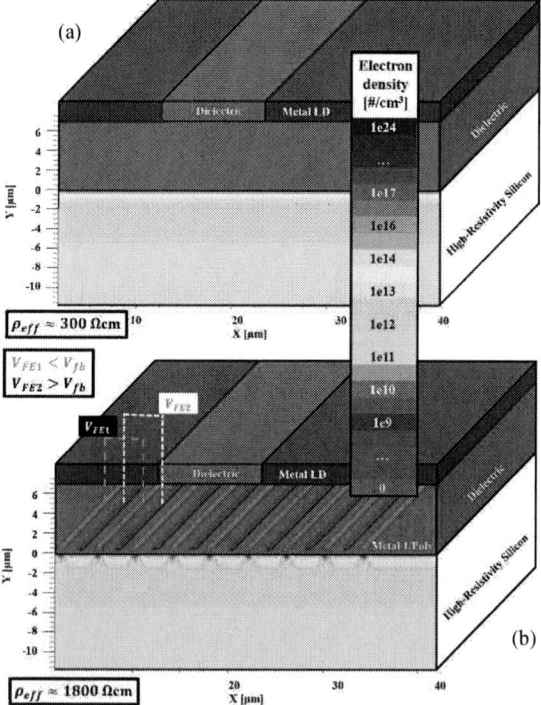

Figure 1. Carrier profile in a HR substrate (with PSC) with (a) and without (b) the FE passivation scheme, below a CPW line in "Metal LD", roughly 13 μm above the Si/SiO2 interface).

This approach is not entirely dissimilar to the PN-implant passivation scheme, which also serves to interrupt the PSC interface layer in HR substrates [5].

In this work, the highly improved quality of CPW lines is demonstrated through the extraction of the ρ_{eff} and linearity metrics by implementing the concept in GLOBALFOUNDRIES' 45RFSOI node.

It should be noted that a similar concept was patented in [6] independently of this research.

For these trials, 45RFSOI was run on HR substrates. Enabling RF performances close to those of TR wafers, this method is compatible with bulk, SOI and FD-SOI technologies.

Finally, a simulation section is included in this work, allowing us to explore potential RF performance results based on the field-effect substrate passivation scheme associated to various biased structures.

II. FIELD-EFFECT PASSIVATION FOR RF

A. Concept: Field-Effect Passivation of PSC

One concept of this passivation is depicted in Fig. 1b, based on an array of adjacent parallel lines implemented in a polysilicon layer, with various bias points applied to the lines. One line in two is biased with V_{FE1}, and the others with V_{FE2}, as illustrated. All lines are biased through large on-chip resistors and are therefore RF-floating nodes (in fact, the poly shapes used to implement them are the core of high-valued RF resistor elements). Through the filed-effect, the applied V_{FE} biases modulate the interface channel (PSC) and can be tuned to interrupt and counter its conductive and lossy nature.

This type of array was implemented physically in GLOBALFOUNDRIES' 45RFSOI technology on HR beneath CPW lines implemented in the topmost metal layer (LD), which is 2.9 μm-thick aluminium at a distance of roughly 13 μm from the Si/SiO₂ interface. The fabricated CPW lines are characterized by a signal-line width of W_c of 35 μm, a gap between signal and ground lines S of 25 μm, and a length L of 2.2 mm.

The field-effect passivation structure is an array of parallel lines of width 0.6 μm, separated by 0.5 μm, and implemented in the polysilicon/M1 layer.

B. RF Measurement Results

These CPWs were measured on-wafer using GSG probes. After pad de-embedding, ρ_{eff} is extracted from the small-signal (S-parameter) data according to [7, 8]. Large-signal measurements performed at 900 MHz enable substrate linearity evaluation, by extracting the power level of the second generated harmonic (at 1.8 GHz) at a DUT fundamental power of +15 dBm.

The measurement results pertaining to the passivated device are plotted in Fig. 2, which illustrates that a fourfold increase in substrate ρ_{eff} (from 420 Ωcm to 1800 Ωcm) is achieved by appropriately biasing the passivation structures to increase substrate impedance by combatting the PSC interface layer.

Linearity is strongly increased simultaneously with ρ_{eff} (expected correlation [9]), as a function of the applied V_{FE} biases, and Fig. 2b demonstrates a 25-30 dB reduction in the device's generated H2 power level by passivating (depleting) the interface through the field-effect scheme.

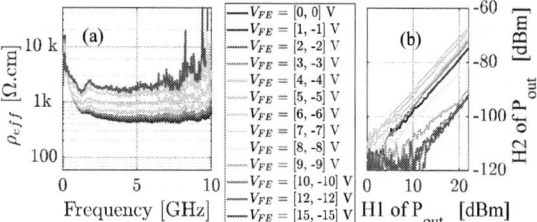

Figure 2. (a) Extracted ρ_{eff} from CPW S-parameter data. (b) Measured second harmonic power (1.8 GHz) vs. fundamental power (900 MHz) at the output of a 2.2 mm-long CPW line.

To highlight these improvements further, Fig. 3 plots the substrate small- and large-signal figures of merit (FoM) as a function of the applied V_{FE}.

When $V_{FE1} = -V_{FE2}$ are large-valued (either significantly higher or lower than the flatband voltage $V_{fb} \approx$ -5 V), electron-rich and hole-rich regions are induced below the adjacent passivation line structures, as illustrated in Fig. 1b. Then, depletion regions are created between these regions of opposite polarity, and serve to substantially increase the resistance of the path along the interface. The PSC effect is then effectively countered, and high values of ρ_{eff} –in the range of 2 kΩcm– and low values of H2 – in the range of -107 dBm– are achieved. Fig. 3 shows these results to be competitive with respect to the state-of-the-art trap-rich SOI RF substrate solution.

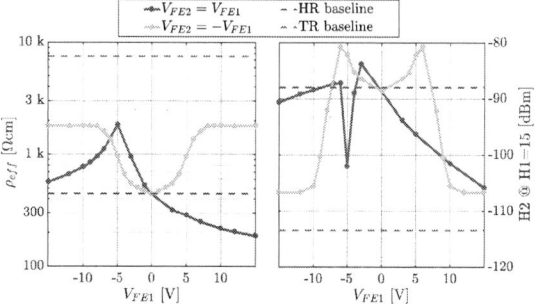

Figure 3. Measured ρ_{eff} (at 5 GHz) and H2 (at H1 = 15 dBm) as a function of the DC bias applied to the passivation structures.

C. Simulation Study

Substrate RF performance on such field-passivated substrates can be evaluated using TCAD tools. By simulating a 2D cross-section of a CPW line on a substrate material stack, the lineic capacitance C and conductance G terms between the central signal electrode and the coplanar ground electrodes of the CPW are obtained. These metrics are the converted into the effective substrate material parameters ρ_{eff} and ε_{eff} [7, 8].

Fig. 4 plots the simulated effective resistivity curves over frequency obtained for several types of field-effect

978-1-6654-3746-2/21 $31.00 © 2021 IEEE

passivation schemes using lower-level electrode layers (DC biased and RF floating).

In the simulations, these lower-level electrodes are implemented using 100 nm-thick polysilicon that has a resistivity of 50 Ωcm (unless explicitly stated otherwise). The poly layer is situated 200 nm from the Si/SiO$_2$ interface, and the CPWs (W_c = 26 μm and S = 12 μm) are defined in a 1 μm-thick aluminum layer that is situated 5 μm from the Si/SiO$_2$ interface.

By applying strong negative/positive biases to an array of poly-Si shapes (each 500 nm wide, with a spacing to each neighboring shape of 1.5 μm), effective passivation is demonstrated. Indeed, ρ_{eff} values in the range of a few kΩcm are achieved, compared to only a few hundred Ωcm when the applied biases are $V_{FE1} = V_{FE2} = 0$ V. These values are in good agreement with the measured data presented in Section II-B.

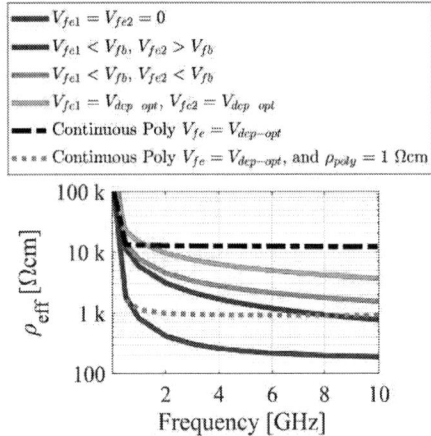

Figure 4. Extracted ρ_{eff}(f) from TCAD-simulated CPW data. Poly-Si thickness is 100 nm, and resistivity is 50 Ωcm.

Fig. 5a presents the electronic configuration in the substrate when an optimal depletion-voltage is applied to the same poly-Si passivation layers. This voltage is defined as the precise condition to apply in order to induce a state of strong depletion at the Si/SiO$_2$ interface beneath these structures. Fig. 4 demonstrates that in that case ($V_{FE1} = V_{FE2} = V_{dep,opt}$) the ρ_{eff} sensed by the CPW line is higher still, as the TCAD results present values in the 5-10 kΩcm-range over the simulated frequencies.

This concept of accurate interface depletion can be pushed further still, by biasing to $V_{dep,opt}$ a continuous-plane of poly-Si. This is illustrated in Fig. 5b, and the simulated ρ_{eff} curve corresponding to this passivation solution is plotted as the black dashed-line in Fig. 4, which demonstrates that values above 10 kΩcm are achievable under these conditions. In this case of a continuous-plane poly-Si when the interface is in strong depletion, the dominating contributing term to the CPW's lineic G term (and hence to ρ_{eff}) is actually the conductance associated to the poly-Si plane itself. The black dashed curve of Fig. 4 illustrates that a 100 nm-thick layer of 50 Ωcm resistivity contributes only to negligible CPW loss (since $\rho_{eff} > 10$ kΩcm), however ρ_{eff} would be lower if the layer's

thickness is increased or its resistivity decreased. This is illustrated by the dashed gray curve of Fig. 4, obtained from a simulation in which the continuous poly-Si plane's resistivity was set instead to 1 Ωcm. This plane becomes somewhat lossy, as an RF voltage drop occurs over its finite associated resistance value, explaining the 10-fold reduction in overall ρ_{eff} sensed by the overlying CPW.

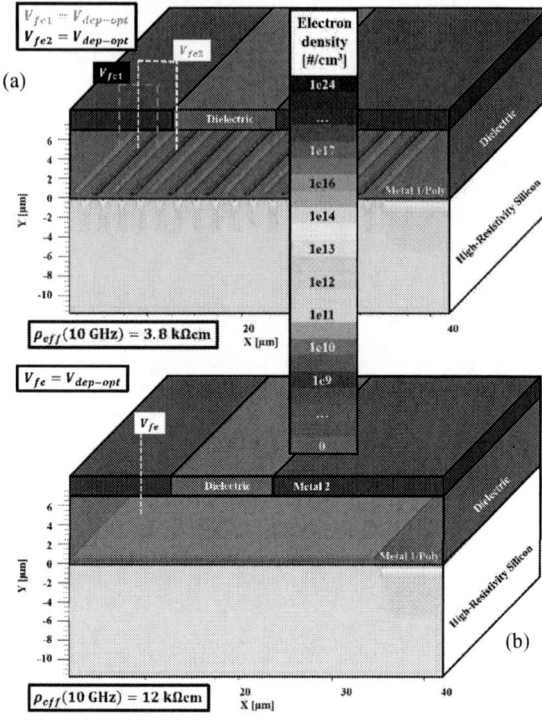

Figure 5. Carrier profile in a HR substrate with (a) the same FE passivation scheme as presented in Fig. 1b but biased to $V_{dep,opt}$ and (b) with a continuous-plane FE passivation structure, also biased to $V_{dep,opt}$.

Practical considerations for implementing such FE-passivation schemes on-chip are discussed in Appendix A.

III. CONCLUSION

In this paper, a novel interface passivation scheme is presented for enabling high-resistivity RF substrate solutions. It employs low-level metal or polysilicon structures to which DC potentials are applied in order to control the interface impedance. Through this passivation technique, measurements of fabricated CPW lines on such passivated substrates achieved an effective resistivity metric of 2 kΩcm, along with a linearity metric of -107 dBm of H2 generated for +15 dBm of fundamental power at 900 MHz. These competitive RF substrate results offer a new path for the passivation of the PSC effect that dominates the quality of low-doped silicon substrates. This technique is compatible with all types of CMOS technologies (Bulk, SOI, FD-SOI, etc.) and does not require any particular substrate engineering, and is directly implementable at the foundry level.

978-1-6654-3746-2/21 $31.00 © 2021 IEEE

ACKNOWLEDGMENT

The authors would like to thank GLOBALFOUNDRIES for chip fabrication and research support. This work is supported by Ecsel JU project "BEYOND5" (grant agreement No 876124) via EU H2020 and Innoviris (Brussels/Belgium) funding.

REFERENCES

[1] M. Rack and J.-P. Raskin, "SOI Technologies for RF and Millimeter Wave Applications", ECS Transactions, vol. 92, no. 4, pp. 79–94, Jul. 2019, DOI: 10.1149/09204.0079ecst.

[2] H. S. Gamble, B. M. Armstrong, S. J. N. Mitchell, et al., "Low-loss CPW lines on surface stabilized high-resistivity silicon", IEEE Microwave and Guided Wave Letters, vol. 9, no. 10, pp. 395–397, Oct. 1999, ISSN: 1558-2329. DOI: 10.1109/75.798027.

[3] D. Lederer and J.-P. Raskin, "New substrate passivation method dedicated to HR SOI wafer fabrication with increased substrate resistivity", IEEE Electron Device Letters, vol. 26, no. 11, pp. 805–807, Nov. 2005, DOI: 10.1109/LED.2005.857730.

[4] M. Rack, F. Allibert and J. . -P. Raskin, "Modeling of Semiconductor Substrates for RF Applications: Part I—Static and Dynamic Physics of Carriers and Traps," in IEEE Transactions on Electron Devices, vol. 68, no. 9, pp. 4598-4605, Sept. 2021, doi: 10.1109/TED.2021.3096777.

[5] M. Rack, L. Nyssens, and J.-P. Raskin, "Low-Loss Si-Substrates Enhanced Using Buried PN Junctions for RF Applications", IEEE Electron Device Letters, vol. 40, no. 5, pp. 690–693, May 2019, ISSN: 1558-0563. DOI: 10.1109/LED.2019.2908259.

[6] O. Gorbachov and E. H. S. Han, "Isolation methods for leakage, loss and non-linearity mitigation in radio-frequency integrated circuits on high resistivity silicon-on-insulator substrates", patent US20150228714A1, RFAXIS, INC. Irvine, CA (US), 13 Aug. 2015.

[7] D. Lederer and J.-P. Raskin, "Effective resistivity of fully-processed SOI substrates", Solid-State Electronics, vol. 49, no. 3, pp. 491–496, 2005, DOI: 10.1016/j.sse.2004.12.003.

[8] L. Nyssens, M. Rack, and J.-P. Raskin, "Effective resistivity extraction of low-loss silicon substrates at millimeter-wave frequencies", International Journal of Microwave and Wireless Technologies, pp. 1–14, 2020. DOI:10.1017/S175907872000077X.

[9] C. Roda Neve and J. Raskin, "RF Harmonic Distortion of CPW Lines on HR-Si and Trap-Rich HR-Si Substrates," in IEEE Transactions on Electron Devices, vol. 59, no. 4, pp. 924-932, April 2012, doi: 10.1109/TED.2012.2183598.

IV. APPENDIX A: DISCUSSION ON PRACTICAL IMPLEMENTATION

A. Applying $V_{dep,opt}$ in practice

The best results are achieved in terms of ρ_{eff} if an optimized depletion voltage $V_{dep,opt}$ can be applied.

- An idea is to implement a voltage-regulation scheme in order to achieve a depletion condition, regardless of PVT variations.
- However, the $V_{dep,opt}$ to be applied to attain high substrate FoMs may be within quite a narrow range, which may be impractical. This range depends on the distance from the FE layer to the Si/SiO$_2$ interface, but also depends on the value of the parasitic interfacial oxide charge density N_{ox}. It would then be necessary for the IC designer wanting to implement such a $V_{dep,opt}$ biasing scheme to have an estimate value of N_{ox}. Depending on these parameters, it is to

be understood that a $V_{dep,opt}$ value beyond a few volts may be impractical.

- In the simulations pertaining to the continuous-plane depletion-inducing arrangement at 200 nm from the Si interface (results of Fig. 4), simulations revealed that there is approximately a 200 mV bias range that allows for the CPW to sense a ρ_{eff} of at least 1 kΩcm.

If the field-effect layer to be biased is closer to the interface, for example 20 nm (which could arise in practice using an SOI film in a UTBB FD-SOI node) then the aforementioned 200 mV range is reduced to only 20 mV.

If the layer is moved further away from the Si interface, the acceptable biasing range to achieve decent ρ_{eff} increases, but so too does the absolute value of $V_{dep,opt}$ (close to the flatband voltage).

We note that $V_{dep,opt}$ may be multi-valued if the fixed oxide charge density is non-uniform over the Si/SiO$_2$ interface. Such consideration may be a source of a lack of robustness of exact $V_{dep,opt}$ -based approaches.

B. Continuous-plane solution vs. array-like solution

While applying $V_{dep,opt}$ (necessary for the continuous-plane solution) may require an accurate control of the applied bias, this condition can be lifted if alternate array-like P and N type regions are defined at the interface through the field-effect (see Fig. 1b).

The continuous-plane solution remains very attractive from a designer's point of view, as it alleviates the co-design between the passivation shapes and the RF device.

It is possible to design a viable array-like solution implemented using metallic layers as the filed-effect arrangements with little effect on the CPW's line parameters, while, for the continuous-plane solution, a layer of significantly higher resistivity must be used (details in the next point below).

C. On the resistivity and thickness of the FE passivation layer when implementing a continuous-plane solution

We have seen that decent results are achievable for a 100 nm thick layer of 50 Ωcm resistivity.

- We point out that if the layer was only 10 nm-thick, then similar RF FoM results would have been achieved using 5 Ωcm resistivity, as both implementations would contribute similarly to the line's absolute G parameter.

Using a 100 nm thick layer of 1 Ωcm significantly deteriorated the RF results (see Fig. 4).

Decreasing the resistivity further will continue to lower the ρ_{eff} parameter, until the slow-wave mode is brought about. Then, an RF equipotential sets up in the layer, and there is no longer any V^2/R type loss within it. The field distributions are significantly changed, and the line becomes highly capacitive, and its characteristic impedance and propagation constant are highly modified. If the plane becomes close to being of metallic properties, inductive coupling (Eddy currents) to the plane will further impact on the line's characteristics (not an accountable-for effect using a semiconductor TCAD tool.

978-1-6654-3746-2/21 $31.00 © 2021 IEEE

A Theoretical Study of Electron Mobility Distribution in FDSOI MOSFET

N. D. Akhavan[1], G. A. Umana-Membreno[1], R. Gu[1], J. Antoszewski[1], L. Faraone[1] and S. Cristoloveanu[2]

[1] Dept. of Electrical, Electronic & Computer Eng., The University of Western Australia, Crawley WA 6009, Australia
[2] IMEP-LAHC, Grenoble INP Minatec, BP 257, 38016 Grenoble, France
e-mail: nima.dehdashti@uwa.edu.au

Abstract—**In this work, we present results of a theoretical study of the electron mobility distribution in long-channel planar fully-depleted silicon-on-insulator (FD-SOI) transistors employing quantum mechanical modelling. The simulation results indicate that, in transistors with 10 nm thick Si channel layer and lengths varying from 50 nm to 200 nm, electronic transport is clearly due to two distinct and well-defined electron populations. These two calculated electron mobility distributions arise from sub-band modulated transport in 10-nm thick Si planar FD-SOI MOSFETs. Self-consistent NEGF–Poisson numerical calculations indicate significant localization of the total electron population near the front interfaces. These finding can be used to better interpret experimental results obtained by magnetic-field dependent geometrical magnetoresistance measurements and mobility spectrum analysis.**

Keywords – theoretical modelling, NEGF-Poission solver, electronic transport, FDSOI MOSFET, mobility spectrum

I. INTRODUCTION

Traditionally, the mobility of a charge carrier in semiconductor is treated as a single valued function of temperature and other semiconductor parameters such as the doping density. This, in part, is due to the fact that common techniques to measure mobility are acquired at a single magnetic field strength, which results in a single value of mobility and carrier concentration. Modern semiconductor devices and structures often consist of multiple semiconductor layers, and thus contain populations of distinct carrier species caused by n- and p-doped regions, two-dimensional conduction of surface and interface layers, and thermally generated minority carriers, among other possibilities [1].

To provide an understanding of mobility spectra, we present the calculation results of the mobility distribution n-channel fully depleted silicon-on-insulator (FDSOI) transistors employing quantum mechanical modelling. A complicated mobility distribution with multiple peaks resulting from electron conductivity in only the Γ conduction band is predicted. The mobility distribution is dependent on the nature of the scattering interactions, position of inversion layer under the gate and therefore is influenced by temperature, doping concentration and thickness of the silicon body. Our numerical modelling presented here, indicate that electronic transport in a 10-nm thick channel transistor is due to two well-defined electron species, which, according to self-consistent NEGF–Poisson calculations, arise from the modulation of carrier populations in the multiple subbands formed within the Si channel region because of strong carrier confinement. The calculated results provide a means to evaluate and interpret experimentally derived mobility spectra.

II. THEORETICAL APPROACH

In order to model the carrier transport in planar n-channel silicon FDSOI-MOSFET, we have followed the simulation approach outlined in [2-4], which has the capability to model carrier transport in mode-space approximation including electron-phonon interactions efficiently. The Hamiltonian is written in the effective mass approximation for the [100] crystal orientation with the values of the effective mass m_x, m_y and m_z obtained from the literature [2]. A schematic of the device in current study has been depicted in Fig. 1.

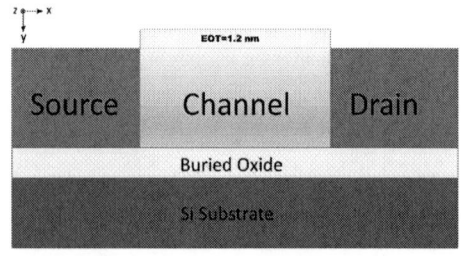

Fig. 1. Schematic representation of the planar SOI-MOSFET used in this study. The gate oxide is HfO_2 with thickness of 1.2 nm, channel thickness of 10 nm and the buried oxide is 10 nm of SiO_2. The silicon substrate is grounded. The carrier transport direction for this study is along x axis.

This work was supported by the Australian Research Council (DP170104555), the Horizon 2020 ASCENT EU project (Access to European Nanoelectronics Network – Project no. 654384), the Western Australian node of the Australian National Fabrication Facility (ANFF), and the Western Australian Government's Department of Jobs, Tourism, Science and Innovation..

978-1-6654-3746-2/21 $31.00 © 2021 IEEE

A. Mode-space and scatteing mechanisms

In order to calculate the mobility spectrum, electron-phonon scattering has to be taken into account. Therefore, the electron–phonon interaction scattering is described via the self-consistent Born approximation, and is included in the NEGF formalism by means of local self-energies. Here we assume bulk band structures for all phonons and neglect effects due to the confinement on the transverse plane [5]. For all phonons, the lesser/retarded self-energy for the *n-th* mode at the *i-th* discrete space site along the transport direction is[6]:

$$\Sigma_{n,n}^{R/<}(x_j, E) = |M_q|^2 \sum_m \overline{G}_{m,m}^{R/<}(x_j, E) F_m^n(x_j) \qquad (1)$$

where *n* and *m* are mode indices. For acoustic phonons,

$$\overline{G}^{R/<}(x_j, E) = G_{j,j}^{R/<}(E) \qquad (2)$$

Whereas for optical phonons:

$$\overline{G}^{<}(x_j, E) = N_{ph} G_{j,j}^{<}(E - \hbar\omega) + (N_{ph}+1) G_{j,j}^{<}(E + \hbar\omega) \qquad (3)$$

$$\begin{aligned}
\overline{G}^{R}(x_j, E) = & N_{ph} G_{j,j}^{R}(E - \hbar\omega) \\
& + (N_{ph}+1) G_{j,j}^{R}(E + \hbar\omega) \\
& + \frac{1}{2}\left[G_{j,j}^{<}(E - \hbar\omega) - G_{j,j}^{<}(E + \hbar\omega) \right]
\end{aligned} \qquad (4)$$

are the lesser and retarded Green's functions for the electron-phonon interactions, respectively. The term N_{ph} is the Bose–Einstein distribution for the phonon energy $\hbar\omega_{ph}$. During numerical evaluation of equations (1) to (4), the principal value of G^R and the real part of the self-energies are generally neglected. It was shown that this approximation introduced no significant error in the transport properties [7-9]. The charge and current criteria of convergence were set to 1%.

B. Mobility definition

Wheren A detailed comparison of different models for the conductivity and mobility based on the Kubo-Greenwood formula, Shur Model and Macroscopic definition is given by Frey for the case of silicon nanowire devices in the presence of electron-phonon scattering [4]. However, all the data presented are limited to the maximum gate length of 50 nm where the ballistic transport is dominant rather than the diffusive transport [10-14]. Comparison of these methods for the Long-channel MOSFET is not the purpose of our study.

The most important common part of calculating mobility in all of the aforementioned methods is the correct definition of inversion charge. The inversion charge density by Frey is defined as the charge density in the middle of the channel which means Q_{inv} at $x=L_G/2$. Niquet has also done an extensive study on the

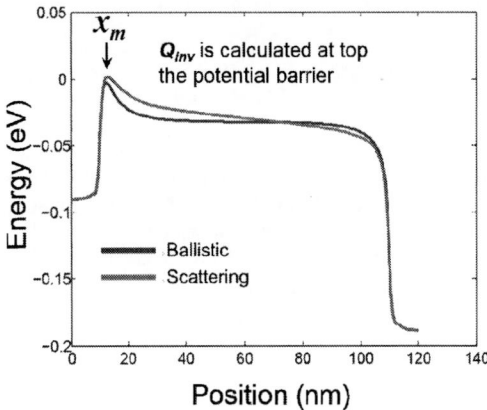

Fig. 2. In the mobility calculation, $Q_{inv}(x_m)$ is defined as the electronic charge per unit length at maximum of the first subband (top of the barrier). This spatial position is often referred to as a "virtual source".

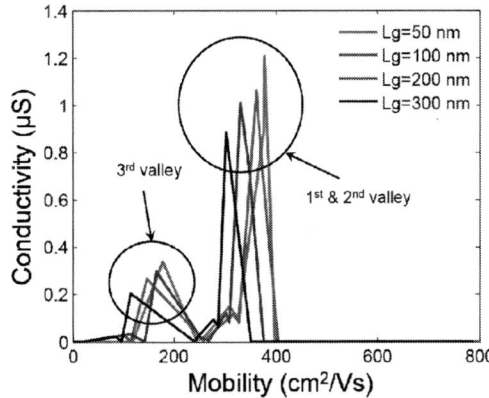

Fig. 3. The mobility distribution $\sigma(\mu)$, calculated using the macroscopic model at T=300 K for different gate length and an inversion charge density of 5×10^{12} cm^{-2} (equal to V_{GS}=0.3 V) in the channel.

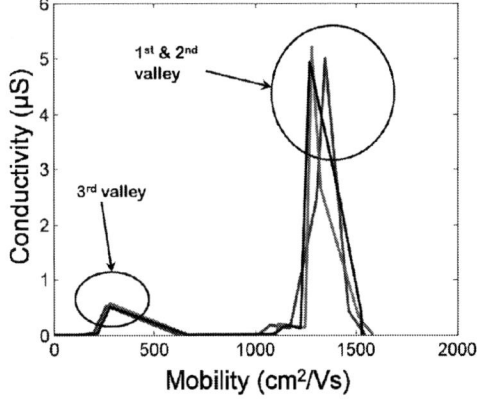

Fig. 4. The mobility distribution, $\sigma(\mu)$, calculated using the kubo-Greenwood formalism for different gate length.

definition of the inversion charge using the same approach [13].

Although this definition is correct for $L_G < 50$ nm, we found out that this definition of charge does not lead to correct values of mobility for larger gate length, whereas the inversion charge density definition by Lundstrom leads to a better definition of Mobility for a wide range of gate length [14]. Therefore, the inversion charge is defined as the carrier density at the top of the source to channel barrier, which is fixed by FET electrostatics.

The effective mobility in the MOSFET from the macroscopic view can be extracted from the current equation as follows [12]:

$$I_D = \mu_{eff} Q_{inv}(x_m) \frac{V_{DS}}{L_G} \qquad (5)$$

Where $Q_{inv}(x_m)$ is the electronic charge per unit length at maximum of the first subband (top of the barrier) in the confinement direction often referred to as a "virtual source" as shown in Fig. 2 [12, 13, 15].

C. Mobility spectrum calculations

In order to calculate the mobility spectrum we use the Kubo–Greenwood formula in order to express the mobility and conductivity as the function of energy. The relative equations can be found in literature and are as follows [4]:

$$\sigma(x_n) = \sum_v \sum_i q^2 \int dE \frac{\tau_{i,v}(x_n, E)}{m_v^*} \frac{[\hbar \, Re(k_i)]^2}{2m_v^*} A_{i,v}(x_n, E) \frac{\partial f(E)}{\partial E} \qquad (6)$$

$$\mu(x_n) = \frac{1}{qn(x_n)} \sum_v \sum_i \sigma_{i,v}(x_n) \qquad (7)$$

where the sum is taken overall valleys v and modes i. The parameter A is the spectral function, is expressed as,

$$A = i[G^R(E) - G^A(E)] \qquad (8)$$

where $G^A = G^{R+}$ and the diagonal elements of G^R correspond to density of states.

III. RESULTS AND DISCUSSION

In this study, we consider a planar SOI-MOSFET with ideal ohmic contact. The source and drain regions have a fixed length of 10 *nm* and donor doping concentration $N_D = 10^{20}$ *cm⁻³*. The short source/drain region allows us to neglect any series in the macroscopic definition of the mobility. The channel is left intrinsic with a variable gate length of 20 nm to 300 nm and a fixed silicon film thickness of 10 nm. HfO₂ layer of 1.2 *nm* thickness is used as the dielectric material. The buried oxide is SiO₂ with a thickness of

10 nm and grounded. A sketch of the device is provided in Fig. 1. Bulk effective masses are considered for Silicon ($m_t = 0.19 m_0$, $m_l = 0.91 m_0$), while the electron effective mass in the oxide is fixed at $0.5 m_0$, with m_0 the free.

The mobility data presented in this section are based on complete simulations as outlined previously and, in order to have a consistent comparison of the mobility, the drain-source voltage (V_{DS}) has been adjusted for different gate length to maintain the low electric field of 10^4 *V/cm* in all devices. No simplifying assumptions are taken in the computation of the barrier height at the device virtual source, and multiple subbands are considered according to the treatment out lined previously.

Fig. 3 shows the mobility distribution , $\sigma(\mu)$, of FDSOI-FET at T=300 K for different gate length and fixed inversion charge density of 5×10^{12} cm⁻² in the channel. The mobility calculated using the macroscopic definition of the mobility. Two distinct mobility peaks are observed in the mobility spectrum. In order to get a better understating of these two distinct mobility peaks, we need to have a closer look at individual subbands and modes in each valley of the silicon. It can be seen that the first peak in the mobility spectrum with lower conductivity comes from the 1st subband of the 3rd valley in silicon. The second peak with higher mobility originates from the 1st subband of the 1st valley and 2nd valley in Silicon. It worth to note that these valleys has the highest mobility due to the lower effective mass. Fig. 4 shows the mobility distribution calculated using the Kubo–Greenwood formula. The detailed definition of this method can be found elsewhere [4]. It is evident that both methods result in a same shape for mobility distribution, however, Kubo-Greenwood formula predicts a higher mobility for the second peak due to approximation involved in the method.

IV. CONCLUSION

Mobility Distribution calculations can be efficiently performed employing NEGF-Poisson solver to predict the broadening of mobility spectrum in the FDSOI MOSFET. Initial simulation results shows the in FDSOI MOSFET two distinct peaks are observed in the mobility spectrum graph. The presented results are consistent with recently reported experimentally data employing high-resolution mobility spectrum analysis (HRMSA) study of the inversion layer electron mobility in fully depleted silicon-on-insulator [16].

REFERENCES

[1] G. A. Umana-Membreno, S. Chang, M. Bawedin, J. Antoszewski, S. Cristoloveanu, and L. Faraone, "Evidence of Sub-Band Modulated Transport in Planar Fully Depleted Silicon-on-Insulator

978-1-6654-3746-2/21 $31.00 © 2021 IEEE

MOSFETs," *IEEE Electron Device Letters,* vol. 35, pp. 1082-1084, 2014.

[2] K. Nehari, N. Cavassilas, F. Michelini, M. Bescond, J. L. Autran, and M. Lannoo, "Full-band study of current across silicon nanowire transistors," *Applied Physics Letters,* vol. 90, p. 132112, 2007.

[3] H. Carrillo-Nuñez, C. Medina-Bailón, V. P. Georgiev, and A. Asenov, "Full-band quantum transport simulation in the presence of hole-phonon interactions using a mode-space k·p approach," *Nanotechnology,* vol. 32, p. 020001, 2020/10/15 2020.

[4] M. Frey, A. Esposito, and A. Schenk, "Computational comparison of conductivity and mobility models for silicon nanowire devices," *Journal of Applied Physics,* vol. 109, p. 083707, 2011.

[5] A. Svizhenko and M. P. Anantram, "Role of scattering in nanotransistors," *IEEE Transactions on Electron Devices,* vol. 50, pp. 1459-1466, 2003.

[6] M. Aldegunde, A. Martinez, and A. Asenov, "Non-equilibrium Green's function analysis of cross section and channel length dependence of phonon scattering and its impact on the performance of Si nanowire field effect transistors," *Journal of Applied Physics,* vol. 110, p. 094518, 2011.

[7] A. Martinez, A. Price, R. Valin, M. Aldegunde, and J. Barker, "Impact of phonon scattering in Si/GaAs/InGaAs nanowires and FinFets: a NEGF perspective," *Journal of Computational Electronics,* vol. 15, pp. 1130-1147, 2016/12/01 2016.

[8] M. Moussavou, M. Lannoo, N. Cavassilas, D. Logoteta, and M. Bescond, "Physically based Diagonal Treatment of the Self-Energy of Polar Optical Phonons: Performance Assessment of III-V Double-Gate Transistors," *Physical Review Applied,* vol. 10, p. 064023, 12/10/ 2018.

[9] M. G. Pala, C. Grillet, J. Cao, D. Logoteta, A. Cresti, and D. Esseni, "Impact of inelastic phonon scattering in the OFF state of Tunnel-field-effect transistors," *Journal of Computational Electronics,* vol. 15, pp. 1240-1247, 2016/12/01 2016.

[10] S. Poli and M. G. Pala, "Channel-Length Dependence of Low-Field Mobility in Silicon-Nanowire FETs," *IEEE Electron Device Letters,* vol. 30, pp. 1212-1214, 2009.

[11] T. Sadi, E. Towie, M. Nedjalkov, C. Riddet, C. Alexander, L. Wang, V. Georgiev, A. Brown, C. Millar, and A. Asenov, "One-dimensional multi-subband Monte Carlo simulation of charge transport in Si nanowire transistors," in *2016 International Conference on Simulation of Semiconductor Processes and Devices (SISPAD),* 2016, pp. 23-26.

[12] E. Gnani, A. Gnudi, S. Reggiani, and G. Baccarani, "Effective Mobility in Nanowire FETs Under Quasi-Ballistic Conditions," *IEEE Transactions on Electron Devices,* vol. 57, pp. 336-344, 2010.

[13] Y.-M. Niquet, V.-H. Nguyen, F. Triozon, I. Duchemin, O. Nier, and D. Rideau, "Quantum calculations of the carrier mobility: Methodology, Matthiessen's rule, and comparison with semi-classical approaches," *Journal of Applied Physics,* vol. 115, p. 054512, 2014.

[14] M. Lundstrom and Z. Ren, "Essential physics of carrier transport in nanoscale MOSFETs," *IEEE Transactions on Electron Devices,* vol. 49, pp. 133-141, 2002.

[15] V. Nguyen, Y. Niquet, F. Triozon, I. Duchemin, O. Nier, and D. Rideau, "Quantum Modeling of the Carrier Mobility in FDSOI Devices," *IEEE Transactions on Electron Devices,* vol. 61, pp. 3096-3102, 2014.

[16] G. A. Umana-Membreno, N. D. Akhavan, J. Antoszewski, L. Faraone, and S. Cristoloveanu, "Inversion layer electron mobility distribution in fully-depleted silicon-on-insulator MOSFETs," *Solid-State Electronics,* vol. 183, p. 108074, 2021/09/01/ 2021.

TCAD Negative Capacitance Ferroelectric Device Modeling for Radiation Detection Applications

Arianna Morozzi[1,*], Michael Hoffmann[2], Stefan Slesazeck[2] and Roberto Mulargia[3]

[1] INFN, Section of Perugia, via A. Pascoli 23c, 06123 Perugia, Italy.
[2] NaMLab gGmbH/TU Dresden, Noethnitzer Str. 64 a, 01187 Dresden, Germany.
[3] INFN, Section of Genova, via Dodecaneso 33, 16146 Genova, Italy.
*arianna.morozzi@pg.infn.it

Abstract— **In this work advanced TCAD (Technology Computer Aided Design) modeling will be used aiming at investigating the potentiality of Negative Capacitance (NC) devices in non-conventional application domains (e.g., radiation detection). A device-level approach to simulate the electrical characteristics of ferroelectric $Hf_{0.5}Zr_{0.5}O_2$ (HZO) has been developed. The validation of the models and of the adopted numerical methods relies on the comparison between simulations and measurements of Metal-Ferroelectric-Metal and Metal-Ferroelectric-Insulator-Metal capacitors. The ferroelectric/dielectric interface could be therefore studied before and after X-ray irradiation. The goal will be to investigate the suitability of innovative NC devices to be used in High Energy Physics experiments detection systems, featuring self-amplified segmented, high granularity detectors.**

Keywords: Ferroelectric devices; TCAD simulation; Numerical models; Radiation damage effects.

I. INTRODUCTION

The negative capacitance (NC) feature of doped HfO_2 has emerged with important technological applications in CMOS nanoscale electronic devices. Indeed, ferroelectricity in HfO_2 does not degrade with the thickness scaling, showing excellent miniaturization properties, different from conventional ferroelectrics. This fosters the HfO_2 integration in the most advanced logic and memory devices. Moreover, the voltage amplification triggered by the ferroelectric material properties, further pushes its use in almost every low-power application.

The main difference between ferroelectric materials and other dielectrics is that ferroelectrics retain their polarization after the applied electric field is removed, while other dielectrics return to an unpolarized state. This property makes ferroelectric materials particularly useful for the design of memories.

In this work the potentiality of Metal-Ferroelectric-Insulators-Metal structures will be explored with an advanced TCAD (Technology Computer Aided Design) simulation approach. All results can be extended to the study of NC-FETs of which radiation hardness will be evaluated after X-ray irradiation. Radiation damage effects can be modeled by considering the increase in the oxide charge density at the ferroelectric/dielectric interface with the dose [1]. The goal is to investigate the suitability of innovative NC devices to be used in High Energy Physics experiments detection systems at future colliders, fostering the fabrication of tracking devices with high spatial resolution, extremely thin layers and capable of detecting signals from noise in harsh radiation environments.

II. MODELING A FERROELECTRIC MATERIAL WITHIN THE TCAD ENVIRONMENT

The devices under test are MFM (Metal-Ferroelectric-Metal) and MFIM (Metal-Ferroelectric-Insulator-Metal) capacitors featuring different thicknesses of the ferroelectric $Hf_{0.5}Zr_{0.5}O_2$ (HZO) and dielectric Al_2O_3 thin films. These structures have been investigated within the Sentaurus Technology CAD (TCAD) environment, by means of ad-hoc customized models [2] to describe the properties of both materials, which are not included within the standard material library of the simulation software.

Al_2O_3 has been implemented as a new insulator material with a relative permittivity of 8. The material HZO has been added in the software material library as a new insulator with a relative permittivity of 25, 5.9 eV of band gap and all its polarization properties have been extracted from experimental measurements, as described in the next sections.

A. Structures under investigation

The simulated MFM and MFIM structures are shown in Fig. 1a and 1b respectively and represent a numerical approximation of real capacitors fabricated on Si substrates. The fabrication process is described in detail in [3]. For both MFM and MFIM devices 7.7 nm and 11.3 nm thin HZO films were considered. Moreover, in MFIM capacitors the thickness of the deposited Al_2O_3 layer ranges from 0.5 to 4 nm.

In ferroelectric materials (FE) the polarization (P) depends nonlinearly on the electric field (E).

978-1-6654-3746-2/21 $31.00 © 2021 IEEE

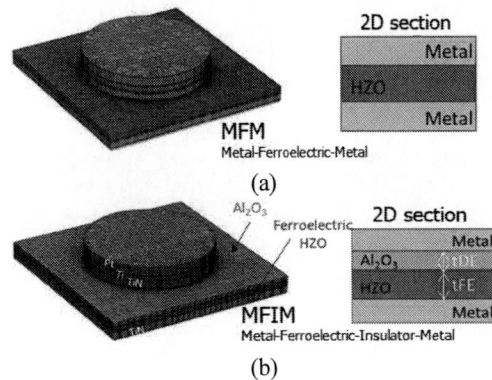

Figure1. Three-dimensional layout of the simulated MFM (a) and MFIM (b) capacitors and their two-dimensional sketches.

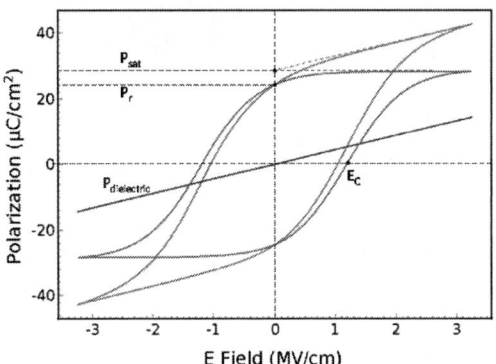

Figure 2. TCAD P-E simulation of a MFM capacitor: ferroelectric polarization properties not activated (blue), activated (magenta). The nominal P-E is red.

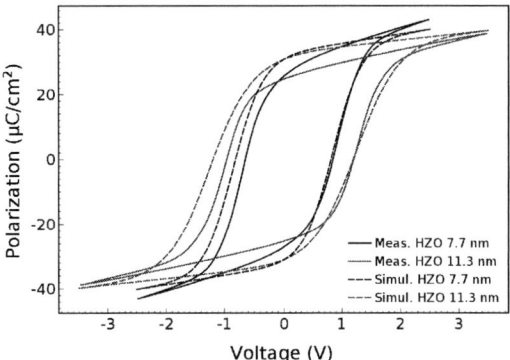

Figure 3. P-V measurements (solid) and simulations (dashed) for MFM capacitors with 7.7 and 11.3 nm thin HZO films.

This dependence is mainly modeled by means of the Preisach and the Ginzburg-Landau-Khalatnikov (GLK) models, within the TCAD environment. Indeed, MFM structures show a hysteretic trend for the P-E characteristic, which can be realistically accounted for by using the Preisach model of hysteresis. GLK equation represents instead, the proper approach to simulate hysteresis-free operation in MFIM capacitors. The latter is an evidence of the charge-boost NC phenomenon which only takes place after proper capacitance matching between the dielectric and the ferroelectric materials [4].

1) Preisach model

The polarization properties of HZO in MFM structures have been mainly described by means of the Preisach TCAD model with three quantities namely the remnant polarization (P_r), the saturation polarization (P_{sat}) and the coercive field (E_C), within the TCAD environment (red curve in Fig. 2). Without specifying these values, the material is considered as a typical insulator (blue curve in Fig. 2). The magenta curve in Fig. 2 represents the overall P-E hysteresis behavior of a MFM capacitor in which not only the polarization properties of the material have been considered, but also its dielectric characteristics (e.g., background permittivity).

From measurements reported in [3] the fabricated MFM capacitors showed excellent ferroelectric properties with coercive fields of ~1 MV/cm and a high P_r of up to 27 $\mu C/cm^2$. The standard polarization-electric field hysteresis was measured by applying triangular voltage signals with 10 kHz frequency. The pursued simulation approach considers the execution of time-variant transitory analyses, with the same characteristics for the voltage source.

These extracted polarization values have been implemented in the Preisach TCAD hysteresis model and the best agreement with experimental data has been achieved by considering 31 $\mu C/cm^2$ of P_r, 33 $\mu C/cm^2$ of P_s and 1.1 MV/cm of E_C for both 7.7 nm and 11.3 nm thin HZO films. The comparison between simulations and measurements in terms of P-Voltage behavior is illustrated in Fig. 3. The good agreement between experimental data and simulations has been obtained independently of the capacitor thickness, thus suggesting a suitable parameterization of the HZO material and its polarization properties within the simulation environment. The developed HZO material model has been used for model and methodologies validation purposes.

2) Ginzburg-Landau-Khalatnikov model

From the simulation point of view, the capacitance enhancement due to the NC effect can be obtained by using the Ginzburg-Landau-Khalatnikov (GLK) equation within the TCAD environment [5], which provides a reliable description of ferroelectric material properties in terms of a free energy (F) expanded as a power series in the FE polarization (P). In the GLK framework, the governing equation for the evolution of P, can be obtained as a gradient flow associated with the free energy (F):

$$\rho \frac{dP_i}{dt} + \nabla_{P_i} F = 0 \tag{1}$$

$$\rho \frac{dP_i}{dt} + 2\alpha_i P_i + 4\beta_i P_i^3 + 6\gamma_i P_i^5 - 2g_{ij}\frac{\partial^2 P_i}{\partial x_j^2} - E_i = 0 \tag{2}$$

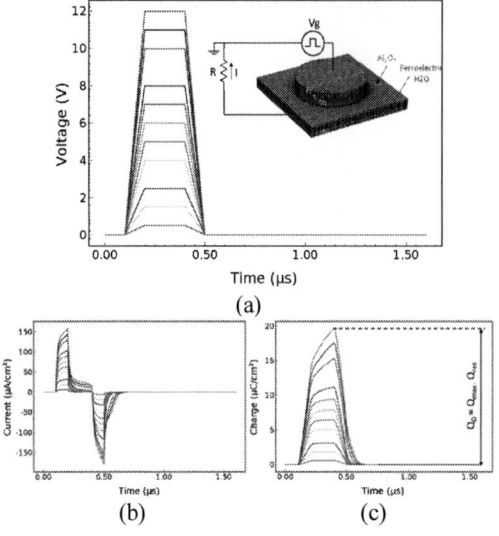

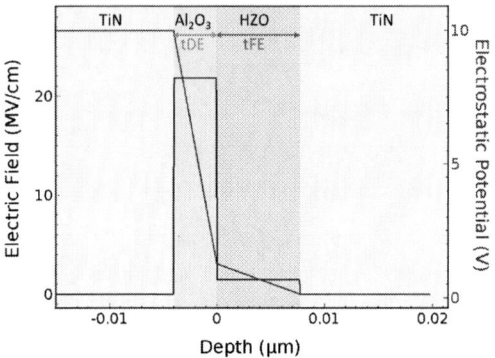

(a)

(b)　　　　　(c)

Figure 4. MFIM 7.7 nm HZO/4 nm Al₂O₃ capacitor. (a) Voltage pulses of different amplitudes, (b) current which flows through the resistor charging/discharging the capacitor, (c) charge on the capacitor as a function of time. Q_D is the released charged calculated as the difference between the maximum Q_{max} and the residual Q_{res} charge.

Figure 5. Electric field and electrostatic potential profiles in MFIM capacitors with 7.7 nm HZO and 4 nm Al₂O₃ (V_{BIAS}= 10 V).

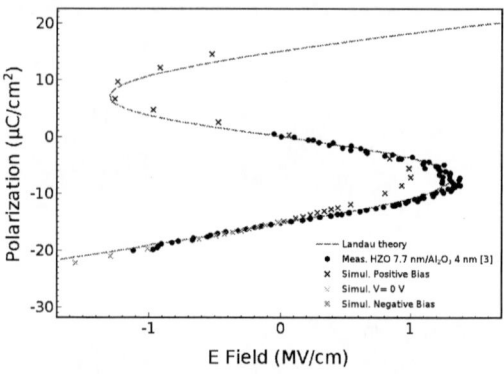

Figure 6. Ferroelectric P-E curve of a MFIM structure 7.7 nm HZO with 4 nm Al₂O₃ extracted from the pulsed signals in Fig. 4, considering P=Q_D+Q_{int} and E_F is the electric field across the ferroelectric material. The NC region corresponds to the negative slope of the S-shaped Landau P-E curve. The theoretical S-shaped Landau curve is in red, measurements in black, while the cross markers represent simulations for positive (blue), negative (orange) and 0 V input voltage applied.

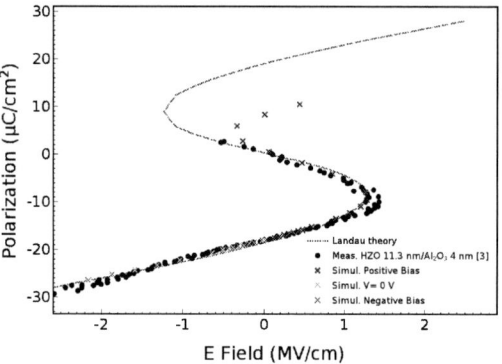

Figure 7. Ferroelectric P-E curve of a MFIM structure 11.3 nm HZO with 4 nm Al₂O₃ extracted from the pulsed signals in Fig. 4, considering P=Q_D+Q_{int} and E_F is the electric field across the ferroelectric material. The NC region corresponds to the negative slope of the S-shaped Landau P-E curve. The theoretical S-shaped Landau curve is in red, measurements in black, while the cross markers represent simulations for positive (blue), negative (orange) and 0 V input voltage applied.

where ρ is the viscosity associated with polarization-switching dynamics, g_{ij} is a coupling coefficient, α, β and γ are the Landau coefficient used to parameterize the energy function. The viscosity has been set to $2.25 \cdot 10^{+04}$ Ωcm, g_{ij} to $1.0 \cdot 10^{-03}$ cm³/F, while α, β and γ are worth $-1 \cdot 10^{11}$ cm/F, $2.5 \cdot 10^{20}$ cm⁵/FC² and $2.0 \cdot 10^{30}$ cm⁹/FC⁴ respectively, in accordance with Landau theory.

Within the experimental framework, pulsed Q-V measurements are necessary to access the ferroelectric NC region during switching by preventing charge injection. Consequently, this procedure has been implemented within the TCAD environment where transient analyses have been carried out to investigate the response of a MFIM capacitor to voltage pulses of increasing amplitudes, as the ones reported in Fig 4a. Fig. 4b reports the current during the capacitor charging and discharging phases. The current integral over time, the charge, is reported in Fig. 4c. The capacitor released charge Q_D is defined as the difference between the

maximum charge Q_{max} and the residual charge Q_{res} after the discharging phase.

Fig. 5 illustrates both the electrostatic potential and the electric field profiles across a MFIM capacitor, at an applied voltage of e.g. 10 V at the top electrode, when the device operates in the NC region. By means of the simulating approach it is possible to easily distinguish between the electric field in the dielectric and in the ferroelectric materials.

Measurements highlighted the presence of a large negative fixed charge (Q_{int}) at the HZO/Al₂O₃ interface of about -15 μC/cm² (HZO 7.7 nm) and -18 μC/cm² (HZO 11.3 nm), which allows to enter the NC region only when applying a large enough positive voltage [3]. Polarization is defined as the sum of Q_D and Q_{int}. The simulated polarization is then plotted as a function of the electric field across HZO (Fig. 6 and 7). For hysteresis-free operation, the dielectric capacitance C_{DE} must be matched to the ferroelectric NC region by following the law $C_{DE} < |C_{FE}|$). This match has been experimentally

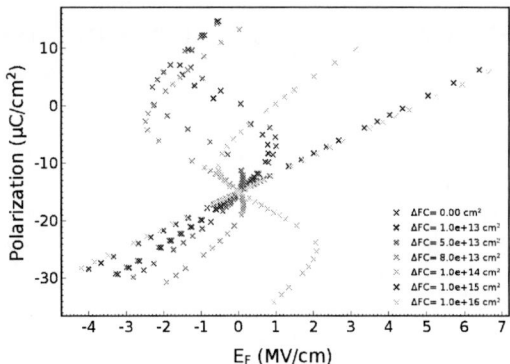

Figure 8. Ferroelectric P-E curve of a MFIM structure 7.7 nm HZO/4 nm Al$_2$O$_3$. The NC region corresponds to the negative slope of the S-shaped Landau P-E curve.

verified for a MFIM structure with 7.7/11.3 nm HZO thin film coupled to 4 nm of deposited Al$_2$O$_3$ [3]. Simulation results are illustrated in Fig. 6 and 7 for and 7.7 and 11.3 nm HZO and 4 nm Al$_2$O$_3$ MFIM capacitors, respectively. Using a Landau-based model, non-hysteretic NC stabilization can be simulated. The stabilized NC region corresponds to the negative slope of the S-shaped Landau P-E curve. A region of NC is observed which starts for positive applied pulses around the coercive field of ~ 1 MV/cm. For negative applied pulses, only a linear dielectric response is observed, which shows that HZO is initially in the negative polarization state due to compensating fixed charges at the HZO/Al$_2$O$_3$ interface. This behavior has been proficiently modeled and the good agreement between measurements and simulations fosters the application of the HZO developed model to the study of more complex devices e.g. NC-FETs.

The radiation hardness of MFIM structures has been investigated with the introduction of additional fixed charge ($\Delta Q_{int}(\phi)$) at the HZO/Al$_2$O$_3$ interface of increasing values, aiming at mimic increasing X-ray doses (ϕ). The overall Q_{int} is the sum of the Q_{int} already present before irradiation (-15/-18 μC/cm^2) and the positive ΔQ_{int} which is the contribution totally induced by the X-ray irradiation ($Q_{int}(\phi) = Q_{int}(0) + \Delta Q_{int}(\phi)$). The purpose will be to verify the suitability of innovative NC devices to be used in High Energy Physics experiments detection systems.

Fig. 8 illustrates the P-E S-shaped curves for different ΔQ_{int} values up to $1.0 \cdot 10^{+16}$ cm^{-2}. For increasing values of dose, hence of Q_{int}, the access to the NC region is granted by applying different voltage signal amplitudes ranges. For not irradiated devices, to enter the NC region is necessary to apply a large enough positive voltage, while for a ΔQ_{int} of $1.0 \cdot 10^{+14}$ cm^{-2} the signal to be applied must be of negative amplitude. However, the access to stabilized NC region is kept for Q_{int} up to $1.0 \cdot 10^{+14}$ cm^{-2} after which the NC region seems to be inaccessible at least in the operating voltage range of the device.

CONCLUSION

The goal of this work was to investigate the suitability of innovative NC devices to be used in High Energy

Physics (HEP) experiments detection systems, featuring self-amplified segmented, high granularity detectors. To cope with the small signal amplitude coming from thin sensors, the internal amplification mechanisms, namely step-up voltage conversion, intrinsic to NC devices allows to exceed the limits imposed by actual CMOS technology and can be used for particle detection in tracking applications for next generation HEP experiments at future colliders.

Advanced TCAD modeling has been used for the purpose at hand, by the introduction of two new insulator materials for the description of the electrical characteristics of Hf$_{0.5}$Zr$_{0.5}$O$_2$ (HZO) and Al$_2$O$_3$. The validation of the developed methods and models relies on the comparison between simulations and measurements of MFM and MFIM capacitors. The TCAD Preisach model turned out to be the proper method to describe MFM structures which show hysteretic P-E characteristics once the remnant and saturation polarization and the coercive field have been set. Instead, using the TCAD GLK model, non-hysteretic NC stabilization can be obtained for MFIM structures, after a proper geometric match between the ferroelectric and dielectric capacitances.

Moreover, the HZO/Al$_2$O$_3$ interface can be studied before and after X-ray irradiation. Surface damage effects induced by radiation can be characterized by considering a fixed charge Q_{int} at the ferroelectric/dielectric interface. Increasing values for Q_{int} accounts for increasing X-ray doses.

When numerical simulations are capable of verify experimental results, they will also gain predictive power, resulting in reduced time and cost in detector design and testing. The good agreement obtained between measured and simulations findings fosters the application of the HZO developed model to the study of more complex devices e.g. NC-FETs, aiming at stabilizing the NC on the wider operation voltage range by achieving a matched design of the ferroelectric layer and the MOS capacitor.

ACKNOWLEDGMENT

This work was financed by the INFN-CSN5, under INFN Young Researcher Grant "NegHEP" and was financially supported out of the State budget approved by the delegates of the Saxon State Parliament.

REFERENCES

[1] A. Morozzi et al., "Polycrystalline CVD diamond device level modeling for particle detection applications", Journal of Instrumentation, vol. 11(12), C12043, 2016.

[2] A. Morozzi et al., "A combined surface and bulk TCAD damage model for the analysis of radiation detectors operating at HL-LHC fluences", Journal of Instrumentation, vol. 11, issue 1214, C12028, 2016.

[3] M. Hoffmann et al., 2018 IEEE International Electron Devices Meeting IEDM, 18-727 (2018).

[4] H. Agarwal et al., "Proposal for Capacitance Matching in Negative Capacitance Field-Effect Transistors," IEEE Electron Device Lett., vol. 40, no. 3, pp. 463-466, 2019.

[5] Sentaurus™ Device User Guide, Version R-2020.09, 2020.

978-1-6654-3746-2/21 $31.00 © 2021 IEEE

Performance of Stacked SOI Nanowires in a Wide Temperature Range

Jaime C. Rodrigues[a], Genaro Mariniello[a], Mikael Cassé[b], Sylvain Barraud[b], Maud Vinet[b], Olivier Faynot[b]
and Marcelo A. Pavanello[a]

[a]Department of Electrical Engineering, Centro Univeritário FEI, São Bernardo do Campo, Brazil [b]CEA-Leti, Grenoble, France
E-mail: jaimear@fei.edu.br

Abstract—This paper investigates the basic electrical characteristics and some analog figures of merit for 2-level vertically stacked nanowire MOSFETs with different fin widths in the temperature range of 93K up to 400 K. Basic electrical parameters such as threshold voltage, subthreshold slope and carrier mobility are evaluated in linear region. On the other hand, analog figures of merit as transconductance, output conductance and voltage gain are evaluated in saturation.

Index Terms—Vertically Stacked Nanowires, Low temperature, Electrical characterization, transcondutance, threshold voltage, subthreshold slope, Multigate transistors

I. INTRODUCTION

The multiple-gate field-effect transistors (MuGFETs) are among the most reliable devices to keep pushing the MOSFET downscaling further into the nanometer size dimensions.[1]

Vertically stacked nanowire MOSFETs, or simply stacked nanowires, are promising devices for future technological nodes that would allow the continuity of the CMOS roadmap, increasing the current density of nanometer-long MOSFETs without increasing the occupied silicon area, exhibiting excellent performance and scalability[1][2].

These MOSFETs consist of two levels of narrow and thin silicon layers (sometimes referred as nanosheets) surrounded by the gate electrode, ensuring excellent electrostatic control of the charges in the channel, leading to reduced the short-channel effects. In case of stacked nanowires fabricated in Silicon-On-Insulator (SOI) substrate, the bottom level is built over the buried oxide and has a Omega-FET shape, whereas the top-level has a Gate-All-Around (GAA) architecture[3]. A TEM image of the cross-section of SOI stacked nanowire MOSFETs fabricated with similar process than those studied in this work is presented in Figure 1[3].

The operation of MOSFETs at low temperatures is known to improve some device electrical characteristics, such as the carrier mobility and subthreshold slope, additionally to the technology scaling[4]. Recently, the operation of MOSFETs in cryogenic regime gained interest because the advent of quantum computing and the need of CMOS circuits to interface electronic circuits operating in higher temperatures with the qbits, which operate in deep-cryogenic mili-Kelvin temperature range.

In this study, the electrical characteristics of 2-level vertically stacked nanowires are presented, in the temperature range

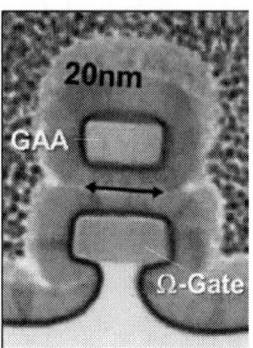

Fig. 1. TEM image of the cross section of Stacked Nanowire MOSFETs fabricated with similar process than those studied in this work[3].

from 93 K up to 400 K. Basic device electrical characteristics at low drain bias as well as some analog figures of merit in saturation are evaluated.

II. DEVICE DESCRIPTION

The devices measured in the work were fabricated in the CEA-Leti, following the process described in ref. [3]. They consist of 2-level n-type silicon stacked nanowire MOSFETs made in SOI wafers with buried oxide thickness of 145nm. Devices with 10 parallel fingers with fixed channel length (L) of 100 nm, fin height (H_{fin}) of 9 nm, and variable fin width (W_{fin}) of 10 nm, 15 nm, 20 nm, 25 nm and 40nm were measured. The body region is undoped (or not intentionally doped p-type material) surrounded by the gate stack, composed of SiO_2-HfSiON and TiN gate metal. The equivalent oxide thickness is about 1.3 nm.

III. RESULTS AND DISCUSSION

A. Basic Electrical Characteristics

The measurements were performed using a Low Temperature Microprobe system from MMR Technologies and a Keysight B-1500 Semiconductor Parameter Analyzer. For the measurements at temperatures lower than 300 K the samples were cooled down to 93 K and biased after the temperature stabilization interval. Then, the temperature is raised to the next temperature until it reaches room temperature again.

978-1-6654-3746-2/21 $31.00 © 2021 IEEE

Figure 2 presents the measured drain current I_{DS} as a function of gate voltage V_{GS}, both in linear and logarithmic scales, of all studied stacked nanowires biased with a drain bias of $V_{DS}=$ 40 mV at different temperatures.

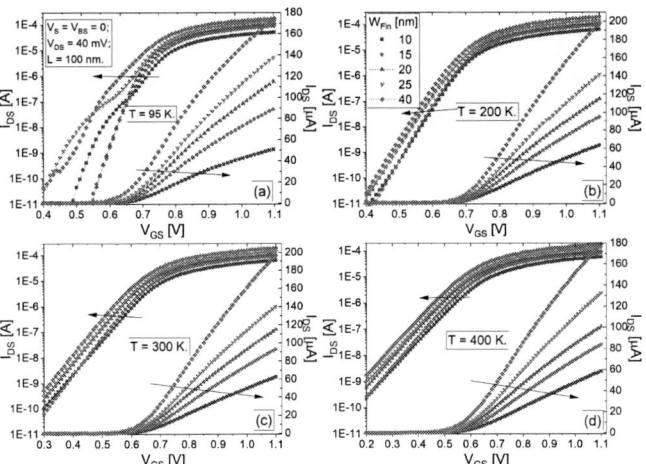

Fig. 2. Measured I_{DS} as a function of V_{GS} for stacked nanowires with fin widths of 10 nm, 15 nm, 20 nm, 25 nm and 40 nm, with $V_{DS}=$ 40 mV, for different temperatures: a) 93K, b) 200 K, c) 300 K, d) 400 K.

Using the data of Figure 2 the threshold voltage (V_{TH}) obtained by dgm/dV_{GS} method [5]. Figure 3 illustrates V_{TH} for the several stacked nanowires at different temperatures. The straight lines presented in Figure 3 represent the linear regression of the measured data for the stacked nanowires with W_{fin} of 10 nn, 20 nm and 40 nm.

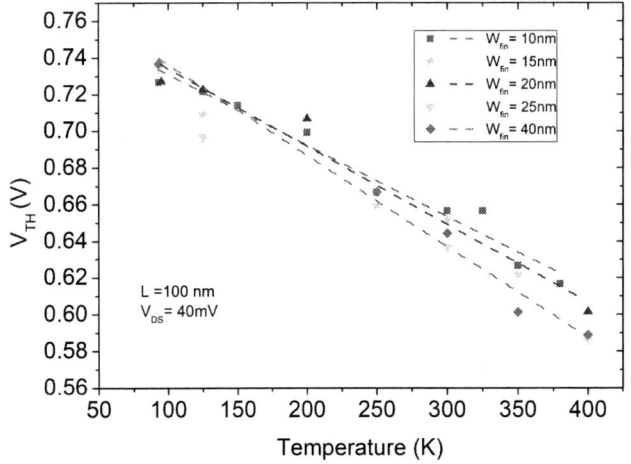

Fig. 3. Threshold voltage as a function of the temperature for transistors with W_{fin} of 10nm, 15nm, 20nm, 25nm and 40 nm.

In the studied temperature range, the V_{TH} increases practically linearly with the temperature reduction, similarly to previously observed for FinFETs[4] and single-level nanowires[4]. Also, it can be noticed in the Figure 3 that the reduction of V_{TH} on temperature is more pronounced for wider stacked nanowires. Table I shows the slopes of the linear regressions

included in Figure 3.[6]

TABLE I
THE THRESHOLD VOLTAGE SLOPE BY TEMPERATURE

$W_{fin}[nm]$	$\delta V_{TH}/\delta T [V/K].10^{-4}$
10	-3,876
20	-4,251
40	-4,966

To get some insight on the obtained results, the analytical model for the threshold voltage and drain current of MuGFETs proposed by Duarte et al. [7][8] has been used. This model proposes a set of universal equations valid for MuGFETs with different geometries such as double-gate, triple-gate, cylindrical Gate-All-Around and rectangular Gate-All-Around. Equation 1 describes the V_{TH}.

$$
\begin{aligned}
V_{\mathrm{TH}} = {} & V_{\mathrm{FB}} - \frac{Q_{d,n}}{C_{g,n}} + 2v_T \ln \frac{N_{\mathrm{si}}}{n_i} \\
& - v_T \ln \left[\frac{C_{\mathrm{ch},n}}{C_{g,n}} \left(1 - e^{\frac{Q_{d,n}}{v_T C_{\mathrm{ch},n}}} \right) \right].
\end{aligned}
\tag{1}
$$

where V_{FB} is the flat band voltage, $Q_{d,n}$ is the depletion charge per unit length, $C_{g,n}$ is the gate oxide capacitance per unit length, $C_{ch,n}$ is the channel capacitance per unit length, $A_{ch,n}$ is the area of the channel, v_T stands for the thermal voltage (kT/q), N_{Si} is the channel doping concentration, q is the electron charge and n_i is the intrinsic carrier concentration.

The $Q_{d,n}$ term of Equation 1 is weakly temperature-dependent in the studied range and has negligible influence on V_{TH}. The reduction of n_i at lower temperatures increases the Fermi potential. The V_{TH} increase with temperature reduction is caused by the Fermi potential increase at lower temperatures. The terms proportional to the channel cross-section of Equation 1 increase with W_{fin}. This increase causes the slight larger variation of V_{TH} for wider transistors observed in Table I.

Figure 4 presents the inverse subthreshold slope *versus* temperature curves for the devices W_{fin}=10 nm, 15 nm, 20 nm, 25 nm and 40 nm.

The inverse subthreshold slope remained close to theoretical limit, obtained for $S = (kT/q).\ln(10)$ [3], in the whole range of temperatures except at T=400 K. Similar inverse subthreshold slope *vs.* temperature behavior has been obtained for all studied devices, wich indicates a body factor close to the unity.

The maximum transcondutance ($g_{m,max}$) and the carrier mobility (μ_n) as a function of temperature for stacked nanowires with W_{fin}=10 nm, 15 nm and 20 nm are presented in Figure 5. For the (μ_n) extraction the Y-Function method has been employed[9] to avoid the series resistance influence.

One can see an almost linear increase of $g_{m,max}$ in the entire temperature range, which is caused by the μ_n increase at lower temperatures. The rate of $g_{m,max}$ increase with temperature is slightly larger for wider W_{fin}, which is also transferred to μ_n. The μ_n is larger for wider devices at any temperature,

978-1-6654-3746-2/21 $31.00 © 2021 IEEE

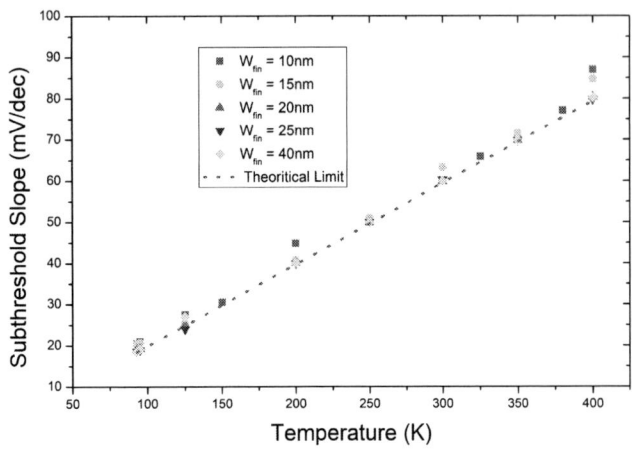

Fig. 4. Inverse subthreshold slope as a function of the temperature for transistors with several W_{fin}.

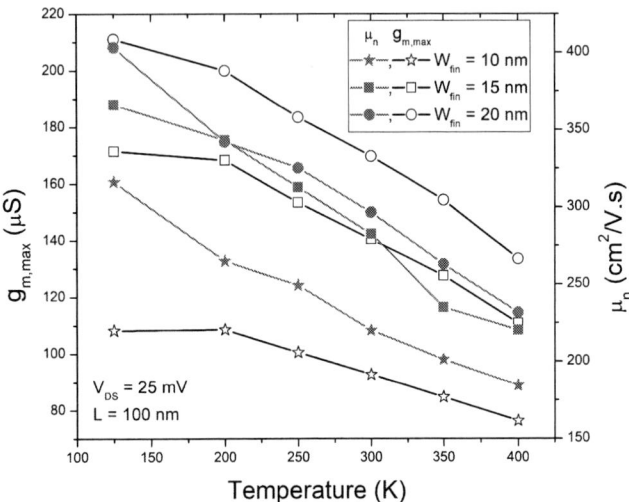

Fig. 5. Maximum transconductance and carrier mobility as a function of the temperature for transistors with W_{fin}=10nm, 15nm and 20 nm

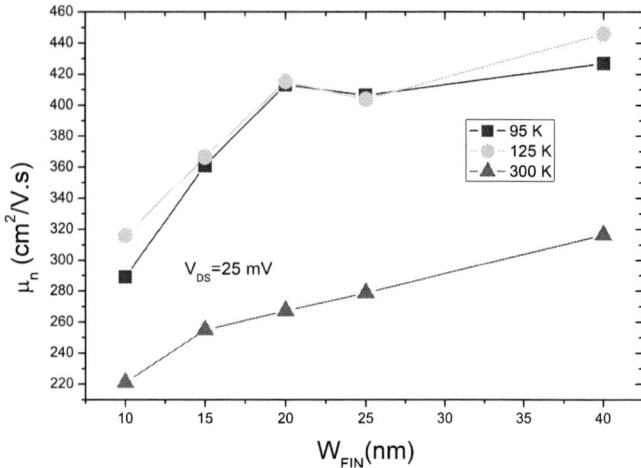

Fig. 6. Carrier mobility at different temperatures as a function of W_{FIN}

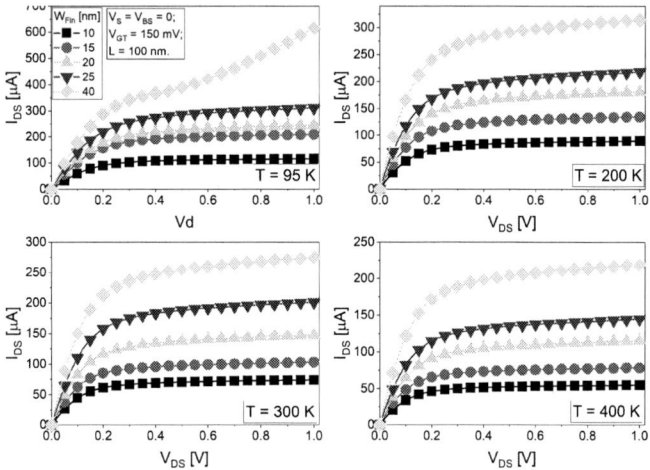

Fig. 7. I_{DS} as a function of V_{DS} for transistors with channel widths of 10 nm, 15 nm, 20 nm, 25 nm and 40 nm, with $V_{GT} = 150$ mV, with different temperatures: a) 93 K, b) 200 K, c) 300 K and d) 400K.

which is associated with the smaller relative conduction in the sidewalls, which have smaller electron mobility due to the crystal orientation, compared with the whole drain current. The influence of W_{fin} on the electron mobility is clarified in Figure 6, where are presented the results of μ_n as a function of W_{fin} extracted at 300 K, 125 K and 95 K.

B. Analog Parameters

The analog parameters of stacked nanowires have been evaluated with the devices biased in saturation and fixed $V_{GT}=V_{GS}-V_{TH}$.

Figure 7 presents the measured I_{DS} as a function of V_{DS} for the stacked nanowires with several W_{fin} at different temperatures.

One can see the monotonous increase of I_{DS} with W_{fin}

due to the larger channel width. At cryogenic temperature the wider device suffers an abnormal increase of I_{DS} at higher V_{DS} in the saturation region, which is consistent with some impact ionization effect at lower temperatures[10].

Figure 8 shows the g_m and output conductance (g_{DS}) as a function of temperature for nanowires with several W_{fin}, biased with V_{GT}=150 mV and V_{DS}=700 mV. It it is possible to notice an increase of g_m and g_{DS} with the increase of W_{fin}. Furthermore, both g_m and g_{DS} decrease when temperature rises from 93 K to 400 K. With the data of Figure 8 the intrinsic voltage gain $A_V=g_m/g_{DS}$ was calculated.

The calculated voltage gain *versus* temperature for the stacked nanowires of several W_{fin}, biased at $V_{DS} = 700$mV and $V_{GT} = 150$mV, is presented on Figure 9.

The voltage gain decreases for wider W_{fin} in any temperature due to the worsened g_{DS}. For the devices with W_{fin} wider than 10 nm the voltage gain decreases at lower temperatures than 250 K. This decrease is in the order of 2.5 dB. In The

978-1-6654-3746-2/21 $31.00 © 2021 IEEE

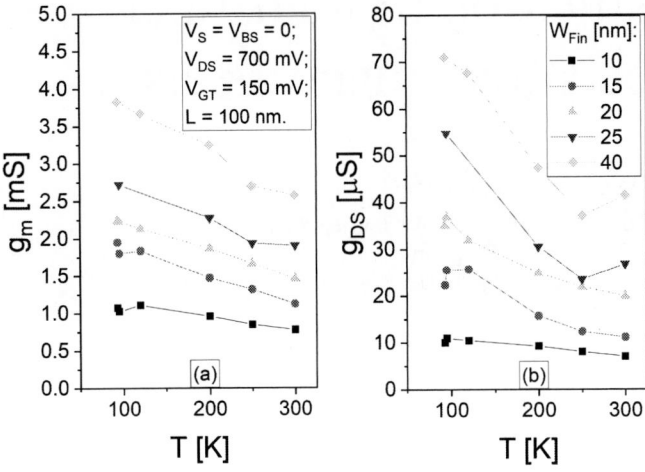

Fig. 8. Calculated g_m and g_{DS} *versus* temperature for the Stacked NWs of several W_{fin} biased at $V_{DS} = 700$mV and $V_{GT} = 150$mV.

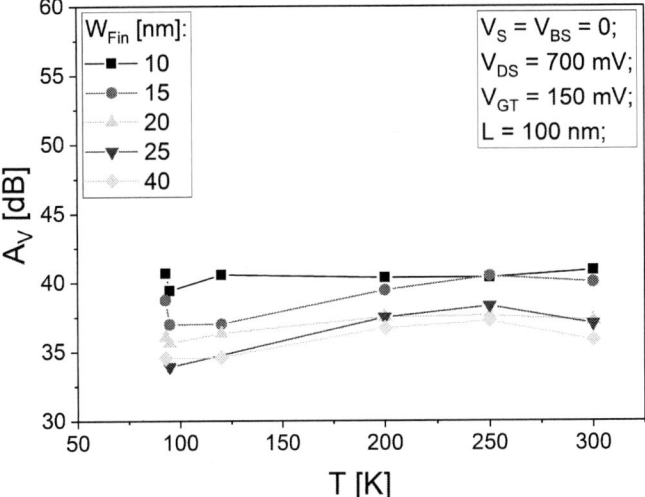

Fig. 9. Calculated A_V *versus* temperature for the Stacked NWs of several W_{fin} biased at $V_{DS} = 700$mV and $V_{GT} = 150$mV.

increase of g_m with temperature is not enough to compensate the degradation of g_{DS} leading to a degradation in the voltage gain.

For the narrow device with $W_{fin} = 10$ nm, the A_V is slight temperature-dependent and decreases about 1.5 dB at 93 K. The smaller degradation of A_V at 93 K with respect to wider devices is associated with the improved device electrostatics that reduces the degradation of g_{DS}.

IV. CONCLUSION

This work studied the performance of vertically stacked nanowires of various fin widths in the temperature range of 93 K up to 400 K. The presented results demonstrated a linear decrease of threshold voltage with temperature for the several investigated stacked nanowires. The threshold voltage reduction on temperature is larger for wider devices. The subthreshold slope is close to the theoretical limit in any temperature except at 400 K, indicating that all devices present a body factor close to the unity regardless the fin width. The carrier mobility increases with temperature decrease and the rate of increase is larger for wider fin widths. This For the analog figures of merit, both g_m and g_{DS} increase at lower temperatures and for wider stacked nanowires. The A_v is weakly temperature dependent in for the stacked nanowires in the investigated temperature range. A maximum variation of 2.5 dB has been found for wider stacked nanowires. For the narrower device, the voltage gain degradation at lower temperatures reduces to 1.5 dB.

REFERENCES

[1] S. Barraud, R. Coquand, M. Casse, M. Koyama, J.-M. Hartmann, V. Maffini-Alvaro, C. Comboroure, C. Vizioz, F. Aussenac, O. Faynot *et al.*, "Performance of omega-shaped-gate silicon nanowire mosfet with diameter down to 8 nm," *IEEE Electron Device Letters*, vol. 33, no. 11, pp. 1526–1528, 2012.

[2] R. Coquand, S. Barraud, M. Casse, P. Leroux, C. Vizioz, C. Comboroure, P. Perreau, E. Ernst, M.-P. Samson, V. Maffini-Alvaro, C. Tabone, S. Barnola, D. Munteanu, G. Ghibaudo, S. Monfray, F. Boeuf, and T. Poiroux, "Scaling of high-k/metal-gate trigate soi nanowire transistors down to 10nm width," in *2012 13th International Conference on Ultimate Integration on Silicon (ULIS)*, 2012, pp. 37–40.

[3] S. Barraud, V. Lapras, M. Samson, L. Gaben, L. Grenouillet, V. Maffini-Alvaro, Y. Morand, J. Daranlot, N. Rambal, B. Previtalli *et al.*, "Vertically stacked-nanowires mosfets in a replacement metal gate process with inner spacer and sige source/drain," in *2016 IEEE International Electron Devices Meeting (IEDM)*. IEEE, 2016, pp. 17–6.

[4] M. A. Pavanello, J. A. Martino, E. Simoen, and C. Claeys, "Cryogenic operation of finfets aiming at analog applications," *Cryogenics*, vol. 49, no. 11, pp. 590–594, 2009.

[5] A. Ortiz-Conde, F. J. Garcíea Sánchez, J. J. Liou, A. Cerdeira, M. Estrada, and Y. Yue, "A review of recent mosfet threshold voltage extraction methods," *Microelectronics Reliability*, vol. 42, no. 4-5, pp. 583–596, 2002.

[6] B. C. Paz, M. A. Pavanello, M. Cassé, S. Barraud, G. Reimbold, M. Vinet, and O. Faynot, "Cryogenic operation of ω-gate p-type sige-on-insulator nanowire mosfets," in *2018 Joint International EUROSOI Workshop and International Conference on Ultimate Integration on Silicon (EUROSOI-ULIS)*. IEEE, 2018, pp. 1–4.

[7] J. P. Duarte, S.-J. Choi, D.-I. Moon, J.-H. Ahn, J.-Y. Kim, S. Kim, and Y.-K. Choi, "A universal core model for multiple-gate field-effect transistors. part i: Charge model," *IEEE Transactions on Electron Devices*, vol. 60, no. 2, pp. 840–847, 2012.

[8] J. P. Duarte, S.-J. Choi, D.-I. Moon, J.-H. Ahn, J.-Y. Kim, S. Kim, and Y. K. Choi, "A universal core model for multiple-gate field-effect transistors. part ii: Drain current model," *IEEE Transactions on Electron Devices*, vol. 60, no. 2, pp. 848–855, 2013.

[9] J. Henry, Q. Rafhay, A. Cros, and G. Ghibaudo, "New y-function based mosfet parameter extraction method from weak to strong inversion range," *Solid-State Electronics*, vol. 123, pp. 84–88, 2016.

[10] R. Cuerdo, Y. Pei, Z. Chen, S. Keller, S. P. DenBaars, F. Calle, and U. K. Mishra, "The kink effect at cryogenic temperatures in deep submicron algan/gan hemts," *IEEE Electron Device Letters*, vol. 30, no. 3, pp. 209–212, 2009.

978-1-6654-3746-2/21 $31.00 © 2021 IEEE

Low temperature investigation of n-channel GAA vertically stacked silicon nanosheets

Bogdan Cretu[1], Anabela Veloso[2], Eddy Simoen[2,3]

[1]Normandie Univ, UNICAEN, ENSICAEN, CNRS, GREYC, 14000 Caen, France
[2]Imec, Kapeldreef 75, B-3001 Leuven, Belgium
[3]Solid-State Physics Department, Ghent University, 9000 Gent, Belgium
e-mail: bogdan.cretu@ensicaen.fr

Abstract — In this work, DC and low frequency noise measurements are performed on n-channel gate all around (GAA) vertically stacked silicon nanosheets (NS) at room and liquid nitrogen temperature. Principal static (DC) parameters such as low field mobility, access resistance and subthreshold swing are estimated. Preliminary low frequency noise studies reveal that the 1/*f* noise may be explained by the carrier number fluctuations mechanism at room and liquid nitrogen temperature operation.

Keywords-GAA NS FET; liquid nitrogen temperature, 1/f noise, carrier number fluctuations

I. INTRODUCTION

It is widely acknowledged that FinFETs are expected to be replaced by GAA nanowire or nanosheet transistors for the next advanced logic technology nodes [1]. The aim of this work is to show preliminary DC and low frequency noise results obtained on GAA NS FETs at low temperature operation. Several studies concerning the low frequency noise in GAA NS FET devices are already reported but only at room or high temperature operation [2-4].

The studied GAA NS n-channel structures, fabricated at imec (Belgium), have two vertically stacked rectangular nanosheets, corresponding to 4 and 22 fins in parallel per device in the floorplan: their width (W_{Fin}) is about 15 nm and each NS thickness (H_{NS}) is around 11 nm. The resulting effective channel width is calculated as 416 nm (4 fins), and 2288 nm (22 fins) from $W_{eff} = 2 \cdot (2 \cdot W_{Fin} + 2 \cdot H_{NS}) \cdot N$, with N being the number of fins per device. The gate stack (SiO_2 interfacial layer + HfO_2) leads to an equivalent oxide thickness of around 0.9 nm. The channel gate length of the studied devices ranges from 28 nm up to 100 nm. Two types of structures are investigated, the main difference being the vertical distance between the stacked nanosheets: 7.5 nm vs. 4.7 nm. More fabrication details may be found in [2].

The DC measurement are performed for devices having different gate length and fixed gate width of 416 nm. The low frequency noise measurements are made on devices having fixed gate length of 100 nm and fixed gate width of 2288 nm.

II. RESULTS AND DISCUSSION

An example of typical drain current transfer characteristics for the first type of devices at different temperatures is shown in Fig. 1. As expected, reducing the temperature leads to increased performances, in particular amelioration of the subthreshold slope or the leakage current etc... As observed from the Table in the inset of Fig. 1, the subthreshold swing takes values very close to the theoretical ones.

The impact of the vertical distance between the nanosheets on the transfer characteristics at 78 K and room temperature operation is illustrated in Fig. 2. Reducing the vertical distance between the nanosheets leads to an increase of the threshold voltage and to a decrease of the low field mobility. It should be noted that this mobility reduction is more pronounced at room temperature operation.

The static parameters are estimated using the Y function methodology [5]. The obtained values for the low field mobility and access resistance at 78 K and 300 K operation, are summarized in Table 1.

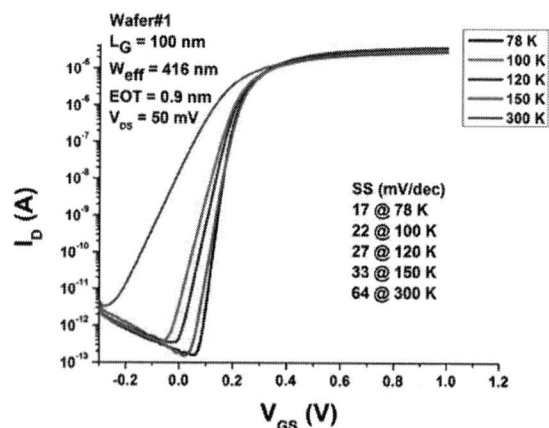

Figure 1. Typical drain current transfer characteristics.

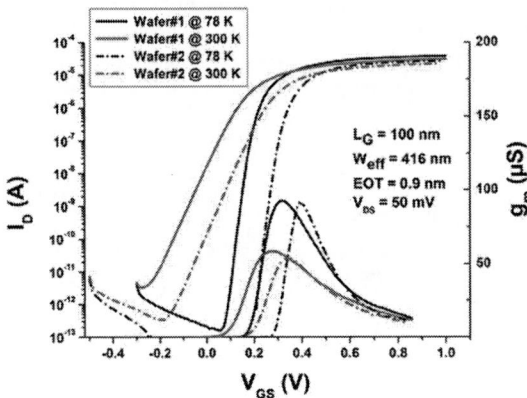

Figure 2. Impact of vertical distance between nanosheets on the tranfers characteristics.

TABLE I.   SUMMARY OF DC PARAMETER ESTIMATION

	Wafer#1: devices with longer vertical distance between NS		Wafer #2: Devices with shorter vertical distance between NS	
	78 K	300 K	78 K	300 K
μ_0 (cm²/V/s)	164	97	145	85
R_{access} (Ω)	151	122	152	116

The access resistance values are quasi-similar at the same temperature operation for both devices. At the same time, reducing the temperature leads to around 25% increase for the access resistance. As expected, a temperature decrease leads to enhanced low field mobility, related to reduced phonon scattering. As may be observed in Fig. 3 at 78 K, from the transconductance behaviour, the low field mobility presents a slight variability of about 12.5%. The same trend is also observed at room temperature

The low frequency noise measurements are conducted at room temperature and at 78 K in several devices having the same geometry: fixed gate length of 100 nm and fixed total gate with of 2288 nm.

To model the flicker noise behaviour versus the applied gate overdrive, information on the low field mobility and the access resistance of each device is needed. Indeed, assuming symmetrical drain and source access regions, in linear operation regime, the 1/f drain current noise spectral density may be expressed by:

$$S_{i_{D,tot}} = S_{i_{ch}} \frac{(r_{tot} - r_{access})^2}{r_{tot}^2} + S_{i_R} \frac{r_{access}^2}{2\, r_{tot}^2} \quad (1)$$

where r_{tot} represent the total resistance (i.e. of I_D/V_{DS}), S_{ich} the channel flicker noise and S_{iR} the access resistance flicker noise.

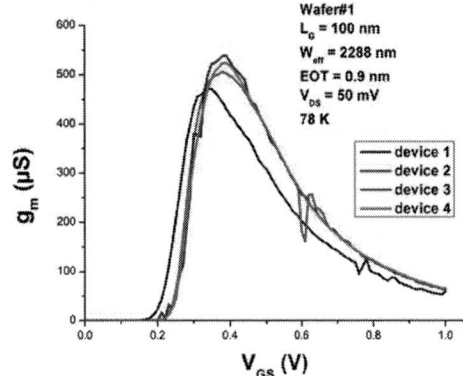

Figure 3. Transconduction versus applied gate voltage at 78 K temperature operation.

As devices present some variability, the static parameters obtained from Table 1 may not be directly applied. It is chosen to estimate the low field mobility individually for each device from the maximum of the transconductance. Moreover, for each device, the effective mobility attenuation function (θ_{eff}) is constructed (eq. 4 from [5]):

$$\theta_{eff}(V_{GS}) = \frac{G_M V_{DS}}{I_D} - \frac{1}{V_{GS} - V_{th} - \frac{V_{DS}}{2}}$$

$$= \theta_1 + \theta_2 \left(V_{GS} - V_{th} - \frac{V_{DS}}{2} \right) (2)$$

The first order mobility attenuation factor may be estimated from the θ_{eff} behaviour with the applied gate voltage. Finally, for each device, an overestimated value of the access resistance is found by neglecting the intrinsic first order attenuation factor from the expression of θ_1.

The overestimated values of the access resistances are in the range of 122-191 Ω at room temperature and in the range of 162-212 Ω at 78 K. These values are in agreement with those estimated for devices having different channel gate length (Table I). One may note that the ratio r_{acces}/r_{tot} is about 0.3 in strong inversion, so that the term $\frac{(r_{tot} - r_{access})^2}{r_{tot}^2}$ may have a strong impact on the 1/f noise modelling (eq. 1).

An example of an input-referred power spectrum density (PSD) versus frequency is illustrated in Fig. 4 at room temperature operation: 1/f, white noise and two generation-recombination (GR) contributions are necessary to model the noise behaviour using the following equation.

$$S_{v_g}(f) = \frac{S_{id}}{g_m^2} = W_n + \frac{K_f}{f} + \sum_{i=1}^{N} \frac{A_i}{1 + \left(\frac{f}{f_{0i}}\right)^2} \quad (3)$$

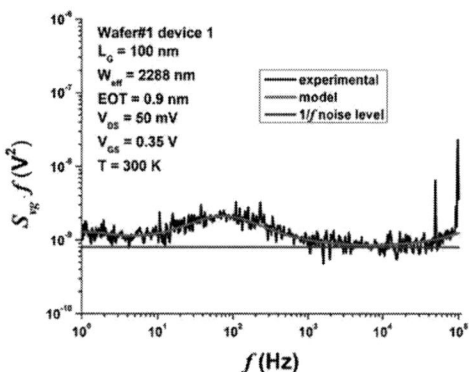

Figure 4. Typical input-referred noise power spectrum. 1/f, white noise and two generation-recombination contributions are necessary to model the noise behaviour

However, for some polarizations, the generation-recombination noise contributions are more pronounced.

In Fig. 5, the S_{id}/I_D^2 (related to the 1/f noise contribution) and $(g_m/I_D)^2$ versus the drain current variation are compared. It may be noted that the carrier number fluctuations may explain the S_{id}/I_D^2 behaviour in weak inversion. Deviation between the S_{id}/I_D^2 and $(g_m/I_D)^2$ in moderate to strong inversion suggest possible correlated mobility and carrier number fluctuations [6].

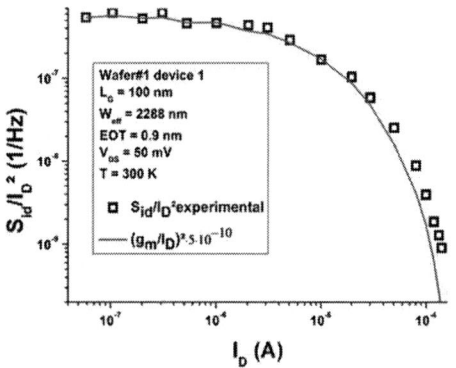

Figure 5. Normalized drain current noise versus the drain polarisation.

From Fig. 6 and 7, the S_{id}/I_D^2 or S_{vg} behaviour may be completely explained in the framework of correlated

mobility and carrier number fluctuations at room and low temperature operation. For some devices, the impact of access resistance noise was also considered at room temperature operation, as shown in Fig. 6 and 7 (eq. 1)).

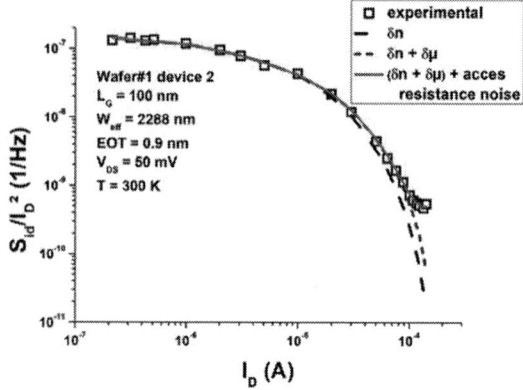

Figure 6. Comparison between the experimental normalized drain current noise versus drain curent and carrier number fluctuations model.

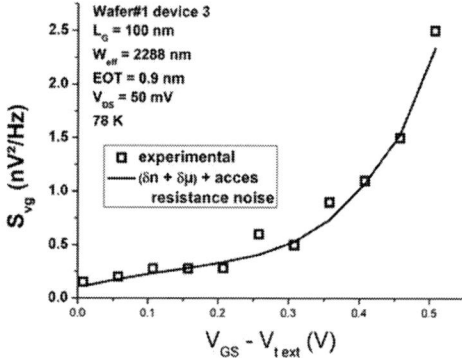

Figure 7. Comparison between the experimental gate voltage noise versus the applied gate overdrive ($V_{GS} - V_{text}$) and the carrier number fluctuations model, including the access resistance noise contribution.

The estimated noise parameters, e.g., the interface trap density (N_{it}), the Coulomb scattering (α_C) coefficient and the low field mobility for each studied device are summarized in Table II.

The obtained values for the interface trap densities show less than one decade variability, in agreement with previously reported ones in similar devices [2-4] and suggest a good insulator-silicon interface quality.

The values of α_C lower than $1 \cdot 10^4$ (Vs/C) may be related to the deeper trap's location in the gate dielectric. Relationship between the N_{it}, α_C and the device mobility may be observed, suggesting an impact of the charged oxide traps on both 1/f noise and low field mobility through remote Coulomb scattering. The inversion layer may be more confined at the interface at 78 K than at 300 K, leading to more effective scattering by charged centers in the oxide leading to an increase of the α_C coefficient.

mobility, subthreshold slope). The low frequency noise measurements confirm that the correlated mobility and carrier number fluctuation mechanism dominates the $1/f$ noise for the studied n-channel GAA vertically stacked silicon nanosheet FETs.

TABLE II.   SUMMARY OF NOISE PARAMETER ESTIMATION

	Devices with longer vertical distance between NS			
	300 K			
	Device 1	Device 2	Device 3	Device 4
N_{it} $(cm^{-3}eV^{-1})$	$3 \cdot 10^{18}$	$7.5 \cdot 10^{17}$	$6 \cdot 10^{17}$	$4.5 \cdot 10^{17}$
α_C (Vs/C)	$1.25 \cdot 10^4$	$0.75 \cdot 10^4$	$0.6 \cdot 10^4$	$0.55 \cdot 10^4$
μ_0 $(cm^2/V/s)$	63	68	70	91
	78 K			
	Device 1	Device 2	Device 3	Device 4
N_{it} $(cm^{-3}eV^{-1})$	$5.8 \cdot 10^{18}$	$2.1 \cdot 10^{18}$	$1.9 \cdot 10^{18}$	$1.56 \cdot 10^{18}$
α_C (Vs/C)	$1 \cdot 10^4$	$0.85 \cdot 10^4$	$1.75 \cdot 10^4$	$1.35 \cdot 10^4$
μ_0 $(cm^2/V/s)$	105	115	119	122

III.   CONCLUSION

Preliminary results on the DC and low frequency noise performances at room and low temperature operation are presented. Predictably, the low temperature operation leads to amelioration of the static device parameters (e.g.

REFERENCES

[1]   https://irds.ieee.org/

[2]   A.V. Oliveira, A Veloso, C. Claeys, N Horiguchi and E. Simoen, "Low-Frequency Noise Assessment of Vertically Stacked Si n-Channel Nanosheet FETs With Different Metal Gates", IEEE Trans. Electron Dev., 2020, 67 (11), pp. 4802-4807, DOI: 10.1109/TED.2020.3024271.

[3]   A.V. Oliveira, A Veloso, C. Claeys, N Horiguchi and E. Simoen, "Low–Frequency Noise in Vertically Stacked Si n–Channel Nanosheet FETs", IEEE IEEE Electron Device Lett., 2020, 41 (3), pp. 317-320, DOI: 10.1109/LED.2020.2968093.

[4]   V. C.P. Silva, W.F. Perina, J.A. Martino, E. Simoen, A. Veloso and P. Agopian. "Analog Figures of Merit of Vertically Stacked Silicon Nanosheets nMOSFETs With Two Different Metal Gates for the Sub-7 nm Technology Node Operating at High Temperatures", IEEE Trans. Electron Dev., 2021, 68(7), pp. 3620-3625, DOI: 10.1109/TED.2021.3077349.

[5]   C. Mourrain, B. Cretu, G. Ghibaudo and P. Cottin, "New Method for Parameter Extraction in Deep Submicrometer MOSFETs" in Proceedings of ICMTS'2000, DOI: 10.1109/ICMTS.2000.844428

[6]   G. Ghibaudo, O. Roux, Ch. Nguyen-Duc, F. Balestra and J. Brini, "Improved Analysis of Low Frequency Noise in Field-Effect MOS Transistors", Phys. Status Solidi (a), 1991;124 (2), pp. 571-581, DOI : 10.1002/pssa.2211240225.

[7]   D. Boudier, B. Cretu, E. Simoen, R. Carin, A. Veloso, N. Collaert, and A. Thean, "Low Frequency Noise Assessment in n- and p-Channel sub-10 nm Triple-Gate FinFETs: Part I: Theory and Methodology," Solid-State Electron, 2017, 128, pp. 102-108, , DOI: 10.1016/j.sse.2016.10.012.

TCAD based Modeling of Sub-surface Leakage in Short Channel Bulk MOSFETs

Harshit Kansal
Department of Electrical Engineering and Computer Science
Indian Institute of Science Education and Research
Bhopal, India
harshit16@iiserb.ac.in

Aditya Sankar Medury
Department of Electrical Engineering and Computer Science
Indian Institute of Science Education and Research
Bhopal, India
adityam@iiserb.ac.in

Abstract— **Aggressive scaling of the channel length of Bulk MOSFETs manifests in higher sub-surface leakage current, which becomes an increasingly significant component of the OFF-State leakage current. By firstly identifying the sub-surface leakage region as one that manifests in a region away from the electrostatic control of the gate, but, within the source/drain junction depth, while being impacted by the drain voltage, we model the sub-surface leakage current using a non-charge-sheet based approach, which is further simplified as a physics-based semi-analytical model similar to the Shockley equation of an ideal 'forward-biased' p-n junction diode. In addition to validating the proposed model, we also show the impact of various structural and electrical parameters, thus demonstrating that the model provides key insights on device behaviour.**

Keywords— **Bulk MOSFET, Non-Charge-Sheet Model, Short Channel, Sub-surface Leakage, TCAD**

I. INTRODUCTION

From an Analog circuit performance perspective, bulk MOSFETs offer superior performance compared to FinFETs, such as higher cut-off frequencies, due to substantially lower parasitic capacitances, despite superior transconductance of FinFETs [1] while FDSOI (Fully Depleted Silicon-on-Insulator) MOSFETs suffer from the self-heating effect and quantum confinement effects at thinner SOI (Silicon-on-Insulator) channel thicknesses [2] and are comparable to bulk MOSFETs [3]. Given these observations, it remains important to investigate and accurately model leakage phenomena [4], which manifest at shorter channel lengths degrading the performance of Bulk MOSFETs; leakage between the source and drain beneath the surface (sub-surface leakage), within the source/drain junction depths, is interesting as it provides an additional path of current conductivity, and has received some attention only recently [5]. In short channel bulk MOSFETs, a region susceptible to sub-surface leakage away from the Si/SiO$_2$ interface has been previously discussed [6], which contributes significantly to the OFF-State leakage current. While authors in [5] have presented a model for the sub-surface leakage, a more detailed physics-based model of the sub-surface leakage region is required to get an insight into the different electrical and structural parameters that are likely to impact the electrostatics in the sub-surface leakage region. It is also important to note that unlike sub-threshold (weak inversion) leakage [7], where the leakage path is close to the oxide/channel interface and is primarily a phenomenon seen as the MOSFET turns 'ON', the sub-surface leakage in contrast manifests farther away from the channel/oxide interface (within the source/drain junction depth) and is an 'OFF' State leakage mechanism. It therefore becomes important to develop an accurate model for this sub-surface leakage current while taking the underlying physics governing this phenomenon into account.

Therefore, in this work, using TCAD Simulations [8], we firstly identify the sub-surface leakage region in short channel (n-channel) Bulk MOSFETs. We then demonstrate that the sub-surface leakage region of finite thickness can be visualized as stacked mono-layers (along the depth of the substrate), where each mono-layer can be modelled as a base contact-less NPN BJT (bipolar junction transistor), with the emitter and collector of the BJT being identical to the source and drain of the bulk MOSFET, as shown in Fig. 1, while the doping of the base is determined to ensure good agreement with the electron density in the lateral direction, at drain-source voltage, $V_{ds} = 0$ V. Also, the collector-emitter voltage (V_{ce}) of each NPN BJT is used to ensure that the current density along the base of each BJT shows good agreement with the current density along a specific mono-layer of the sub-surface leakage region. We model the sub-surface leakage current, as a net sum of the current densities of each NPN BJT (each mono-layer), while also taking the thickness of the sub-surface leakage region into account. As discussed earlier, given that the current densities of different mono-layers is different over the sub-surface leakage region, by modeling the entire sub-surface leakage region with an average current density (average current density of various mono-layers of the sub-surface leakage region) with an equivalent collector-emitter voltage ($V_{ce(eq)}$) and showing that the sub-surface leakage current resembles the physical behaviour of a forward-biased p-n junction diode (similar to Shockley equation), the non-charge-sheet model can be significantly simplified. Finally, besides showing good agreement with TCAD Simulation results, various dependencies of the proposed model on electrical and structural parameters are also taken into account.

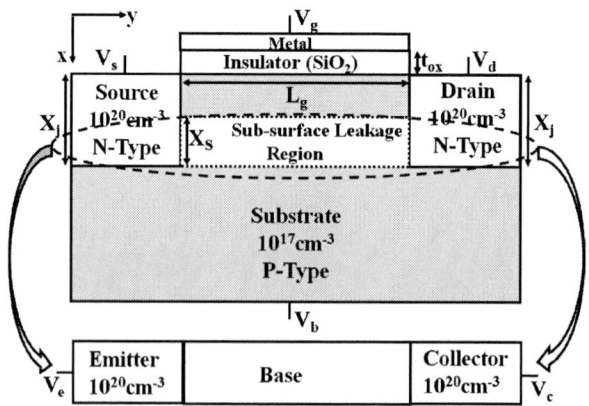

Figure 1: Schematic of short-channel Bulk MOSFET having Source and Drain regions of N-Type (10^{20} cm^{-3}) along with Substrate region of P-type (10^{17} cm^{-3}), showing a sub-surface leakage region (Xs) away from the Si/SiO$_2$ interface. The leakage path is identical to a stack of base contact-less NPN BJT with Emitter and Collector regions identical to Source and Drain region respectively, while the doping in the base region (for each mono-layer) is varied along the depth of the substrate.

978-1-6654-3746-2/21 $31.00 © 2021 IEEE

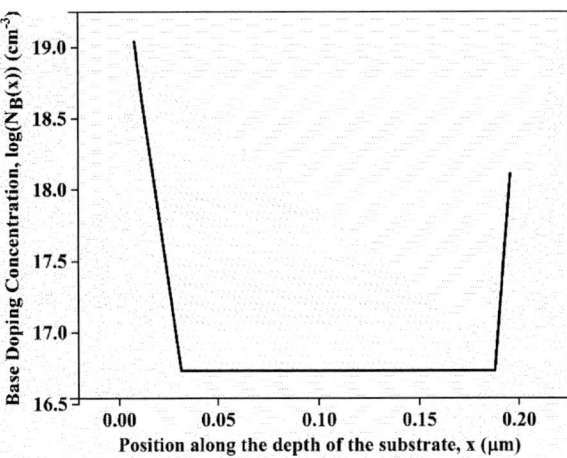

Figure 2: Variation of Base doping concentration along the depth of the substrate, so as to get good agreement in electron density between discrete mono-layers in MOSFET and the corresponding base contact-less NPN BJT, shown in Fig. 3.

II. SIMULATION METHODOLOGY

Sentaurus 3-D TCAD is used to simulate the short-channel Bulk MOSFETs, at room temperature, considering semi-classical behaviour, through the self-consistent solution of the 2-D Poisson's equation, with the drift-diffusion model and the electron/hole continuity equation. Also, constant electron mobility model, along with Shockley Read-Hall statistics is considered.

III. RESULTS AND DISCUSSION

As a first step towards developing a more detailed understanding of this sub-surface leakage phenomenon, after having identified the sub-surface leakage region [6], as shown in Fig. 1, we visualize the sub-surface leakage region, as a discretized stack of mono-layers along the depth of the substrate where each mono-layer is essentially a lateral cut from source to drain through the sub-surface leakage region. Each mono-layer of this discretized stack can further be visualized as an NPN BJT, where the base of the BJT models the sub-surface leakage region of a particular mono-layer while the emitter and collector are identical to the source and drain (both in terms of length and doping) of the MOSFET. We model the 1-D electrostatics of each mono-layer as a base contact-less NPN BJT, whose structure is shown in Fig. 1, with the base doping of each BJT uniquely representing the electrostatics of a particular mono-layer, as shown in Fig. 2. In this base contact-less NPN BJT, the base doping of each BJT is determined for $V_{ds}= 0$ V ($V_{ce} = 0$ V), through TCAD simulations, so as to obtain good agreement with the electron density of each corresponding mono-layer of the MOSFET in the sub-surface leakage region, as seen in Fig. 3. Thus, for $V_{ds} = 0$ V, each discrete NPN BJT emulates a particular mono-layer of the sub-surface leakage region of the short channel MOSFET (as clearly demonstrated in Fig. 3) and cumulatively the sub-surface leakage region can be considered as a stack of these NPN BJT's with the base doping concentration varying along the depth of the substrate, as seen in Fig. 2. An extended region of flatness and lowered

DST, (SERB), Government of India (Grant No: ECR/2-17/000011)

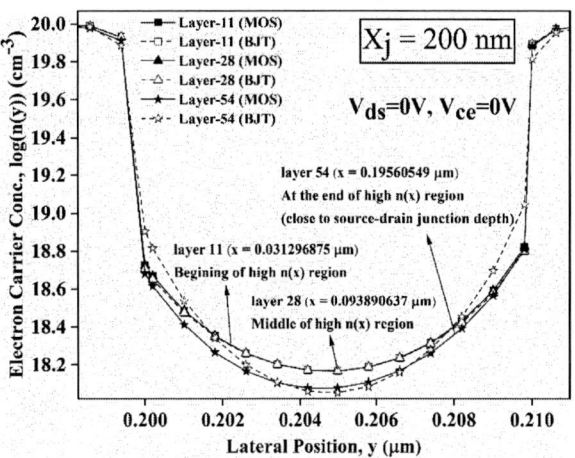

Figure 3: Comparison of Electron density (n(y)) (in log scale) along the channel (Lateral Direction (y)), between discrete mono-layers in MOSFET and the corresponding base contact-less NPN BJTs for $V_{ds}=0$V.

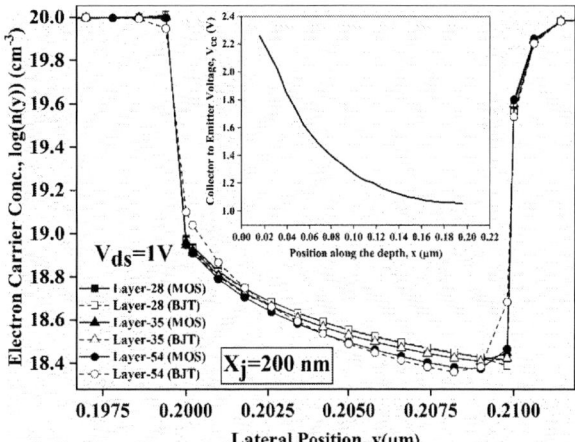

Figure 4: Comparison of Electron density (n(y)) (in log scale) along the channel (Lateral Direction (y)), between discrete mono-layers in MOSFET and the corresponding base contact-less NPN BJT (identical to obtained for $V_{ds}=0$V) for $V_{ds}=1$V and for varying V_{ce} values along the depth. Inset plot shows the variation of V_{ce} along the depth of the substrate.

base doping in Fig. 2 suggests that this region coincides with a region of high electron concentration in the sub-surface.

By using this same base doping concentration ($N_B(x)$) (as shown in Fig. 2) but by applying a suitable collector-emitter voltage (V_{ce}), we are able to show that the same NPN BJT is now also able to emulate the electron density profile of the corresponding mono-layer of the MOSFET, for $V_{ds} = 1$ V, as shown in Fig. 4. This variation of V_{ce}, along the depth of the sub-surface leakage region, is shown in the inset plot in Fig. 4 and enables good agreement of the electron current densities in each NPN (base contact-less) BJT with the corresponding mono-layer of the sub-surface leakage region in the short channel Bulk MOSFET, as shown in Fig. 5. Collector-emitter voltage (V_{ce}) showing a sharp increase in the upper layers of the sub-surface leakage region, while saturating to a value closer to V_{ds} in the layers closer to the source-drain junction depth (where the electrostatics becomes more 1-D in nature, controlled by the drain). Therefore, given the visualization of the sub-surface Leakage region as being a stack of the NPN BJT's (where each BJT emulated the electrostatics of an individual mono-layer)

978-1-6654-3746-2/21 $31.00 © 2021 IEEE   151

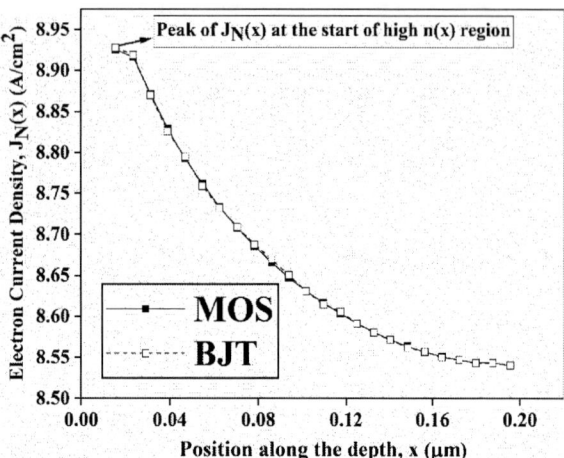

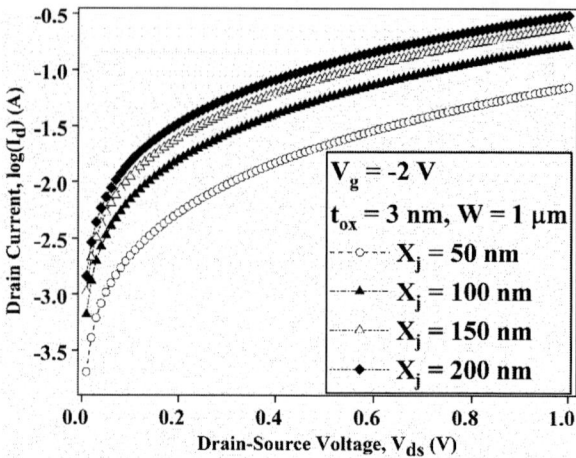

Figure 5: Electron Current Density ($J_N(x)$) along the depth of the Sub-surface leakage region in the MOSFET showing good agreement with results obtained from the base contact-less BJT Model.

along the depth of the substrate, the total current density can be considered as being the cumulative effect of each of these mono-layers (also each of these NPN BJT's model the individual mono-layers), as shown in (1):

$$ J = J_1 + J_2 + J_3 + \cdots + J_N = \sum_{i=1}^{N} J_i \qquad (1) $$

In equation (1), J_1 represents the current density in the top-most mono-layer of the sub-surface region, while J_N represents the lowest mono-layer (close to the source/drain junction depth).

By simplifying the cumulative electron current density of the sub-surface leakage region, shown in (1), in terms of an average electron current density over the thickness of the sub-surface leakage region, a model for the sub-surface leakage current is proposed in (2):

$$ I_{model} = J_{avg} W X_S \qquad (2) $$

where, J_{avg} is the average electron current density over the sub-surface leakage region of thickness, X_s and W is the width of the sub-surface channel. This simplified model for the sub-surface leakage current is a non-charge-sheet based model, as it takes the thickness of the sub-surface leakage region into account.

In Fig. 6, the drain-source current (I_{ds}), under strong accumulation bias ($V_{gs} = -2$ V), which is essentially the sub-surface leakage current, is shown as a function of the drain voltage, for different structural parameters, in log-scales, clearly demonstrating similar characteristics to a forward-biased p-n junction. This phenomenological understanding of the sub-surface leakage current, along with the modeling of each mono-layer of the short channel MOSFET as a base contact-less NPN BJT (where it may be recalled that by applying a certain V_{ce} to the NPN BJT, the equivalent electrostatics of a specific mon-layer of the MOSFET were accurately modeled for $V_{ds} = 1$ V) described earlier, can be used to simplify the model in (2), as shown in (3) below:

$$ I_{model} = J_0 W X_S \left(exp\left(\frac{qV_{ce(eq)}}{nkT}\right) - 1 \right) $$

Figure 6: Drain current (I_d) (in log-scale) as a function of V_{ds}, where the drain current under deep accumulation bias, essentially due to the sub-surface leakage phenomenon, closely resembles the I-V characteristics of a forward-biased diode.

$$ I_{model} = I_0 \left(exp\left(\frac{qV_{ce(eq)}}{nkT}\right) - 1 \right) \qquad (3) $$

where I_0 represents the voltage independent sub-surface leakage current in the MOSFET, $V_{ce(eq)}$ represents the equivalent collector-emitter voltage over the entire sub-surface leakage region and 'n' is a fitting parameter obtained from TCAD Simulations, similar to the ideality factor seen in the Shockley Diode equation. The model for the sub-surface leakage current, shown in (3), is compared with the results obtained from TCAD Simulations and is shown in Fig. 7, where very good agreement is seen over a wide range of source/drain junction depths (X_j) and oxide thicknesses (t_{ox}). The parameter I_0, in (3), is strongly dependent on the sub-surface leakage region thickness (X_s). Thus, the strong dependencies of X_s on L_g and X_j, translates in terms of similarly strong dependency of I_0 on X_j and L_g, as shown in Fig. 8.

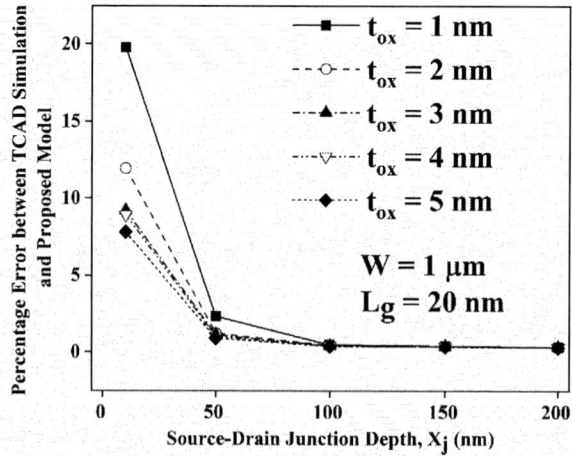

Figure 7: Percentage Error between the model proposed for the sub-surface leakage current, shown in equation (1), and the TCAD Simulation results, as a function of source-drain junction depth, when $L_g = 20$ nm, $V_{gs} = -2$ V (deep accumulation bias).

978-1-6654-3746-2/21 $31.00 © 2021 IEEE

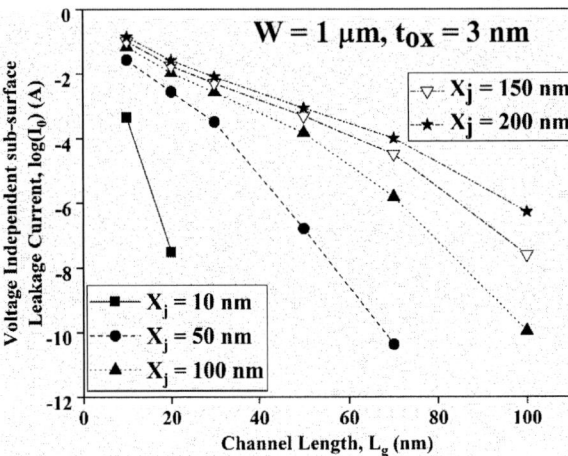

Figure 8: Logarithm of I_0 as a function of L_g for different source-drain junction depths, demonstrating that I_0 decreases sharply with increasing L_g, for different source-drain junction depths.

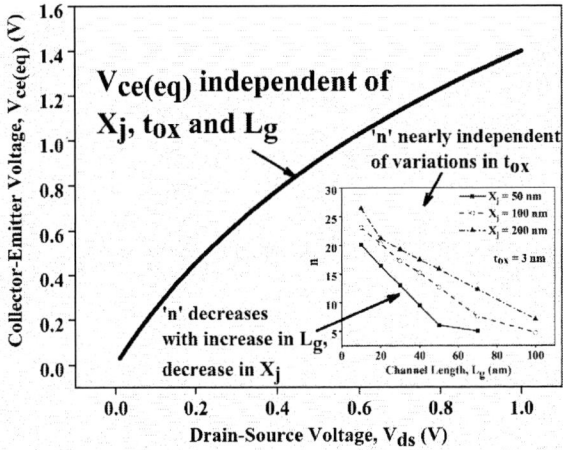

Figure 9: Equivalent Collector-Emitter Voltage, $V_{ce(eq)}$ (a key model parameter, shown in (3)), independent of process parameter variations, shown as a function of Drain-Source Voltage (V_{ds}), where $V_{ce(eq)}$ increases with increasing V_{ds}. In the inset plot, the parameter 'n' decreases with increasing L_g for different source/drain junction depths.

Figure 8, shows that I_0 decreases with increasing L_g, decreasing X_j, as the electrostatic control of the drain over the channel, including the sub-surface leakage region, gets weaker relative to the gate. Furthermore, Fig. 9, shows $V_{ce(eq)}$ as a function of V_{ds}, where $V_{ce(eq)}$ is independent of other structural parameter variations, while the inset plot shows the ideal factor 'n' which is strongly dependent on structural parameters such as X_j, L_g, while being weakly dependent on the oxide thickness, t_{ox} and being independent of variations of V_{ds}.

Thus, the ratio, $\frac{V_{ce(eq)}}{n}$, clearly is a function of V_{ds} for different channel lengths showing that as channel length increases, the effect of V_{ds} becomes more significant. This, along with a much-reduced I_0, with increasing L_g, shows that while I_{ds} (sub-surface leakage current) reduces significantly with increase in L_g, but at the same time with increasing L_g, the relative effect of V_{ds} on the sub-surface leakage current also increases. It may be recalled that the sub-surface leakage

current, showed smaller values of the current at larger channel lengths with larger variations in current values with increase in V_{ds}, which is both qualitatively and quantitatively explained by our model.

IV. CONCLUSION

In this work, we have clearly identified a region away from the Si/SiO2 interface, where due to poor electrostatic control from the gate, in short channel Bulk MOSFETs, there tends to be a significant flow of current from drain to source, termed as the sub-surface leakage current. Using TCAD Simulations, we develop an understanding of the parameters which are most likely to impact this sub-surface leakage current and identify that this current is essentially not controlled by the gate voltage, while being strongly controlled by the drain voltage and impacted by structural parameters such as channel length and source/drain junction depth. Through modelling the electrostatics of the sub-surface leakage region as a stack of base contact-less NPN BJT's with appropriate base doping and collector-emitter voltage, we are able to propose a non-charge sheet model for this phenomenon. Furthermore, by showing that this sub-surface leakage current has a forward-biased diode like dependence on the drain voltage, we are able to make simplifications to the non-charge sheet model, while also clearly demonstrating the accuracy of the proposed model through detailed comparisons with TCAD simulation results. Finally, we also clearly show how the proposed model is able to take all the critical structural and electrical parameter dependencies into account.

ACKNOWLEDGMENT

We acknowledge financial support from DST, (SERB), Government of India (Grant No: ECR/2-17/000011)

REFERENCES

[1] V. Passi, and J. P. Raskin, "Review on analog/radio frequency performance of advanced silicon MOSFETs," *Semiconductor Science and Technology 32 no. 12*, p. 123004, 2017.

[2] A. S. Medury, K. N. Bhat and N. Bhat, "Impact of carrier quantum confinement on the short channel effects of double-gate silicon-on-insulator FINFETs," Microelectronics journal 55, pp. 143-151, 2016.

[3] M. M. Arshad, V. Kilchytska, M. Emam, F. Andrieu, D. Flandre, and J. P. Raskin, "Effect of parasitic elements on UTBB FD SOI MOSFETs RF figures of merit," *Solid-State Electronics 97*, pp. 38-44, 2014.

[4] K. Roy, S. Mukhopadhyay and H. Mahmoodi- Meimand, "Leakage current mechanisms and leakage reduction techniques in deep-submicrometer CMOS circuits," Proceedings of the IEEE 91 no.2, pp. 305-327, 2003.

[5] Yen-Kai Lin, Sourabh Khandelwal, Aditya Sankar Medury, Harshit Agarwal, Huan-Lin Chang, Yogesh Singh Chauhan, and Chenming Hu, "Modeling of Subsurface Leakage Current in Low V_{TH} Short Channel MOSFET at Accumulation Bias." *IEEE Transactions on Electron Devices 63*, no. 5, pp. 1840-1845, 2016.

[6] H. Kansal, and A. S. Medury, "Short-Channel Effects and Sub-Surface Behavior in Bulk MOSFETs and Nanoscale DG-SOI-MOSFETs: A TCAD Investigation," *Silicon Nanoelectronics Workshop*, pp. 1-2, IEEE, June 2019.

[7] W. M. Elgharbawy and M. A. Bayoumi, "Leakage sources and possible solutions in nanometer CMOS technologies", IEEE Circuits and Systems magazine 5, no.4, pp. 6-17, 2005.

[8] "Technology Computer-Aided Design.", 1986, [online] Available:http://www.synopys.com/Tools/TCAD/Pages/default.aspx.

978-1-6654-3746-2/21 $31.00 © 2021 IEEE

An artificial synaptic thin-film transistor based on 2D MXene–TiO$_2$

Y X Cao[1-3], C Zhao[1-3,*], I Z Mitrovic[2], Y N Liu[4,5], L Yang[6,7], H van Zalinge[2], C Z Zhao[1-3]

[1] School of Advanced Technology, Xi'an Jiaotong-Liverpool University, Suzhou, China.
[2] Department of Electrical Engineering and Electronics, University of Liverpool, Liverpool L69 7ZD, UK.
[3] AI University Research Centre (AI-URC), Xi'an Jiaotong-Liverpool University, Suzhou, China.
[4] Department of Applied Mathematics, Xi'an Jiaotong-Liverpool University, Suzhou, China.
[5] Department of Applied Mathematics, University of Liverpool, Liverpool L69 7ZD, UK.
[6] Department of Chemistry, Xi'an Jiaotong-Liverpool University, Suzhou, China.
[7] Department of Chemistry, University of Liverpool, Liverpool L69 7ZD, UK.
*E-mail: Chun.Zhao@xjtlu.edu.cn

Abstract—**MXenes, a new class of two-dimensional transition metal carbides and nitrides, has the potential as a floating gate in storage devices due to its inherent characteristics such as two-dimensional structure, high density of states and high work function. Based on synthetic MXene and TiO$_2$ generated by surface oxidation, this study used a low-energy, pollution-free and low-cost aqueous solution method to fabricate artificial synaptic thin-film transistors. Moreover, we tested its synaptic properties under light stimulation. Successful simulations include excitatory postsynaptic current, spike number dependence plasticity, spike frequency dependence plasticity and spike width dependence plasticity, and proposed a potential application: high-pass filter. This MXene-based artificial synaptic device serves as a data storage medium to inspire the application of future storage devices.**

Keywords—*MXene, synaptic, thin-film transistor*

I. INTRODUCTION

As a new type of two-dimensional metal carbide, carbonitride or nitrogen nitride, MXenes have the general formula of $M_{n+1}X_nT_x$ (n = 1-4). Due to its good metal conductivity [1], water dispersion [2], high optical transparency [3], adjustable work function [4, 5], electromagnetic interference shielding [6], the photothermal effect [7], good mechanical properties and other physical and chemical properties [8] they have attracted significant attention. This work demonstrates an artificial synapse thin-film transistor (TFT) processed by an aqueous solution method based on synthetic MXene and TiO$_2$ prepared by surface oxidation. Furthermore, its synaptic properties under light stimulation were tested. Due to the high optical transparency of MXene [3], adjustable work function [4,5] and the existence of TiO$_2$ tunnelling layer, successful simulations include excitatory postsynaptic current (EPSC), spike number dependence plasticity (SNDP), spike frequency dependence plasticity (SFDP) and spike width dependence plasticity (SWDP), leading to a potential application: high-pass filter.

II. EXPERIMENTAL SECTION

For the solution preparation, the InO$_x$ precursor solution with a concentration of 0.15 M was prepared by dissolving indium nitrate hydrate (In(NO$_3$)$_3$·xH$_2$O, Aladdin) in deionized water (DI water). Subsequently, the prepared precursor solutions were ultra-sonicated and filtered through a 0.45 µm poly (ether sulfone) (PES) syringe to obtain transparent solutions. For MXene preparation: stir 2 g of lithium fluoride (LiF, 99.99% metal base, Aladdin) and 40 ml of hydrochloric acid (HCl, AR 36.0~38.0%, Sinopharm Chemical Reagent

Co., Ltd.) in a polytetrafluoroethylene (PTFE) beaker for 30 minutes. Add 2g of titanium aluminium carbide MAX (MAX-

(a)

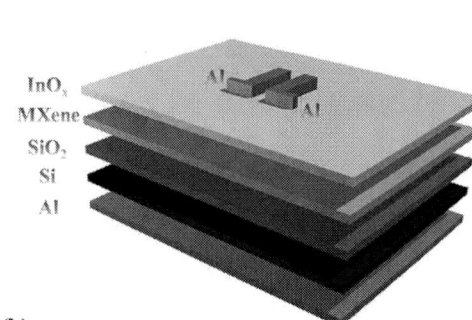

(b)

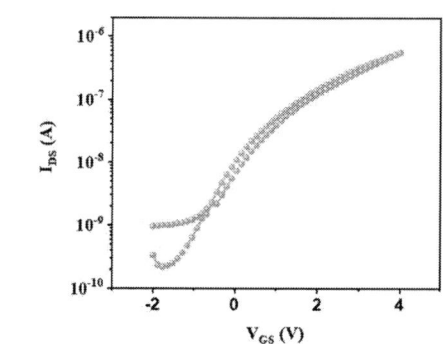

(c)

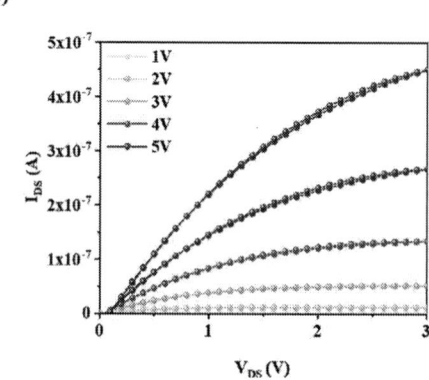

Figure 1. (a) The cross-sectional structure , (b) the transmission characteristics and (c) the output characteristics of the artificial synapse thin- film transistor based on Mxene-TiO$_2$.

Ti$_3$AlC$_2$, 98%, 11 Technology Co., Ltd.) slowly into the first step beaker, adjust the reaction temperature to 35°C, and continue stirring for 24h in a fume hood. Subsequently, the resulting solution was centrifuged (3500 rpm, 10 min), the supernatant was poured out, 40 ml of deionized (DI) water was added to the centrifuge tube sediment, the sediment was shaken by hand to mix with deionized water, ultrasonic treatment centrifuge the tube in a high-power ultrasonic machine (750 W) for 15 minutes. Then repeat the above centrifugation and ultrasonic treatment until the pH of the supernatant decanted after centrifugation is 5, and then add 40ml of ethanol (CH$_3$CH$_2$OH, AR ≥ 99.7%, Sinopharm Chemical Reagent Co., Ltd.) to the centrifuge tube and ultrasound for 1.5 h (with the function of intercalator), centrifugation (10000 rpm, 10 min). Next, add 20 ml of deionized water to the centrifugal sediment, shake well, and sonicate for 20 minutes. At last, centrifuged it again at 3500 rpm for 3 min to obtain the black-brown few-layer dispersion of about 0.7 mg/ml. For TFTs fabrication, the heavily doped n-type silicon substrates with 100 nm thermally grown silicon oxide (SiO$_2$) were performed as the gate electrodes and the dielectrics of the TFT and a 30 min-air plasma process was applied on the SiO$_2$ surfaces to improve the hydrophilicity. The MXene was then spin-coated at 5000 rpm for 30 s on the surfaces of SiO$_2$. After that, the InO$_x$ precursor solution was spun onto the surface of the MXene and then annealed at 250°C for 30 minutes in an ambient atmosphere. Finally, the Al source and drain electrodes were deposited onto the semiconductor layer through thermal evaporation. The channel length (L) and width (W) were 10 and 150 μm, respectively, with a thickness of 100 nm.

III. RESULTS AND DISCUSSION

The brain contains a large number of neurons to receive interactive signals (such as electrical and optical signals) in different ways, and realize cross-modal neuromorphic calculations in the multi-sensory association area. The synapse is the connection point between two adjacent neurons and plays a vital role in the transmission of neural information (Figure 1a) [9]. Inspired by the brain and nervous system, we tried to demonstrate a photoelectric artificial synaptic thin-film transistor with synergistic electrical and optical plasticity. The schematic diagram of the photoelectric artificial synapse TFT is shown in Figure 1a. In the operation of the synaptic transistor, UV light pulses, which act as the pre-synaptic input, are shone onto the InO$_x$ channel of the transistor with the gate terminal floated and the drain voltage (Vd) fixed at 4V. Current and channel conductance is similar to postsynaptic current and synaptic weight, respectively. Under the stimulation of ultraviolet pulses, the capture and release of photo-generated holes in the InO$_x$-MXene/SiO$_2$ interface and/or SiO$_2$ layer are similar to the role of neurotransmitters in regulating synaptic strength in biological synapses. Since the bandgap of the InO$_x$ film is 2.8 eV and the photon energy of ultraviolet light is 3.4 eV, many electron-hole pairs are generated in the InOx channel. Therefore, under the influence of the drain voltage, a photocurrent flowing between the source and drain of the transistor is generated, increasing the postsynaptic current. Some light-generated holes are trapped in the InOx-MXene/SiO$_2$ interface and/or SiO$_2$ layer and are gradually released during the non-period of the ultraviolet light pulse. The trapped holes induce conduction electrons in the InO$_x$ channel layer. Therefore, the postsynaptic current

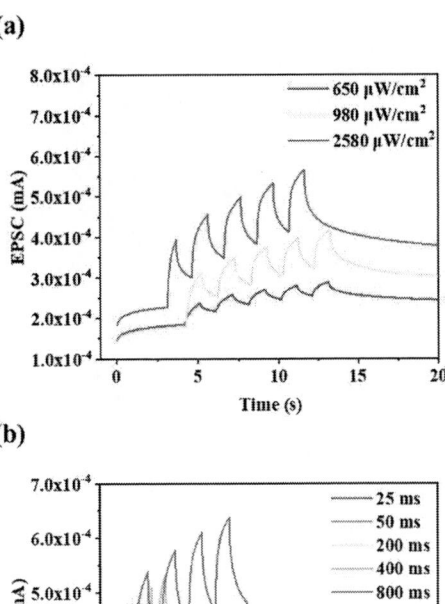

(a)

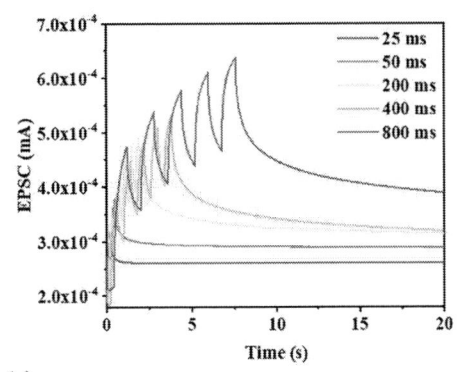

(b)

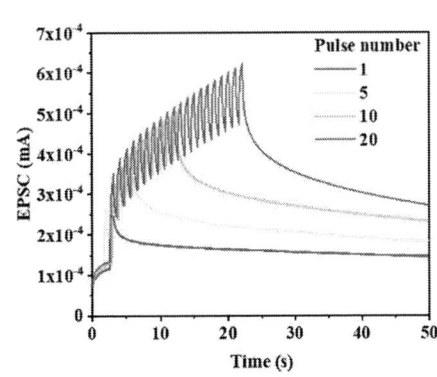

(c)

Figure 2. (a) EPSC under different light intensity, (b) different illumination widths and (c) different amounts of light.

will not disappear immediately after the ultraviolet radiation is turned off. Figure 1b shows its transmission characteristic curve. The transfer curves of the transistors were measured by varying the gate voltage from -2 V to 4 V and then back to -2 V. A constant voltage V$_{DS}$ = 4 V was applied between the source-drain electrodes. The transfer curves exhibited a clear clockwise hysteresis loop and indicated that the channel conductance could be reversibly manipulated between the high and low conductance states. The associated output characteristic curve is depicted in Figure 1c and it shows clear pitch-off characteristics and good current saturation.

978-1-6654-3746-2/21 $31.00 © 2021 IEEE

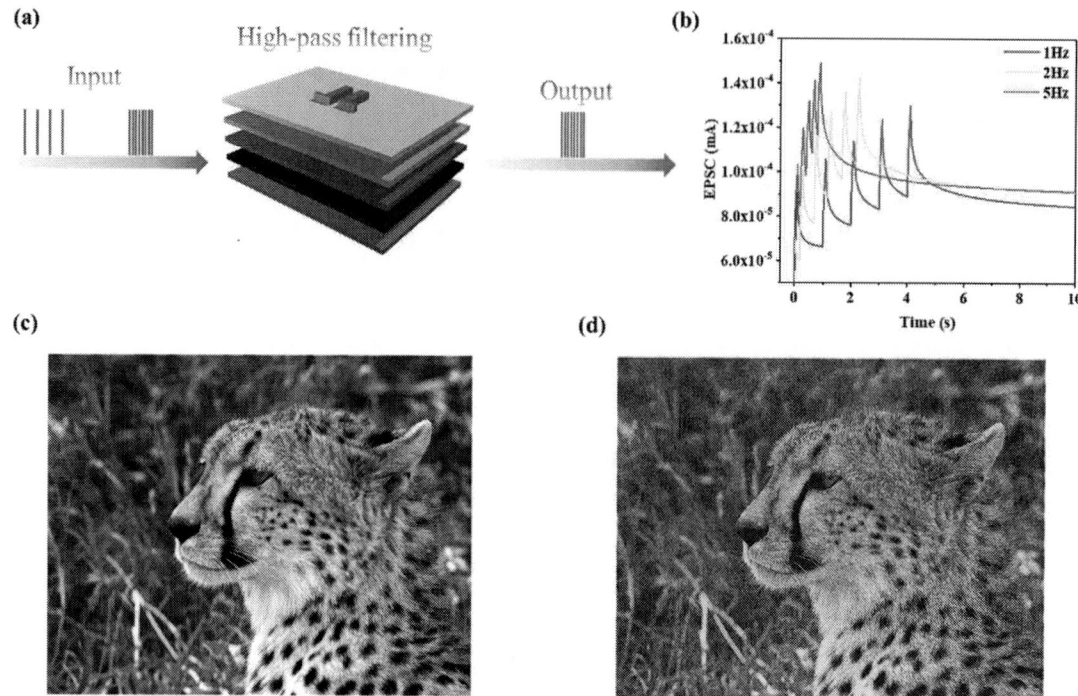

Figure 3. (a) Schematic diagram of the application of a high-pass filter based on MXene's artificial synaptic TFT. (b) EPSC under different illumination frequencies. (c) The original image of a leopard. (d) The image of the leopard after sharpening with the high-pass filtering function.

The EPSC levels of five consecutive presynaptic spikes at different light intensities with Δt of 1s are shown in Figure 2a. With the increase of light intensities from 650 μW/cm² to 2580 μW/cm², the amplitude changing of EPSCs was also increased. Figure 2b shows the SWDP characteristics. When the width of the light pulse increased from 25 ms to 800 ms, the EPSC peaks of the TFT have more than doubled. The peak value of the postsynaptic current was affected not only by the magnitude of pulses' width but also by the number of light pulses. Figure 2c shows the EPSC levels of presynaptic light spikes at different numbers (5 - 50). It could be obtained that a higher number of light spikes leads to a promotion of the EPSC. These results indicate that the proposed synaptic TFT has synaptic plasticity for various presynaptic light spikes, just like the characteristics of biological nerves. The stronger the stimulus, the stronger the response [10, 11].

Biologically, synapses with a low probability of neurotransmitter release have a high-pass filtering effect [9, 12]. The high-pass filter allows high-frequency signals exceeding a certain cut-off value to pass and greatly attenuates low-frequency signals [13]. Figure 3a depicts a schematic diagram of a high-pass filter used for signal processing. In order to explore the filtering characteristics of MXene-based artificial synaptic TFTs, three light pulse sequences of different frequencies were used. As shown in Figure 3b, when the frequency increases from 1 Hz to 5 Hz, the EPSC amplitude increases sharply. To demonstrate more vividly the high-pass filtering function of the artificial synaptic TFT based on MXene, the image of the leopard is taken as an example to simulate the filtering process. The sharpened image 3d is significantly improved compared to the original image 3c. Therefore, this MXene-based artificial

synaptic TFT has been proven to have the potential as a dynamic high-pass filter for signal processing.

IV. CONCLUSION

In summary, we demonstrated an artificial synaptic TFT based on MXene as a floating gate and TiO₂ on the surface as a tunnelling layer. Based on the adjustable work function brought by MXene and its excellent light synaptic transmittance, the device has good photosynaptic characteristics as demonstrated by the simulated EPSC, SWDP, SNDP and SFDP. Moreover, the high-pass filter application is proposed.

ACKNOWLEDGMENT

This research was funded in part by the Natural Science Foundation of the Jiangsu Higher Education Institutions of China Program (19KJB510059), Natural Science Foundation of Jiangsu Province of China (BK20180242), IZM acknowledges British Council UKIERI F.No.184-1/2018(IC) project, the Suzhou Science and Technology Development Planning Project: Key Industrial Technology Innovation (SYG201924), and the Key Program Special Fund in XJTLU (KSF-P-02, KSF-T-03, KSF-A-04, KSF-A-05, KSF-A-07, KSF-A-18).

REFERENCES

[1] B. Anasori, M. R. Lukatskaya, and Y. J. N. R. M. Gogotsi, "2D metal carbides and nitrides (MXenes) for energy storage," vol. 2, no. 2, pp. 1-17, 2017.

[2] M. Ghidiu, M. R. Lukatskaya, M.-Q. Zhao, Y. Gogotsi, and M. W. J. N. Barsoum, "Conductive two-dimensional titanium carbide 'clay' with high volumetric capacitance," vol. 516, no. 7529, pp. 78-81, 2014.

[3] K. Hantanasirisakul *et al.*, "Fabrication of Ti3C2Tx MXene transparent thin films with tunable optoelectronic properties," vol. 2, no. 6, p. 1600050, 2016.

[4] Y. Liu, H. Xiao, and W. A. J. J. o. t. A. C. S. Goddard III, "Schottky-barrier-free contacts with two-dimensional semiconductors by surface-engineered MXenes," vol. 138, no. 49, pp. 15853-15856, 2016.

[5] B. Lyu *et al.*, "Large-area MXene electrode array for flexible electronics," vol. 13, no. 10, pp. 11392-11400, 2019.

[6] F. Shahzad *et al.*, "Electromagnetic interference shielding with 2D transition metal carbides (MXenes)," vol. 353, no. 6304, pp. 1137-1140, 2016.

[7] H. Lin, X. Wang, L. Yu, Y. Chen, and J. J. N. l. Shi, "Two-dimensional ultrathin MXene ceramic nanosheets for photothermal conversion," vol. 17, no. 1, pp. 384-391, 2017.

[8] Z. Ling *et al.*, "Flexible and conductive MXene films and nanocomposites with high capacitance," vol. 111, no. 47, pp. 16676-16681, 2014.

[9] L. Abbott and W. G. J. N. Regehr, "Synaptic computation," vol. 431, no. 7010, pp. 796-803, 2004.

[10] D. Choi, M.-K. Song, T. Sung, S. Jang, and J.-Y. Kwon, "Energy scavenging artificial nervous system for detecting rotational movement," *Nano Energy,* vol. 74, 2020.

[11] Y. Kim *et al.*, "A bioinspired flexible organic artificial afferent nerve," *Science,* vol. 360, no. 6392, pp. 998-1003, 2018.

[12] P. Feng *et al.*, "Printed Neuromorphic Devices Based on Printed Carbon Nanotube Thin‐Film Transistors," vol. 27, no. 5, p. 1604447, 2017.

[13] W. Huang *et al.*, "Zero-power optoelectronic synaptic devices," vol. 73, p. 104790, 2020.

Silicon and hafnia thin film transfer on c-plane sapphire: effect of substrate thickness on the ferroelectric hafnia properties

Valentin Antonov
Laboratory of Silicon Material Science,
Rzhanov Institute of Semiconductor
Physics SB RAS,
Novosibirsk, Russia
ava@isp.nsc.ru

Sergey Tarkov
Laboratory of Silicon Material Science,
Rzhanov Institute of Semiconductor
Physics SB RAS,
Novosibirsk, Russia
ser-tarkov@yandex.ru

Vladimir Popov
Laboratory of Silicon Material Science,
Rzhanov Institute of Semiconductor
Physics SB RAS,
Novosibirsk, Russia
popov@isp.nsc.ru

Andrey Miakonkikh
Laboratory of Microstructuring and
Submicron Devices,
Valiev Institute of Physics and
Technology RAS,
Moscow, Russia
miakonkikh@ftian.ru

Andrey Lomov
Laboratory of Microstructuring and
Submicron Devices,
Valiev Institute of Physics and
Technology RAS,
Moscow, Russia
lomov@ftian.ru

Konstantin Rudenko
Laboratory of Microstructuring and
Submicron Devices,
Valiev Institute of Physics and
Technology RAS,
Moscow, Russia
rudenko@ftian.ru

Abstract— Silicon-on-sapphire (SOS) substrates with Si and hafnia nanolayers are investigated depending on the substrate thickness. It is shown that the stress decrease in hafnia leads to an increase in the coercive field needed for the ferroelectric pseudo-MOSFET hysteresis.

Keywords— hydrogen transfer; silicon-hafnia-on-sapphire; interlayer stress; ferroelectricity

I. INTRODUCTION

Silicon layers of nanoscale thickness transferred to sapphire wafers (silicon-on-sapphire (SOS) strucrures) have serious prospects for the use as substrates in the production of new generations of microcircuits for subtepahertz receivers and "digital radio" for mobile phones. But, at present, HR-TR SOI substrates are used for the production of such devices, since their cost is noticeably lower [1]. To reduce the cost of SOS substrates, the method of hydrogen-induced transfer of Si and HfO_2 layers at elevated temperatures was used [2]. Annealing, which is necessary to remove defects, promotes the formation of interlayer silicon dioxide (IL) and, as a consequence, the appearance of vacancies in sapphire. Thus, a large positive built-in charge is formed at the silicon-sapphire interface [3] and a very high negative threshold voltage $V_{t,n}$ appears, and leads to high electronic currents in the SOS pseudo-MOSFET. Even with the substrate thickness of 100 µm or less, this voltage becomes more than 10 kV. To compensate the above-described built-in charge effect, we proposed in [3], instead of SiO_2, to use 20 nm thick HfO_2 layers providing $V_{t,n}>-0.5$ kV. The silicon dioxide layer elimination is due to the fact that already 50 nm thick SiO_2 layers gave an unacceptably high modulus $V_{t,n}<-6.0$ kV [4]. In addition, the presence of an amorphous SiO_2 layer sharply worsens the heat removal from the device Si layer. The aim of this study was to develop a method for suppressing the oxygen diffusion from sapphire and reducing the positive charge using interlayer hafnium dioxide. This interlayer was formed by the hafnia plasma-enhanced atomic layer deposition (PEALD) onto the silicon donor wafer before the hydrogen irradiation and a subsequent bonding after it.

II. EXPERIMENT

A 20 nm thick PEALD HfO_2 layer was grown on monocrystalline silicon wafers at the temperature of 250 °C. Then these wafers were subjected to the ion implantation of H_2^+ at room temperature with the fluence of $3 \cdot 10^{16}$ cm^{-2} and the energy of 130 keV. To transfer silicon and hafnium dioxide layers, we used the technique described earlier in [4]. The SOS structures obtained in this way were annealed in stages at the temperatures of 450-1100 °C.

To reveal the perfection degree of silicon/buried dielectric/sapphire substrate heterointerfaces, the cross-sections of SOS samples were prepared using the ion-beam cut method. The first measurements by scanning electron microscopy (SEM) confirmed the presence of uniform thickness of the device layer of silicon and a hidden dielectric, as well as the perfection of the interfaces in the areas, the length of which is orders of magnitude greater than that of the declared thicknesses (Fig. 1).

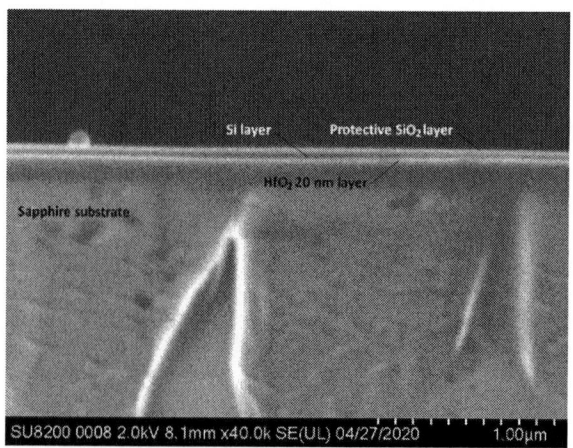

Fig. 1. Image obtained by scanning electron microscopy of a cross-section of the SOS-structure with a hidden 20 nm HfO_2 layer.

The electric properties of the SOS structures were evaluated by the drain-gate current-voltage characteristics measured by the method of SOS-pseudo-MOS transistors. They are presented in the Table 1 for the SOS structures with thick and thin silica [3, 4] and for a thin hafnia interlayer. In this case, the measured structure was placed on a copper metal gate by the substrate, and tungsten probes with the tip radius of 20 μm, pressed against the device silicon layer with the force of 60 g at the distance of 100 μm from each other, played the role of the source and drain. We compared the properties of SOS structures with the 20 nm thick hafnia IL with N$^+$-implanted (w NI) into sapphire substrates and without the N$^+$ implantation (w/o NI). X-ray diffraction studies were performed to investigate the hafnia film crystallinity evolution after annealing (Fig. 2). The measurements of pseudo-MOS transistors in a quasi-static electric mode were carried out using a stepwise gate voltage (up to 5500 V) sweep by decreasing and increasing it at the rate of 500, or 2 V/s relative to the voltage points with a minimal source-drain conductivity (Fig. 3).

TABLE 1. Charge carrier mobility μ, fixed charge Q_{ox} and interface state density D_{it} for electrons (e) / holes (h) in the Si layers for the different SOS structures with thick and thin silica [3, 4] or thin hafnia interlayer in this work.

SOS structure	μ_e / μ_h, cm²/(Vs)	Q_{ox}, cm^{-2}	$D_{it,e}$ / $D_{it,h}$, cm^{-2}eV^{-1}
Thick SiO$_2$ 310 nm	250 / 50	4.7·10^{11}	6.3·10^{11} / 4.1·10^{11}
Thin SiO$_2$ 50 nm N$^+$, 50 keV	105 / 37	2.1·10^{11}	1.3·10^{12} / 3.8·10^{11}
Thin HfO$_2$ 20 nm	230 / 33	1.2·10^{12}	7.0·10^{11} / 2.4·10^{12}

The same stand was used to determine the relaxation by measuring the current 50 times for 1, 2, or 4 s at each stage with a change in the sweep speed from 3 to 300 V/s. The sample with a thick SiO$_2$ interlayer showed that the hysteresis window width remained unchanged when the gate voltage range was changed. With an increase in this voltage, a change in the ratio of the maximum currents for slow and fast sweeps was found, and it is apparently associated with the presence of various types of charge carrier traps at the SiO$_2$ interface.

III. RESULTS AND DISCUSSION

Monocrystalline hafnium and aluminum oxide (hafnia and sapphire) are a promising dielectric substrate that provides high dielectric permittivity (high-k ε ~11.5 and 25, respectively) and thermal conductivity. The X-ray diffraction patterns from all samples treated at 450-1100°C demonstrate a lot of diffraction peaks (Fig. 2). Most of them have not chemically removed Si layer, but if it is present, clear peak shifts are observed at 35° for the [200]m P2$_1$/c phase. A qualitative analysis of crystalline peaks indicates the presence of crystallites in two phases of the substance: nonpolar monoclinic phase P2$_1$/c and orthorhombic polar phase Pca2$_1$. The presence of centers of charge trapping from silicon and charge transfer in the oxide, the role of which is played in oxides with a high-k by protons and oxygen vacancies, leads to the appearance of hysteresis in the I$_{DS}$–V$_G$ dependences, similar to the ferroelectric switching of PE. But, in our case, hysteresis is associated with the formation of a double electric layer (DEL), with the recharging of trapping centers, or re-polarization of dipoles [5, 6]. But it should be noted that we often observe hysteresis as a

function of polarization *versus* field in our structures with high-k dielectric layers, similar to FEFET ferroelectric transistors (Fig. 3).

Monocrystalline C-surface sapphire with a high thermal conductivity and dielectric constant ε ~ 11.5 is a promising substrate with dielectric properties. The presence of protons and oxygen vacancies in oxides with a high-k, which are the centers of trapping and charge transfer in the oxide with the previous charge capture from silicon, leads to the appearance of hysteresis in the IDS-VG dependences, similar to the ferroelectric switching of PE [5, 6]. In SOS structures containing a hafnium dioxide interlayer, these high-k dielectrics exhibit a typical hysteresis for ferroelectric transistors FEFETs as a function of polarization P from the electric field $E = V_G / d$, where d is about the sapphire thickness (Fig. 3).

Only by measuring the charge Q as a function of the applied electric field E, we practically cannot distinguish between the ferroelectricity and polarization P due to the spatial charge region (SCR). Therefore, it is important for us to obtain additional data that unambiguously characterize the material used as a ferroelectric [7]. For example, such differences can be established by measuring the dielectric constant ε.

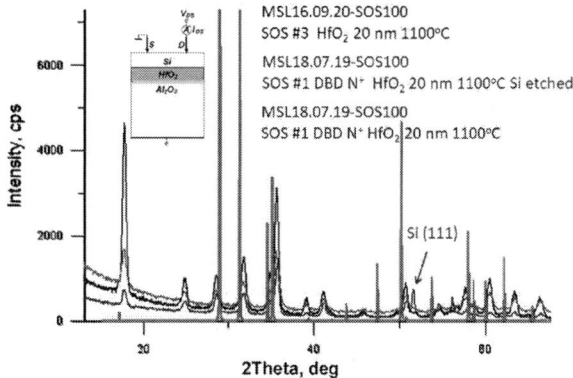

Fig. 2. GIXRD pattern of SOS samples with thick 20 nm hafnia w or w/o NI. In the inset is the scheme of SOS pseudo-MOS-FET measurements and SOS cross-section.

The temperature dependence of ε in the ferroelectric case should be described by the Curie-Weiss law. In addition, in the low-frequency range, there should be no pronounced dependence of ε on the ferroelectric thickness, which is recorded in the DEL of ionic conductors.

In addition, the SCR polarization has a saturation value of P, which depends on the thickness, while, in the absence of an inactive layer near the electrode, the ferroelectric polarization does not depend on the thickness. Experiments were carried out at different temperatures with different thicknesses in SOS structures. The results obtained demonstrate the presence of mechanical stress at the silicon/hafnium dioxide interface, as well as the presence of the compressive biaxial stress for the Si/HfO$_2$ interface providing the conductivity switching at the weak electric field $E < 10^5$ V/cm [8, 9].

For electrons in SOS-structures with an intermediate HfO$_2$ layer, relaxation measurements showed a slow increase in the current at the steps of forward branch, but at a positive bias, and, therefore, not because of protons, but because of polarization or the transfer of electrons trapped by volume hafnia traps to the HfO₂/Al₂O₃ interface with the substrate. (Fig. 4).

The much larger drop in the current manifested itself at the steps with a decrease in the gate potential due to the depolarization, rather than the capture of electrons by traps at the HfO₂/Al₂O₃ interface. This means that the main contribution to the hysteresis for the sample with the HfO₂ IL comes from polarization/depolarization. For the hole conductivity in this pseudo-MOSFET SOS-structure, a constant current was observed at the steps of the forward branch, which indicates the absence of proton drift from the bulk of hafnia traps towards the sapphire substrate and the displacement of electrons trapped in HfO₂ to the Si/SiO₂ interface, as well as a weak current drop with a decrease in the negative gate potential modulus due to the depolarization of ferroelectric domains in HfO₂. A drop in the current across the steps with a decrease in the gate potential can also manifest itself due to the capture of electrons ejected from traps at the HfO₂/Al₂O₃ interface into the bulk of HfO₂, but this requires an electron drift against the field [10].

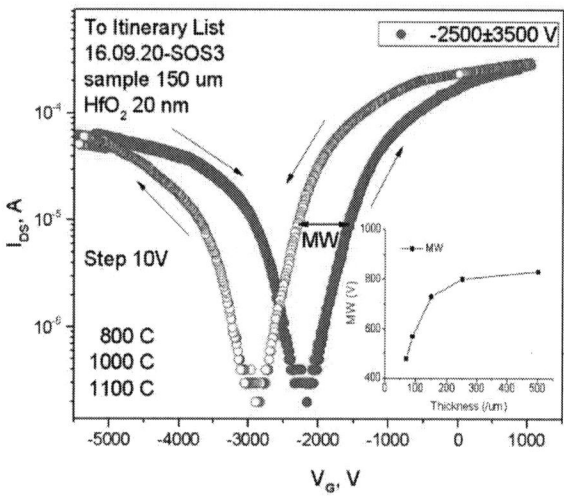

Fig. 3. Drain-gate characteristics of SOS pseudo-MOS-FET with thick 20 nm hafnia w/o NI. In the inset is the MW dependence on the sapphire thicness.

In order to reveal the traps contribution to the substrate, the sapphire plates were irradiated with N₂⁺ (NII) ions. Such implantation created the inclusions of oxygen vacancies and aluminum nitride in sapphire, which were the centers of trapping of charge carriers [11]. The AlN conduction band edge is located near the similar silicon band bottom. After the implantation, a silicon layer with a thermally grown SiO₂ layer was transferred onto sapphire using the SmartCut technology.

The annealing of defects at 1000 °C led to a large positive charge formation in the latent silicon dioxide of the SOS structure. This fact became an obstacle to achieving the depletion in the silicon layer even when the maximum possible (pre-breakdown) gate bias voltage of -6 kV was reached. If the sapphire substrate was treated with N2+ ions, the depletion voltage would increased to -400 V. In this case, the predominance of charge carrier trapping centers near the SiO2/Al2O3 interface and in sapphire led to the hysteresis

loop direction a change to the opposite for both cases with NII [12].

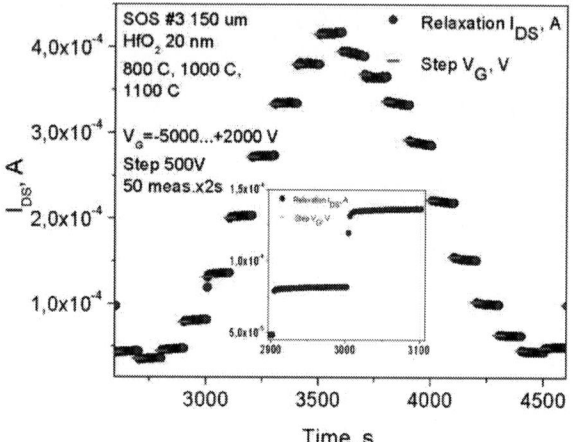

Fig. 4. Transient characteristics for the electron current in SOS pseudo-MOS-FET with a thin 20 nm HfO₂ layer. In the inset is the current change during two up-steps (+500 V) of voltage sweeping.

According to the remarks above we can suggest that the main reason of the memory window decrease is an increase in the coercive field with the sapphire thickness. This decrease leads to the decrease in the interlayer biaxial stress that was as high as ~1 GPa after the 1100°C annealing measured by the single phonon Raman peak shift of +3.5 cm⁻¹ in the 30 nm Si layers in the SOS structure with hafnia ILs (Fig. 5). The decrease of biaxial stress in hafnia leads to an increase in the coercive field with the decreasing sapphire thickness needed for the polarization switching of ferroelectric pseudo-MOSFETs.

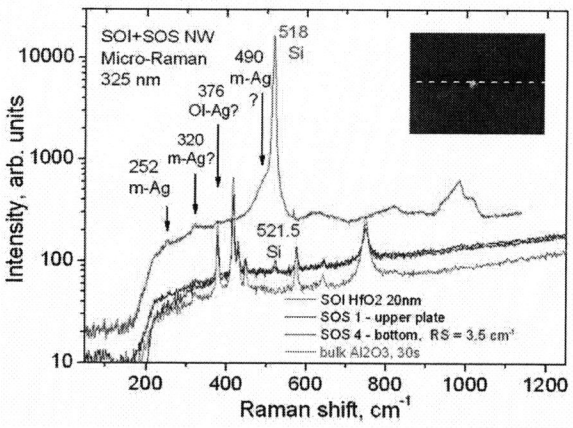

Fig. 5. a-micro-Raman spectra taken in the UV laser beam direction along the Si layer surface along the axis [110] for SOI (upper spectrum) and SOS (lower spectra) of 30 nm layers after annealing at 1100°C. In the inset is a micrograph of the measurement geometry of two identical halves (SOS1 and SOS4) of the structure folded with the Si layers facing each other and a UV laser spot with the diameter of 2 microns mainly on the lower half (SOS4) of the SOS structure, as well as a UV laser spot at a distance of 50 microns from its surface (bulk Al₂O₃).

IV. CONCLUSIONS

These results give a clear evidence that the physical origin of the hysteresis switching in the SOS pseudo-MOSFETs with hafnia interlayers is the compressive biaxial stress is due to the large difference in the thermal expansion coefficients between silicon and sapphire transferred to the HfO_2 ILs during the thermal treatments. This stress stimulates the ferroelectric behavior in hafnia nanolayers.

The charge trap density in these ILs is small enough and does not camouflage the ferroelectric switching behavior as in the case of nitrogen-implanted sapphire.

ACKNOWLEDGMENT

The authors thank A.V. Sheremetiev, N.V. Dudchenko, and E.D. Zhanaev from ISP SB RAS for their help in producing SOS-structures. All authors and co-authors declare no competing financial interest. This work was funded by the RFBR grant no. 19-42-543012 and also was supported by grants no. 066-2019-0004 and 0242-2021-0003 of the Ministry of Science and Higher Education of Russia.

References

[1] G. Scheenab, R.Tuyaertsa, M.Racka, L.Nyssensa, J.Rassona, M. Nabeta, J-P.Raskin. "Post-process porous silicon for 5G applicationss" Sol. State Electron., 2019, doi:10.1016/j.sse.2019.107719.

[2] V.P. Popov, E.D. Zhanaev, N.V. Dudchenko, V.A. Antonov, and A.I. Popov, "Method for silicon-on sapphire formation". RF Patent No. 2538352, Byul. Izobret. No. 1 (2015). In Russian.

[3] V.P. Popov, V.A. Antonov, and V.I. Vdovin "Positive Charge in SOS Heterostructures with Interlayer Silicon Oxide". Semiconductors, vol. 52, 10, pp. 1341–1348, October 2018.

[4] V. Antonov, S. Tarkov, V. Popov. Current hysteresis in SOS Heterostructures with Interlayer Silicon Oxide. 2020 Joint International EUROSOI Workshop and International Conference on Ultimate Integration on Silicon, EUROSOI-ULIS 2020. 9365505.

[5] N. Miyata. "Low temperature preparation of HfO_2/SiO_2 stack structure for interface dipole modulation". Appl. Phys. Lett., vol. 113, 251601, December 2018.

[6] Yongli He, Sha Nie, Rui Liu, Yi Shi, Qing Wan. "Indium–Gallium–Zinc–Oxide Schottky Synaptic Transistors for Silent Synapse Conversion Emulation". IEEE Electron. Device Lett., vol.40, pp. 139-142, January 2019.

[7] H.Cliem and B.Martin. "Pseudo-ferroelectric properties by space charge polarization". J. Phys.: Condens. Matter, 20, 321001, 2008.

[8] A. Kruv, S.R.C. McMitchell, S. Clima, O.O. Okudur, N. Ronchi, G. Van den bosch, M. Gonzalez, I. De Wolf, and J.Van Houdt. "Impact of mechanical strain on wakeup of HfO2 ferroelectric memory". IEEE International Reliability Physics Symposium (IRPS), EEE Xplore, 2021, doi: 10.1109/IRPS46558.2021.9405159

[9] Yefan Liu, Sergiu Clima, Gaspard Hiblot, Philippe Matagne, Mihaela Loana Popovici, Ben Kaczer, Dimitrios Velenis, Ingrid De Wolf. "Investigation of the Impact of Externally Applied Out-of-Plane Stress on Ferroelectric FET". IEEE Electron Device Letters, vol. 42(2), pp. 264-267, february 2021

[10] Md. Nur Kutubul Alam, Ben Kaczer, Lars-Åke Ragnarsson, Mihaela Popovoci, Gerhard Rzepa, Naoto Horiguchi, Marc Heyns, Jan Van Houdt. "On the Characterization and Separation of Trapping and Ferroelectric Behavior in HfZrO FET". J. Electron Devices Society, vol. 7, pp. 855-862, 2019, doi: 10.1109/JEDS.2019.2902953.

[11] Shujing Zhao, Fengbin Tian, Hao Xu, Jinjuan Xiang, Tingting Li, Junshuai Chai, Jiahui Duan, Kai Han, Xiaolei Wang, Wenwu Wang, Tianchun Ye. "Experimental Extraction and Simulation of Charge Trapping during Endurance of FeFET with TiN/HfZrO/SiO$_2$/Si (MFIS) Gate Structure". arXiv:2106.15939, 2021.

[12] C. Jin, C.J. Su, Y.J. Lee, P.J. Sung, T. Hiramoto, M. Kobayashi. "Study on the Roles of Charge Trapping and Fixed Charge on Subthreshold Characteristics of FeFETs". IEEE Transactions on Electron Devices, vol. 68(3), pp. 1304-1312, march 2021.

Operational Transconductance Amplifier Design with Gate-All-Around Nanosheet MOSFET using Experimental Lookup Table Approach

Júlia C. S. Sousa[1*], Welder F. Perina[1], Eddy Simoen[2], Anabela Veloso[2],
Joao A. Martino[1], Paula G. D. Agopian[1,3]

[1]LSI/PSI/USP, University of Sao Paulo, Sao Paulo, Brazil
[2]imec, Leuven, Belgium
[3]UNESP, Sao Paulo State University, Sao Joao da Boa Vista, Brazil
email*: julia.cristina@usp.br

Abstract— This paper presents the design of an Operational Transconductance Amplifier (OTA) with Gate-All-Around Nanosheet MOSFETs (GAA-NSH). The circuit simulation was performed using an experimental Lookup Table (LUT) approach. The experimental drain current and gate capacitance were extracted and used in a Verilog-A model in order to design the OTA for different transistor efficiency (gm/ID) values. The results present a compromise between power consumption (PC), voltage gain (Av) and the Gain-Bandwidth-Product (GBW). For gm/I_D of 8 V^{-1} an Av of 71.8 dB is obtained for a GBW of 361.3 MHz. These results were compared with other OTA designs using FinFET and TFET devices. The NSH OTA presents higher GBW, and considering the Av and PC, while NSH present better behavior than FinFETs, the behavior is worse than TFET OTA circuit for strong inversion operation.

Keywords - Nanosheet (NSH); Operational Transconductance Amplifier; Transistor Efficiency (gm/I_D); Lookup Table; Analog Circuit Design.

I. Introduction

The exponential growth of devices scaling and computational power, as well as decreasing power consumption and cost of devices unit per dollar are the main drivers of the semiconductor business into advanced and novel device research. New technological challenges, such as the 5G protocol, aim to decrease latency times, improve the integration of several IoT applications and services to the internet, and allow a higher data trafficking [1]. This will influence and bring innovative technologies onto several areas of application, ranging from preventive health to urban surveillance. These applications require new devices that consumes even less power and are able to operate at higher frequencies and with higher bandwidths [2].

Gate-All-Around (GAA) devices are MOS transistors composed of thin channels fully surrounded by gate material. This device architecture optimizes the gate-channel electrostatic coupling and decreases the body factor influence on current and subthreshold slope, becoming nearly ideal. GAA-Nanosheet (NSH) FETs are widely regarded as the next main devices to be introduced in the logic scaling roadmap, set to replace triple-gate FinFET for advanced nodes beyond the 5nm technology node [3]. GAA-NSH are already shown to have excellent performance on digital applications [4].

In order to evaluate the performance of the GAA-NSH in analog devices, this work presents a design of an operational transconductance amplifier (OTA) under different bias conditions. The OTA was chosen for being a traditional circuit in analog and mixed signal applications, due to its versatility in signal treatment paths for amplification and buffering.

II. Device Overview

The GAA-NSH MOS transistors used in this work, shown in Figure 1, were fabricated at imec, in Belgium. The fabrication process resembles to a large extent he FinFET fabrication flow [5], with several added key steps namely: Si/SiGe multi-layer fins formation and Si channels released by selective removal of the sacrificial SiGe layers, in-between Si sheets, at replacement metal gate (RMG) module [4], followed by gate stack deposition.

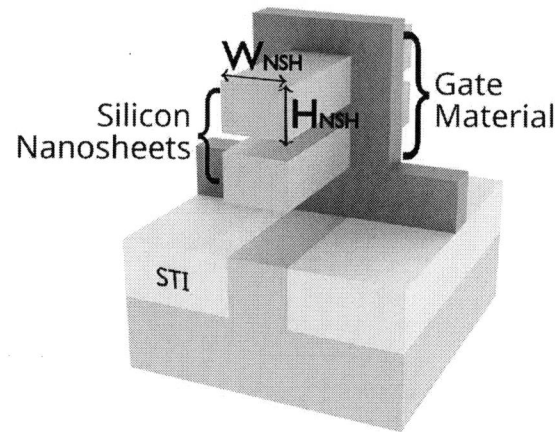

Fig. 1. GAA-NSH schematic.

The schematic shows the presence of two vertically stacked nanosheets with W_{NS} of 15nm and H_{NS} of 11nm, with a thickness of 7.5nm for the metal gate. A single fin in layout corresponding to two vertically stacked sheets has a total effective width (W_{eff}) of 104nm. The gate dielectric consists of an interfacial layer of SiO_2 covered by HfO_2 layer resulting in an effective oxide thickness (EOT) of 0.9nm. The device lengths (L) vary within 200nm~28nm, from which devices of 100nm were chosen to characterize and further design the circuit, due to being relatively immune to short channel effects. Only for capacitance measurements, a device L of approximately 1μm was used in order to create a larger gate area that allowed measuring the extremely small gate capacitance values.

III. Lookup Table Approach and Validation

In order to create a simulation model, Verilog-A was used along with a lookup table (LUT) obtained from characterizations performed on a fabricated device. Afterwards, simulations were made using the Spectre solver within the Cadence Virtuoso environment.

The GAA-NSH $I_D \times V_G$ curves were obtained with NMOS-NSH and PMOS-NSH with an L of 100nm and 22 fins and a total width (W_{eff}) of 2.288µm. The curve extraction utilizes a voltage step of 10mV within a range of -0.5V to 1V, while being parametrized for V_{DS} with a 50mV step. The $I_D \times V_D$ curves were also obtained using a sweep from 0V to 1V with a 10mV precision, while parametrized in V_{GS} with a 50mV step.

Figure 2 shows that the comparison between the measurements and simulations for both curves overlap and thus the device behavior is well represented within the simulator.

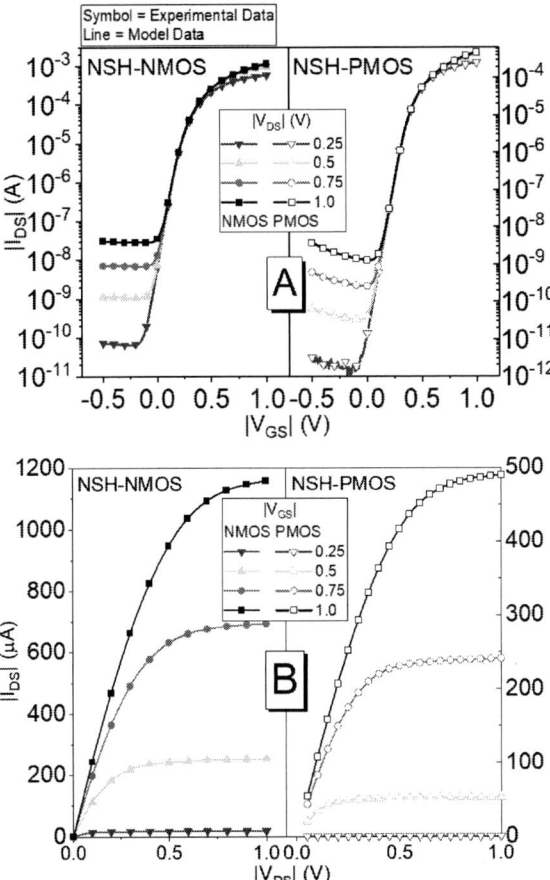

Fig. 2. Drain current as a function of gate voltage (A) and drain voltage (B) for NSH PMOS and NMOS devices.

The transconductance, g_m, and transistor efficiency, g_m/I_D, were extracted from the experimental curves by taking the first derivative of the $I_D \times V_G$ and dividing the result by current I_D, respectively. Figure 3 shows the g_m/I_D curves as a function of the current density $I_D/(W/L)$. The transistor efficiency values are important to define transistors sizing methodology, due to being more accurate, for advanced devices, than quadratic equations, while dealing with the main tradeoffs between voltage gain, frequency, sizing and parasitic capacitances.

Furthermore, the gate capacitance per unit area, c_{gg}, was extracted from PMOS-NSH and NMOS-NSH with an L of 1µm and the same W_{eff} of 2.288µm, and normalized by dividing by the channel area. Gate voltage sweeps were

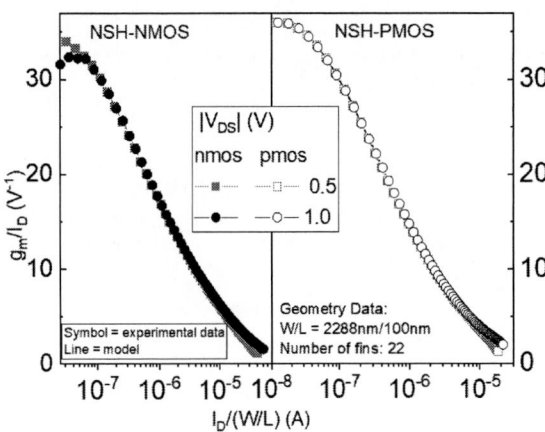

Fig. 3. GAA-NSH g_m/I_D curves.

performed from -0.51V to 1.05V with a step of 15mV, while applying a small voltage 100kHz sinusoidal signal. The obtained capacitance values was then scaled for L = 100nm and by the number of fins by using W_{eff} = 104nm. The LUT was build assuming all transistors would operate in the saturation region and thus $c_{gs} \approx 2/3\ c_{gg}$ was adopted.

IV. ANALYSIS OF CIRCUIT PERFORMANCE

The OTA, shown in Figure 4, was designed with three transistor efficiency (g_m/I_D) conditions for transistors $M_{1,4}$ and M_6: 11 V^{-1}, 8 V^{-1} and 5 V^{-1}. In order to achieve the same transistor efficiency values for NSH-PMOS and NSH-NMOS, the sizing was defined in order to compensate the different current densities associated with a given g_m/I_D, that arise from the different mobility values between the transistor types. The current mirror devices $M_{5,7,8}$ were designed to ensure operation in the strong inversion region, with low g_m/I_D values. A current gain of two is used between the reference current and the first stage, and it is six fold for the second stage. A load capacitance of 200 fF and a miller compensation capacitance of 100 fF were used.

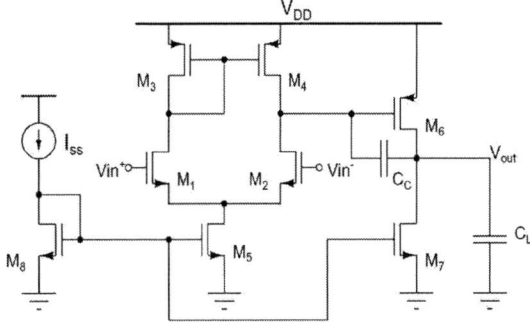

Fig. 4. OTA circuit

Table I shows the device sizing for the different bias conditions as multiples of the number of fins, with unit size defined by L and W_{eff}. NSH-PMOS $M_{3,4}$ are twice the size as the NSH-NMOS $M_{1,2}$ to compensate for the different current levels. The transistors for the current mirror use minimum sizing to increase current density ($I_D/(W/L)$) and decrease

978-1-6654-3746-2/21 $31.00 © 2021 IEEE

g_m/I_D. NSH-NMOS have measured currents on saturation region up to 50µA/fin, and for the defined V_{DS} availability of 2.1V across three transistors, devices are already close to triode operation. And thus, transistors $M_{5,7,8}$ are doubled in size for the $g_m/I_D = 5$ V^{-1} project, that requires a larger reference current. Transistor M_6 was sized as $6*M_{3,4}$, and M_7 was sized in order to place the output common voltage at around ½ V_{DD}. V_{DD} was defined as 2.1V and provides extra drain voltage headroom, allowing for an evaluation on GAA-NSH at with a unit gain frequency (f_T) performance.

TABLE I. GAA-NSH SIZING

Device	Number of Fins	
	Project using gm/I$_D$ = 5	*Other gm/I$_D$ projects*
$M_{1,2}$	5	5
$M_{3,4}$	10	10
M_5	4	2
M_6	60	60
M_7	12	6
M_8	2	1
Length (L)	100nm	
Width per fin (W$_{eff}$)	104nm	

Table II shows the OTA performance across transistor efficiency values, and Figure 5 presents the bode plot with Phase and Gain frequency response. For higher g_m/I_D, we see lower power consumption and currents, higher gain and lower

TABLE II. OTA PERFORMANCE FOR DIFFERENT BIASINGS

Specification	Transistor Efficiency – g$_m$/I$_D$		
	5 V^{-1}	*8 V^{-1}*	*11 V^{-1}*
VDD	2.1V		
CL	200 fF		
CC	100 fF		
Ref. Current	71 µA	32 µA	15 µA
Power	1.205mW	544.5µW	259.5µW
Gain	65.6 dB	71.8 dB	89.1 dB
GBW	496.7MHz	361.3MHz	255.9MHz
Phase Margin	64°	63°	63°

bandwidth.

For a defined W_{eff}/L, current levels and transconductance decreases with increasing transistor efficiency as it can be derived from Figure 3, however, the current decreases much faster [6], and so does the output conductance of the device. The circuit gain is mainly dependent on the devices intrinsic voltage gain (g_m/g_{DS}) ratios, which increase with g_m/I_D. This tradeoff between transistor efficiency, frequency of operation and power consumption is characteristic of MOSFET devices.

The bandwidth of the circuit can be estimated using the device f_T, which decreases for higher g_m/I_D. f_T is a function of the gate capacitances and device transconductance, and so, decreases for larger dimensions due to the increasing of

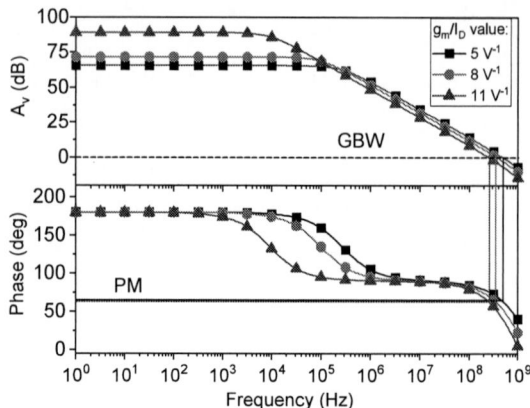

Fig. 5. OTA Frequency Responses

parasitics. For fixed dimensions, the frequency response will be mainly defined by g_m levels, which decrease with g_m/I_D. While designing an OTA, these tradeoffs must be taken into consideration to find the optimal frequency-gain compromise for the application. For many applications, the GAA-NSH presents a good frequency-gain coefficient, which could be sacrificed in order to obtain operations at a lower V_{DD} or higher transistor efficiencies. The Phase Margin (PM) for all the designs is around 60°, and thus the transient response is stable and has no overshoots.

Table III presents a performance comparison across different technologies [7,8]. Analysis of the DC parameters shows that the GAA-NSH OTA presents a significant power reduction while presenting an increase of gain for the same transistor efficiency condition and supply voltage as the FinFET counterpart, despite the lower GBW value, marking an overall improvement over the previous industry adopted technology. The GAA-NSH OTA also presents competitive results against the TFET device, which exhibits a high gain for ultra-low power consumption. This comes at the expense of a very low gain bandwidth product and lower load capacitance. Table III shows that it is also possible to achieve the same gain as the TFET with a larger GBW for a higher g_m/I_D, however with a much larger power consumption cost.

TABLE III. OTA PERFORMANCE FOR DIFFERENT DEVICES

Specification	Devices		
	FinFET [8]	*TFET* [7]	*GAA-NSH (This Work)*
V_{DD}	2.1 V	4.5 V	2.1 V
C_L	100 fF	30 fF	200 fF
C_C	30 fF	100 fF	100 fF
g_m/I_D	8 V^{-1}	7 V^{-1}	8 V^{-1}
Power	1.41 mW	9 µW	544.5 µW
Gain	67.61 dB	88.8 dB	71.8 dB
GBW	880 MHz	718 kHz	361.3 MHz
Phase Margin	63°	63°	63°

Table IV gives further information about the design choices for the FinFET [8] and GAA-NSH OTAs. Both designs utilize the same number of fins for the input transistors ($M_{1,2}$). The second stage of the FinFET uses a tenfold increase of current in relation to the first stage branches, while the GAA-NSH utilizes a sixfold increase, which allowed for a larger gain while decreasing the power consumption. The analog parameters for g_m and g_{ds} hint at an increase for GAA-NSH gain due to increase in transconductance, while output conductance suffers little change. From the Early Voltage (V_{EA}) values, for transistors $M_{1,2}$ it can be noted that the GAA-NSH is slightly more immune to short channel effects, while presenting smaller dimensions than the FinFET, for both

TABLE IV.   FINFET VS. GAA-NSH PROJECT SIZING COMPARISON

Parameter	Devices	
	FinFET	**GAA-NSH** **(This Work)**
Analog Parameters		
$g_{m1,2}$	243 µS	261 µS
$g_{DS3,4}$	3.5 µS	3.2 µS
$V_{EA1,2}$	10.52 V	11.4 V
Device Geometry		
Fins for $M_{1,2}$	5	5
Fins for $M_{3,4}$	7	10
Fins for M_6	70	60
W_{FIN} (FinFET), W_{NS} (nanosheet)	20nm	15nm
L	150nm	100nm

W_{NSH} and L.

V. CONCLUSIONS

The OTA, shown in Figure 4, was designed with three transistor efficiency conditions for transistors $M_{1,4}$ and M_6: 11 V^{-1}, 8 V^{-1} and 5 V^{-1}. The current mirror devices $M_{5,7,8}$ were designed in order to ensure operation in the strong inversion region, with low g_m/I_D values. Table I shows the device sizing for the different bias conditions as multiples of the number of fins, with unit size defined by L and W_{eff}.

The OTA design with Gate-All-Around Nanosheet MOSFETs presents an improvement in power consumption and voltage gain when compared with a FinFET OTA, while

a tradeoff between the smaller power consumption and the final GBW for the GAA-NSH circuit is observed. Comparing with a TFET OTA, although the NSH OTA presents lower voltage gain (71.8dB vs TFETs 88dB) and higher power consumption, a much higher GBW (361 MHz against to 718 kHz) is obtained.

The NSH is a good device choice for high voltage gain and high bandwidth requiring applications, presenting a comparable gain to the TFET device while having a much larger Bandwidth.

ACKNOWLEDGMENT

The authors acknowledge CNPq and CAPES for the financial support. The devices have been processed in the frame of imec's Core Partner Program on Logic Devices.

REFERENCES

[1] 5G PPP Architecture Working Group, Version 3.0, 2019. Available: https://5g-ppp.eu/wp-content/uploads/2019/07/5G-PPP-5G-Architecture-White-Paper_v3.0_PublicConsultation.pdf

[2] M. Mezzavilla et al., IEEE Communications Surveys & Tutorials, vol. 20, no. 3, pp. 2237-2263, thirdquarter 2018, doi: 10.1109/COMST.2018.2828880.

[3] A. Veloso et al., 2021 5th IEEE Electron Devices Technology & Manufacturing Conference (EDTM), 2021, pp. 1-3, doi: 10.1109/EDTM50988.2021.9420942.

[4] A. Veloso, et.al., Solid-State Electronics, Volume 168, 2020. https://doi.org/10.1016/j.sse.2019.107736

[5] A. Veloso, et.al., ECS Transactions, vol. 72, no. 2, pp.85--95, may, 2016, doi: 10.1149/07202.0085ecst

[6] P. Jespers, The gm/ID Methodology, A Sizing Tool for Low-voltage Analog CMOS Circuits. Boston, MA: Springer US, 2010.

[7] Nogueira, Alexandro, et. al. EUROSOI 2020, 1-5. 10.1109/EUROSOI-ULIS49407.2020.9365287.

[8] Sousa, Bruna, et al, Semiconductor Science and Technology. 36. 10.1088/1361-6641/abd349, 2020

978-1-6654-3746-2/21 $31.00 © 2021 IEEE

In-situ recovery of on-membrane PD-SOI MOSFET from TID defects after gamma irradiation

Amor Sedki
Institute of Information and Communication Technologies, Electronics and Applied Mathematics (ICTEAM)
Louvain-la-Neuve, Belgium
Electronics and Micro-Electronics Laboratory, Physics Department Faculty of Sciences of Monastir
Monastir, Tunisia
sedki.amor@uclouvain.be

Valeriya Kilchytska
Institute of Information and Communication Technologies, Electronics and Applied Mathematics (ICTEAM)
Louvain-la-Neuve, Belgium.
valeriya.kilchytska@uclouvain.be

Farès Tounsi
Institute of Information and Communication Technologies, Electronics and Applied Mathematics (ICTEAM)
Louvain-la-Neuve, Belgium.
fares.tounsi@uclouvain.be

Nicolas André
Institute of Information and Communication Technologies, Electronics and Applied Mathematics (ICTEAM)
Louvain-la-Neuve, Belgium.
nicolas.andre@uclouvain.be

Laurent A. Francis
Institute of Information and Communication Technologies, Electronics and Applied Mathematics (ICTEAM)
Louvain-la-Neuve, Belgium
laurent.francis@uclouvain.be

Denis Flandre
Institute of Information and Communication Technologies, Electronics and Applied Mathematics (ICTEAM)
Louvain-la-Neuve, Belgium)
denis.flandre@uclouvain.be

Abstract— This paper demonstrates a procedure for total in-situ recovery of on-membrane n-type MOSFET from Total Ionizing Dose (TID) defects, due to the exposure to gamma radiation. After a total dose of 348 krad (Si), several annealing steps were applied using an integrated micro-heater with a maximum temperature of 364 °C. The electrical characteristics of the transistor are recorded initially in normal conditions, after irradiation and then after each step of the thermal annealing. The electro-thermal annealing of the transistor allowed a total recovery of the original characteristics after a major shift due to radiation-induced defects. Power Spectral Density (PSD) of noise measurements showed a clear domination of the Random Telegraph Noise (RTN) behavior due to the creation of oxide defects after irradiation. After annealing, the RTN behavior vanishes with a further important decrease of flicker noise. Low-frequency noise measurements of the transistor confirmed the neutralization of oxide defects after annealing.

Keywords— Silicon On Insulator (SOI); MOSFET; Gamma irradiation; Total Ionizing Dose (TID); In-Situ Thermal Annealing; Flicker noise.

I. INTRODUCTION

Silicon-on-insulator (SOI) process has long been suitable for radiation environment applications. SOI made devices benefit from an intrinsic immunity to transient ionizing radiation effects thanks to the buried oxide (BOX) layer. The BOX reduces Single Event Effects (SEE) by blocking the induced ionizing particles in the substrate [1]. However, SOI devices remain more sensitive to Total Ionizing Dose (TID) effects, as the thick BOX forms an additional container for radiation-induced positive charges that directly impact the electrical characteristics of the

MOSFETs [2]. Various techniques are used for the mitigation of SEEs such as radiation-hardened design, redundancy, shielding, etc. [3]. However, TID effects remain more challenging to mitigate due to the cumulative creation of fixed positive charges in the SiO_2 insulating layers as well as interlayers interface [4]. This issue produces problematic robustness and reliability considerations for CMOS circuits. Recently, an increasing interest is put on in-situ electro-thermal annealing techniques for the recovery of oxide defects begotten by TID [5-6]. This technique is based on the thermal hydrogen passivation of oxide traps [7]. The current paper demonstrates an in-situ recovery of On-Membrane Partially Depleted (PD)-SOI MOSFET by electro-thermal annealing after being exposed to ^{60}Co gamma radiation. The electrical characteristics of the MOSFET are remarkably shifted after a total dose of 348 krad (Si). A severe shift of the threshold voltage (V_{th}) is observed in addition to a decrease of the transconductance (gm). Low Frequency Noise (LFN) measurements are performed before and after irradiation, as well as after annealing in order to demonstrate the creation and neutralization of oxide defects. The noise PSD shows a clear dominance of a strong Random Telegraph Noise (RTN) behavior after irradiation. A remarkable decrease of the noise PSD is observed after annealing which confirms the neutralization of oxide defects.

II. MEASUREMENTS SETUP

The Device Under Test (DUT) is a Body-tied 6μm-wide 1μm-long n-MOSFET implemented in a circular membrane with 600μm diameter. The device is located in the close vicinity of a Tungsten-based micro-heater, used for the in-situ thermal annealing by Joule heating. This design is fabricated using a

978-1-6654-3746-2/21 $31.00 © 2021 IEEE

1μm PD CMOS SOI technology (Fig. 1). Two PIN diodes are implemented underneath the micro-heater and next to the transistor side for temperature monitoring. An isolation process is used to reduce the flow of leakage current through the silicon substrate and thus the crosstalk among different transistors, which can eventually affect the logic state (On-Off) of each device. All the devices are embedded on-membrane after etching the back-silicon substrate with a deep reactive ion etching (DRIE) process. This design is widely used in gas sensors to minimize thermal dissipation through the substrate. The device showed initially a reliable characteristic in room temperature as well in high temperatures up to 280°C. The calibration of thermo-diodes and the temperature monitoring technique are described in a previous work [8]. The device was packaged in a Dual In-Line (DIL) ceramic package and connected to a pre-designed printed circuit board for direct characteristics recording. Next, the DUT was installed in a ^{60}Co gamma panoramic irradiator (in the Cyclotron Research Center at the Université catholique de Louvain, Belgium). Two sources of ^{60}Co are used providing a maximum dose rate of 1.2krad/h each. A HP4145 Semiconductor analyzer and a laptop with a pre-defined IC-CAP script to control measurements were installed in a shielded room and connected to the device through coaxial cables. A positive bias V_{gs} of 3V is applied to the transistor's gate during irradiation. The other contacts are connected to the ground. This configuration is usually used to separate the electron-hole pairs and increase the radiation sensitivity. A total dose of 348krad (Si) is obtained. LFN measurements were performed using a low-frequency noise analyzer (LFNA–*Keysight E4727A*). The RTN and flicker noise measurements were performed in saturation regime with V_{ds}=3V and by applying a constant drain current I_{ds}=10μA. The time trace measurements were recorded for 5s with a time step Δt of 1μs. This time resolution was sufficient to extract the emission and capture time constants. The PSD measurements were performed from 1Hz to 1MHz.

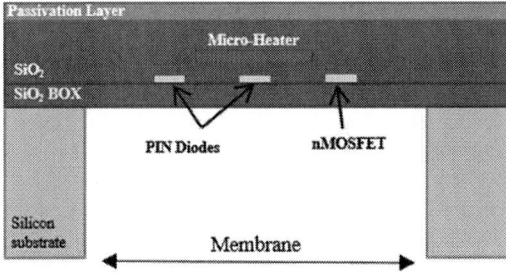

Fig. 1. Cross section schematic of the DUT.

III. RESULTS ANALYSIS

A. I-V measurements:

Fig. 2. presents the subthreshold characteristics of the MOSFET in linear regime with V_{ds} = 50mV, measured in fresh condition and at different doses of gamma radiation up to a total dose of 348krad (Si) during 286 hours. In fresh condition, a non-exponential behavior is observed in the subthreshold slope. This is could be explained by the post-process defects, created by the back DRIE etching [9]. The limited leakage current of this process-induced defects is modeled by a parasitic transistor

in the subthreshold region [10]. At lower radiation doses, a remarkable degradation of the subthreshold slope is manifested due to the cumulative creation of positive traps at the side walls of the MOSFET. After a total dose of 348 krad (Si), the leakage current is drastically increased by more than one decade. The I-V measurements reveals an important shift of the threshold voltage (ΔV_{Th} = 850mV) after irradiation. The maximum g_m value at V_{ds} = 50mV decreased by 13% after the same total dose. These degradations are mainly due to the creation of positive charge traps in the BOX and at the Si/SiO$_2$ interface. The threshold voltage is directly dependent on radiation-induced charges Q_{total} as expressed in following model, by [11]:

$$V_{Th} = \Phi_{MS} + 2\,\Phi_t + \frac{\sqrt{2\,q\,\varepsilon_{si}\,N_a\,2\,\Phi_f}}{C_{ox}} - \frac{Q_{total}}{C_{ox}} \qquad (1)$$

where q is the electron charge, ε_{si} is the dielectric permittivity of silicon, N_a is the substrate doping, C_{ox} is the gate oxide capacitance per unit area, Φ_{MS} the metal–semiconductor work function difference, Φ_f is the Fermi level potential, and Q_{total} is the total gate oxide charge density per unit area at threshold. The negative shift of the threshold voltage could be expressed by:

$$\Delta V_{Th} = -\frac{Q_{total}}{C_{ox}} \qquad (2)$$

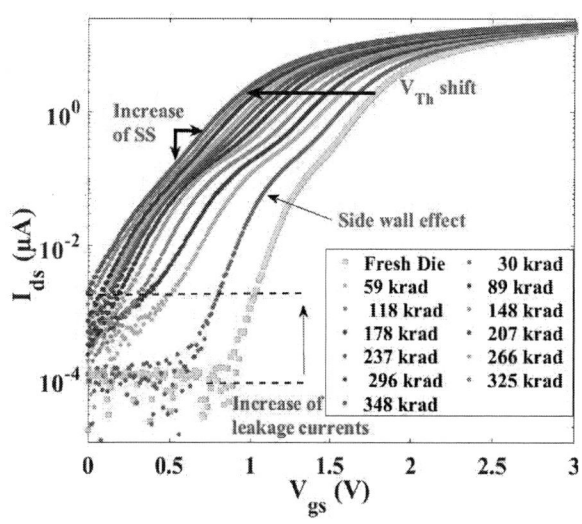

Fig. 2. I_{ds} versus V_{gs} measurements in the linear regime (V_{ds} = 50mV) extracted at different doses of gamma irradiation.

The thermal annealing is applied after the exposure to gamma radiation with a total dose of 348krad (Si). Fig. 3(a) presents the V_{Th} variations after exposure to gamma radiation and after each annealing step of 30 s at different annealing temperatures. For the same annealing temperature at 265°C, the threshold voltage slightly increased after each annealing step reaching a limit of V_{th}=1.33V after 16 annealing steps. The typical/pre-radiation threshold voltage is recovered after a maximum thermal annealing at 364°C. The evolution of the maximum tranconductance with thermal annealing is presented in Fig. 3(b) where it shows a total recovery of g_m at 265°C annealing temperature.

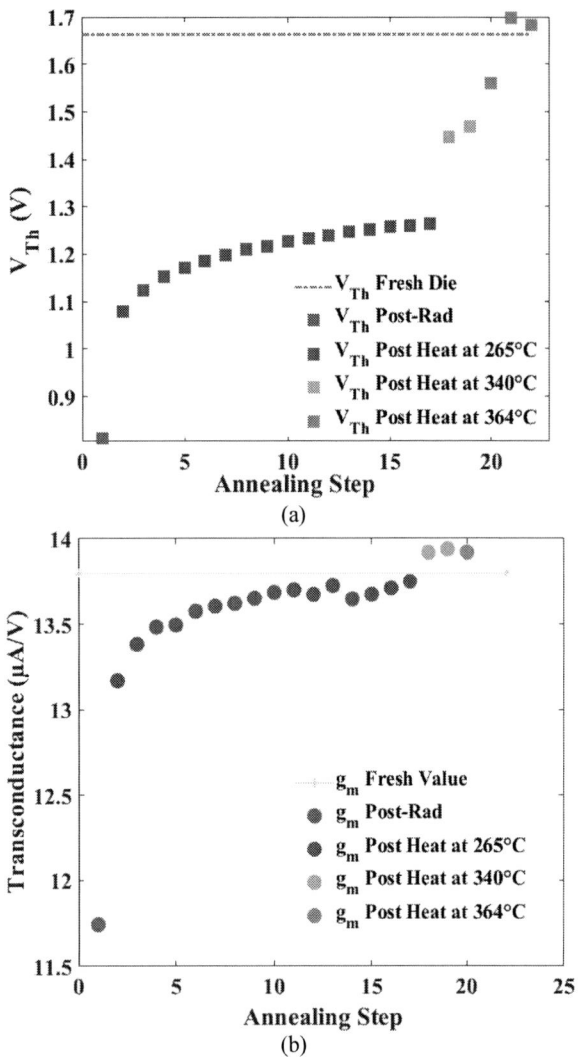

Fig. 3. (a) Threshold voltage of the MOSFET, extracted in linear regime after irradiation and after each annealing step, (b) Maximum transconductance of the MOSFET, extracted in linear regime after irradiation and after each annealing step.

B. Noise measurements

Fig. 4 presents the noise PSD, measured at a fixed drain current I_{ds}=10μA that corresponds to a voltage overdrive ($V_O = V_{gs} - V_{th}$ = 220mV) with V_{ds} = 3V. Only low frequencies between 1Hz and 1kHz are considered due to the influence of output capacitances at high frequencies. The fresh die results are compared to the measurements after gamma irradiation and after the full-recovery annealing. The noise PSD measurement of the fresh die shows a 1/f flicker-noise behavior. After irradiation, a clear Lorentzian-like plateau and $1/f^2$ decrease are observed. This indicates the domination of RTN behavior, explained by the creation of oxide traps due to the ionizing radiation. After annealing, the RTN behavior vanished with an important decrease of the flicker noise related to the

neutralization of defects caused by the irradiation, as well as caused by the DRIE process [9]. This can be quantified using the Lorentzian model (Eq. 3) and by extracting the capture and emission time constants (τ_c and τ_e) from the time trace $I_d(t)$ measurements, with τ is the equivalent time constant.

$$S_{id,RTN}(f) = \Delta I_d^2 \frac{4\,\tau^2}{\tau_e + \tau_c} \frac{1}{1 + (2\pi f \tau)^2} \qquad (3)$$

The Lorentzian model is presented by the purple curve in Fig. 4. We clearly see the fair match between the model and the flicker noise measurements. This confirms the contribution of the RTN behavior to the PSD of noise and thus the ontribution of radiation-induced defects to the PSD of noise.

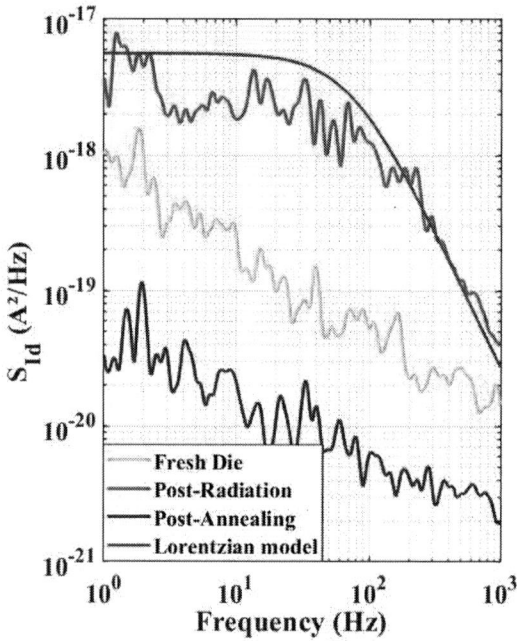

Fig. 4. Power spectral density of noise measurements before radiation, after radiation and after thermal annealing (Measured in saturation regime at fixed Ids=10μA and Vds=3V).

IV. CONCLUSION

This work studied the benefits of in-situ thermal annealing to recover an nMOSFET I-V characteristic after a strong gamma irradiation. After a TID of 348krad (Si), an important degradation of the transistor electrical characteristics such as the threshold voltage and the transconductance is observed due to the creation of positive charges in the BOX and gate-oxide layers. After an in-situ thermal annealing with a maximum temperature of 364°C, a total recovery of the initial electrical characteristics is achieved. Low-frequency noise measurements demonstrated the contribution of radiation-induced defects to the noise PSD. The strong RTN behavior observed after irradiation confirms the creation of oxide defects. After in-situ thermal annealing, the pre-radiation MOSFET characteristics are recovered with a further remarkable decrease of the flicker noise.

978-1-6654-3746-2/21 $31.00 © 2021 IEEE

REFERENCES

[1] J. Alvarado, V. Kilchytska, E. Boufouss, B. S. Soto-Cruz and D. Flandre, "A Compact Model for Single Event Effects in PD SOI Sub-Micron MOSFETs," in IEEE Transactions on Nuclear Science, vol. 59, no. 4, pp. 943-949, Aug. 2012, doi: 10.1109/RADECS.2011.6131345.

[2] B. Ning, D. Bi, H. Huang, Z. Zhang, Z. Hu, M.Chen and S. Zou, "Bias dependence of TID radiation responses of 0.13μm partially depleted SOI NMOSFETs", Microelectronics Reliability, Vol. 53, pp. 259-264, 2013, doi: 10.1016/j.microrel.2012.08.005.

[3] Calligaro. C, "Rad-Hard Mixed-Signal IC Design, Theory and Implementation", Springer International Publishing, pp. 273-297, 2020, doi: 10.1007/978-3-030-25267-0_17.

[4] M. L. Alles, R. D. Schrimpf, R. A. Reed, L. W. Massengill, R. A. Weller, M. H. Mendenhall, D. R. Ball, K. M. Warren, T. D. Loveless, J. S. Kauppila and B. D. Sierawski, "Radiation hardness of FDSOI and FinFET technologies," IEEE 2011 International SOI Conference, 2011, doi: 10.1109/SOI.2011.6081714.

[5] S. Amor, N. André, V. Kilchytska, F. Tounsi, B. Mezghani, P. Gérard, Z. Ali, F. Udrea, D. Flandre and L. A. Francis, "In-situthermal annealing of on-membrane silicon-on-insulator semiconductor-based devices after high gamma dose irradiation", Nanotechnology, vol. 28, pp 184001, 2017, doi: 10.1088/1361-6528/aa66a4.

[6] S. Amor, V. Kilchytska, D. Flandre and P. Galy, "Trap recovery by in-situ annealing in fully-depleted MOSFET with active silicide resistor,"

in IEEE Electron Device Letters. IEEE Electron Device Letters, vol. 42, no. 7, pp. 1085-1088, July 2021, doi: 10.1109/LED.2021.3079244.

[7] W. Markus, M. Masuaki, F. Katsuyuki, L. Ziyuan, A. Koichi, K. Yoshiya and F. Shinji," Influence of H2-annealing on the hydrogen distribution near SiO2/Si(100) interfaces revealed by in situ nuclear reaction analysis", Journal of Applied Physics, vol. 92, pp. 4320-4329, 2002.

[8] S. Amor, N. André, P. Gérard, S. Z. Ali, F. Udrea, F. Tounsi, B. Mezghani, L.A. Francis and D. Flandre, "Reliable characteristics and stabilization of on-membrane SOI MOSFET-based components heated up to 335 °C", Semiconductor Science and Technology, vol. 32, pp. 014001, 2016, doi: 10.1088/1361-6641/32/1/014001.

[9] S. Amor, L. V. Brandt, V. Kilchytska, M. Machhout, L. A. Francis and D. Flandre, "Low-Frequency Noise Analysis Of On-Membrane Mosfet And In-Situ Thermal Annealing," DTIP, 2020, doi: 10.1109/DTIP51112.2020.913.

[10] K. O. Petrosyants, I. A. Kharitonov and L. M. Sambursky, "SOI/SOS MOSFET universal compact SPICE model with account for radiation effects," EUROSOI-ULIS 2015: 2015 Joint International EUROSOI Workshop and International Conference on Ultimate Integration on Silicon, 2015, pp. 305-308, doi: 10.1109/ULIS.2015.7063834.

[11] M. M. Pejovic and A. B. Jaksic, "Contribution of fixed oxide traps to sensitivity of pMOS dosimeters during gamma ray irradiation and annealing at room and elevated temperature", Sensors and Actuators, vol. 174, pp. 85-90, 2012, doi: 10.1016/j.sna.2011.12.011

New 10V to 1V level shifter based on new N/PMOS high voltage in FDSOI technology

Ph. Galy
STMicroelectronics
Crolles France
philippe.galy@st.com

Sebastien Haendler
STMicroelectronics
Crolles France
sebastien.haendler@st.com

Abstract— **The main purpose of this study is to introduce a new way to build a high voltage (HV) level shifter based on new HV N/P MOS transistors. The approach is to develop and to characterize dual HV MOS devices in thin silicon film on the 28 nm UTBB FDSOI technology . Afterwards, a basic level shifter design allows to translate 10V on input signal to 1V on the output signal for a direct application in the standard thin oxide MOS device(s). The HV N/P MOS devices and the level shifter characterizations demonstrate that this new concept provides an easy and efficient solution to design at layout level this kind of devices or IP block with the standard process. Thus, this solution is without process extra cost and it is an adaptative design solution for dedicated applications.**

Keywords—FDSOI , HV NMOS, level shifter

I. Introduction

Today, the 28nm high k metal ultra-thin body and buried oxide (UTBB) FDSOI technology allows to design with standard thin and thick oxide MOS transistors. According to the structure of these devices, the power voltages are 1V and 1.8V whatever the type of threshold voltage (regular & low V_T). The fine tune of V_T is possible thanks to poly & back gate bias. To address high voltage application (3V, 5V .. 10V) some design technics are used, especially the stacked MOS transistors solution. In case of high voltage magnitude, the number of stacked devices could be important. By the way, the silicon area is a penalty and moreover the design could be difficult. In this study, we will introduce the concept to build, in the standard process, a adaptative HV MOS device and the simple design of the HV level shifter. A silicon demonstrator is performed in 28 nm UTTB FDSOI and all measurements are at room temperature at silicon probes level.

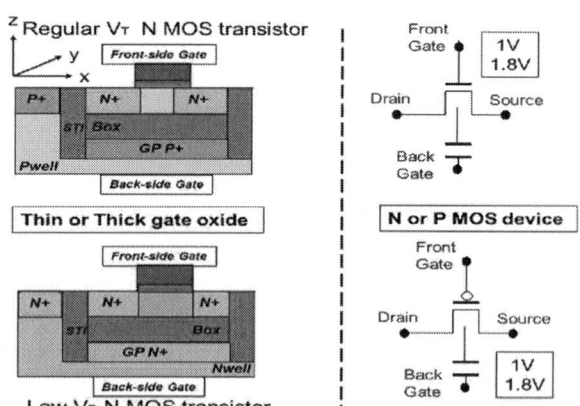

Fig. 1. NMOS device structures in 28nm FDSOI and N/P design symbols.

II. HV N/P MOS devices solution

A. HV device approach in standard FDSOI technology

In this part II, we introduce the concept to design a new high voltage MOS transistor in the thin silicon film in the standard 28 nm FD-SOI technology node [1]. First of all, the MOS core transistor is a sequence of N/P/N (and its dual for PMOS) doping as indicated in figure 2. The gate oxide is performed by the STI process brick. The thickness of the STI allows to adjust the V_T threshold voltage at layout step. Afterwards, on the right and left sides, two doped fingers of silicon active are abutted to perform the left/right gates. When, we applied this approach in 28 nm FDSOI technology, the PMOS is a junction less transistor due to the fact that only Pint body is possible in this standard process. Moreover, it is interesting to note that is possible to modify the V_T flavor of a N/P MOS transistor during the layout step by adjusting the STI thickness and also the active doping value by choosing the proper mask.

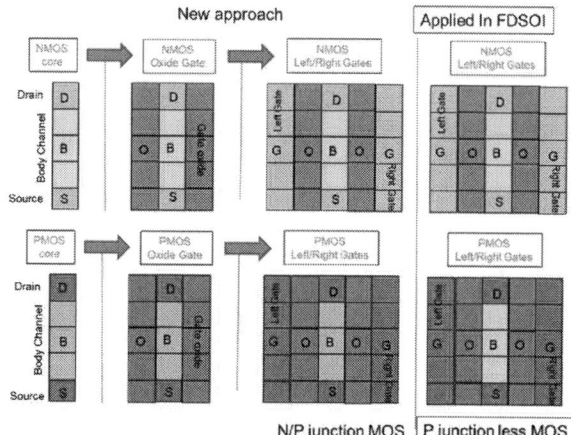

Fig. 2. High voltage N/P MOS transistors in thin silicon film and an application in 28 nm FDSOI technology.

Based on this approach, it is possible to develop other type of devices as to know: a BIMOS transistor, a silicon controlled rectifier (SCR), or a Qubit devices which are describe in [1].

Thus, the next section is a dedicated focus on high voltage N and P MOS transistors fabricated in the 28nm UTBB FDSOI. The silicon demonstrator follows the standard process and all I-V characterizations are performed at wafer level and at room temperature.

978-1-6654-3746-2/21 $31.00 © 2021 IEEE

B. HV N/P MOS in 28 nm FDSOI characterization

Figure 3 depicts the double gates N/P MOS transistors symbol and topology layout in 28nm FDSOI. Based on this technology, dimensions are for gate oxide thickness $T_{ox} = W_{STI} = 50,4nm$, for the MOS width is $W = T_{epi} = (2 \times 18 \text{ nm})$ and for a length of $L = 2,7\mu m$. The back-gate could be with N_{well} or P_{well} doping to have two types of device. The buried oxide thickness is $T_{BOX} = 25nm$ which leads to a threshold voltage $V_T = 4V$ for the back gate. The design layers are: N/P drain/source/gate doping and N_{well}/P_{well} masks for back gates. Also, an un-silicide area in operated in the channel region to avoid a full short between the drain and the source.

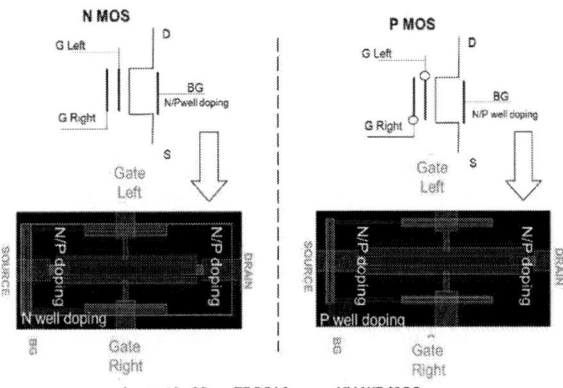

Fig. 3. Electrical schematics and MOS topologies performed in 28nm UTBB FDSOI technology.

The electrical measurements for $I_d(V_g)$ curves are in the range for : $V_g = [-10V, 10V]$, for $Vbg = [-2V, 0V, 2V]$ and for $Vd = 1V$. Here, the two gates are connected together to simplified the electrical exploration ($Vg = Vgr = Vgl$). Figures 4 reports the measurements on the NMOS which demonstrates a good control with a threshold voltage at $V_T = +2V$ for $V_{bg} = 0V$. This threshold voltage is also well modulated by the back gate voltage in forward and reverse bias condition. On this device, the drain leakage current at room temperature is 1pA @ $V_g = 0V$ (blue curve). Afterwards, the curve $I_d(V_d)$ for the range $V_d = [0V, 6V]$ and for $V_g = [0V, 10V]$ are also given. It appears clearly that the front gate sustains 10V without degradation. For $V_g = 0V$ the drain breakdown voltage is around 6V as represented on figure 4.

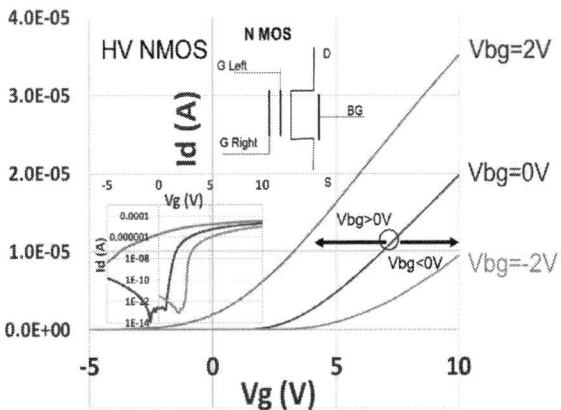

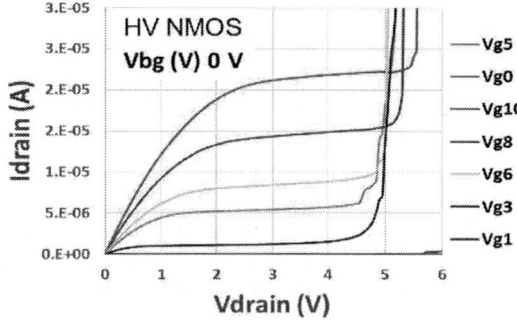

Fig. 4. Measures: $I_d(V_g)$ for Vd=1V & V_{bg} and $I_d(V_d)$ for #V_g on the HV NMOS transistor in 28nm FDSOI

The same type of measurements is carried out on its dual HV PMOS transistor. Extractions show that: the front gate sustains -10V and -30V for the drain breakdown voltage. This high breakdown voltage is due to the fact that the PMOS is a junction-less transistor. The drain leakage current at room temperature is 0,1pA @ $V_g = 0V$. The threshold voltage is -2V and the drain current is less than the NMOS by 3 times. Once again, a V_T modulation is possible thanks to the forward and reverse back gate bias.

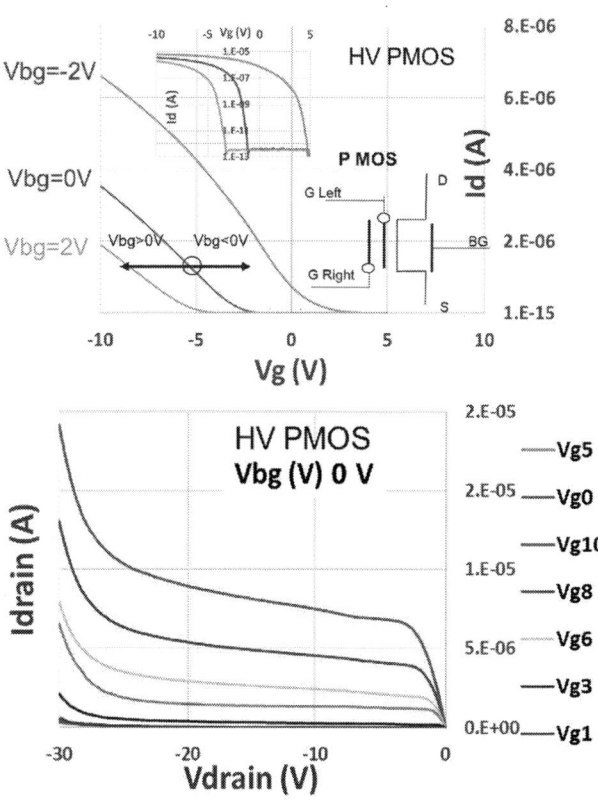

Fig. 5. Measures: $I_d(V_g)$ for Vd=1V & V_{bg} and $I_d(V_d)$ for #V_g on the HV NMOS transistor in 28nm FDSOI

To use this kind of HV devices in a product, the intrinsic ESD robustness is an important point to be addressed and are presented in [2]. Preliminary results on these HV transistors in 28nm FDSOI are also introduced and discussed in [3].

978-1-6654-3746-2/21 $31.00 © 2021 IEEE

III. NEW 10V TO 1V/1.8V LEVEL SHIFTER

These encouraging results obtained on the HV N/PMOS transistors allows to imagine the possibility to design simply a 10V to 1V level shifter in the standard process flow. A simple extension from 10V to 1.8V will be introduce too.

A. Level shifter : design and topology

Basically, the high voltage level shifter is an inverter where the input signal will be a +/-10V excursion, the power voltage Vdd is fixed at 1V to be compliant to the thin oxide MOS device (or could be 1.8V for the thick oxide gate). The output signal will be directly link to Vdd bias (the Vss terminal is the ground). Thus, the design is to use two N&P MOS connected in an inverter function. All front gates are connected to the Vin terminal. The PMOS is the pull up transistor and the NMOS the pull down one. The N_{well} back gate is common for both transistors. The objective is to have one extra bias control. Figure 5 gives the symbol, the design with the two HV MOS and the final layout is also reported.

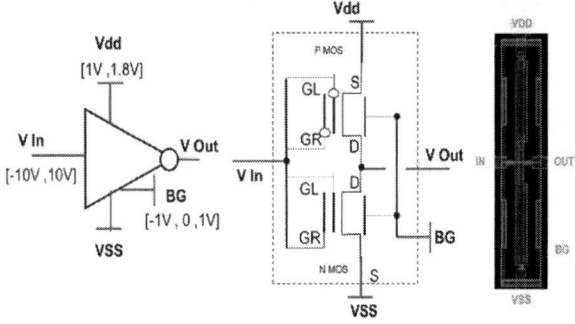

Fig. 6. a) Design of the high voltage level shifter with HV N/P MOS in thin silicon film based on 28 nm FDSOI.

B. From 10V to 1V Level shifter

In this section, the design block on silicon demonstrator is biased at V_{dd}=1V and a -10V to + 10V voltage is applied on the input gate of the level shifter. For these measurements, the back gate voltage is set up at -1V, 0V and +1V to evaluate the transfer function shift according to the previous measurements obtained on N/P MOS transistor (see figure 4 & 5). Two curves are measured on sample and reported on figure 7. The first curve is the V_{out} (V_{in}) transfer function where Vin is in the range [-10V, 10V] and for three values of the back gate. The second reported curve is the current I $_{vdd}$ (V_{in}).

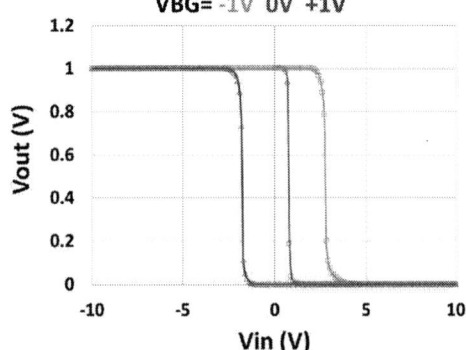

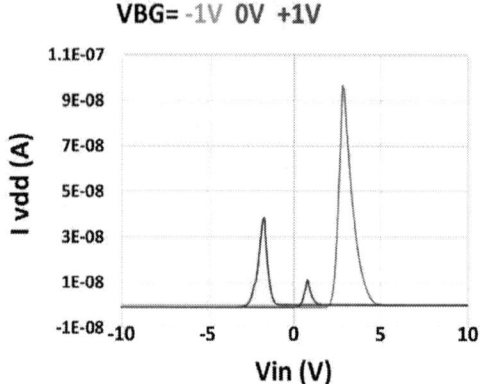

Fig. 7. 10V / 1V level shifter V_{out} (V_{in}) & I_{Vdd} (V_{in}) curves

For V_{bg}=0V, the threshold voltage is at 0.8V which is compliant for a direct application on the core MOS. It appears, by measurement, that for a bias of the back gate between -1V to 1V the threshold voltage of the level shifter is modulated from -1.8V to 2.8V. Moreover, the nominal current peak is 10 nA compared to 38 nA and 94 nA for Vbg=+/-1V. These measurements are performed on several IP blocks and same values are confirmed and no major dispersion is observed.

C. Level shifter bias extension

A natural bias extension for this design is to change the value of the power voltage V_{dd} from 1V to 1.8 V to be compliant with the voltage range of a standard thick oxide gate MOS transistor in this 28 nm FDSOI technology. Figure 8 depicts the two typical curves on this HV level shifter.

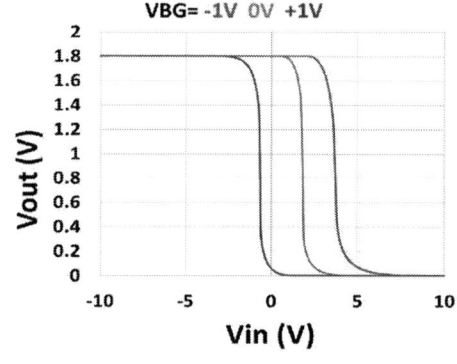

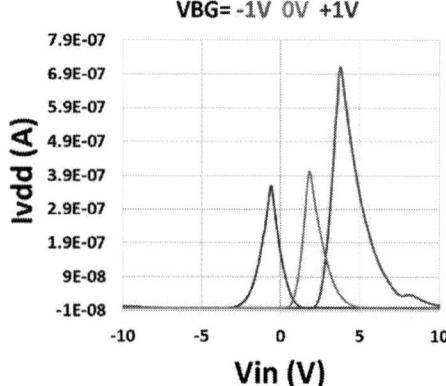

Fig. 8. 10V / 1.8V level shifter V_{out} (V_{in}) & I_{Vdd} (V_{in}) curves

For the transfer function, the measured threshold voltages are: V_T = -0.8V, 1.7V and 3.8V respectively for V_{bg} = -1V, 0V and +1V. It is interesting to notice that the current peak I_{Vdd} is more higher than previously case and leads to 338 nA, 393 nA and 680 nA. for V_{bg} = -1V, 0V and 1V. This solution remains still functional and could be deeply investigated in term of V_{dd} overdrive and back bias control.

IV. CONCLUSION

This study introduces a new high voltage level shifter fabricated in 28 nm UTBB FD-SOI technology. This new solution is based on new high voltage N/P MOS transistors developed in the standard process. All DC measurements are performed at wafer level and at room temperature.

The high voltage N/PMOS devices are within the thin silicon film. The silicon measurements demonstrate good I-V performances and where the back gate control is always operational in the range of [-1V, 1V]. Moreover, the front gate voltage sustains +/- 10V and the threshold voltage V_T could be adjusted by the STI thickness during the design step.

A simple inverter design is proposed to make the 10V to 1V high voltage level shifter. The two N/P MOS are used and connected with common front gates. The silicon measurements show that the transfer function is indeed operational and where the threshold voltage could also be modulated by the back bias too. The monitoring of the current peak is also reported and the maximum current peak is 94 nA at room temperature.

By simply changing the V_{dd} bias of this level shifter, the transfer function is modified and a 10V to 1.8V. It is obtained without extra effort. Also, this power overdrive leads to increase the current peak of I_{Vdd} at 680 nA The next step is to investigate the design optimisation by connecting the back gate to the Vin. Obviously, these approaches could be adapted for the FDSOI next generation node.

REFERENCES

[1] Ph. Galy, I. Kreikouki, S. Rochette, D. Drouin, M. Pioro-Ladrière ; « New 2D/3D integration for device/design applications, preliminary results in 28 nm UTBB FD-SOI at room and cryogenic temperature«. MAM 2020.

[2] Voldman, S., W. Anderson, R. Ashton, M. Chaine, C. Duvvury, T.J. Maloney, E. Worley. "A strategy for characterization and evaluation of ESD robustness of CMOS semiconductor technologies". 2001, Microelectronics Reliability 41(3):335-348. DOI: 10.1016/S0026-2714(00)00236-5

[3] Ph. Galy, B. Jacquier, S. Haendler " Intrinsic ESD robustness on new high voltage N/PMOS devices in 28nm FDSOI UTBB CMOS technology through TLP/VFTLP characterizations" . ESREF 2021

Use of CMOS Image Sensor for early detection of ischemic and haemorrhagic stroke

G. Pignataro[1], P. Cepparulo[1], O. Cuomo[1], A. Cusano[2], R. Rao[4], M. Ruvo[3], F. Palma[4]

[1] Division Pharmacology, Department Neuroscience, University of Naples "Federico II", Napoli, Italy
[2] CeRICT srl, Benevento, Italy
[3] Istituto di Biostrutture e Bioimmagini, CNR, Napoli, Italy
[4] Università di Roma La Sapienza, DIET, Rome, Italy, fabrizio.palma@uniroma1.it

Abstract—We present the development of a lab-on-chip system potentially able to determine specific miRNA levels that enable a differential diagnosis between ischemic and hemorrhagic stroke, through the specialization of CMOS Image Sensors. In particular, the system allows investigations on the photoluminescence of samples of biological liquid to be analyzed (plasma, lysate, biological fluid) following the capture of the specific miRNA by an antisense set of ad hoc designed Peptide Nucleic Acids (PNA) that confers the biological specificity and sensitivity. The CMOS Image Sensor-biochip is modified with a first PNA that captures the target miRNA. A second PNA bringing a fluorescent tag binds the target miRNA enabling detection of the 3-component complex by the CMOS.

Keywords-System on –chip, fluorescence, image sensors, stroke.

I. Introduction

The goal of the study is the development of a biosensing system able to detect specific sets of miRNA whose differential levels can discriminate between ischemic and hemorrhagic stroke. Ideally, the biosensor would provide a low-cost tool for the rapid diagnosis of stroke events outside hospital structures.

Silicon-based Optical Sensors are already key components for medical applications, specifically for diagnosis [1]. They will become more and more important being enablers of many other applications and are used in many development projects, like for example in lab-on-chip or improved endoscopy.

The main enabling technologies are: the classical CMOS Image Sensors (CIS), able capture a scene converting it into a digital 2D format, Photo-diode for X-ray detection, optical sensors for single photon detection, like APD, SPAD and its arrayed architecture SiPM.

II. The need of fast Stroke diagnosis

The triggering event of stroke results from the interruption of blood flow to a portion of the brain, caused

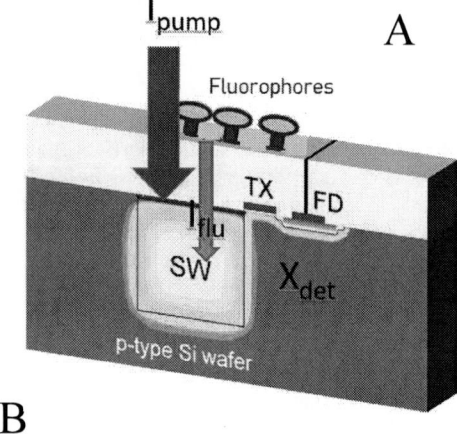

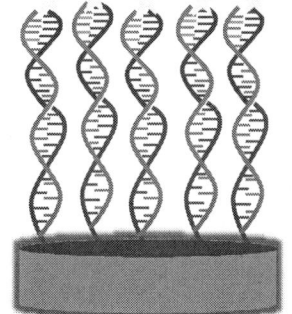
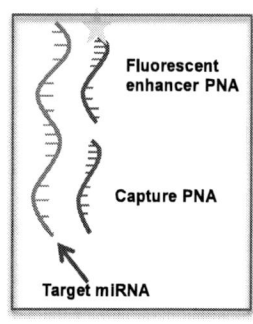

Figure 1. **A**. Sketch of the filtering structure in a Pinned Photo Diode. violet curve depicts the absorption of the radiation of the short wavelength pump, while green curve indicates the photo-emitted radiation, which can reach the SW collection region. **B**. Schematic of the PNA-miRNA sandwich array used to detect the target miRNAs.

by two possible events: hemorrhagic or ischemic stroke [2, 3]. The first case is due to the rupture of a cerebral vessel and accounts for 15% of all strokes. On the other side, in the ischemic stroke a thrombus or embolus are responsible for the obstruction of the vessel that supplies the brain, accounting for 85% of cases.

According to the Global Burden of Disease Study, performed from 1990 to 2013, stroke is the second main cause of death (representing the 11.8% of all deaths

978-1-6654-3746-2/21 $31.00 © 2021 IEEE

worldwide) and the third leading cause of disability-adjusted life years (DALYs) worldwide [4-6].

To date, therapeutic interventions acting on the activity and/or expression of individual genes have yielded negative results, and the only effective treatment for ischemic stroke is limited to recombinant tissue plasminogen activator (tPA) (National Institute of Neurological Disorders and Stroke rt-PA Stroke Study Group, 1995). Unfortunately, treatment benefit is time-dependent, because of the short therapeutic time window, related to stroke pathophysiology and penumbra restoring; therefore the majority of patients can't receive this therapy. Currently, it's considered to be a good prospect of success acting within 4.5 hours after the onset of symptoms [7].

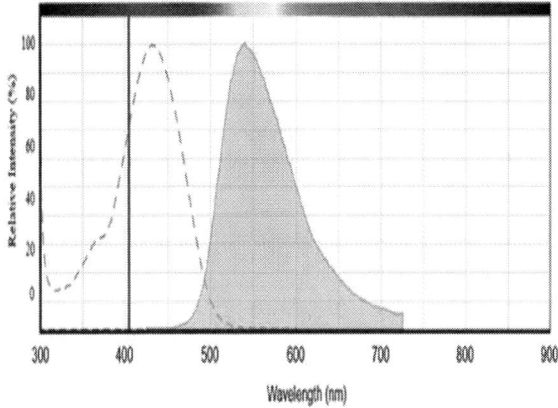

Figure 2. Absorption (dashed line) and emission (continuous line) bands of Alexa Flour 430 (xxx). The violet line represents the emission of 405 nm laser diode. Reproduced with permission [15]

Modern neuroimaging tools, such as computed tomography (CT) or magnetic resonance imaging (MRI), are now used to diagnose a stroke and identify its subtype [8]. However, some hospitals do not yet provide MRI and CT service and many patients cannot undergo these techniques. Moreover, time required to reach the medical center and to prepare patients for imaging tests often does not match with the urgency of an early diagnosis for an immediate therapy. In addition, the application of these imaging tools weighs on stroke care costs [9]. For all these reasons, the search of biomarkers is critically important to speed up the diagnosis of stroke and selectively distinguish cerebral ischemia from the hemorrhagic damage, in order to ensure therapeutic interventions in a very short time frame from the onset of disease.

MicroRNAs (miRNAs) are evolutionarily conserved non-coding RNA in size of 20-22 nucleotides, synthesized through processing by both nuclear and cytosolic proteins [10], and are important downregulators of gene expression via mRNA cleavage or translational repression [10].

Over the last 10 years, the role of miRNAs in stroke has been widely discussed and evaluated, focusing attention on the regulation of stroke risk factors [11] and mechanisms activated and elicited by the ischemic insult [3].

The presence of microRNA in blood and the ability to measure their levels in a non-invasive way has opened new doors in the search for peripheral biomarkers for the diagnosis and prognosis of diseases such as brain ischemia. Since the recommended therapeutic window is very limited, biomarkers for stroke have the potential to expedite diagnosis and institution of treatment. Moreover, in the last decade it has been evidenced that expression levels of miRNAs in blood are reproducible and indicative of several diseases [12].

Goal of the developing system is to allows to make a differential diagnosis between ischemic and hemorrhagic stroke with the aim of providing a low-cost tool for obtaining rapid diagnosis even outside of important hospital structures.

Silicon-based Optical Sensors are already key components for medical applications, specifically for diagnosis. They will become more and more important being enablers of many other applications and are used in many development projects, like for example in lab-on-chip or improved endoscopy.

III. THE LAB-ON-CHIP STRUCTURE

The structure of a modern pinned photodiode used in CMOS image sensors is shown in Figure 1A. The main elements are: an n-type buried signal charge storage well (SW) region, sandwiched between a lower p-type layer and a $p+$ pinning layer at the top surface in contact with the lower active layer, a transfer gate, TX, and an $n+$ output floating diffusion, FD.

In order to show the concept, we describe the case of front side illumination (FSI) architecture; the light enters the top surface and is absorbed in the pnp layers in accordance with the wavelength dependent absorption coefficient. The violet light of the pump, I_{pump}, is blocked by the $p+$ pinning layers, above the SW. The depth of the SW, X_{det}, plays in this case a relevant role. The fluorescence light, I_{flu}, on the contrary is substantially absorbed in the pinning layer, and carriers generated can be collected into the SW by diffusion and drift without appreciable recombination and signal loss. Green light is absorbed with good quantum efficiency and collection efficiency, though the doping can be tailored.

The backend of CMOS integrated circuits is usually covered by a top layer of silicon dioxide. This makes the surface particularly suitable for functionalization with biomolecules In particular the process include Peptide Nucleic Acids deposited following a pre-treatment for the introduction of anchoring chemical groups (See Figure 1B).

978-1-6654-3746-2/21 $31.00 © 2021 IEEE

Surface modification using (3-aminopropyl) triethoxysilane (APTES), producing a terminal amine group (-NH₂), has been found to be useful for covalent coupling of protein to the surface of the silica materials [13, 14]. The process involves Piranha cleaning H2SO4:H2O2(3:1), silanization with APTES in Et-OH/DI water solution (95/5, v/v), washing step in Et-OH and sonication plus soft baking at T = 110 °C for 1h. The fluorophore bound to the enhancer PNA is brought to the CMOS surface only in presence of the target miRNA. We find as particularly suitable fluorophore the Alexa Fluor 430. Absorption band has a maximum at 430 nm, nevertheless a common InGaN laser diodes. at 405 nm, undergoes a relatively large absorption, while it ensures a large spectral separation from the emission band.

As shown in Figure 2, the large gap between the spectrum of the laser and the mission of fluorescence, permits to design the sequence of the semiconductor and dielectric layer on the back end of the CIS such as to insure a strong rejection of detected photocurrent arising from the pump radiation. Figure 3 reports the photo current, simulated in a pinned structure with 1 mW cm⁻³ radiation. It shows a very good rejection of the pump light at 405 nm, with respect to the fluorescence light peaked at 540nm.e 2, the large gap between the spectrum of the laser and the mission of fluorescence, permits to design the sequence of the semiconductor and dielectric layer on the back end of the CIS such as to insure a strong rejection of detected photocurrent arising from the pump radiation. Figure 3 reports the photo current, simulated in a pinned structure with 1 mW cm⁻³ radiation. It shows a very good rejection of the pump light at 405 nm, with respect to the fluorescence light peaked at 540nm.

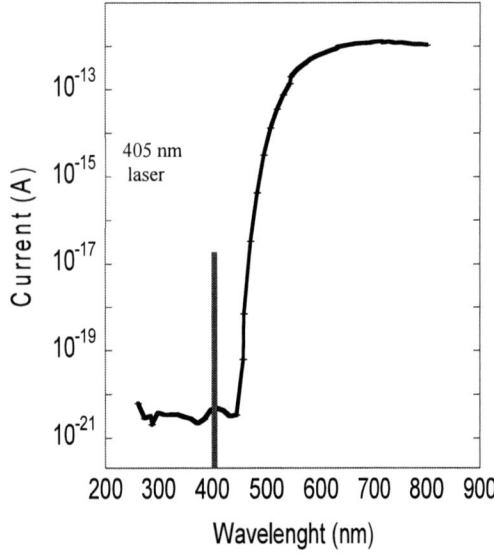

Figure 3. Simulated photocurrent with 1 mW radiation obtained in a Pinned Photo Diode structure. Line indicates the laser emission.

TCAD Sentaurus Device [16] performs simulations using the hydrodynamic equations, based on the work of Stratton [17] and Bløtekjær [18], adopting a simplified approach of six Partial Differential Equation [19]. The transistor source and drain diffusions have phosphorous peak doping concentration $1.4 \times 10^{+20}$ cm⁻³, with a Lightly Doped Diffusion implantation peak $4. \times 10^{+19}$ cm⁻³. Polysilicon gate is phosphorous doped 10+20 cm⁻³.

CONCLUSIONS

We presented the structure and the design of a lab-on-chip system for make a differential diagnosis between ischemic and hemorrhagic stroke, through the specialization of CMOS Image Sensors. The structure combines the high sensitivity, characteristics of CIS with the filtering capability obtained by a proper design of the doping structure.

Financial support. The support from project NEON, grant n° ARS001_00769 is acknowledged.

REFERENCES

[1] E.R. Fossum, D.B. Hondongwa, "A Review of the Pinned Photodiode for CCD and CMOS Image Sensors," Electron Devices Society, IEEE Journal of the, vol. 2, no. 3, pp. 33-43, 2014.

[2] Donnan GA, Fisher M, Macleod M, Davis SM. Stroke. Lancet. 2008 May 10;371(9624):1612-23.

[3] Khoshnam SE, Winlow W, Farzaneh M, Farbood Y, Moghaddam HF. Pathogenic mechanisms following ischemic stroke. Neurol Sci. 2017 Jul;38(7):1167-1186

[4] Hankey GJ. Stroke. Lancet. 2017 Feb 11;389(10069):641-654.

[5] Feigin VL, Barker-Collo S, Krishnamurthi R, Theadom A, Starkey N. Epidemiology of ischaemic stroke and traumatic brain injury. Best Pract Res Clin Anaesthesiol. 2010 Dec;24(4):485-94.

[6] Feigin VL, Norrving B, Mensah GA. Global Burden of Stroke. Circ Res. 2017 Feb 3;120(3):439-448.

[7] Emberson J, Lees KR, Lyden P, Blackwell L, Albers G, Bluhmki E, Brott T, Cohen, G, Davis S, Donnan G, Grotta J, Howard G, Kaste M, Koga M, von Kummer R, Lansberg, M, Lindley RI, Murray G, Olivot JM, Parsons M, Tilley B, Toni D, Toyoda K, Wahlgren N, Wardlaw J, Whiteley W, del Zoppo GJ, Baigent C, Sandercock P, Hacke W; Stroke Thrombolysis Trialists' Collaborative Group. Effect of treatment delay, age, and stroke severity on the effects of intravenous thrombolysis with alteplase for acute ischaemic stroke: a meta-analysis of individual patient data from randomised trials. Lancet. 2014 Nov 29;384(9958):1929-35.

[8] Rudkin S, Cerejo R, Tayal A, Goldberg MF. Imaging of acute ischemic stroke. Emerg Radiol. 2018 Dec;25(6):659-672.

[9] Burke E, Dobkin BH, Noser EA, Enney LA, Cramer SC. Predictors and biomarkers of treatment gains in a clinical stroke trial targeting the lower extremity. Stroke. 2014 Aug;45(8):2379-84.

[10] Hammond SM. An overview of microRNAs. Adv Drug Deliv Rev. 2015 Jun 29;87:3-14

[11] Koutsis G, Siasos G, Spengos K. The emerging role of microRNA in stroke. Curr Top Med Chem. 2013;13(13):1573-88.

[12] Viswambharan V, Thanseem I, Vasu MM, Poovathinal SA, Anitha A. miRNAs as biomarkers of neurodegenerative disorders. Biomark Med. 2017 Feb;11(2):151-167

[13] J Gass, T.; Ligler, F. S. Immobilized Biomolecules in AnalysissA Practical Approach; Oxford University Press: New York, 1998.

[14] Subramanian, A.; Kennel, S. J.; Oden, P. I. Enzyme Microb. Technol. 1999, 24, 26-34. I. S. Jacobs and C. P. Bean, "Fine

particles, thin films and exchange anisotropy," in Magnetism, vol. III, G. T. Rado and H. Suhl, Eds. New York: Academic, 1963, pp. 271–350.

[15] AAT Bioquest, Inc. https://www.aatbio.com/spectrum/Alexa_Fluor_4.

[16] Sentaurus Device, Synopsys User Manual, Version Y-2006.06, 2006.

[17] Stratton, R. "Diffusion of hot and cold electrons in semiconductor barriers", Physical Review, 126(6), 1962, 2002-2013. DOI: 10.1103/PhysRev.126.2002

[18] Blotekjaer, K. "Transport equations for electrons in two-valley semiconductors" IEEE Transactions on Electron Devices, 17(1), 1970, pp. 38-47. DOI: 10.1109/T-ED.1970.16921

Szeto, S., & Reif, R. "A unified electrothermal hot-carrier transport model for silicon bipolar transistor

High performance silicon-based substrate using buried PN junctions towards RF applications

Maxime Moulin[#1], Martin Rack[*], Thibaud Fache[#], Zdenek Chalupa[#], Christophe Plantier[#], Yves Morand[#], Joris Lacord[#], Frédéric Allibert[$], Fred Gaillard[#], Jose Lugo[#], Louis Hutin[#] and Jean-Pierre Raskin[*]

[#]CEA, LETI, Minatec Campus, Université Grenoble Alpes, F-38000, Grenoble, France

[*]Institute of Information and Communication Technologies, Electronics and Applied Mathematics (ICTEAM), Université Catholique de Louvain, Belgium

[$]SOITEC, Bernin, France

[1]maxime.moulin@cea.fr

Abstract — **This paper shows the potential of buried PN junctions as a substrate interface passivation solution to increase the effective resistivity (ρ_{eff}) figure of merit of a High-Resistivity (HR) substrate suffering from Parasitic Surface Conduction layer (PSC). We characterize Coplanar Waveguides (CPW) in order to monitor the substrate frequency response. We demonstrate that this method can be implemented using an industrial process with an effective resistivity reaching 2kΩ.cm with 0.1 dB/mm loss at 6 GHz for a HR+PN substrate. This method is suitable for local PSC passivation, targeting advanced SoC (System-on-Chip) in FD-SOI technology for next wireless communications generations.**

Keywords — **PN junctions; HR-silicon; parasitic surface conduction; RFSOI; CMOS; substrate resistivity; CPW.**

I. INTRODUCTION

Growing complexity and demand for power efficiency at a lower cost challenge the integration of all electronic functions on a single chip. Silicon substrate losses and non-linearity are limiting characteristics of active and passive devices [1]. HR-Silicon substrates aim to reduce RF losses [2]. The presence of positive fixed charges in the top oxide layer [3] bend the conduction band close to the Fermi level, creating an inversion layer named as Parasitic Surface Conduction (PSC), leading to a drastic reduction of the substrate effective resistivity (ρ_{eff}) [4]. The introduction of a polysilicon layer, known as trap-rich layer, between the top oxide layer and the Si solves this problem [5], since the many traps can ionize to compensate the fixed oxide charge while ensuring an interface Fermi level deep in the band gap. In the case of the co-integration of hybrid/logic/active components in FDSOI technology, substrate interface passivation must be localized to avoid performance degradation of active and hybrid components. Recently, a novel passivation method was introduced, which consists in creating a chain series of PN junctions beneath the insulator. This approach generates highly resistive depleted regions, which interrupt the PSC and thus increase the effective resistivity, mitigating substrate losses and reducing harmonic distortion in HR-Si substrates [6-7]. In this paper, we introduce a smaller implantation pitch to achieve higher RF-substrate performance that can be comparable to trap-rich wafers [8], We consider that we have an HR-SOI substrate from which we remove SOI and BOX (Buried Oxide), giving conditions mimicking fully

processed NOSO-like (Not On Silicon Oxide) areas with a 28nm process flow.

Our paper is organized as follows, Section II introduces the test structure definitions which describe how the CPW devices used to characterize the substrate performance were built, then in Section III we discuss about small-signal simulation and experimental results with different conditions and PN junctions pattern that show a very good enhancement on RF Figures of merit (RF FoMs), based on this approach.

II. TEST STRUCTURES DEFINITION

Starting from wafers of different nominal resistivity $\rho_{nominal}$ (10-15 Ω·cm, 2.2-2.5 kΩ·cm) and implantation conditions of alternating P+ (B at 15 and 20 keV) and N+ areas (P at 35 and 45 keV) with a dose variation, respectively of, $5 \cdot 10^{12}$, $1 \cdot 10^{13}$, $2 \cdot 10^{13}$ and $4 \cdot 10^{13}$ cm^{-2} were performed through a 25 nm silicon dioxide layer and a lateral periodicity of 324nm. A 1050°C spike activation anneal was performed and a stack of dielectrics representative of a Back-End-Of-Line (BEOL) process with 2 Metal Layers in FDSOI 28nm was deposited; the CPW devices were then defined on top of this dielectric.

III. SMALL-SIGNAL SIMULATION AND EXPERIMENTAL RESULTS

A TCAD deck was built on Synopsys Sentaurus software, lumping the dielectric stack into an equivalent single layer of thickness 1.75μm and relative permittivity 3.9 (Fig.1).

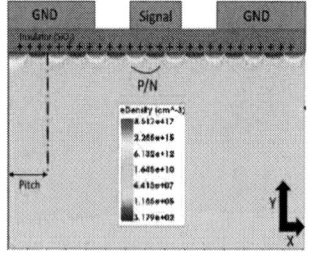 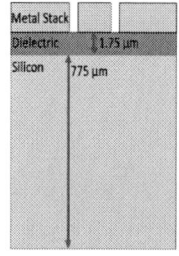

Fig. 2. Representation of the TCAD 2D electron concentration profile in a HR-Si substrate with buried PN junctions and its equivalent stacked materials (not to scale).

The formation of a series of PN junctions and their associated depletion regions lead to locally enhanced resistivity

978-1-6654-3746-2/21 $31.00 © 2021 IEEE

that interrupts the PSC layer as depicted in Fig. 2. We leverage this effect in order to have the highest depletion zone density at the top-Si. We compute the local resistivity of such a structure including local PN junctions underneath the oxide with different pitches using equation (1). We obtain an average local resistivity of 2.4 kΩ.cm (2µm) and 13.7 kΩ.cm (324 nm) multiplying $\rho_{local,avg}$ by 5 for a pitch reduced by 6.

$$\rho = [q(n\mu_n + p\mu_p)]^{-1} \qquad (1)$$

where, ρ is the nominal resistivity of the silicon, obtained by the inverse of adding the product of the electronic charge, q, the carrier mobility, and the density of carriers of each carrier type.

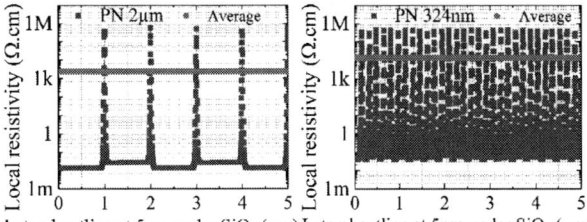

Fig. 2. Local resistivity from lateral cutline at 5 nm underneath the oxide with two different pitches.

A. *SMALL-SIGNAL PERFORMANCE*

We extracted resistivity profiles from our previously described TCAD structure in order to perform small-signal AC simulations. The substrate effective resistivity (ρ_{eff}) was computed from the measured line RLCG parameters by making use of Heinrich's CPW conformal mapping model [9]. On-wafer small-signal measurements of CPW lines were performed using a Signatone prober, a PNA-X and a pair of GSG Infinity Probes (pitch 100 µm). The setup was calibrated on an Impedance Standard Substrate (ISS) through an LRRM (Line Reflect Reflect Match) correction. From the measured S-parameters we extracted the effective resistivity ρ_{eff} [4] and CPW losses α from 100 MHz up to 12 GHz using a de-embedding method proposed in [10]. Results are shown below.

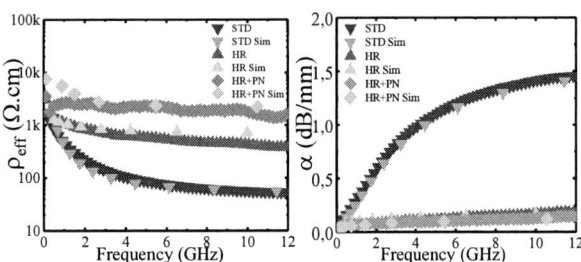

Fig. 3. Electrical characteristics extracted from S-parameters measurements of CPW lines, compared to TCAD simulations. The HR+PN measurements and simulations correspond to the following ion implantation conditions: (2.10^{13} cm^{-2}, 15 keV) for B and (2.10^{13} cm^{-2}, 35 keV) for P.

Due to the PSC located at the interface between the substrate and the dielectric layer, the HR substrate presents in general a reduced performance: the effective resistivity drops under 1

kΩ.cm for frequencies higher than 1 GHz, down to 400 Ω.cm at 10 GHz, against a nominal resistivity of 2.5 kΩ.cm therefore mitigating the potential of this type of substrate. However, the HR+PN substrate with a high density of PN junctions offers a major enhancement of the effective resistivity, keeping its value around 2 kΩ.cm for frequencies up to 10 GHz compared to [9] which decreases when the frequency increases. The performance is multiplied by 4 and 30 at 6 GHz compared to HR and STD substrates, which is a key frequency for 5G applications.

B. *FREQUENCY STABILITY AND DIELECTRIC DEPENDANCE*

The densification of PN junctions serves on the one hand to increase the RF FoMs but on the other hand to stabilize the frequency of the RF FoMs, pushing back the resistivity drop at higher frequencies [6-7]. This can be understood in terms of a higher count of depletion capacitances between the CPW and substrate lowering the overall capacitive coupling. However, the relative importance of this effect should also vary with the thickness of the BEOL dielectric T_{ox}. As shown on Fig. 4, a hypothesis of T_{ox}=200nm as in [6] indicates a substantial improvement of ρ_{eff} at high frequency can be expected by scaling the PN junctions pitch to sub-micron values. In our case however with thicker dielectric (T_{ox}=1.75µm), while buried PN junctions still significantly contribute to improving ρ_{eff}, the pitch dependence appears significantly weakened.

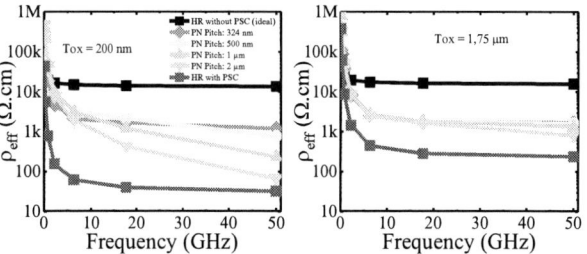

Fig.4. Electrical characteristics extracted from S-parameters TCAD of CPW lines with different dielectric thicknesses in simulation, ideal HR, HR suffering from PSC and HR+PN with varying pitch. Left: Tox = 200 nm as in [6]. Right: Tox = 1.75 µm.

C. *SMALL-SIGNAL PERFORMANCE VS. DOSE*

In the previous sub-section, we showed that small-signal performance was enhanced for the HR+PN substrate for given doping conditions. Frequency measurements of ρ_{eff} for various ion implantation doses are shown on Fig. 5.

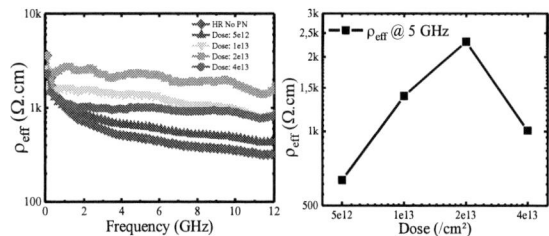

Fig.5. Left: Electrical characteristics extracted from S-parameters measurements of CPW lines with varying implanted dose. The dose is the same for B and P, with energies respectively of 15 keV and 35 keV. Right: Effective resistivity at 5 GHz versus implantation dose.

978-1-6654-3746-2/21 $31.00 © 2021 IEEE

Qualitatively, one should expect ρ_{eff} to keep increasing as the dose is lowered within this range, because the depletion region of the PN junctions becomes wider. Yet, the graph shows non-monotonous behavior with an optimum at $2 \cdot 10^{13}$ cm^{-2}. To explain this discrepancy, 2D TCAD-based process simulation was used (Fig. 6), highlighting the importance of doping geometry. Indeed as the P and N implantation depths become comparable to half of the implantation pitch, a mismatch between depth profiles may cause the conductive pockets of one polarity to laterally merge underneath the others. This is visible on Fig. 6 (a) and (c), where reducing the dose to $5 \cdot 10^{12}$ cm^{-2} counter-intuitively leads to a moderately doped n-type channel forming under the p-type regions. Fig. 6 (b) and (d) illustrate that this effect disappears by relaxing the pitch to 1µm.

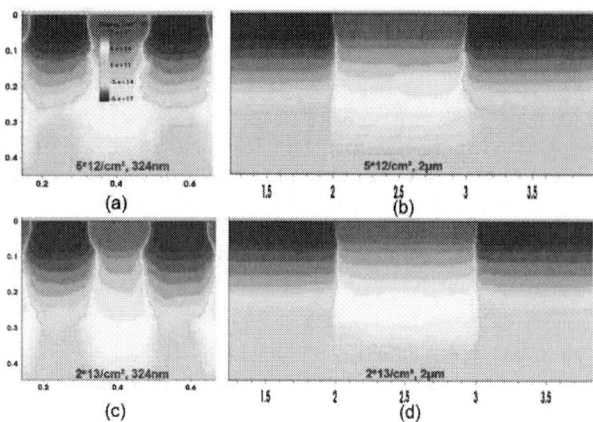

(a)
(b)
(c)
(d)

Fig.6. 2D-TCAD Process PN junctions simulation view. (a) 324 nm PN region pitch with a $5 \cdot 10^{12}$ cm^{-2} dose. (b) 2 µm PN region pitch with as $5 \cdot 10^{12}$ cm^{-2} dose. (c) 324 nm PN region pitch with a $2 \cdot 10^{13}$ cm^{-2} dose. (d) 2 µm PN region pitch with a $2 \cdot 10^{13}$ cm^{-2} dose.

D. SMALL-SIGNAL PERFORMANCE VS. PN JUNCTIONS PATTERNS

The patterns serve to obstruct the electrical field propagation in unwanted directions. In this subsection, we show the quantitative impact of matching the PN pattern to the geometry of the passive component for improving the RF performance. In the case of a CPW, we seek to minimize substrate-mediated resistive coupling between signal and ground. It is therefore intuitive to orient PN stripes parallel to the signal line (Fig. 7 (a)). On the other hand, we can expect that drawing PN stripes perpendicular to the CPW direction should be suboptimal (Fig. 7 (b)). Finally, we can consider another type of configuration decoupled from the linear geometry of the component above, such as a checkerboard (Fig. 7 (c)).

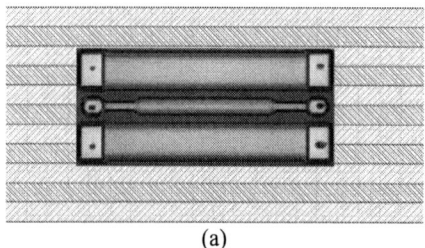

(a)

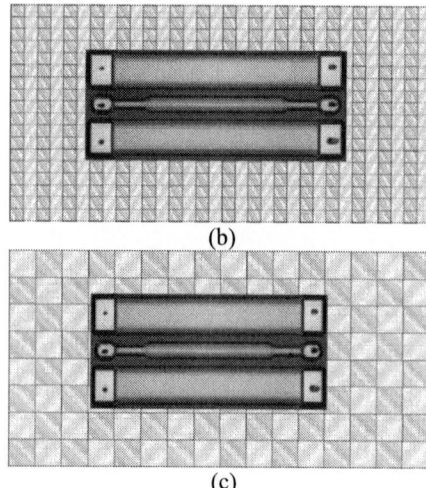

(b)

(c)

Fig.7. PN junctions patterns under CPW device. (a) Parallel PN junctions, (b) Perpendicular PN junctions, (c) Checkerboard PN junctions. These figures are not to scale. On our samples, the ratio between CPW gap and PN pitch is above 80.

The small-signal performance of these different patterns are detailed in Fig. 8.

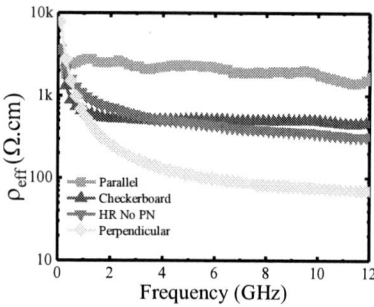

Fig.8. Electrical characteristics extracted from S-parameters measurements of CPW lines with a different PN junctions patterns. The dose is the same for B and P, with energies respectively of 15 keV and 35 keV.

As expected, the parallel configuration yields the best performance. The perpendicular junction design, however, has a catastrophic impact on ρ_{eff} at high frequency, making the doping approach counterproductive. Unfortunately, the checkerboard shows little to no improvement compared to an undoped HR substrate.
In conclusion, the benefits of this approach is strongly layout-dependent, and the patterning of PN junctions should be advantageously adapter to that of the overarching passive component.

IV. CONCLUSION

We have shown a major substrate enhancement technique enabling high quality silicon substrate in terms of RF FoMs by using a process flow at 28nm design rules. PN passivation offers a local trap-rich like effect, suitable for PD-SOI, FD-SOI and bulk technologies. These results extend work by Rack *et al.* on a more aggressive scale and validate the relevance of the

978-1-6654-3746-2/21 $31.00 © 2021 IEEE

approach. Nevertheless, experimental results proved that for a passive design, PN junctions must be oriented in the signal-ground conduction path in order to take full advantage of this substrate passivation technique.

ACKNOWLEDGMENT

The authors acknowledge the support of European Commission, French State and Auvergne-Rhône Alpes region through the funding of ECSEL project BEYOND5 part of IPCEI (Important Project of Common European Interest) microelectronics and French Nano2022 program.

REFERENCES

[1] F. Allibert, L. Andia, Y. Morandini, C. Veytizou, M. Rack, L. Nyssens, Jean-Pierre Raskin, E. Augendre, "Engineering SOI Substrates for RF to mmWave Front-Ends", in Microwave Journal, Oct. 2020.

[2] J.-P. Raskin, A. Viviani, D. Flandre and J. -. Colinge, "Substrate crosstalk reduction using SOI technology," in IEEE Transactions on Electron Devices, vol. 44, no. 12, pp. 2252-2261, Dec. 1997, doi: 10.1109/16.644646.S. Zhang, C. Zhu, J. K. O. Sin, and P. K. T. Mok, "A novel ultrathin elevated channel low-temperature poly-Si TFT," *IEEE Electron Device Lett.*, vol. 20, pp. 569–571, Nov. 1999.

[3] Yunhong Wu, S. Gamble, B. M. Armstrong, V. F. Fusco and J. A. C. Stewart, "SiO/sub 2/ interface layer effects on microwave loss of high-resistivity CPW line," in IEEE Microwave and Guided Wave Letters, vol. 9, no. 1, pp. 10-12, Jan. 1999, doi: 10.1109/75.752108.

[4] D. Lederer and J.-P. Raskin, "Effective resistivity of fully-processed SOI substrates. Solid-State Electronics", 2005, vol. 49, no 3, p. 491-496.

[5] D. Lederer and J. -. Raskin, "New substrate passivation method dedicated to HR SOI wafer fabrication with increased substrate resistivity," in IEEE Electron Device Letters, vol. 26, no. 11, pp. 805-807, Nov. 2005, doi: 10.1109/LED.2005.857730.

[6] M. Rack, L. Nyssens and J. -. Raskin, "Low-Loss Si-Substrates Enhanced Using Buried PN Junctions for RF Applications," in IEEE Electron Device Letters, vol. 40, no. 5, pp. 690-693, May 2019, doi: 10.1109/LED.2019.2908259.

[7] M. Rack, L. Nyssens and J. -. Raskin, "Silicon-substrate enhancement technique enabling high quality integrated RF passives," 2g019 IEEE MTT-S International Microwave Symposium (IMS), 2019, pp. 1295-1298, doi: 10.1109/MWSYM.2019.8701095.

[8] K. Ben Ali, C. Roda Neve, A. Gharsallah and J. -. Raskin, "RF SOI CMOS technology on commercial trap-rich high resistivity SOI wafer," 2012 IEEE International SOI Conference (SOI), 2012, pp. 1-2, doi: 10.1109/SOI.2012.6404404.

[9] W. Heinrich, "Quasi-TEM description of MMIC coplanar lines including conductor-loss effects," in IEEE Transactions on Microwave Theory and Techniques, vol. 41, no. 1, pp. 45-52, Jan. 1993, doi: 10.1109/22.210228.

[10] A. M. Mangan, S. P. Voinigescu, Ming-Ta Yang and M. Tazlauanu, "De-embedding transmission line measurements for accurate modeling of IC designs," in IEEE Transactions on Electron Devices, vol. 53, no. 2, pp. 235-241, Feb. 2006, doi: 10.1109/TED.2005.861726.

[11] Martin Rack, PhD Thesis, "Modelling of Advanced Silicon-Based Substrates for RF and mm-Wave Applications", 2021.

978-1-6654-3746-2/21 $31.00 © 2021 IEEE

Si/Si$_{0.7}$Ge$_{0.3}$ A2RAM nanowires fabrication and characterization for 1T-DRAM applications

J. Lacord[1], F. Tcheme Wakam[1,2], Z.Chalupa[1], J.-M. Hartmann[1], P. Besson[1], V. Loup[1], C. Vizioz[1], L. Brevard[2], F. Aussenac[1], X. Mescot[2], K. Lee[2] and M. Bawedin[2]

(1) Univ. Grenoble Alpes, CEA, LETI, 38000 Grenoble, France
(2) Univ. Grenoble Alpes, IMEP-LAHC, Grenoble INP Minatec, CNRS, F-38000 Grenoble,
joris.lacord@cea.fr

Abstract—**A2RAM devices are fabricated using an adaptation of Si-Nanowire process flow. They include a Si-SiGe heterostructure to improve memory performance. Even the device structure is not exactly what we expect, we succeed to evidence 1T-DRAM programming.**

Keywords: 1T-DRAM, A2RAM, electrical characterization;

I. Introduction

1T-DRAM has been proposed more than 25 years ago [1]. In general, 1T-DRAMs take advantage of the floating-body effects usually considered in CMOS technology as parasitic and detrimental. A2RAM is one of them: it is a transistor on SOI with a doped buried layer (the bridge) short circuiting the source and the drain (Fig.1). The '1' state is defined by the presence of an excess of majority carriers in the undoped part of the silicon film (the body) while the '0' state corresponds to a body empty of charge. The charges stored in the body modulates the current flow through the bridge: a high amount of charge leads to '1' state and high read current (I1). Conversely, in '0' state, the bridge tends to be fully depleted and the current I0 is negligible. Programming '1' is usually achieved by charge generation via band-to-band tunneling (B2B) at drain-body junction [2]. Erasing is performed through capacitance coupling that evacuates the holes from the body. Its concept has been experimentally demonstrated in [3] in planar SOI devices. The aim of this paper is to improve the A2RAM performance previously shown thanks to a more advance process flow: scaled dimensions to improve gate electrostatic control and Si$_{0.7}$Ge$_{0.3}$ body

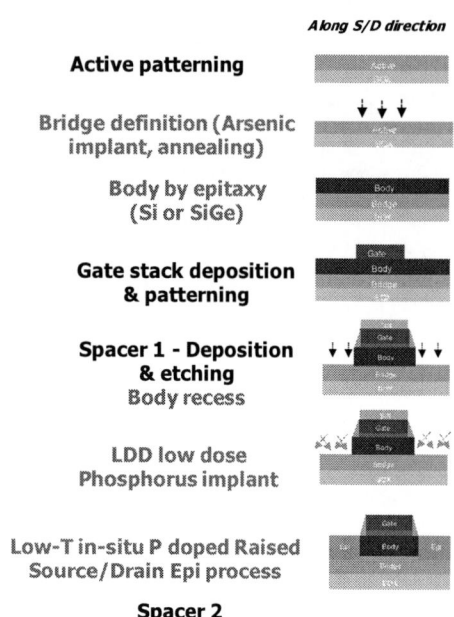

Figure 2: A2RAM process flow based on nanowire process flow [3]. Process steps in red are specific to A2RAM.

and Si-bridge heterostructure to have a deeper potential well to enhance charge storage in the body [4].

II. Device Fabrication

A2RAM devices are fabricated using an adaptation of Leti's nanowire 300mm process flow [5]: A2RAM specific process steps, mainly to build the bridge (in red on Fig. 2) had all been defined by TCAD simulations. After active patterning, bridge implantation and activation was performed, targeting a uniform doping at 1e18cm^{-3}. Indeed, in our previous study, we demonstrated by TCAD that the I1/I0 ratio is maximized for a bridge uniformly doped at 1e18cm^{-3} [6]. Then body was formed by Si$_{0.7}$Ge$_{0.3}$ epitaxy. Fig. 3 shows a TEM cross section of the SiGe body – Si bridge stack. Our previous studies [6, 7] show the major impact of body and bridge thicknesses. We therefore fabricated A2RAM devices with 3 different

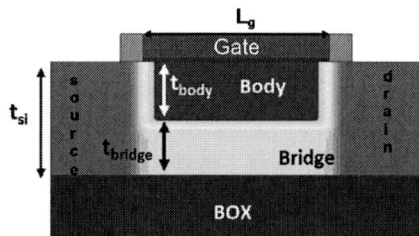

Figure 1: A2RAM device sketch with the main technological parameters.

978-1-6654-3746-2/21 $31.00 © 2021 IEEE

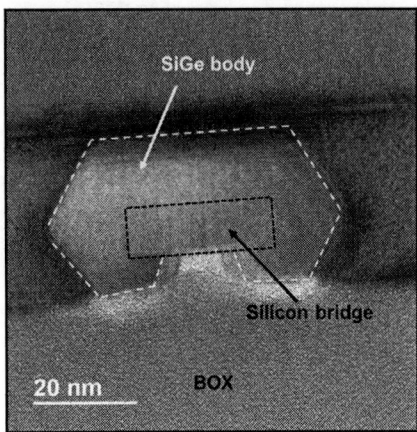

Figure 3: A2RAM TEM cross section just after body epitaxy

body-bridge stacks (10nm/10nm, 10nm/15nm and 15nm/10nm). To control the bridge doping profile (mandatory for A2RAM performance [6]), we had to limit dopant diffusion, so the thermal budget after the bridge/body stack formation. That was why source and drain are fabricated in 2 steps after the High-K/TiN/Polysilicon gate fabrication. First, a recess plasma etching of SiGe-body selectively stopped on the Si-bridge was conducted. Then, a low temperature (725°C) in-situ phosphorus doped silicon selective epitaxy was performed.

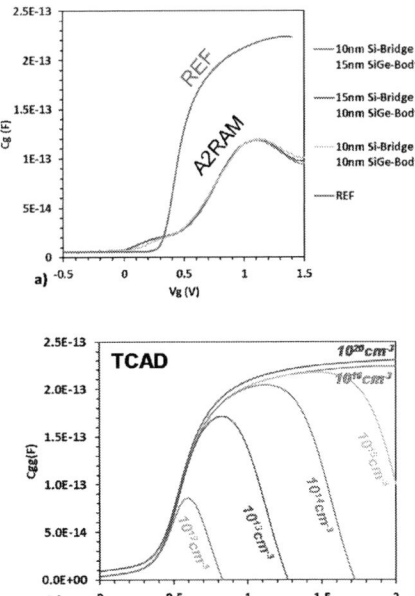

Figure 4: C_{gg}-V_g measured on A2RAM devices and reference CMOS device (W=10μm and L_g=1μm) b)) C_{gg}-V_g from TCAD for various polysilicon doping above gate metal.

III. CV MEASUREMENTS: EXPERIMENTAL BRIDGE EVIDENCE

To evidence that the body-bridge stack was properly fabricated, we intend to use the method described in [7] based on C_{gg}-V_g measurements There is in Fig.4-a a hump in the depletion regime (around V_g=0~0.5V) which is characteristic of conduction in the bridge. The A2RAM strong inversion regime capacitance is otherwise half the reference one while it should be the same [7]. Measurements show that this is not due to gate leakage. We finally demonstrate that the polysilicon on gate TiN of the fabricated A2RAM devices is undoped thanks to C_{gg}-V_g curves from TCAD simulations, with a doping of polysilicon on top of metal gate varying from 1e20cm^{-3} (standard) to 1e10 cm^{-3}(undoped) (Fig.4-b). This behavior is due to the depletion of the undoped polysilicon between gate silicide and gate TiN occurring when gate voltage increases. This is evidenced by Fig.5 which shows the electron density in the polysilicon of A2RAM structure for three different gate voltages. Indeed, in the MOS capacitance formed by S/D contact – spacer – gate polysilicon, positive gate voltage applied on gate silicide produces a rounded electric field from spacer to the top of the gate (illustrated by the arrow on Fig. 5) which is stronger if gate voltage is more positive. This explained why the depletion layer formed on spacer edge under positive gate voltage is thicker for larger gate voltage. If gate voltage is large enough, depletion layers on both sides are merged, leading to a depleted region between gate silicide and TiN. In this condition, the gate acts as two capacitances in series: the gate oxide capacitance (which is constant) and the polysilicon capacitance (which decreases when depletion layer, so gate voltage, increases), what is consistent with CV measurement of Fig. 4-b.

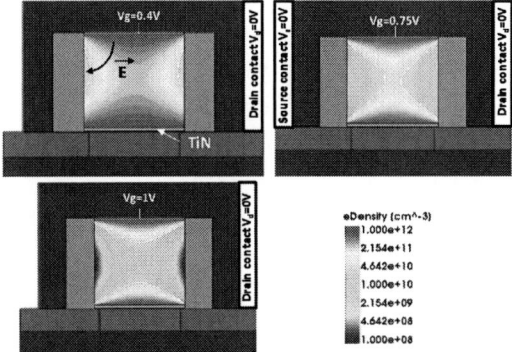

Figure 5: electron density in polysilicon (1e12cm^{-3} doping) for 3 different gate voltages saved during CV TCAD simulation of Fig.4-b. The white line represents the limit of depletion layer.

978-1-6654-3746-2/21 $31.00 © 2021 IEEE

IV. MEMORY PERFORMANCE

To evaluate the impact of the undoped polysilicon on memory performance with a '1' state written by GIDL, we used the sequence: programming (W1)-read (R)-erase (E)-read (R) where each step occurred during 1µs. Fig.6-a shows the corresponding variation of drain current with time obtained by TCAD with doped and undoped polysilicon (Fig.6-a). We evidence that erasing operation was impossible with a '0' state read current I_0 ~10 times superior to '1'state read current I_1 in case of the undoped polysilicon. This is due to the complete depletion of the polysilicon produced by erasing operation ($V_g > 0$, as in Fig. 5) and preserved by the negative gate voltage applied during hold and read operations. In this condition, the polysilicon behaves as an insulator. It is evidenced by the Fig.6-b which shows the linear increase of potential in the polysilicon from the top of the gate to the TiN layer during '0' state reading. Obviously, the potential is constant for a doped polysilicon which acts as a metal. As a consequence, potential is largely higher in TiN layer in

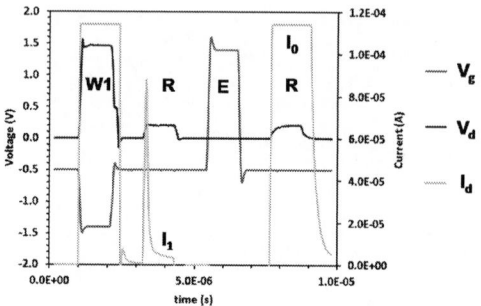

Figure 7: measurement of the memory sequence programming (W1) read (R) erase (E) read (R) on one A2RAM device. Voltages (left axis) and drain current (right axis).

undoped polysilicon case. This produces an increase of the potential in the device silicon film sufficient to form an inversion layer and to activate the front channel conduction (see horizontal potential distribution during '0' state reading on Fig.6-c). So, for undoped polysilicon the '0' state read current is equal to current flowing through the inversion layer, which is larger than the '1' state read current only flowing through the bridge. This behavior is confirmed by the measurement of the same memory sequence on the fabricated A2RAM devices (Fig. 7) where we can see that $I_1 \gg I_0$. Finally, we also remarked on Fig.6-a that '1' state read current is reduced by a factor of ~5 because of the undoped polysilicon gate. This is mainly due to the lower amount of charges store in the body during programming.

As the initial state of A2RAM device, so the equilibrium state is close to '0' state, we will use the sequence R-W1-R-R to assess the programing ability and the memory performances (Fig.8). The first read operation provided the read current of equilibrium state I_{DC} (with $I_{DC} > I_0$ [3,6]) and the two last read operations yielded I_1 and a first evaluation of retention ability. Fig. 8 demonstrates that the programming operation is functional because the read current after programming is higher than the current measured during the first read operation. In the following, to improve the accuracy of the characterization, I_{DC} is measured with a lower compliance current than I_1 and so, I_{DC} and I_1 are presented on different plots. Fig. 9 shows the

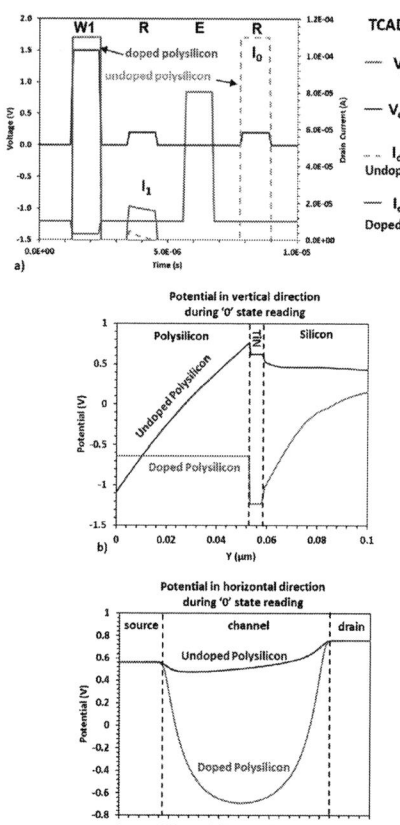

Figure 6: a) TCAD Drain current vairation corresponding to the memory sequence programming (W1) read (R) erase (E) read (R) on A2RAM device. Potential distribution during '0' state reading extracted from TCAD simulation. b) vertical distribution in the middle of the gate. c) horizontal distribution from source to drain.

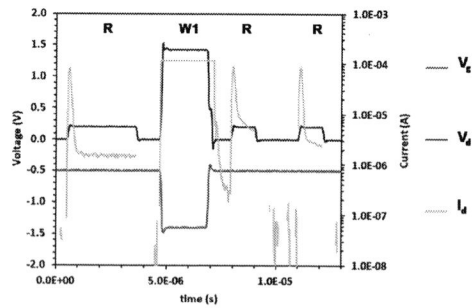

Figure 8: measurement of the memory sequence read (R) programming (W1) read (R) read (R) on one A2RAM device. Voltages (left axis) and drain current (right axis).

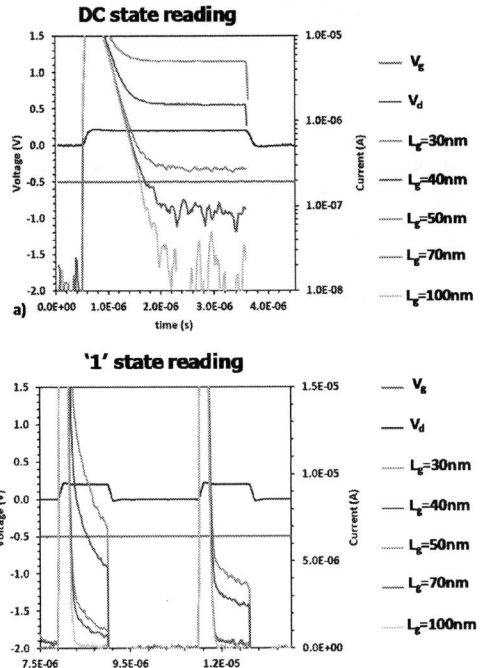

Figure 9: DC (a) and '1' state (b) read current of A2RAM device (t_{bridge}=15nm, t_{body}=10nm) for various gate length L_g. Drain (blue) and gate (red) voltages are also shown.

measurement I_{DC} (a) and I_1 (b) versus time of A2RAM devices (t_{bridge}=15nm, t_{body}=10nm) for various gate lengths (30 to 100nm). I_{DC} is constant with time during the read operation and decreases for longer gates because short channel effect are reduced. I_1 is lower for shorter gates because the bridge resistance is lower (during '1' state reading, current is flows through the bridge). We also observe that I_1 is decreasing during the read operations and is close to I_{DC} value in the end of the last read operation. This indicates a very low retention ability. Finally, for all cases, I_1>I_{DC}, demonstrating the ability to program and to read the '1' state. Similarly, Fig.10 shows I_{DC} and I_1 for L_g=40nm for the different body-bridge stacks. On Fig.10-a, we observe that I_{DC} is mainly impacted by the total silicon thickness and is lower for thinner stack. This is due to the short channel effect reduction for thinner films. Fig.10-b shows that A2RAM with 20nm stack does not provide relevant '1' state. In 25nm thick stack device, we observe I_1>I_{DC} and so we demonstrate the A2RAM programming operation.

CONCLUSION

We demonstrated A2RAM device fabrication with advanced technology. The bridge is evidenced by capacitive measurement and the '1' state programming by

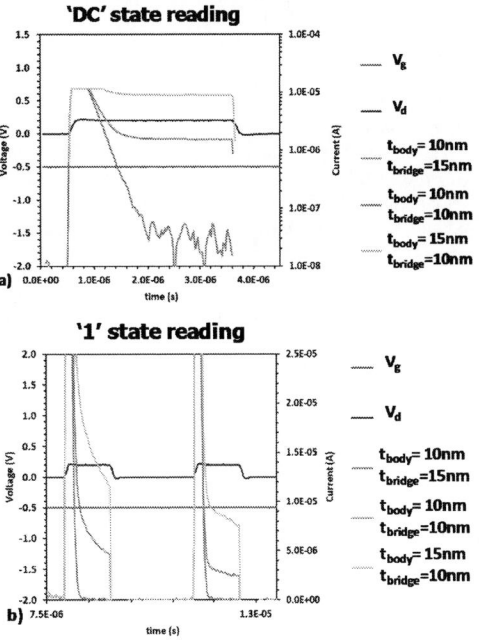

Figure 10: DC (a) and '1' state (b) read current of A2RAM device (L_g=40nm) for the 3 available body/bridge stacks.

transient measurement. The limited A2RAM performance demonstrated here are mainly due to the undoped polysilicon in the gate stack. Thanks TCAD simulations, we detailed why the undoped polysilicon reduces the A2RAM performance and demonstrated that we can expect '1' state read current improvement by a factor of 5 with a doped polysilicon. We finally demonstrated that the best A2RAM structure is L_g=40nm / t_{si}=25nm.

ACKNOWLEDGMENT

The research leading to these results has received funding from the European Union's Horizon 2020 research and innovation program under grant agreement No 687931 REMINDER.

REFERENCES

[1] H.-J. Wann et al., IEDM 93.

[2] M. Bawedin et al. "Floating-Body SOI Memory: The scaling tournament," Springe Science p 393-421, edited by A. Nazarov et al., 2011.

[3] N. Rodriguez et al. EDL 2012.

[4] J. S. Shin et al, EDL 2012.

[5] S. Barraud et al., VLSI 2013

[6] F. Tcheme Wakam et al. SISPAD 2017

[7] F. Tcheme Wakam et al. SSE 2019

Thermal Stability of Ferroelectricity in Hafnia-Zirconia-Alumina Buried Oxide Stacks

Fedor Tikhonenko
Laboratory of Silicon Material Science
Rzhanov Institute of Semiconductor
Physics SB RAS
Novosibirsk, Russia
ftikhonenko@gmail.com

Valentin Antonov
Laboratory of Silicon Material Science
Rzhanov Institute of Semiconductor
Physics SB RAS
Novosibirsk, Russia
ava@isp.nsc.ru

Vladimir Popov
Laboratory of Silicon Material Science
Rzhanov Institute of Semiconductor
Physics SB RAS
Novosibirsk, Russia
popov@isp.nsc.ru

Andrey Miakonkikh
Laboratory of Microstructuring and
Submicron Devices
Valiev Institute of Physics and
Technology RAS,
Moscow, Russia
miakonkikh@ftian.ru

Konstantin Rudenko
Laboratory of Microstructuring and
Submicron Devices
Valiev Institute of Physics and
Technology RAS,
Moscow, Russia
rudenko@ftian.ru

Abstract— The thermal stability of 20 nm buried oxide (BOX) with ferroelectric (Fe) stacks grown by plasma enhanced atomic layer deposition (PEALD) was investigated in bonded silicon-on-insulator (SOI) structures after the RTA treatment at 600-1000 ^{o}C for using them in silicon-ferroelectric-silicon (SFS) optical switches or FeFET memories. Various stacks had different ferroelectric thermal stabilities, where the highest residual polarization P_r = 5.5 μC/cm^2 was demonstrated for HfO$_2$:Al$_2$O$_3$ (10:1) deposited monolayer stacks and annealed up to 900 ^{o}C.

Keywords— *silicon on insulator; buried hafnia zirconia alumina stack; ferroelectricity; thermal stability*

I. INTRODUCTION

The need to increase the computing power of electronic devices has faced in recent years significantly increased difficulties in the transition to nanometer topological nodes. The solution is to use alternative architectures based on quantum computing or neuromorphic systems. The latter are particularly attractive due to the relative ease of implementation and the possibility of scaling by the CMOS technology [1]. For the functioning of neuromorphic computing systems, it is necessary to ensure the storage and processing of information in a single logical element. A possible variant for such neuromorphic units can be a ferroelectric field effect transistor (FeFET) with a gate dielectric based on the recently discovered hafnium-zirconium dioxides HfO$_2$/ZrO$_2$ (HZO) with a high dielectric permittivity (high-k) [2]. The advantages of HZO FeFET are speed, simplicity of design and compatibility with the CMOS technology. The disadvantage of HZO FeFET is a relatively low (<700 ^{o}C) thermal stability of ferroelectric HZO properties [3]. The FeFET implementation requires ferroelectric layers with specified parameters, such as the coercive field E$_c$ and the residual polarization P$_r$, and to ensuring the thermal stability of these characteristics. It is known that the physical properties of thin layer ferroelectric capacitors based on HZO strongly depend on the electrode material [4]. The properties of metal-ferroelectric-metal (MFM) and metal-ferroelectric-semiconductor (MFS) capacitors are mainly studied. However, the semiconductor-ferroelectric-semiconductor structures (SFS) have not been practically studied, despite their huge potential for applications from optical SFS switches to neuromorphic

processors [5, 6]. In this report, the ferroelectric properties of the SFS capacitors and SOI pseudo-MOSFETs are investigated with different high-k BOXs based on hafnia-zirconia-alumina (HZAO) multilayers after a rapid thermal annealing (RTA) at 600-1000 ^{o}C.

II. EXPERIMENT

The SFS structures were made by the direct bonding and hydrogen induced transfer of thin Si layers on the ultrathin high-k dielectric films (DeleCut method) covering n- or p-type silicon wafers with the (001) orientation and the resistivity of 0.1 – 20 Ohm·cm [7]. 20 nm thick high-k dielectric layers were deposited before bonding to the Si substrates by PEALD with the supercycles of monolayers HfO$_2$:ZrO$_2$:Al$_2$O$_3$ at the ratio HO:ZO:AO = m:n:p, where m=5, 10, 200; n=5, 0, 0; p=1, 1, 0, respectively, at the substrate temperature of 250 ^{o}C. The HO and HZO reference samples did not contain alumina layers (p=0). The Si substrates were subjected to the N$_2$/NH$_3$ plasma treatment forming the silicon oxide free SFS interfaces with a smaller interface state density D$_{if}$.

The electric properties of SFS mesastructures of the 3 mm in diameter were determined from the drain-gate characteristics of pseudo-MOSFETs with the tungsten needle tips with the radii of 20 μm and the pressing force of 60 g at the 100 μm distance, used as drain-source contacts, and Si substrates as a gate electrode [7]. Pulsed C-V and I-V positive-up-negative-down (PUND) impulses and triangle P-V hysteresis measurements on the SFS mesastructures with the Keithley 4200 SCS unit provided the residual ferroelectric polarization P_r and the determination of coercive field E_c.

The obtained SFS-structures with a buried layer consisting of alternating of HfO$_2$, or HfO$_2$/ZrO$_2$ and Al$_2$O$_3$ layers, were subjected to a successive RTA treatment in nitrogen during 30 s and the analysis of transient pseudo-MOSFET characteristics (Fig. 1, 2). At each stage, by measuring the current-voltage characteristics of pseudo-MOSFETs, the mobility of charge carriers in the cut-off layer and the density of states D_{if} are estimated at the silicon-dielectric interfaces. The threshold voltage shift dependences on the gate voltage-sweep and pulsed PUND rates are performed for the determination of the charge carriers mobility $\mu_{n,p}$, D_{if}, E_c and P_r (Fig. 1-4).

III. RESULTS AND DISCUSSION

The HAO pseudo-FeFET hysteresis after the thermal treatments at 650-800 °C during 30 s due to ferroelectric polarizations in the whole range of annealing temperatures is presented in Figure 1. The increase of T > 800 °C leads

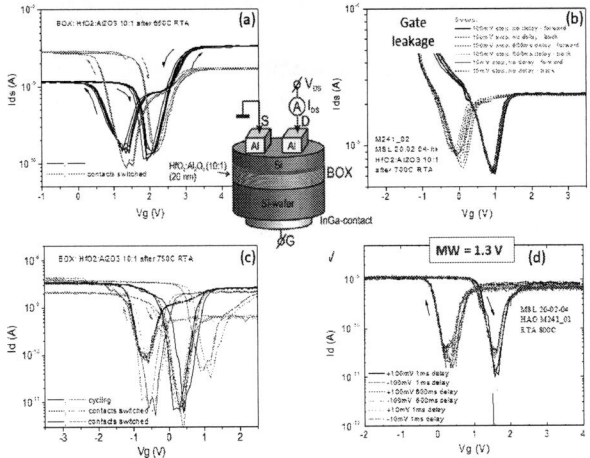

Fig. 1. HAO pseudo-FeFET transfer I_{ds}-V_g curves after RTA at 650-800 °C. In inset the measurement sceme.

to the decrease in the memory window (MW = ΔV_{th}) from 1.3 V (Fig. 1d) to 0.5 V (Fig. 3a). These memory windows occur due to the polarization switching P-up (↑) and P-down (↓) in the threshold voltage V_{th} by the applied gate voltage V_g:

$$\Delta V_{th} = V_{th}\uparrow - V_{th}\downarrow = 2\,P_r t_{BOX}/(\varepsilon_0\varepsilon_{BOX}) \qquad (1)$$

In Figure 2 is the HZAO pseudo-FeFET hysteresis after the RTA at 750-800 °C. Here the hysteresis occurs due to the D_{if} recharges (750-700 °C) and HZAO polarization switching (750-800 °C). The HZAO buried oxide hysteresis shows a more complex behavior due to the competitions of leakage currents, BOX bulk, interface charges and polarization.

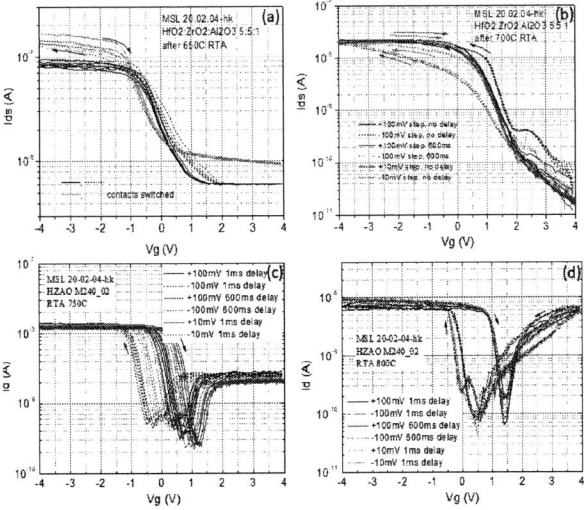

Fig. 2. HZAO pseudo-FeFET transfer I_{ds}-V_g curves after the RTA at 650-800 °C

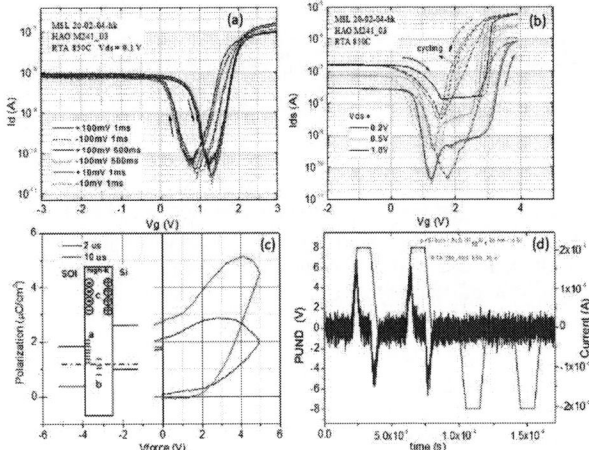

Fig. 3. HAO pseudo-FeFET transfer I_{ds}-V_g curves for different sweeping rates (a) and drain voltages (b), P-V curves with the SFS band diagram inset, where "a" is the interface states, "b" means the bulk traps and "c" is the P-down polarization charge (c), PUND pulse voltage and current peaks (d) after the RTA at 850 °C

The HAO pseudo-FeFET threshold voltage instabilities with the drain voltage V_{ds} increase due to the charge trapping at the interface states D_{if} after the RTA treatment at 850 °C. But their charges do not decrease the polarization switching hysteresis substantially. It dominates for the whole temperature range up to 950 °C, for all scanning rates and on the drain voltages $V_{ds} \leq 1.5$ V.

The P-V curves and PUND pulse measurements confirm the ferroelectric nature of the HAO hysteresis in the SFS structure after the RTA at 650-850 °C is demonstrated in Figures 3a, b. The current amplitude is weakly dependent on the voltage sweep rate. The value of $2P_r \leq 11$ µC/cm² is observed for the HAO BOX ferroelectric even after the RTA annealing at 850 °C (Fig. 3c). The pulse current at the negative voltage pulses is small due to the space charge region extension at the backward biasing of the n-SOI/BOX/p-Si substrate structure [7, 8]. A decrease in the polarization switching is possible, but the difference between the first and the second positive current pulses are clearly observed (Fig. 3d).

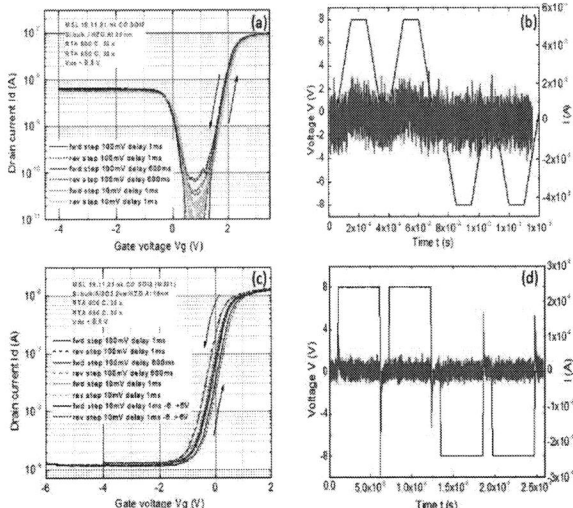

Fig. 4. HZAO pseudo-FeFET transfer I_{ds}-V_g curves for different sweeping rates (a) and the PUND pulse voltage and current peaks for the

HZAO SFS structure after RTA at 850 °C (b), HZAO/AO pseudo-FeFET transfer I_{ds}-V_g curves for different sweeping rates (c) and the PUND pulse voltage, and current peaks for the HZAO/AO SFS structure after the same RTA treatment at 850 °C (d).

The threshold voltage shift dependences on the scanning speed were also studied for HZAO BOX, or HAO/AO BOX and HZAO/AO BOX with a 2 nm alumina layer in the SFS structures. The direct P_r determination by Equation 1 was provided also using the PUND pulse measurements (Fig. 4, 5). It was found that the interface charge states have a significant effect on the pseudo-FeFETs characteristics.

On the one hand, their recharge can cause a change in the width and direction of the drain-gate transfer hysteresis (Fig. 4a, c). The density of states depends on the structure properties. It reaches the highest values in the samples in which there is a direct contact of the 2 nm Al_2O_3 layer with Si substrates. But, on another hand, it allows saving the ferroelectric hysteresis even for the HZAO BOX layer with the much lower thermal stability than that of HAO BOX (Fig. 5a, b). Indeed, the transfer I_{ds}-V_g hysteresis at the larger fields is smaller or corresponds to the capture on the states in the BOX bulk and at the interface boundaries for HZAO BOX, or HAO/AO BOX and HZAO/AO BOX with 2 nm alumina layers. These data are confirmed by two equal pulse amplitudes in the PUND sequences (Fig. 4b).

Thus, the presence of nanometer thick amorphous intermediate layers leads to an increase in the concentration of charge carrier capture centers in BOX and to the compensation of ferroelectric hysteresis by their charges. Nitriding the silicon surface before the high-k dielectric deposition and bonding reduces the thickness of the amorphous interlayers and leads to an increase in the ferroelectric hysteresis and its thermal stability. It occurs, possibly, due to the decreases in the silicon oxidation and high-k oxygen vacancies generation during followed thermal treatments. The obtained results should be taken into account when designing and manufacturing devices based on the SFS structures and two gate FeFET transistors (Fig. 5).

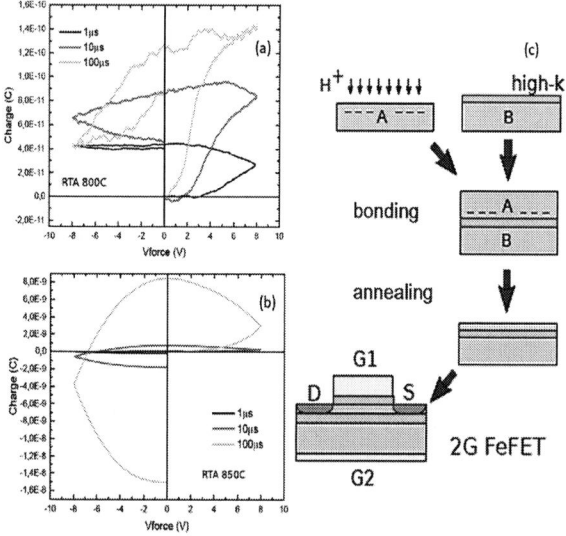

Fig. 5.   HZAO P-V polarization curves for different sweeping rates after the RTA at 800 °C (a) and the same after the RTA at 850 °C (b), suggested SFS-structure and the two gate FeFET manufacturing process with the HAO BOX as a bottom ferroelectric and the HZAO layer as an upper ferroelectric (c).

IV. CONCLUSIONS

1. The thermally stable ferroelectric buried oxide (up to 850°C) was formed by the PEALD HfO_2:Al_2O_3 (10:1) 20 supercycle stack deposition and the hydrogen induced silicon nanolayer transfer above resulting in a high quality SFS-structure.

2. The pseudo-MOSFET drain-gate transfer characteristics with ferroelectric BOX have a characteristic hysteresis.

3. The magnitude and direction of this hysteresis can be changed by the rapid thermal annealing at temperatures of 800-950°C.

4. The HfO_2: ZrO_2:Al_2O_3 (5:5:1) ferroelectric BOX was thermally stable only up to 800°C.

5. To observe the ferroelectric repolarization effects, it is necessary to reduce the influence of charge states at the boundary and in the HZAO high-k bulk by optimizing the SFS manufacturing technology, e.g., nitriding the silicon surface before the high-k dielectric deposition.

6. It is possible to implement a two-gate field-effect transistor with full depletion and two ferroelectrics on a SOI substrate using HAO BOX as a bottom and HZAO as an upper ferroelectric in the last gate process.

ACKNOWLEDGMENT

This work was partially funded by RFBR grant no.19-29-03031 and by the Ministry of Science and Higher Education of Russia, programs no. 0242-2021-0003 and no. 0066-2019-0004.

References

[1] T. S. Böscke, J. Müller, D. Bräuhaus, U. Schröder, U. Böttger. "Ferroelectricity in hafnium oxide thin films". Appl. Phys. Lett. vol.99, 1029033, 2011.

[2] T. Mikolajick, U. Schroeder, S. Slesazeck. "The past, the present, and the future of ferroelectric memories." IEEE TED, vol.67, pp. 1434-1443, April 2020.

[3] K. Toprasertpong, K. Tahara, T. Fukui, (...), M. Takenaka, S. Takagi. "Improved ferroelectric/semiconductor interface properties in Hf0.5Zr0.5O2 ferroelectric FETs by low-temperature annealing." IEEE EDL, vol.41, 9177059, pp.1588-1591, October 2020.

[4] L. Yao, X. Liu, Y. Cheng, B. Xiao. "A synergistic interplay between dopant ALD cycles and film thickness." Nanotechnol., vol.32, 215708, March 2021.

[5] P. Bidenko, J. Han, J. Song, S. H. Kim. "Study on charge-enhanced ferroelectric SIS optical phase shifters utilizing negative capacitance effect." IEEE Journ. Quant. Electron. vol.56, 9154724, December 2020.

[6] M. S. Tarkov, A. N. Leushin, F. V. Tikhonenko, I. E. Tyschenko, V. P. Popov. "Logic elements and crossbar architecture based on SOI two-gate ferroelectric transistors." 2020 Joint International EUROSOI Workshop and International Conference on Ultimate Integration on Silicon, EUROSOI-ULIS 2020, 9365362, 2020.

[7] V. Popov, F. Tikhonenko, V. Antonov, I. Tyschenko, A. Miakonkikh, S. Simakin, K. Rudenko. "Diode-like current leakage and ferroelectric switching in silicon SIS structures with hafnia-alumina nanolaminates." Nanomaterials, 2021, vol.11(2), 291, p. 1-14,

[8] V. P. Popov, V. A. Antonov, F. V. Tikhonenko, S. M. Tarkov, A. K. Gutakovskii, I. E. Tyschenko, et al. "Robust Semiconductor-on-Ferroelectric Structures with Hafnia-Zirconia-Alumina UTBOX Stacks Compatible with the CMOS Technology" Journ. Phys. D: Appl. Phys., vol. 54, 225101, 2021.

Si thickness influence on subthreshold currents at high temperatures in FDSOI CMOS

Mattias Ekström, Laura Zurauskaite and Per-Erik Hellström

Affiliations: School of Electrical Engineering and Computer Science,
KTH Royal Institute of Technology, Stockholm, Sweden
mekstr@kth.se, lauraz@kth.se, pereh@kth.se

Abstract—Fully depleted silicon-on-insulator (FDSOI) CMOS with thick buried oxide (BOX) can operate at higher temperatures compared to bulk CMOS. This work demonstrates, both experimentally and through simulations, that the subthreshold characteristics (off-state leakage current, I_{leak} and subthreshold swing, SS) are greatly improved at high temperatures by reducing the Si thickness (t_{Si}) in FDSOI CMOS. Fabricated N and PFET devices exhibit low $I_{leak} < 300$ pA/µm and close to ideal subthreshold swing (SS<132 mV/dec) at 300 °C. TCAD simulations closely match measured data and show that electrostatic control of the Si layer is key to achieve close to ideal SS and low I_{leak}. With proper gate electrodes FDSOI CMOS can achieve an $I_{off} < 1$nA/µm at 300 °C for both P and NFETs. This result shows that FDSOI CMOS can find use as low power control logic at high temperatures.

Keywords-FDSOI CMOS, High temperature, subthreshold swing

I. INTRODUCTION

High temperature electronics could find applications in gas sensing [1], data logging for geothermal energy drilling [2], and monitoring industrial processes. Operation of bulk MOSFETs is limited to ~200 °C due to increasing leakage currents, subthreshold swing (SS) and the shift of the threshold voltage (V_T) [3]. On the other hand, fully depleted silicon-on-insulator (FDSOI) MOSFETs have been shown to stay operational above 200 °C [3]–[5]. FDSOI has the advantage of near-ideal subthreshold swing and avoids parasitic drain/body pn-junctions that leak at high temperatures. However, it has been observed that SS deviates from the near-ideal value of $k_BT\ln(10)/q$ at high temperatures [5], resulting in rapidly increasing off-state leakage current with increasing temperatures above 200 °C. In this work, we show that, at high temperatures, both the off-state leakage current and the subthreshold swing are effectively controlled by reducing the Si thickness. We demonstrate that properly designed FDSOI CMOS can achieve an $I_{off}<1$nA/µm at 300 °C, implying that static power consumption can be kept low even at high temperatures.

II. EXPERIMENTS

FDSOI MOSFETs were fabricated on SOI substrates from SOITEC with 145 nm buried oxide (BOX) and Si thickness (t_{Si}) thinned down to 22 nm by thermal oxidation and HF stripping. After active area definition, a gate stack of 5 nm thermally grown SiO_2 gate oxide (GOX), 11.5 nm atomic layer deposited (ALD) TiN and 80 nm n^+ *in-situ* doped poly-Si was formed. The source/drain (S/D) was implanted with BF_2 and As for PFETs and NFETs respectively. Spacers were formed by 12 nm ALD SiO_2 and 50 nm plasma enhanced chemical vapor deposited SiN. After S/D activation by rapid thermal anneal at 1000 °C for 10 s, two level Al metallization was employed followed by forming gas anneal for 30 min at 400 °C. High temperature reliability could be improved by replacing Al with a refractory metal, like W, and by adding a final passivation layer. A cross section schematic of fabricated FET is displayed in Fig. 1.

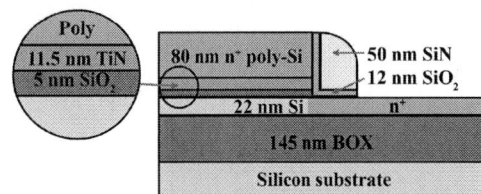

Figure 1.   Schematic of fabricated FDSOI NFET with t_{Si}=22 nm.

Sentaurus TCAD 2D-simulation of NFETs with t_{Si}=10, 22, 55 and 85 nm was performed by creating the transistors through process flow simulation. The BOX (140 nm) and active thicknesses were set exactly, while the GOX was grown by diffusion-reaction simulation, giving a 4 nm thickness. The final active layer thickness was ~2 nm thinner due to the consumption of silicon. Ion implanted profiles were generated by Monte-Carlo implantation and diffusion-reaction simulation. Both the substrate and active region were set to have a nominal boron doping of 10^{15} cm^{-3}. The TiN work function was set to 4.785 eV, negligible interface state density was assumed (10^{10} cm^{-2} eV^{-1}) and drift-diffusion I_{DS}-V_{GS} sweeps were performed at different temperatures

978-1-6654-3746-2/21 $31.00 © 2021 IEEE

neglecting quantum mechanical effect. The Boltzmann approximation was used globally, as it often improved numerical convergence at many different temperatures and biasing conditions for the multiple thicknesses.

III. RESULTS AND DISCUSSION

Measured transfer characteristics of 1 μm gate length MOSFETs at room temperature (RT) displays (Fig. 2) a low global variability of the threshold voltage (V_{Tp}=0.678 ± 0.013 V, V_{Tn}=0.305 ± 0.017 V) and subthreshold swing SS~66 mV/dec. for both PFETs and NFETs (102 devices of each type were measured over a wafer at RT).

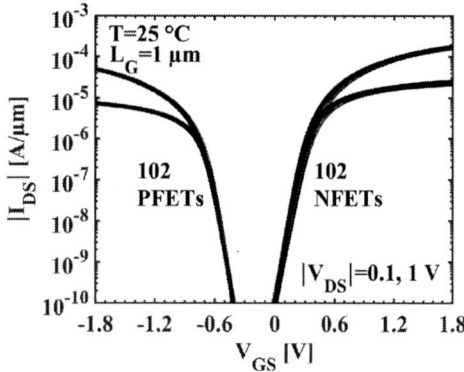

Figure 2. I_{DS}-V_{GS} characteristics of 102 PFETs and 102 NFETs at room temperature showing ideal subthreshold swing and low global variability of V_T (σ_{VT}<20 mV) and I_{DS} over the wafer.

At elevated temperatures a few devices of each type were measured and the results are displayed (Fig. 3A and 3B) for one device for clarity. As the temperature increases, V_T decreases and SS increases giving rise to an increased subthreshold current for both NFETs and PFETs (Fig. 3A and B). The leakage current in off-state, I_{leak}, (V_{GS}<-0.5 V for NFETs and V_{GS}>0 V for PFETs) is also increased with temperature. However, even at 300 °C, I_{leak} is less than 300 pA/μm in contrast to other SOI devices presented in literature [3], [4]. TCAD simulated I_{leak} (not shown here) closely matched the measured data indicating that off-state leakage current in fabricated devices, at elevated temperatures, is representative for Si devices and determined by the electrostatics in the Si layer. Simulations show that at RT the Si layer is fully depleted both in weak inversion and in the off-state. However, at elevated temperature and biasing in off-state, a conduction channel (electrons for NFETs and holes for PFETs) is formed in the Si layer at the back interface close to the BOX (see Fig. 4). The simulated electron distribution in the active region when the device is totally off (V_{GS} = -0.5 V) is approximately triangular with the peak at the silicon/BOX interface. The peak electron concentration increases with Si layer thickness. Therefore, the formation of the conduction channel at the

back interface can effectively be suppressed by reducing the Si thickness.

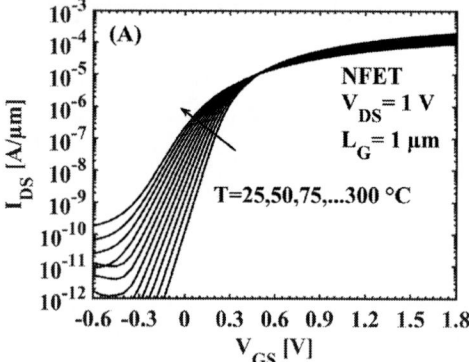

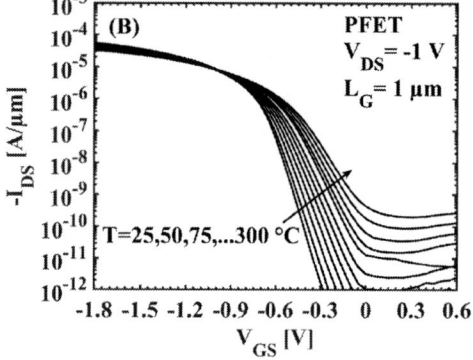

Figure 3. I_{DS}-V_{GS} characteristics of (A) n- and (B) p-channel MOSFETs at |V_{DS}|=1 V in the temperature range of 25-300 °C. Subthreshold swing and leakage current increases with temperature.

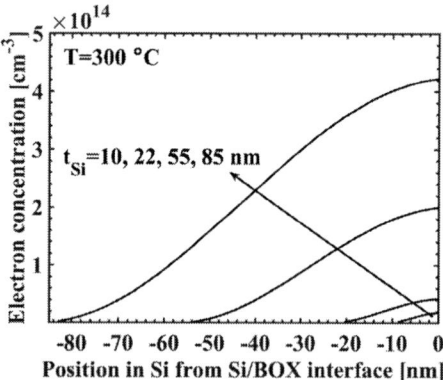

Figure 4. Simulated electron concentration distribution through the active region of NFETs at V_{GS} = -0.5 V (totally-off) at 300 °C. The back-channel peak concentration at the silicon/BOX interface increase with thickness, and the distribution is approximately triangular. The leakage current is proportional to the the area under the curves, and consequently the leakage current is a superlinear function of thickness at high temperature.

Due to the same gate electrode on fabricated NFETs and PFETs, the measured off-current I_{off} at $V_{GS}=0$ V is 216 nA/µm and 435 pA/µm at 300 °C for N and PFET, respectively. However, exploiting work function engineering and increasing NFET work function +0.35 eV an $I_{off}<1$ nA/µm can be achieved at 300 °C for both N and PFETs with $t_{si}=22$ nm in thick BOX FDSOI.

Figure 5 shows experimental SS (this work and [5]) together with TCAD simulations as a function of temperature. The simulations match the experimental result. Although SS is increased compared to the ideal SS at elevated temperatures, reducing t_{si} effectively suppresses the increase. Similar SS dependence on t_{si} has been shown by simulation in [6]. At $t_{si}=22$ nm and 300 °C, SS for fabricated PFETs and NFETs is <132 mV/dec., which is only 12 % higher than ideal SS.

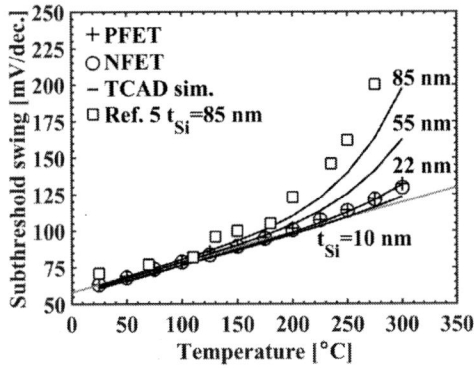

Figure 5. Subthreshold swing dependence on temperature in FDSOI MOSFETs with $t_{si}=22$ nm (this work) compared to $t_{si}=85$ nm [5]. The grey line displays the ideal linear temperature dependence while the black lines show TCAD simulation results for thick BOX FDSOI NFETs with $t_{si}=10$, 22, 55 and 85 nm.

The reason for the deviation of simulated SS from the ideal value ($SS_{ideal} = k_BT \ln(10)/q$) was investigated. The textbook derivation, starting from Pao-Sah's double integral and the charge-sheet approximation, gives the subthreshold swing linked to the surface potential ψ_s,

$$SS = SS_{ideal} \frac{\partial V_{GS}}{\partial \psi_s}, \qquad (1)$$

This is based on an approximation where the electron channel is infinitely narrow, and has a negligible voltage drop across it. The potential variation in silicon with gate voltage was extracted, as shown in Fig. 6. The subthreshold swing can be estimated from Eq. 1 by sampling $\Delta V_{GS}/\Delta\psi_s$ as shown in Fig. 6C. Fig. 7 displays the I_{DS}-V_{GS} subthreshold swing and the estimated SS from (1). It shows good correspondence at RT, but poor correspondence at high temperature (300 °C).

Careful analysis of the devices revealed that at high temperature there are three subtle mechanisms that cause the deviation from (1). First, the assumption of an infinitely narrow channel is invalid in weak-inversion, where it instead is more spread out through the active

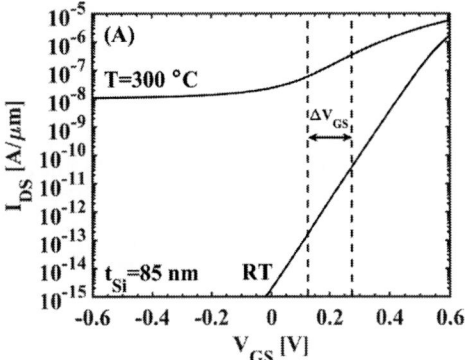

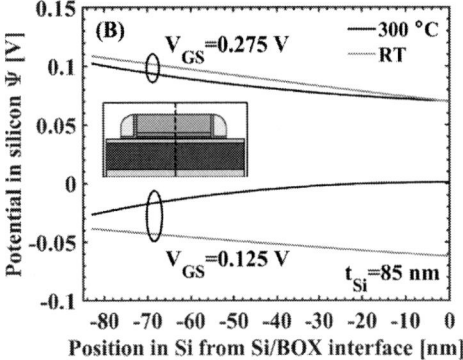

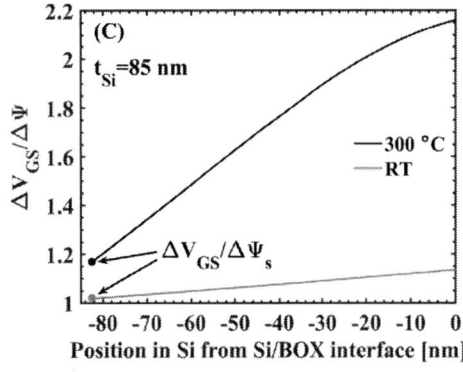

Figure 6. Extracting subthreshold swing from I_{DS}-V_{GS} and estimating it from internal electrostatics. (A) I_{DS}-V_{GS} at two different temperatures for the simulated 85 nm device. The subthreshold swing is extracted between the gate voltages 0.125 V and 0.275 V. (B) The internal electrostatic potential of silicon is extracted at the two different gate voltages by taking a 1D-cut along the active region, center of the gate. (C) The reciprocal variation of potential with gate voltage is estimated from the difference of the curves in (B).

region. Secondly, the weak-inversion channel is broader at higher temperature than at RT (consider Fig. 4, the electron channel exist at 300 °C even when fully turned off). This result requires a new interpretation of SS from internal electrostatics. We consider that there are multiple channels through the active region that conduct in parallel. Each of these channels have different SS (because $(\Delta\psi/\Delta V_{GS})^{-1}$ increases with depth), and SS

978-1-6654-3746-2/21 $31.00 © 2021 IEEE

extracted from I_{DS}-V_{GS} is a measure of the average of $(\Delta\psi/\Delta V_{GS})^{-1}$. We found that

$$SS = SS_{ideal} \langle \frac{\partial V_{GS}}{\partial\psi} \rangle, \quad (2)$$

which is the estimate from the average value of $(\partial\psi/\partial V_{GS})^{-1}$, gave a better correspondence to I_{DS}-V_{GS} at both RT and 300 °C than estimating SS from (1), as seen in Fig. 7.

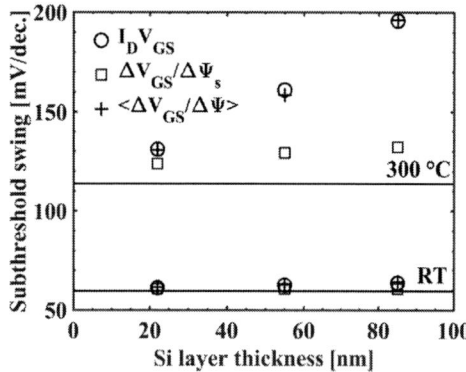

Figure 7. Simulated subthreshold swing estimated from I_{DS}-V_{GS} and internal electrostatics. The ideal values SS_{ideal} are drawn as straight lines.

The third mechanism is the increase of $(\partial\psi/\partial V_{GS})^{-1}$ at high temperature. The surface potential estimate of (1) can be considered an approximation or special case of (2), and is generally good at RT because $(\partial\psi/\partial V_{GS})^{-1}$ varies little with depth. This is because the active region is almost charge-free, and $\psi(x)$ is almost linearly decreasing with depth (see Fig. 6B), RT-curves). The voltage drop is given by Poisson's equation, which depends on the total charge density ρ. The total charge density of the MOSFET is given by

$$\rho(\psi(x)) = q(N_d^+ - N_a^- + p(\psi(x)) - n(\psi(x))) \quad (3)$$

where all terms have their usual meaning. (3) can be approximated several ways at weak-inversion, but very simplified,

1. Charge-free (linear voltage drop)
2. Doping (N_d-N_a term) dominates (quadratic voltage drop)
3. Intrinsic charge (p-n term) dominates (non-linear voltage drop)

It can be seen in Fig. 6B) that the voltage drop is almost linear at RT, indicating that the device is almost charge-free (the net doping, ~10^{15} cm^{-3}, is low enough that the quadratic behavior is suppressed for thin active layer thicknesses). The doping does not change with temperature (100 % dopant ionization at any temperature), but the voltage drop shows non-linearity at 300 °C. The total charge density was found to be approximately independent of gate voltage (0.125-0.275 V) at RT (doping dominates), but varied greatly at 300

°C, indicating that intrinsic charges dominated, even in what should be depletion. Thus, at very high temperature, the voltage drop increases because the total charge density increases with increasing intrinsic charge carriers. The larger voltage drop means that the gate voltage modulates poorly the silicon potential, as seen in Fig. 6C). The average value of $(\partial\psi/\partial V_{GS})^{-1}$ increases as a result, leading to a higher average SS. Thinner active silicon leads to lower voltage drop, and consequently a lower average value of $(\partial\psi/\partial V_{GS})^{-1}$ and lower SS. This result is in general agreement with [5].

IV. CONCLUSION

Fabricated FDSOI CMOS displayed low leakage current in off state and close to ideal subthreshold swing at elevated temperatures. Reducing t_{si} was shown to efficiently reduce the subthreshold leakage currents. It was found through TCAD simulation that the subthreshold swing is related to the depth-average silicon potential change with gate voltage change because the weak-inversion channel is spread out through the active region at high temperatures. The average value is degraded by the voltage drop over the active region due to intrinsic charge carrier greatly contributing to the total charge density at high temperatures, and thinner active silicon thickness gives smaller voltage drops. The totally-off leakage current is due to the formation of an inversion channel at the silicon/BOX interface, and the peak charge carrier concentration increases with increasing active layer thickness. FDSOI CMOS with proper gate electrodes can achieve an I_{off} < 1 nA/μm at 300 °C for both P and NFETs.

ACKNOWLEDGMENT

Prof. Mikael Östling and Prof. Carl-Mikael Zetterling are acknowledged for supporting the work and PhD Ganesh Jayakumar for part of the device processing.

REFERENCES

[1] T. Usagawa, K. Ueda, A. Nambu, A. Yoneyama, Y. Kikuchi, and A. Watanabe, "Pt–Ti–O gate silicon–metal–insulator–semiconductor field-effect transistor hydrogen gas sensors in harsh environments," *Jpn. J. Appl. Phys.*, vol. 55, no. 6, p. 67102, Jun. 2016.

[2] J. D. Cressler and H. A. Mantooth, *Extreme Environment Electronics*. CRC Press, 2013.

[3] D. Flandre, "Silicon-on-insulator technology for high temperature metal oxide semiconductor devices and circuits," *Mater. Sci. Eng. B*, vol. 29, no. 1–3, pp. 7–12, 1995.

[4] Francis, Terao, Gentinne, Flandre, and Colinge, "SOI technology for high-temperature applications," in *International Technical Digest on Electron Devices Meeting*, 1992, pp. 353–356.

[5] T. Rudenko, V. Kilchytska, J. P. Colinge, V. Dessard, and D. Flandre, "On the high-temperature subthreshold slope of thin-film SOI MOSFETs," *IEEE Electron Device Lett.*, vol. 23, no. 3, pp. 148–150, 2002.

[6] V. Kilchytska, N. Collaert, M. Jurczak, and D. Flandre, "Specific features of multiple-gate MOSFET threshold voltage and subthreshold slope behavior at high temperatures," *Solid. State. Electron.*, vol. 51, no. 9, pp. 1185–1193, 2007.

RF performances at cryogenic temperatures of inductances integrated in a FDSOI technology

Quentin Berlingard,
CEA Leti,
Université Grenoble Alpes,
F-38000, Grenoble, France,

Jose Lugo-Alvarez,
CEA Leti,
Université Grenoble Alpes,
F-38000, Grenoble, France,
jose.lugo@cea.fr

Lauriane Contamin,
CEA Leti,
Université Grenoble Alpes,
F-38000, Grenoble, France,

Cédric Durand,
STMicroelectronics,
850 rue Jean Monnet,
38920 Crolles, France

Philippe Galy,
STMicroelectronics,
850 rue Jean Monnet,
38920 Crolles, France

Andre Juge,
STMicroelectronics,
850 rue Jean Monnet,
38920 Crolles, France

Silvano De Franceschi,
CEA-IRIG,
Université Grenoble Alpes,
MINATEC, 38054 Grenoble, France

Maud Vinet,
CEA Leti,
Université Grenoble Alpes,
F-38000, Grenoble, France,

Tristan Meunier,
CNRS, Institut Néel,
Université Grenoble Alpes,
38042 Grenoble, France

Mikaël Cassé,
CEA Leti,
Université Grenoble Alpes,
F-38000, Grenoble, France,

Fred Gaillard,
CEA Leti,
Université Grenoble Alpes,
F-38000, Grenoble, France,

Abstract − **This paper investigates the RF performances of octagonal inductances integrated in a Fully Depleted Silicon On Insulator (FD-SOI) technology. 0.5nH to 4 nH inductors were first characterized vs frequency down to 4.2K and then modelled thanks to lumped elements including induction, coil proximity and skin effects. At low temperature, the quality factor is improved by a factor two because of the metal resistivity reduction, in agreement with DC measurements. Such performances and RF characterizations are mandatory to enable cryogenic CMOS RF integrated circuits design.**

Keywords − **RF modelling, inductance, FDSOI, CryoCMOS, cryogenic RF measurements**

I. INTRODUCTION

There is an increasing range of applications for Radio-Frequency (RF) electronics at cryogenic temperatures, such as control electronics for quantum computing and aerospace electronics [1][2]. Control circuits for quantum computers are needed to operate as low as 1K in the Super High Frequency (SHF) band. Such circuits will contain passive filters, attenuators and amplifiers, among other RF components. For this reason, we need to assess performances of the above circuits and develop component models at cryogenic temperatures to enable circuits design. Other works have focused on RF circuits at 77K or lower temperatures such as Low Noise Amplifiers (LNA) [3], Voltage-Controlled Oscillators (VCO) [4], transformers [5]. Here, the focus is made on the performances of inductances integrated in a FD-SOI technology, a technology of choice for those applications.

Indeed, FD-SOI technologies are good candidates to design low temperature RF electronics, because of the high performance of FD-SOI transistors, thanks to high electrostatic integrity and low variability [6].

In this paper, RF characterization of inductances fabricated in the 28nm FDSOI technology from STMicroelectronics [6] at 300K, 200K, 100K, 77K, 50K, 20K and 4K are presented. Then a model to analyse the measurements is proposed and discussed. We observe a gain in RF performance mainly due to the reduction of the metal

series resistance, confirmed by DC measurements is discussed. Please note that our measurements are limited to 4K because the test bench does not have the option to pump on the helium to reach lower temperatures.

II. EXPERIMENTAL SET-UP AND MEASURED INDUCTANCES

We used a cryogenic probe station Lakeshore EMTTP4 to perform 1-port RF measurements from 500 MHz to 20 GHz. The temperature is measured at three different points: under the sample holder, in the main chamber and in the external chamber. The sample are held on the sample holder with silver paint, this will also help in the thermalization of the sample and we assume that the sample is at the same temperature as the sample holder. A 1-port SOL (Short, Open, Load) calibration is realized at each temperature on a CS-5 Impedance Standard Substrate (ISS). According to [7], it is better to perform the calibration at the temperature of interest to obtain accurate and repeatable S-parameters. We have additionally used open and short structures to de-embed the raw measurements. Inductances are fabricated using the two thickest metals (top copper metal and alucap) available in the Back-end-of-Line (BEOL) of the considered 28nm-FDSOI technology. We then measured 18 inductances (like the one in Figure 1) with different number of turns, metal track width and inner diameter. Note that due to the 1-port test structure configuration, inductance output is short-circuited to the ground.

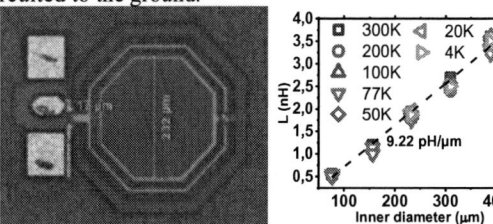

Figure 1. (Left) Example of one octagonal FDSOI inductance with two turns, an inner diameter of 237.6 µm and a metal track width of 13.95 µm. (Right) Extracted value of the two-turn measured inductances as a function of the inner diameter at the different temperatures.

978-1-6654-3746-2/21 $31.00 © 2021 IEEE

Temperature	300 K	200 K	100 K	50 K	20 K	4 K
C_{res}	0.15 pF	0.16 pF	0.15 pF	0.15 pF	0.16 pF	0.16 pF
L_{res}	1.3 nH	1.2 nH	1.3 nH	1.3 nH	1.3 nH	1.4 nH
R_{res}	51 Ω	50 Ω	51 Ω	48 Ω	48 Ω	48 Ω
L_{var}	0.46 nH	0.40 nH	0.39 nH	0.29 nH	0.21 nH	0.21 nH
R_{dc}	1.72 Ω	1.10 Ω	0.48 Ω	0.21 Ω	0.15 Ω	0.15 Ω

Table 1. model parameters extraction based on measurement.

III. RESULTS AND DISCUSSION

a. Measurements results

The extracted values from the inductance measurements are presented in Figure 1 (Right) for temperatures from 300 K down to 4K. All inductances have a similar RF response, analysis will be detailed for one device. The analysed inductance has an inner diameter of 237.6 µm, a metal track width of 13.95 µm, two turns and a short-circuit at output port. Figure 2 shows the inductance value vs frequency (from 300 K down to 4K) and associated quality factor (Q) evolution (only available at 300K and 77K).

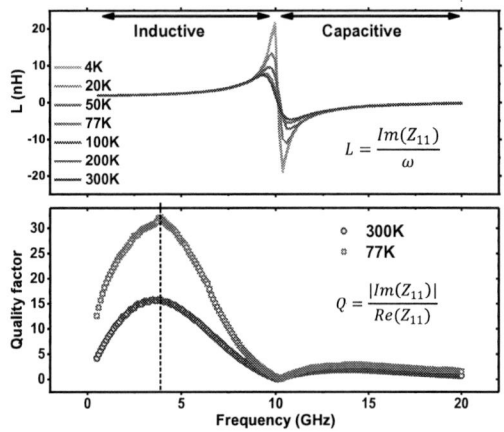

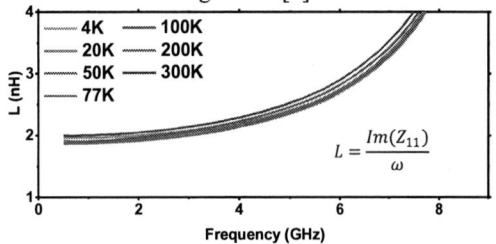

Figure 2. (Top) Inductance value vs frequency and (Bottom) quality factor value vs frequency. Inductance with inner diameter of 237.6 µm, a metal track width of 13.95 µm and two turns.

As expected, the inductance has an inductive response at low frequency and then becomes capacitive at high frequency after the Self Resonant Frequency (SRF) [8]. The inductance value at low frequency remains the same regardless of the temperature and its value is 1.9 nH as it is shown in Figure 3. At 3.9 GHz, maximum Q increases from 15 at 300K to 33 at 77K. These observations stem from the change on the metal resistivity with the temperature, while the geometric capacitance and inductance do not vary, as described in the following model [9].

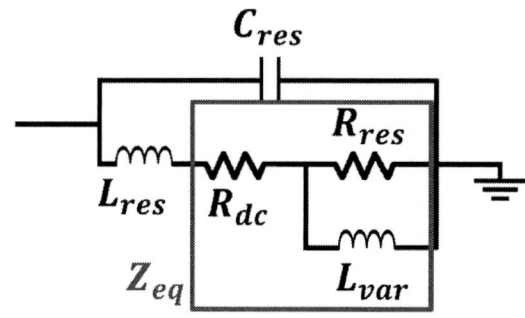

Figure 3. Zoom on the inductance value vs frequency at low frequency.

b. Proposed Model

Figure 4. Proposed RF model for the inductances.

Figure 4 represents the proposed model. It is composed of one main $R_{dc}L_{res}C_{res}$ loop describing the self-resonant behaviour, and one $R_{res}L_{var}$ loop for the auto induction effect, coil proximity effect and the skin effect. L_{res} and R_{dc} represent the inductance and DC resistance introduced by the metal track. C_{res} corresponds to the parasitic capacitance in the inductance such as the capacitance between the different metals composing the inductance and to the pattern ground shield. L_{res} and C_{res} define the resonance frequency. R_{res} and L_{var} aim to capture the auto induction effect, the coil proximity effect and the skin effect by producing a resistance frequency dependency [9] (L_{var} contribution to resonance frequency being negligible). The skin effect changes the current densities at high frequency, resulting in an increased resistance of the metal tracks. This frequency-dependent resistance leads to a non-symmetrical peak of $Re(Z_{11})$ at the resonance frequency.

Fitting the Z_{11} impedance yields the values presented in table 1.

This model is in good agreement with measurements as shown in Figure 5. It shows the real and imaginary part of the analyzed inductance.

We observe a SRF at 9.8 GHz and it is independent of the temperature. We are able to reproduce the change in the signal through the lumped model presented in section III.b.

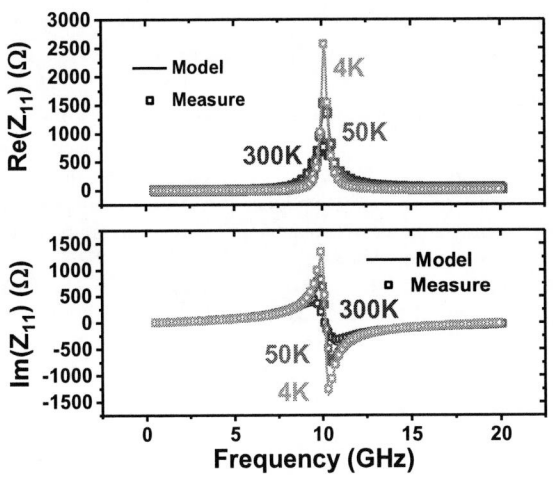

Figure 5. Inductance real and imaginary Z_{11} at ambient temperature and at 77K. Measurements (dot) and model (dash). Inductance with inner diameter of 237.6 µm, a metal track width of 13.95 µm and two turns.

c. Metal Resistivity temperature dependence

As expected the capacitance C_{res} and the inductance L_{res} do not vary with the temperature (see Table 1), since they only depend on the inductance geometrical dimensions. On the other hand, R_{dc} decreases with temperature as expected thanks to [10]. This resistance reduction is the main contributor of the difference between the 300K and 4K results (Figure 2 and 5). Moreover, the resistivity of the metal has a direct impact on the current density and it modifies the skin effect. Therefore, R_{res} and L_{var} vary with temperature too.

$$Re(Z_{eq}) = R_{dc} + \frac{R_{res}L_{var}^2\omega^2}{R_{res}^2 + L_{var}^2\omega^2} \quad (1)$$

$$Im(Z_{eq}) = \frac{R_{res}^2 L_{var}\omega}{R_{res}^2 + L_{var}^2\omega^2} \quad (2)$$

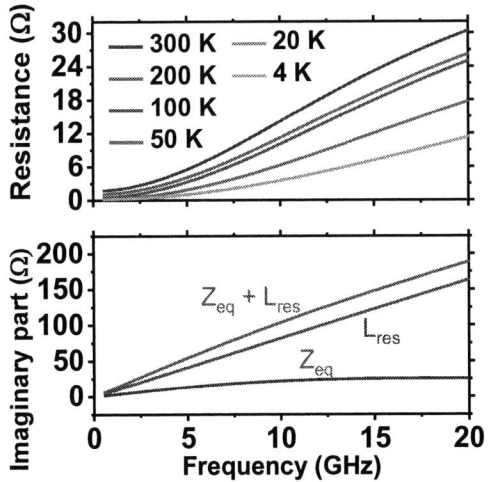

Figure 6. (Top) Effective resistance introduced by the skin effect modelled by R_{dc}, L_{var} and R_{res} as a function of frequency. (Bottom) Contribution of L_{res} and Z_{eq} to the imaginary part of Z11 at 300 K

Figure 6 (top) shows the variation of the effective resistance R_{eq} originating from R_{dc}, L_{var} and R_{res}. The formula is given in (1). When frequency increases, the resistance gets larger leading to a smaller skin depth. Besides, when temperature is lowered, R_{eq} decreases since the resistivity of the metal drops. Figure 6 (bottom) presents the Contribution of L_{res} and Z_{eq} to the imaginary part of Z11 at ambient temperature. The formula of the imaginary part of Z_{eq} is given in (2). This figure shows that the imaginary part introduced by the inductance L_{var} and the resistance R_{res} is negligible in comparison with the imaginary part of the inductance L_{res} (which is $L_{res}\omega$). This observation is available at all measurements temperatures.

To understand more precisely the variation of the resistivity of top copper metal and alucap used in the inductance, we have performed DC measurements from 10K to 300K, as shown in Figure 7. We have used a low DC power to avoid the line self-heating, and we have performed a four-point probe measurement to de-embed contact resistance. The temperature of the device is checked with previous resistance measurements as a reference. The resistivity increases linearly with the temperature between 50K and 300K as represented by the two fits, which can be used for modelling. Below 50K, the decrease in resistivity saturates because there is no more phonon scattering. The observation are the same with the DC resistance of the inductance. The DC resistance extracted with this method is integrated in the model of section III.b, its value is presented in the table 1.

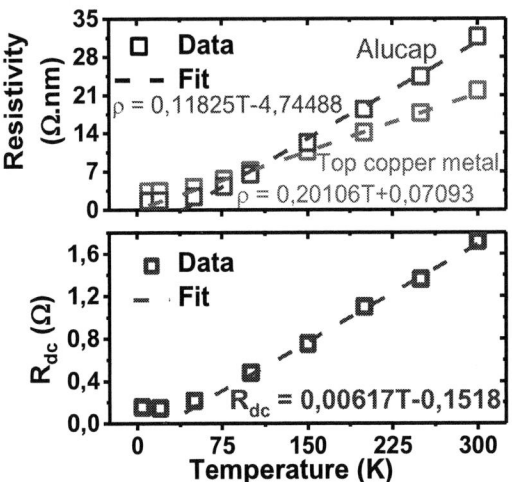

Figure 7. (top) DC resistivity of top copper metal and alucap as a function of temperature and (bottom) DC resistance of the studied inductance (stack of both metals).

IV. MODEL AND MEASUREMENTS LIMITS

In the previous section, we compared the model with the input impedance and not with the quality factor or the value of the inductance. As we can see on Figure 8 (bottom), the model does not fit well the quality factor. The quality factor is very sensitive to the $Re(Z_{11})$ in particular at low frequency where it reaches its maximum.

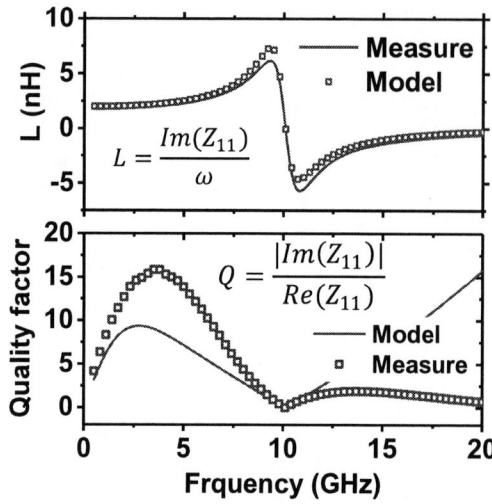

Figure 8. (Top) Inductance value and (bottom) quality factor at 300K. Measurements (dot) and model (dash).

According to our measurement, at low frequency, $Re(Z_{11})$ is less than 2 Ω around 1GHz. The proposed model seems limited as it could not fit the resonance peak and have an input impedance real part smaller than 2 Ω. An error of a few ohms at low frequency will degrade the model quality factor and will not be able to fit the maximum of quality factor.

The second uncertainty is based on the measurements. Indeed, the inductance input resistance is approximately 1 Ω. This value is very small and sensitive, in the range to the RF measurement accuracy we performed. To illustrate this limitation, Figure 9 shows the real part of the short circuit used to de-embed.

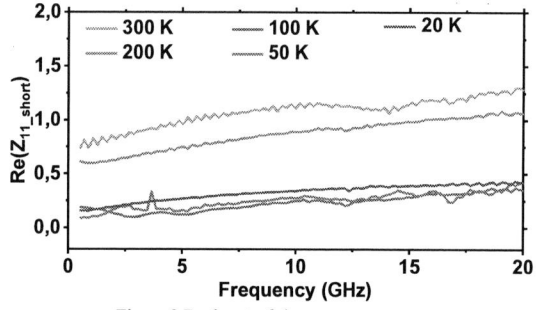

Figure 9 Real part of short measurement

The variations between the different measurements should be linear with the temperature until 50K in agreement with the resistivity measurement. But we can see that the gap is bigger between 200 K and 100 K than between 300 K and 200 K. Under 100 K, the short measurements reach its minimum and even if we go down in temperature, the value of the short is still the same. From that, we can deduce that the resistance accuracy is in the range of 0.3 ohms explaining the difficulties to extract low serial inductance resistances at low temperature and frequencies. Therefore, after applying the de-embedding it will be impossible to extract the inductance DC resistance. To obtain this value we propose to perform a 4-point probes DC measurement as in III.c.

V. CONCLUSION

In this work, we presented RF performances of a set of inductances characterized up to 20GHz at cryogenic temperatures. The proposed RLC model fits at ambient and low temperature, and it includes the frequency dependency of the inductance resistance. At 77K, the maximum quality factor is doubled compared to its 300K value thanks to the resistivity reduction at cryogenic temperatures. This effect must be considered when using passive devices at cryogenic temperatures.

VI. ACKNOWLEDGEMENTS

The authors acknowledge Spintec and especially Cécile GREZES and Laurent VILA for their help with the experimental setup. This work benefited from the facilities and expertise of the OPE)N(RA characterization platform of FMNT (FR 2542,fmnt.fr) supported by CNRS, Grenoble INP and UGA. We would like also to thank STMicroelectronics for 28FDSOI samples delivery.

VII. REFERENCES

[1] Dijk, J. Van, Hart, P., Kiene, G., Overwater, R., Padalia, P., & Staveren, J. Van. (2020). Cryo-CMOS for Analog / Mixed-Signal Circuits and Systems.
[2] Charbon, E. (2019). Cryo-CMOS Electronics for Quantum Computing Applications. ESSCIRC 2019 - IEEE 45th European Solid State Circuits Conference, 1–6.
[3] Li, M., Zhang, X., Cai, K., Jin, Shichao, Wei, B., & Cao, B. (2007). Design and fabrication of cryogenic low noise amplifier in low RF band. 2007 International Conference on Microwave and Millimeter Wave Technology, ICMMT '07, 8–11.
[4] Gulec, H. O., & Yelten, M. B. (2019). A cryogenic LC VCO utilizing cryogenic models of active devices. Proceedings - 2019 6th International Conference on Electrical and Electronics Engineering, ICEEE 2019, 220–224.
[5] Patra, B., Mehrpoo, M., Ruffino, A., Sebastiano, F., Charbon, E., & Babaie, M. (2020). Characterization and Analysis of On-Chip Microwave Passive Components at Cryogenic Temperatures. IEEE Journal of the Electron Devices Society, 8(April), 448–456.
[6] Planes, N., Weber, O., Barral, V., Haendler, S., Noblet, D., Croain, D., Bocat, M., Sassoulas, P. O., Federspiel, X., Cros, A., Bajolet, A., Richard, E., Dumont, B., Perreau, P., Petit, D., Golanski, D., Fenouillet-Béranger, C., Guillot, N., Rafik, M., Haond, M. (2012). 28nm FDSOI technology platform for high-speed low-voltage digital applications. Digest of Technical Papers - Symposium on VLSI Technology, 33(4), 133–134.
[7] J. Laskar, J. J. Bautista, M. Nishimoto, M. Hamai, and R. La. (1996). Development of Accurate On-Wafer, Cryogenic Characterization Techniques. IEEE TRANSACTIONS ON MICROWAVE THEORY AND TECHNIQUES, 44(I), 5–10.
[8] Bantas, S., Koutsoyannopoulos, Y., & Liapis, A. (2004). An inductance modeling flow seamlessly integrated in the RF IC design chain. Proceedings -Design, Automation and Test in Europe, DATE, 3, 39–43.
[9] Cao, Y., Groves, R. A., Zamdmer, N. D., Plouchart, J., & Wachnik, R. A. (2002). Frequency-Independent Equivalent Circuit Model. 7–10.
[10] Duthil, P. (2014). Material properties at low temperature. CAS-CERN Accelerator School: Superconductivity for Accelerators - Proceedings, 77–95.

978-1-6654-3746-2/21 $31.00 © 2021 IEEE

Combined effects of BTI, HCI and OFF-State MOSFETs Aging on the CMOS Inverter Performance

A. Crespo-Yepes
Dept. of Electronic Engineering
Autonomous University of Barcelona
Bellaterra, Spain
albert.crespo@uab.es

C. Nasarre
Autonomous University of Barcelona
Bellaterra, Spain

N. Garsot
Autonomous University of Barcelona
Bellaterra, Spain

J. Martin-Martinez
Dept. of Electronic Engineering
Autonomous University of Barcelona
Bellaterra, Spain
javier.martin.martinez@uab.es

R. Rodriguez
Dept. of Electronic Engineering
Autonomous University of Barcelona
Bellaterra, Spain
rosana.rodriguez@uab.es

E. Barajas
Dept. of Electronic Engineering
Polytechnic University of Catalunya
Bacelona, Spain
enrique.barajas@upc.edu

X. Aragones
Dept. of Electronic Engineering
Polytechnic University of Catalunya
Bacelona, Spain
xavier.aragones@upc.edu

D. Mateo
Dept. of Electronic Engineering
Polytechnic University of Catalunya
Bacelona, Spain
diego.mateo@upc.edu

M. Nafria
Dept. of Electronic Engineering
Autonomous University of Barcelona
Bellaterra, Spain
montse.nafria@uab.es

Abstract— In this work, the degradation of the transistors in a CMOS inverter under the various biasing configurations in a complete operation cycle and their impact on the circuit performance are experimentally studied. The relationships between transistors parameters (threshold voltage and mobility) and circuit specifications shifts (peak current and inversion voltage) are explained in terms of the different device aging mechanisms that are active depending on the voltages at the circuit terminals. Moreover, the combined effects of the different aging mechanisms sequentially activated (such as BTI, HCI and OFF-state), at device and circuit levels, emphasizing the role of the OFF-state degradation, are also discussed.

Keywords— Aging, degradation, CMOS inverter, BTI, Hot-Carrier Injection, Off-state stress.

I. INTRODUCTION

Device aging mechanisms such as Bias Temperature Instabilities (BTI) and Hot-Carrier Injection (HCI) degradation are of increasing importance in deep CMOS scaled technologies, significantly reducing device and circuit reliability [1]. Customary, BTI and HCI aging are studied in stand-alone devices [2-7] and aging models are formulated for device and circuit lifetime predictions [7-10]. However, to develop realistic and accurate device aging models, it is necessary to take into account the actual operation conditions of the devices when integrated in a circuit. This is not a simple task, since normally it is difficult to access the device terminals in a circuit to observe the aging impact on the device electrical parameters, so that it can be correlated to circuit performance changes. But also, because the stress conditions of the device in the circuit dynamically change by the aging suffered by the devices themselves, what is difficult to reproduce in stand-alone experiments. The effect of two different stress configurations that sequentially activate two aging mechanisms (one after the other) in an isolated device has been analyzed previously in the literature [11-13]. However, device aging in a circuit may differ from that actually experienced in this kind of tests. For example, the relevance of the OFF-state degradation (of growing interest in the last years) may not be fully revealed [14-19]. Other works analyze the sequential effect of several aging mechanisms by applying pulsed stress conditions, which are closer to circuit

operation. However, these experiments do not allow the observation of the device and circuit parameters shifts during the stress, but only at the end of the stress [20-22]. This can be useful for circuit lifetime prediction, but relevant information on the aging suffered by the devices in the circuit as a consequence of the different activated aging mechanisms is lost, as well as how they could be mutually compensated. In this regard, experimental studies of the circuit performance degradation are rather limited and device-circuit aging correlations are still unclear [21-25].

In this work, the aging of the transistors in a CMOS inverter during a complete cycle of operation and their impact on the circuit performance is experimentally studied. Several stress configurations have been considered, to evaluate the impact of NBTI, PBTI, HCI and OFF-state aging mechanisms, and their interdependences, on the pMOSFET and nMOSFET threshold voltages (V_{TH}) and mobilities (μ), as well as the circuit performance degradation. The role of the OFF-state degradation in combination with the sequential activation of BTI and HCI is also characterized and discussed. Finally, the correlation of the device aging and the CMOS inverter performance shifts is studied.

II. DEVICE AND STRESS PROCEDURE DESCRIPTION

The samples used in this work were 2.45 GHz wideband Power Amplifiers (PA) that are operated as a CMOS inverter (see Figure 1). They were fabricated on a commercial 65-nm technology with 1.2 V nominal operation voltage [22]. pMOSFET and nMOSFET W/L dimensions were 180μm/60nm, and the gate was split in 45 fingers. In order to exactly fix the inversion point (i.e, the input voltage at which output and input are equal), the structure includes an external resistive feedback between the circuit input and output that can be connected or disconnected. When connected, it provides self-biasing (see Fig. 1.B) [22] of the structure, emulating the transitions during the inverter operation. When disconnected, the structure behaves as a standard CMOS inverter. This topology was purposely chosen to allow access to all the transistor terminals, so that the aging of the individual devices, as well as their impact on the PA performance under RF conditions could be evaluated [23]. However, in this work, these circuits will be operated as

978-1-6654-3746-2/21 $31.00 © 2021 IEEE

CMOS inverters, to analyze the shifts of the device/circuit performances in more detail, using a sequence of constant voltages at the gates (input terminal) that emulates the inverter operation.

Different electrical stress configurations of the devices were identified in a complete inverter operation cycle (Figures 1 A, B and C). When a low voltage (logic '0') is applied to the input (case A), a negative voltage drop between the gate-source and gate-drain terminals of the pMOSFET is forced, so that pMOSFET suffers NBTI stress. The nMOSFET is OFF, but a non-uniform electric field is applied to the gate dielectric, due to the large V_{DS} forced by the '1' logic output, so that actually the nMOSFET suffers OFF-state aging [14-19]. Note that usually this stress mode is overlooked during stand-alone device reliability tests but is actually acting in an inverter. When the input voltage is switched to '1' (high input voltage), the output voltage shifts from '1' to '0'. During the transition, both transistors are ON and a significant current flow through their channels, being subjected to HCI stress (case B). The worst case HCI stress corresponds to the inversion voltage (V_{INV}) where $V_{IN}=V_{OUT}$ ($V_{DS}=V_{GS}$), for this short channel devices [26]. Finally, case C represents the situation when a logic '1' (i.e. high voltage) is applied to the input, so that the nMOSFET suffers from PBTI and the pMOSFET from OFF-state aging (case C, complementary to case A).

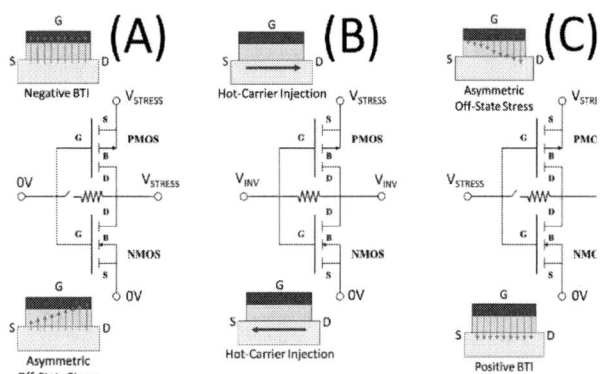

Figure 1: CMOS inverter voltage configuration and active aging mechanisms at (A) low input, (B) inversion point, and (C) high input. The vertical electric field at the gate dielectric during BTI and OFF-State aging (blue arrows) and the currents for HCI (red arrows), for the different inverter stress conditions, are also depicted.

This stress sequence has been emulated using the voltage sequence 'S0V1' in Fig. 2. Note however, that the actual fast '0' to '1' (or vice versa) state transitions have been replaced by a constant voltage at V_{INV} (Input/Output voltages at the inversion point), thanks to the self-biasing). To induce a measurable degradation, the supply voltage was increased to 2.4 V. The high (2.4 V) and low (0 V) input stresses are applied during 300 seconds and the stresses at V_{INV} during 30 seconds. The stress effects on the devices and circuit parameters have been evaluated by measuring the transistors curves (I_D-V_{GS} and I_D-V_{DS}, to extract V_{TH} and μ) and the CMOS inverter characteristics (transfer curve and peak current) before any stress (pristine samples) and after each stress phase (marked with orange flags in Fig. 2. To explain the results obtained during this stress sequence, simpler stress configurations have been considered. To evaluate the effects of NBTI, PBTI, OFF-state and HCI aging on the transistors and the inverter performance, only low input (S0) and only

high input (S1) have been considered. To analyse the effects of the sequential activation of mechanisms, low-high inputs (S0-1) have also been alternatively applied. To allow comparisons, the duration of all the stress phases in the sequence has been kept equal (300 s). The stress sequences have been applied several times (namely, iterations), to evaluate the evolution of the degradation with time. Figure 3 shows typical CMOS inverter transfer curves of the pristine device (continuous lines) and after an accelerated stress test (dashed lines) [27]. As observed, the inversion voltage is slightly shifted to higher values, but also a larger reduction of the current peak is measured.

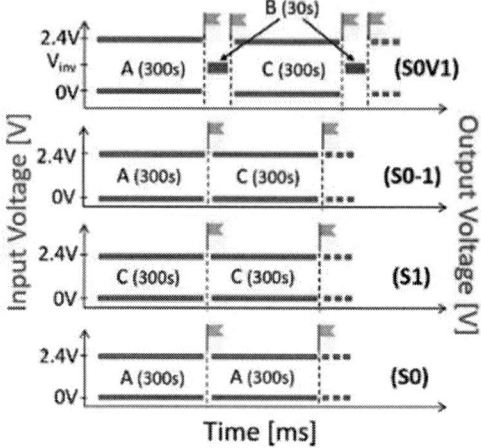

Figure 2: Applied input (blue) and output (red) voltages for each stress sequence. V_{STRESS}=2.4 V in all cases. Labels indicate the stress voltage configurations of the transistors: (S0V1) alternation of low and high inputs, with and intermediate 30 seconds stage where input and output voltages are coincident thanks to the inverter self-bias, (S0-1) alternation of low and high inputs, (S1) high input, and (S0) low input. Flags indicate the times at which the electrical characterization of the CMOS inverter and MOSFET transistors is performed in the cycled scheme.

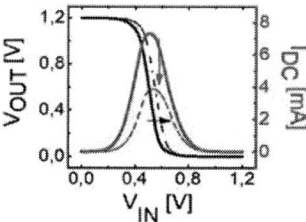

Figure 3: Typical inverter transfer curve (black) and peak current (red) for a pristine (solid lines) and damaged (dashed) CMOS inverter.

III. RESULTS AND DISCUSSION

In order to analyze and correlate the device aging with the variations of the circuit performance, the shifts of V_{TH} and μ of the nMOSFET and pMOSFET, V_{INV} in the inverter transfer curve, and inverter peak current (I_{DC}) as a function of the stress time (i.e. #iteration) for the different stress sequences have been evaluated. Figure 4 shows the evolution of the relative variation of V_{INV} and I_{DC} of the CMOS inverter (Fig. 4A and 4B, respectively) and V_{TH} and μ of the nMOSFET and pMOSFET transistors (Fig. 4C-4E, and Fig. 4D-4F, respectively) as a function of the number of stress iterations.

For S0 (low input, open triangles), the nMOSFET is under OFF-State stress, whereas the pMOSFET is under NBTI. Then, as expected for this technology, V_{TH} shift of the pMOSFET is larger than for the nMOSFET, (~1.2% vs

~0.8%). However, surprisingly, ΔV_{TH} and $\Delta\mu$ of the nMOSFET are relevant, especially for $\Delta\mu$ (~7% versus ~3% for the pMOSFET), which must be attributed to the OFF-state stress [14, 16]. At circuit level, ~8% I_{DC} reduction and ~0.8% inversion point shift are observed. These results suggest that (i) I_{DC} variation is mostly controlled by $\Delta\mu$, whereas the transfer curve shift is mainly determined by ΔV_{TH} of the transistors, and (ii) in this electrical situation the inverter degradation is mainly controlled by the OFF-state nMOSFET aging. Therefore, the NBTI degradation of the PMOS is less detrimental on the circuit performance shifts than the OFF-state of the nMOSFET, though the last is not usually considered during circuit reliability analysis.

For S1 (high input, solid triangles), no significant variations of the inverter performance are observed (Fig. 4A and 4B), clearly related to the negligible variations of μ and V_{TH} of both transistors. Regarding the nMOSFET, this can be explained by the negligible effect to PBTI stress. Moreover, results also suggest that, contrarily to the nMOSFET, the OFF-state stress in the pMOSFET is not as relevant (~0.3%). Therefore, a low input (S0) is much more detrimental than a high input (S1), from the circuit perspective (open triangles vs solid triangles in Fig. 4). However, unexpectedly, this is because the device aging mechanisms activated when a low input (S0) is applied produce larger device degradations, and therefore, larger circuit performance shifts.

The results obtained when sequence S0-1 (solid circles) is applied, i.e., there is an alternation of high and low inputs, are depicted in Fig. 4 with red circles. For the nMOSFET, small fluctuations of ΔV_{TH} around 0% are registered, what suggests that the application of a high voltage (PBTI) after a low voltage (OFF-state) mostly compensates the effects of the previous OFF-state stress on the V_{TH}. However, this does not hold for $\Delta\mu$, since a large decrement of μ is observed, following a similar trend as in S0. On the other hand, μ and V_{TH} of the pMOSFET are clearly degraded during low input stage (NBTI), but largely relaxed during the next high input phase (OFF-state) [15], as observed in Fig. 4D and 4F, producing oscillating variations of the pMOSFET parameters. All those changes at device level are observed as small variations of the inversion voltage (< ~0.2%) around the fresh value, but ~6% variation of I_{DC} (as for sequence S0) due to the large μ reduction of both transistors. The large pMOSFET relaxation (during high input stress) is also observed in the peak current evolution, which oscillates, although it tends to decrease as observed for S0. These results seem to directly relate the μ variations of both transistors with peak current shifts, and V_{TH} changes with inversion voltage variations.

Finally, results corresponding to S0V1 (open squares), where the intermediate HCI stress is introduced, are shown with open squares in Fig. 4. For the pMOSFET parameters, degradation and relaxation are alternatively observed again, and the self-bias stage (HCI) simply acts as an intermediate degradation state between the maximum and minimum values. Therefore, during inverter operation, in this technology, NBTI is the main aging mechanism for the pMOSFET, which is largely relaxed when the voltage difference at its terminals changes (with or without current flow), having HCI a negligible effect. Regarding the nMOSFET parameters, similar trends as for S0 are observed for S0V1. For ΔV_{TH}, since the alternation of high and low voltages compensates the V_{TH} shift (S0-1 in Fig. 4C), and the inclusion of the HCI state provokes a degradation of V_{TH}, this suggests that HCI aging mechanism plays an important role in the V_{TH} degradation of the NMOS transistors (and,

consequently, in the variation of inversion voltage). Then, the damage becomes permanent, so that a similar trend as for S0 is observed, also at circuit level (i.e., large increase of ΔV_{INV}). As far as $\Delta\mu$ is concerned, no remarkable difference is observed with respect to the case S0-1, and only similar fluctuations as for S0 are measured. Consequently, no significant differences are observed for I_{DC} in the S0, S0-1 and S0V1 cases, being dominated by the nMOSFET $\Delta\mu$ (fig. 4B).

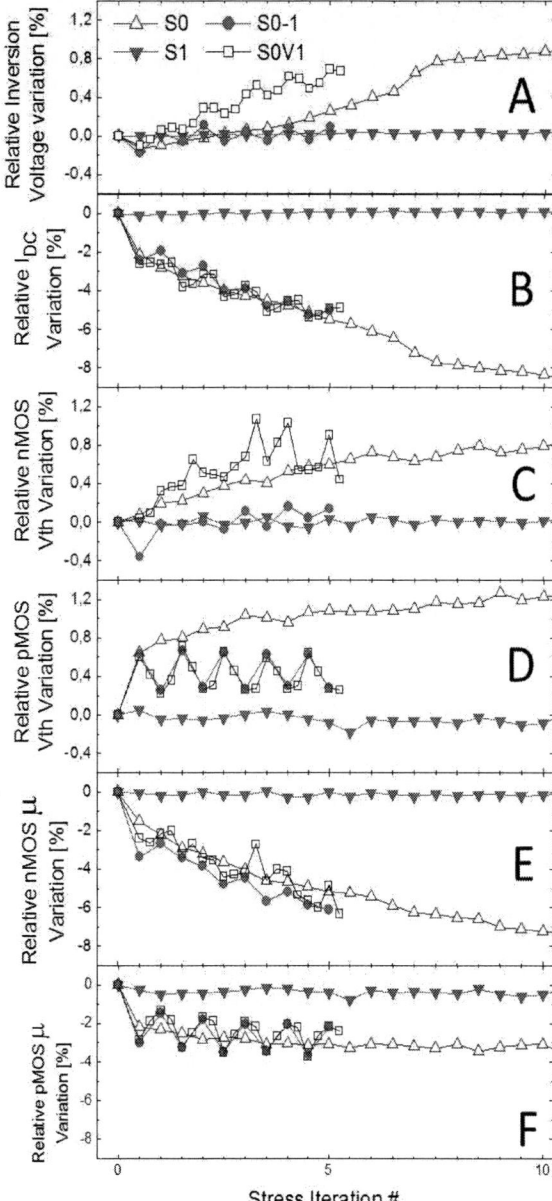

Figure 4. Relative variations of the inversion voltage (A), peak current at the inversion point (B) of the CMOS inverter, threshold voltage (C and D) and carrier mobility (E and F) of the nMOS (C and E) and pMOS (D and F).

IV. CONCLUSIONS

Correlations between devices and circuit degradations are experimentally analysed in a CMOS inverter. To unravel the impact of the degradation of both nMOSFET and pMOSFET transistors on the CMOS inverter performance, several stress-measurement sequences have been defined that, on the one hand, emulate the voltages applied during a complete '0-1' input cycle and, on the other, consider particular stress cases (at device level) along the sequence. Measurements confirm that the shift of the inversion voltage in the CMOS inverter transfer curve is mainly determined by the V_{TH} shifts of the transistors, whereas the peak current is mostly related to the degradation of µ.

The device degradation is strongly dependent on the 'stress history', which depends on its actual operation in the circuit. Degradation can become permanent or compensated, depending on the order that aging mechanisms are activated. Then, lifetime prediction procedures must take into account these aging mechanisms interdependencies.

The results highlight the relevance of the nMOSFET OFF-state degradation on the circuit performance. Though its effects on the inversion voltage may be compensated by a subsequent nMOSFET PBTI stress, cumulative effects are observed for the case of the peak current. So that, on the one hand, models for this mechanism will have to be included in aging prediction CAD tools and, on the other, it may have to be accounted for when designing idle or off-state configurations in digital and analog circuits (i.e. high inputs may be less detrimental than low ones).

ACKNOWLEDGMENT

This work was supported by projects PID2019-103869RB / AEI / 10.13039/501100011033 and TEC2016-75151-C3-R (AEI/FEDER, UE).

REFERENCES

[1] E. Maricau, G. Gielen, Analog IC reliability in nanometer CMOS. New York: Springer-Verlag, 2013.

[2] M. Denais, V. Huard, C. Parthasarathy, G. Ribes, F. Perrier, N. Revil, A. Bravais,"Interface trap generation and hole trapping under NBTI and PBTI in advanced CMOS technology with a 2-nm gate oxide," IEEE Trans. On Device and Materials Reliability, vol. 4, no. 4, pp. 715-722,Desember 2004.

[3] R. Degraeve, M. Aoulaiche, B. Kaczer, P. Roussel. T. Kauerauf, S. Sahhaf, G. Groeseneken, "Review of reliability issues in high-k/metal gate stacks," Physical and Failure analysis of integrated circuits (IPFA), 2008.

[4] D. Gao, C. Liu, Z. Gan, P. Ren, C. Zhan, W. Wong, Z. Chen, Y. Xia, "The study on the variation of NBTI degradation in high-scales FinFET technology," IEEE International Conference on Solid-State and Integrated Circuit Technology (ICSICT), 2018.

[5] A. Hokazono, S. Balasubramanian, K. Isimaru, H. Ishiuchi, C. Hu, T-J. K. Liu, "MOSFET Hot-Carrier Reliability improvement by Forward-Body Bias," IEEE Electron Device Letters, vol. 27, no.7, pp. , 2006.

[6] E. Amat, T. Kauerauf, R. Degraeve, R. Rodriguez, M. Nafria, X. Aymerich, G. Groeseneken, "Gate Voltage Influence on the Channel Hot-Carrier Degradation of High-k-Based Devices," IEEE Transactions on Device and Materials Reliability, vol. 11, no. 1, 2011.

[7] V. Huard. M. Denais, C. Parthasarathy, "NBTI degradation: from physical mechanisms to modelling," Microelectrics Reliability, vol. 46, pp. 1-23, 2006.

[8] I. Messaris, S. K. Goudos, S. Nikolaidis, C: A. Dimitriadis, " A software Tool for Aging Analysis of the CMOS Inverter based on Hot Carrier Degradation modeling," 5th International Conference on Modern Circuits and Systems Technologies (MOCAST), 2016.

[9] B. Tudor, J. Wang, Z. Chen, R. Tan, W. Liu, F. Lee, " An accurate and scalable MOSFET aging model for circuit simulation," 12th International Symposium on Quality Electronic Design, 2011.

[10] M. Miura-Mattausch, H. Miyamoto, H. Kikuchihara, D. Navarro, T. K. Maiti, N. Rohbani, C. Ma, H. J. Mattausch, A. Shiffmann, A. Steinmair, E. Seebacher, "Modeling of dynamic trap density increase for aging simulation of any MOSFET circuits," 47th European Solid-State Device Research Conference (ESSDERC), 2017.

[11] A. Bravaix; M. Saliva; F. Cacho; X. Federspiel; C. Ndiaye; S. Mhira; E. Kussener; E. Pauly; V. Huard, "Hot-carrier and BTI damage distinction for high performance digital application in 28nm FDSOI and 28nm LP CMOS nodes", IEEE 22nd International Symposium on On-Line Testing and Robust System Design (IOLTS), 2016.

[12] E. Bury; A. Chasin; M. Vandemaele; S. Van Beek; J. Franco; B. Kaczer; D. Linten, "Array-Based statistical characterization of CMOS degradation modes and modeling of the time-dependent variability induced by diferent stress patterns in the {VG, VD} bias space", International Reliability Physics Symposium, 2019.

[13] B. Ullmann; M. Jech; K. Puschkarsky; G. Rott; M. Waltl; Y. Illarionov; H. Reisinger; T. Grasser, "Impact of Mixed Negative Bias Temperature Instability and Hot Carrier Stress on MOSFET Characteristics", IEEE Transaction on Electron Device, 2019.

[14] D. Varghese; H. Kufluoglu; V. Reddy; H. Shichijo; S. Krishnan; M. A. Alam, "Universality of Off-Sate degradation in Drain extended NMOS transistors", International Electron Device Meeting, 2006.

[15] N-H. Lee, H. Kim, B. Kang, "Effect of OFF-State Stress and Drain Relaxation Voltage on Degradation of a Nanoscale pMOSFET at High Temperature," IEEE Electron Device Letters, vol. 32, no. 7, pp. 856-858, 2011.

[16] J. K Kim, N, H, Lee, G. J. Kim, Y. Y. Lee, J. E. Seok, Y. S. Lee, " Effect of OFF-State stress on reliability of nMOSFET in SWD circuits of DRAM," Microelectronics Reliability, vol. 88, pp. 183-185, 2018.

[17] N-H. Lee, H. Kim, B. Kang, "Impact of OFF-State Stress and Negative Bias Temperature Instability on degradation of Nanoscale pMOSFET," IEEE Electron Device Letters, vol. 33, no. 2, pp. 137-139, 2012.

[18] A. S. Teng K. W. Lai, R. Tu, M. Y. Lee, A. Kuo, Y. H. Chao, C. W. Lin, K. W. Liu, W, J. Tsai, C. Y. Lu, "Gate Bias Temperature Stress-Induced Off-State Leakage in nMOSFETs: Mechanism, Lifetime Model and Circuit Design Considerations," IEEE International Reliability Physics Symposium, 2014.

[19] J. Trommer; V. Havel; T. Chohan; F. Mehmood; S. Slesazeck; G. Krause; G. Bossu; W. Arfaoui; A. Mühlhoff; T. Mikolajick, "Off-state Impact on FDSOI Ring Oscillator Degradation under High Voltage Stress", International Integrated Reliability Workshop (IIRW), 2018

[20] M. Jech; B. Ullmann; G. Rzepa; S. Tyaginov; A. Grill; M. Waltl; D. Jabs; C. Jungemann; T. Grasser, "Impact of Mixed Negative Bias Temperature Instability and Hot Carrier Stress on MOSFET Characteristics", IEEE Transactions on Electron Device, 2019.

[21] J. Martin-Martinez, S. Gerardin, E. Amat, R. Rodriguez, M. Nafría, X. Aymerich, A. Paccagnella, G. Ghidini, "Channel-Hot-Carrier Degradation and Bias Temperature Instabilities in CMOS Inverters," IEEE Transactions on Electron Devices vol. 56, no. 9, pp. 2155-2159, 2009.

[22] E. Barajas, D. Mateo, X. Aragones, A. Crespo-Yepes, R. Rodriguez, J. Martin-Martinez, and M. Nafria, "Design of a Broadband CMOS RF Power Amplifier to establish device-circuit aging correlations," ICMTS conference, 2017.

[23] X. Aragones, M. Mateo, E. Barajas, A. Crespo-Yepes, R. Rodriguez, J. Martin-Martinez, and M. Nafria, "Aging in CMOS RF Linear Power Amplifiers: Experimental Comparison and Modelling," IEEE International Symposium on Circuits and Systems, 2019.

[24] A. Chenouf, B. Djezzar, A. Benadelmoumene, H. Tahi, "Deep Experimental Investigations of NBTI Impact on CMOS inverter Reliability," 24th International Conference on Microelectronics (ICM), 2012.

[25] A: Kerber, P. Srinivasan, S. Cimino, P. Paliwoda, S. Chandrashekhar, Z. Chbili, S. Uppal, R. Ranjan, M. I. Mahmud, D. Signh, P. P. Manik, J. Johnson, F. Guarin, T Nigam, B. Parameshwaran, " Device reliability metric for end-of-life performance optimization based on circuit level assessment," IEEE International Reliability Physics Symposium (IRPS), pp. 2D-3.1 – 2D-3.5, 2017.

[26] E. Amat, T. Kauerauf, R. Degraeve, A. De Keersgieter, R. Rodriguez, M. Nafria, X. Aymerich, G. Groeseneken, "Channel Hot-Carrier Degradation in Short-Channel Transistors with High-k/Metal Gate Stacks," IEEE Transactions on Device and Materials Reliability, vol. 9, no. 3, 2009.

[27] J. Diaz-Fortuny, P. Saraza-Canflanca, A. Toro-Frias, J. Martin-Martinez, E. Roca, R. Rodriguez, F. V. Fernandez, and M. Nafria, "A Model Parameter Extraction Methodology Including Time-Dependent Variability for Circuit Reliability Simulation," 15th Int. Conf. on Synthesis, Modeling, Analysis and simulation Methods and Applications to Circuit Design (SMACD), 2018.

978-1-6654-3746-2/21 $31.00 © 2021 IEEE

Charge-based modeling of field effect transistors,

Make it easy.

Jean-Michel Sallese
Ecole Polytechnique Fédérale de
Lausanne
1015 Lausanne - Switzerland
jean-michel.sallese@epfl.ch

Abstract—**In this presentation, we revisit some charge-voltage dependences for different architectures of field effect transistor, emphasizing on compactness and simplicity while maintaining a close link with physics, which makes these models predictive and accurate for general purposes of compact modeling.**

Keywords—Compact model, MOSFET, JFET, Junctionless, HEMT, nanowire, double gate, EKV.

I. INTRODUCTION

In the playground of electron devices modeling, and in particular field effect transistors modeling, compact models aim at bridging devices with physics in a simple manner. Compact models stand midway between exact and empirical descriptions of a device, and the objective is to end up with 'simple' relations that can be easily and effectively implemented in electrical simulators for instance. Additionally, these can be used for 'hand' calculations and/or educational purposes. To cope with simplicity, we need to introduce some assumptions at the device and equations levels. At the level of the structure, the architecture needs to be idealized to focus on major features, leaving besides non-fundamental aspects. For instance, the device in treated in one dimension, the mobility is assumed constant, as for the doping densities, and the geometry is long enough to avoid solving complex 2D electrostatics, at least at the early stage. This aims at shaping the corner stone of the model given that all relevant effects can be included afterwards, as corrections to the main set of equations.

Once the process of device idealization is done, the next step is to seek for approximate solutions of non-linear equations governing the physics of the device. Ideally, a compact model should give an intuitive picture of how things work, i.e. they should avoid intricate relationships with many model parameters. Even though some empirical approaches can do the job, compact modeling should focus on the physics of the device first. Ideally, compact models should also predict how technological parameters may impact the device characteristics, which is out of reach with empirical solutions. Therefore, a good compact model should have as few parameters as possible, but the devil is in the details. Practically, many more 'empirical' parameters pop up into the model when comes the time to simulate 'second order' effects. Almost all non-ideal aspects of the devices such as non-uniformity in the doping, short channel effects resulting from electrostatics in two dimensions, field dependent mobility, to cite the mains, are treated as corrections to the set of core equations.

When deriving a compact model, one objective is to reach a balance between accuracy and simplicity, and not necessarily to find the exact analytical solution.

In this presentation, we review very broadly charge-based compact models of four architectures of field effects transistors which were developed at EPFL. For compactness, only ideal devices will be considered, leaving behind second order effects, even though these are important to fine tune the simulations. Starting from the model of the bulk MOSFET, known as the EKV model, we will go through core equations of multigate Field Effect Transistors, Junctionless Field Effect Transistors and the old JFET, to end with the HEMT.

II. THE EKV MODEL OF THE BULK MOSFET.

The first version of the EKV model was published in 1995 by Enz et. al. [1]. The aim was to explore and simulate analytically the region laying between the weak inversion and the strong inversion, which is the moderate inversion. Today, there is a clear evidence that moderate inversion has many advantages for analog circuits in terms of power consumption, speed and reliability, including for RF applications. The model was definitely design oriented, in the sense that it was dedicated to analogue designer. To this purpose, different features were introduced such as the pinch-off voltage, the normalization of potentials and currents, and the key design parameter known as the inversion coefficient. However, the relation linking the current to potentials was somehow empirical, but nevertheless explicit. Note that the bulk is used as reference, whereas most of compact models are using the source. Despite the unquestionable advantages in terms of compactness and ease of use, the model in its first version was lacking some physics, especially the choice of the smoothing function between current and voltages, and the way the pinch-off voltage was defined.

To overcome these issues, the model was revisited introducing more physics in its formulation. All starts with the linearization of the mobile charge density of the channel with respect to the surface potential [2]. A glance at the plot of figure 1 highlights the almost linear dependence between these variables, given that non-linearity comes from the depletion charge 'only'. A linear relation between Q_{inv} and ψ_S seems reasonable then. In the new version of the EKV model, this will be the only approximation introduced in a long channel MOSFET, but it will be very significant in many aspects.

First, it defines a new quantity which is the pinch-off surface potential ψ_p [2] The pinch-off surface potential has a clear physical meaning: when the surface potential reaches the pinch-off surface potential, the mobile charge density

978-1-6654-3746-2/21 $31.00 © 2021 IEEE

cancels. This value is shown by red dots in figure (1). The pinch-off surface potential varies with the gate voltage. The pinch-off voltage initially defined in the EKV model [1] is in fact a definition, $V_P = \Psi_P - 2\varphi_F$ where φ_F is the Fermi potential in the substrate. For simplicity, the pinch-off voltage can also be approximated by $V_P = (V_G - V_T)/n$ where n is the slope of these straight lines.

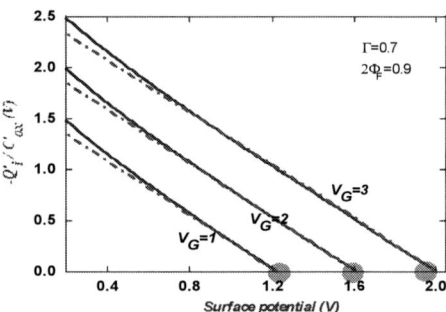

Figure 1. Inversion charge linearization versus the surface potential for different gate voltages.

But the most striking is how this simplifies the heart of the model. Powerful normalizations factors for the current $I_{SP} = 2n\mu C_{ox} U_T^2 W/L$ and charge $Q_{SP} = 2nC_{ox} U_T$ come out 'naturally', and squeezes the model of the MOSFET in two equations:

One between charges and potentials (at source and drain):
$$Ln(q_{S,D}) + 2q_{S,D} = v_P - v_{S,D} \quad (1)$$
The other between current and charges:
$$i = (q_S^2 + q_S) - (q_D^2 + q_D) \quad (2)$$
This means that for given pinch-off, source and drain potentials, all normalized quantities are known. Only after technological parameters are given the real current and the gate voltage are obtained. Interestingly, in saturation, the normalized current defines an inversion coefficient *IC*. The last but not least, g_m/I is technology invariant when plotted versus the inversion coefficient, and defines unambiguously the onset of moderate inversion see figure (2).

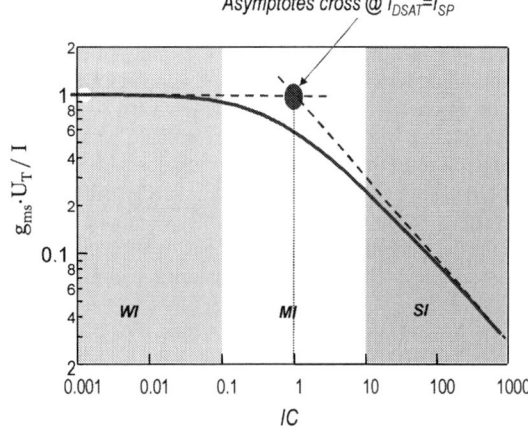

Figure 2. The g_m/I invariant versus the inversion coefficient IC. The operation modes of the MOSFET are clearly defined.

In addition, the model was generalized to simulate non-quasi static operation in MOSFETs [3]. Recently, these DC and non- quasi static models were extended to simulate UTBSOI FETs [4] and validated up to 100 GHz (28 nm channel) [5].

III. DOUBLE GATE, GATE ALL AROUND, ARBITRARY WRAPPED GATE FETs.

A. Double Gate FET

Unexpectedly, from a modeling point of view, the undoped double gate MOSFET has an exact solution regarding the charge versus potential dependence, and this was published by Taur [6], but still, the model is quite complex and not so intuitive. Instead of using the exact solution, looking for an approximate solution will be 'better' [7]. Following a completely different approach, we inherit from interesting features that belong to the bulk MOSFET, such as normalizations, pinch-off voltage, g_m/I invariant. Here, there is no point to introduce any kind of charge linearization scheme, even though this is correct since there is no depletion charge, but we prefer to focus on an integration constant of the differential Poisson-Boltzmann equations, and extract its value in the 'below threshold' regime where it has the greatest impact. As for the bulk MOSFET, normalizations apply to the variables, and in particular a specific current comes out 'naturally', $I_{SP} = 4\mu C_{ox} U_T^2 W/L$.

Interestingly, as for the bulk MOSFET, only two very simple relations are necessary to simulate the DG FET under symmetric operation, including normalization. At the drain we have (at the source we should keep v_P only):

$$v_P - v_D = -2 \cdot q_{m-D} + \ln(-(q_{m-D}/2)) + \ln(1 - q_{m-D} \cdot C_{ox}/2C_{Si}) \quad (3)$$

$$i = -q_m^2 + 2 \cdot q_m + 2 \cdot C_{Si}/C_{ox} \cdot \ln(1 - q_m \cdot C_{ox}/2C_{Si})\Big|_{q_m(S)}^{q_m(D)} \quad (4)$$

Where q_m is the mobile charge density. The main deviation with respect to the bulk MOSFET, apart some factor 2 in scaling, lays in an additional logarithmic quantity $\ln(1 - q_m \cdot C_{ox}/2C_{Si})$. This term holds for the coupling between the gates and depends only on the oxide to silicon capacitances ratio.

There is a subtle difference between the exact solution in [6] and the approximate one in [7]: the exact solution predicts that the potential at the center of the channel, across the gates, will saturate when the device is biased above the threshold. This is expected since the electric field from the gate will be screened by the large amount of charges in the channel. Interestingly, it happens that the main effect of the approximation is to let this center potential free to shift when the gate voltage goes increasing. Still, the error introduced is of minor importance since above the threshold most of the charge density is located at the Si/SiO₂ interface, and not at the center.

Now, what if the channel gets doped, i.e. p-type doped for a n-channel, which was used to tune the threshold voltage in DG FETs ? Again, the integration constant can be rebuilt from the 'below threshold' regime, and this gives the same expression provided an apparent silicon thickness is used [8]. This apparent thickness depends on the doping. It decreases when the doping is increased, figure (3).

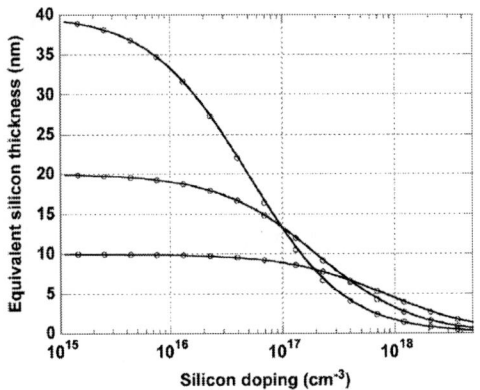

Figure 3. The apparent silicon thickness with respect to the doping density in a double gate FET.

Therefore, equations (3) and (4) can be used for inversion mode doped double gate FET *where no analytical solution exists*. Looking for an approximate solution for the undoped DG FET became an asset to solve the case of the doped channel. We will see that this goes even beyond.

B. Arbitrary wrapped FET

It is quite common to consider the FinFET as a double gate MOSFET, but in reality it is restrictive. First, the cross section of the silicon channel is never an ideal rectangle, secondly the oxide layer is scarcely uniform. The question is how can we simulate a *real* FinFET, and more generally, can we find a simple way to simulate a wrapped FET where both the shape of the Fin and the gate capacitance are not ideal ?

Noting that above the threshold, charges in silicon 'see' the interface, whereas below the threshold they 'see' a volume, we can alleviate all the burden coming from complex 2D analysis of arbitrary geometries.

The 'trick' of the integration constant that we already used will be mapped on this problem as well [9]. After some manipulation, we conclude that a wrapped FET can be simulated as a symmetric double gate FET provided an equivalent silicon thickness is used. In addition, this thickness takes a very simple form $T_{Eq} = 2\,S/P$, i.e. it is

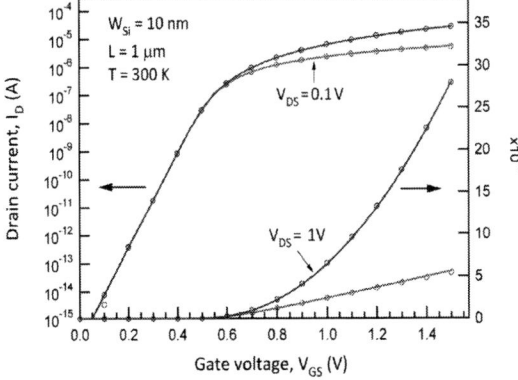

Figure 4. I_D versus V_G for a triangular FET using the equivalent silicon thickness. (dot: TCAD, line: model)

twice the cross section of the channel over the perimeter [9], and the equivalent width (in a double gate 'sense') is simply

half of the perimeter $W_{Eq} = P/2$. Figure (4) shows a triangular channel of 10 nm sides (model and TCAD).

Now, what about the nanowire which is the 'worst rectangular shape' ? This is maybe the most striking outcome: inserting the equivalent thickness in relations (3) and (4) gives back the exact solution for the Poisson Boltzmann equation in cylindrical coordinate [10]…

Finally, the equivalent capacitance reverts to the mean value of the capacitance taken along the interface, a very intuitive picture. Following these simple rules, accurate results were obtained without the need to introduce any fitting parameters.

IV. MODELING JUNCTIONLESS FETs AND JFETs.

Now comes the turn of the junctionless field effect transistor (JLFET), a device proposed by Colinge [11] that is midway between the Junction FET and the MOSFET. Some approaches attempted to model the JLFET imposing the full depletion approximation or using inversion mode MOSFET models. These choices are not justified since the JLFET is none of these devices, and clearly asks for a dedicated solution. To be convinced, we can look at figure (5) of an ideal JLFET with a silicon channel is 20 nm thick. Different doping densities have been used. For relatively low doping densities, at first glance the mobile charge seems to follow the same dependence with V_G as for inversion mode MOSFET, i.e. a linear variation in strong inversion, but when the doping goes beyond some 10^{18}, non-linear shapes become visible. Obviously, there is no way to catch them with regular MOSFET models.

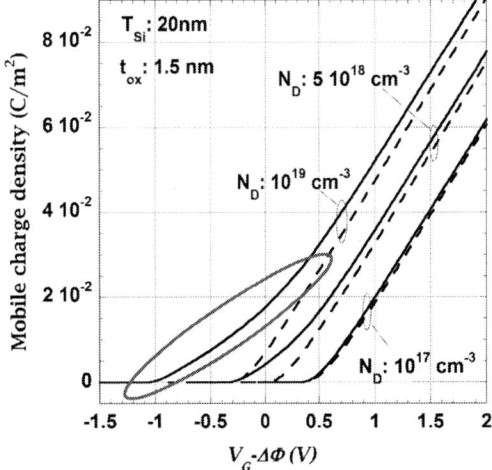

Figure 5. Mobile charge density versus V_G for a junctionless FET highlighting the quadratic dependence of relation (5).

To this purpose, the Poisson-Boltzmann differential equation was solved by finite difference built on three points, two at the Si-SiO$_2$ interface, and one at the center of the slab [12] (to our knowledge, this equation had no analytical solution). JLFET can operate in accumulation and depletion, so here we will focus on depletion only. The approximate solution of the charge versus voltages in depletion mode is [12]:

$$V_{GS} - V_{FB} =$$

$$-\frac{Q_{SC}^2}{8 \cdot q \cdot N_D \cdot \varepsilon_{Si}} - \frac{Q_{SC}}{2 \cdot C_{ox}} + U_T \cdot \ln\left(1 - \left(\frac{Q_{SC}}{q \cdot N_D \cdot T_{SC}}\right)^2\right) \quad (5)$$

Still, normalization doesn't come out easily for the JLFET, but the analytical expression is easy to use and highlights the device parameters. Three regimes are spotted: the logarithmic dependence for below threshold operation, the linear term for MOSFET like operation, and the quadratic one that accounts for the depletion. Now, what if we consider a nanowire, as for instance those used in a variety of label free biosensor, which are likely uniformly doped structures? The problem has no analytical solution (at least we could not find it), but we don't need it: just divide par a factor of 2 the doping density and the density of states of the semiconductor, take the silicon thickness as twice the nanowire radius, and the width as the perimeter [13].

Now, imagine we remove the linear term in relation (5).., then we get a model for the JFET, which now include the exponential decay of the charge at low gate voltages. Indeed, a JFET looks a JL FET without oxide, i.e. with an infinite gate capacitance ... If so, a model of the JFET is the asymptotic model of the JLFET [14]:

$$V_{GS} - V_{FB} = -\frac{Q_{SC}^2}{8 \cdot q \cdot N_D \cdot \varepsilon_{Si}} + U_T \cdot \ln\left(1 - \left(\frac{Q_{SC}}{q \cdot N_D \cdot T_{SC}}\right)^2\right) \quad (6)$$

For nanowire JFETs, the same transformations stated before applies. We will not discuss other aspects of modeling here, but complete AC models have also been derived for these kinds of devices based on the DC models without additional parameters [15].

V. Modeling HEMT.

So far, we modelled field effect transistors assuming explicitly a three-dimensional density of states for the semiconductor (even if the channel is at the Si-SiO$_2$ interface) and Boltzmann statistics. The validity of these assumptions are acceptable given the quite large density of states of silicon and large effective masses. But when modeling AlGaAs HEMTs these hypotheses break down. HEMT is a genuine 2D quantum well device, and GaAs has a low effective DOS of about 410^{17}cm^{-3} and a low conduction band effective mass. As for the MOSFET, a depletion charge exists in HEMT. Then, why not relying on the charge linearization concept introduced for MOSFETs while imposing 2D DOS and Fermi-Dirac statistics ? Following this idea [16], we can inherit from important features which are the pinch-off surface potential, the pinch-off voltage, charge and current normalizations, the inversion coefficient so useful for analog design, and the invariant g$_m$/I characteristics. The normalized current is still given by relation (2) but the barrier capacitance is used in place of the oxide capacitance in the normalization factors I_{SP} and Q_{SP}.

The main difference is between the carrier density and the potentials which writes :

$$U_T \ln\left[\exp\left(\frac{n_{ch}}{\mathrm{DoS}_{2D} U_T}\right) - 1\right] + \frac{q n_{ch}}{C_b n_q} + u_1 \sqrt[3]{n_{ch}^2} = \psi_p - V$$

The logarithmic term holds for below threshold, the linear for above threshold, and the power term handles the shift of the ground state in the quantum well with the surface electric field. As for the MOSFET, a complete AC model follows [17].

VI. Acknowledgements

I would like to thank F. Jazaeri, C. Lallement, W. Grabinski, B. Iniguez and M. Bucher for their constructive interactions.

VII. References

[1] Enz C, Vittoz E, Krummenacher F, 'An analytical MOS transistor valid in all regions of operation and dedicated to low-voltage and low-current applications', Aalog Integ. Cir. And Sign Proc., Vol. 8, Iss 1, pp. 83-114 (1995).

[2] Sallese JM, Bucher M, Krummenacher F, et al., 'Inversion charge linearization in MOSFET modeling and rigorous derivation of the EKV compact model', Solid State Electronics 47 (4), pp. 677-683, 2003.

[3] Sallese JM, Porret AS, 'A novel approach to charge-based non-quasi-static model of the MOS transistor valid in all modes of operation', Solid State Electronics 44 (6), pp. 887-894, 2000.

[4] El Ghouli, Salim; Rideau, Denis; Monsieur, Frederic; Patrick Scheer; Gilles Gouget; André Juge; Thierry Poiroux; Jean-Michel Sallese; Christophe Lallement 'Experimental g(m)/I Invariance Assessment for Asymmetric Double-Gate FDSOI MOSFET ' Vol. 65, Issue 1, Pages: 11-18 (2018).

[5] El Ghouli, Salim, J.M Sallese, Jean-Michel, A. Juge, P. Scheer, C. Lallement, 'Transadmittance Efficiency under NQS Operation in Asymmetric Double Gate FDSOI MOSFET', IEEE TED Volume: 66, Issue 1, Pages: 300-307, 2019.

[6] Taur Y 'Analytic Solutions of Charge and Capacitance in Symmetric and Asymmetric Double-Gate MOSFETs', IEEE Transactions on Electron Devices Vol. 48, Issue 12, pp. 2861-2869 (2001).

[7] Sallese JM, Krummenacher F, 'A design oriented charge-based current model for symmetric DG MOSFET and its correlation with the EKV formalism', Solid State Elect. 49, pp.485-489 2005.

[8] Sallese JM, Chevillon N, Pregaldini F, Lallement C. Iniguez B.'The Equivalent-Thickness Concept for Doped Symmetric DG MOSFETs' Transistors' IEEE Transactions on Electron Devices Vol. 57, Issue 1, pp. 2917-2924 (2010).

[9] Chevillon N., Sallese J.M. et. al. , 'Generalization of the Concept of Equivalent Thickness and Capacitance to Multigate MOSFETs Modeling' IEEE Transactions on Electron Devices Vol. 59, Issue 1, pp. 60-71 (2012).

[10] D. Jiménez, B. Iñíguez, J. Suñé, E, L. F. Marsal, J. Pallarès, J. Roig, and D. Flores 'Continuous Analytic I–V Model for Surrounding-Gate MOSFETs', IEEE Transactions on Electron Devices Vol. 25, Issue 8, pp. 571-573 (2004).

[11] Colinge JP et. al. 'Nanowire transistors without junctions' Nature Nanotechnology, Vol. 5, Iss. 3, pp. 225-229 (2010).

[12] Sallese JM, Chevillon N, Lallement C, Iniguez B, Pregaldini F. 'Charge-Based Modeling of Junctionless Double-Gate Field-Effect Transistors' IEEE Transactions on Electron Devices Vol. 58, Issue 8, pp. 2628-2637 (2011).

[13] Sallese JM, Jazaeri F., Barbut L, Chevillon N.. Lallement C. 'A Common Core Model for Junctionless Nanowires and Symmetric Double-Gate FETs', IEEE Transactions on Electron Devices Vol. 60, Issue 12, pp. 4277-4280 (2013).

[14] F. Jazaeri, N. Makris, A. Saeidi, M. Bucher, J.M. Sallese 'Charge-based Model for Junction FETs' IEEE TED Volume: 65 Issue: 7 Pages: 2694-2698, 2018.

[15] Modeling nanowire and double-gate junctionless field-effect transistors, Farzan Jazaeri, Jean- Michel Sallese, Cambridge University Press 2018.

[16] Jazaeri F, Sallese JM, 'Charge-Based EPFL HEMT Model' in IEEE Transactions on Electron Devices, vol. 66, no. 3, pp. 1218-1229 (2019).

F. Jazaeri, M. Shalchian and J.M. Sallese, "Transcapacitances in EPFL HEMT Model," in IEEE Transactions on Electron Devices, vol. 67, no. 2, pp. 758-762 (2020).

Conductance modulation in Al/SiO₂/n-Si MIS resistive switching structures

Piotr Wiśniewski
Centre for Advanced Materials and Technologies CEZAMAT
Warsaw University of Technology
Warsaw, Poland
Center for Terahertz Research and Applications (CENTERA)
Institute of High-Pressure Physics, Polish Academy of Science
Warsaw, Poland
e-mail: piotr.wisniewski@pw.edu.pl

Jakub Jasiński
Institute of Microelectronics and Optoelectronics
Warsaw University of Technology
Warsaw, Poland
e-mail: jakub.jasinski@pw.edu.pl

Andrzej Mazurak
Institute of Microelectronics and Optoelectronics
Warsaw University of Technology
Warsaw, Poland
e-mail: andrzej.mazurak@pw.edu.pl

Abstract—**In this work, we discuss the small-signal conductance modulation in the Metal-Insulator-Semiconductor structure. We present the DC measurement results of Al/SiO₂/n++-Si device exhibiting resistive switching phenomena. We show that the application of millisecond voltage pulse train results in conductance modulation. Furthermore, we show conductance modulation effect obtained with other bias schemes.**

Keywords— resistive switching; RRAM; MIS structure; silicon oxide; pulse measurements

I. INTRODUCTION

For the last years, a lot of work has been carried out regarding resistive switching (RS) phenomena in Metal-Insulator-Metal (MIM) structures. Such a device can be used as a Resistive Random-Access Memory (RRAM) [1]. RRAMs can also be used in other than memory applications, e.g., hardware security, neuromorphic computing [2], [3]. The latter one is intensively studied nowadays due to the potential ease of scaling MIM structures and construct systems able to mimic the behavior of neurons and synapses, and solve some class of problems in a very efficient way [4]–[7]. Some works present the analysis of silicon oxide as a material for MIM resistive switching device [8]–[15]. This work shows that a well-known Metal-Insulator-Semiconductor (MIS) structure can exhibit resistive switching properties. We present the measurement results of Al/SiO₂/n++-Si MIS device and discuss the results of small-signal conductance modulation by pulse train schemes, which can be potentially interesting to imitate some synaptic plasticity. We also show the DC bias impact on the device's conductance.

II. STRUCTURE AND MEASUREMENT

The considered structure was fabricated using standard CMOS compatible processes: RCA cleaning, thermal oxidation, wet etching, optical lithography, magnetron sputtering. We used highly doped n-type wafers (As dopant, resistivity < 0.005 Ωcm) from Siegert Wafer Gmbh. Gate oxide was prepared using dry oxidation process in a pure oxygen atmosphere in a horizontal furnace system (Thermco HTR system), for 10 minutes at 820°C. It resulted in ~5 nm of silicon oxide, as measured with the use of spectroscopic ellipsometry. After the fabrication process structures were annealed in H₂/Ar atmosphere in 400 °C for 30 minutes.

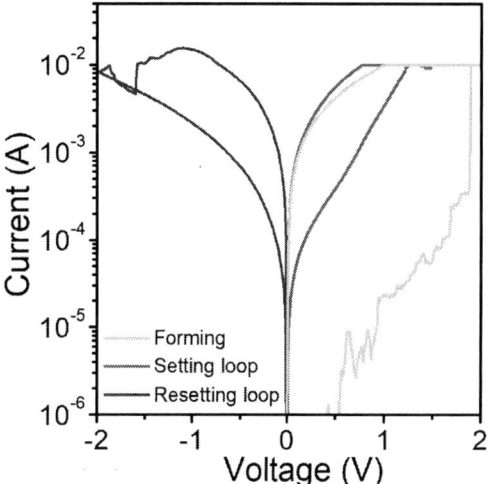

Figure 1. Measured current-voltage characteristics of Al/SiO₂/n++ Si device (CC=10 mA).

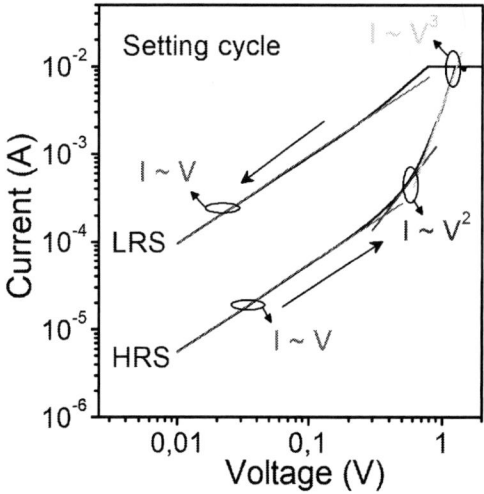

Figure 2. Current-voltage characteristics of Al/SiO₂/n++ Si device at SET cycle with fitted curves of different slope.

978-1-6654-3746-2/21 $31.00 © 2021 IEEE

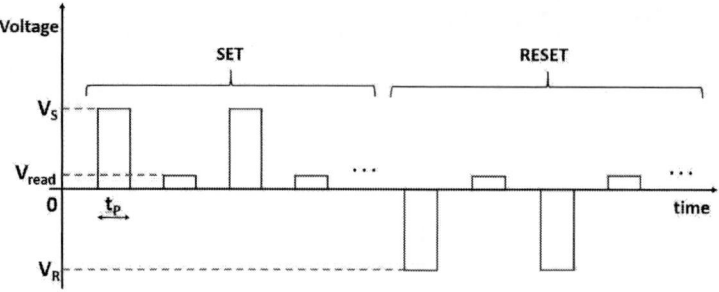

Figure 3. Pulse sequence used for conductance modulation during SET and RESET cycles.

In order to perform electrical characterization of fabricated structures, we used Keithley 4200-SCS Semiconductor Characterization System and Süss MicroTec PM8 low noise probe shield. All measurements were carried out at room temperature. To generate pulse sequence and study the modulation of device's conductance, we used high-resolution, static source-measurement unit (SMU) typically used for standard current-voltage measurements combined with the small-signal, capacitance-voltage unit (CVU).

III. RESULTS AND DISCUSSION

In Figure 1 we present the measured current-voltage characteristics of the MIS RS device with gate electrode diameter of 106 μm. Initial electroforming run and set cycle were measured with a compliance current (CC) of 10 mA. In most of the tested structures we can observe that electroforming, set and reset voltages are below 2 V and current-voltage hysteresis is visible (see Figure 1). We analyzed the I-V curves and tried to fit to the different transport mechanisms. We identified the Space Charge Limited Current (SCLC) as the main transport mechanism [16][17]. In Figure 2, we can observe that at low voltage values, we have ohmic conduction. Going forward to higher voltages at high resistance state (HRS), we observe Child's square law region ($I \sim V^2$) and a high field region. At low resistance state (LRS) we observe ohmic conduction. A deeper analysis of I-V characteristics of these devices and study of transport mechanisms is a scope of a different paper (in preparation).

In Figure 3 we introduced the scheme of pulse measurement scenario. We applied a sequence of positive pulses with V_S = 2 V in SET cycle. After every pulse, we performed a small-signal measurement procedure to obtain conductance value. Bias voltage during this procedure was set at V_{read}= 0.1 V. A similar scheme with V_R = -2 V was used in the RESET cycle. The width of a voltage pulse was t_p = 7 ms. In Figure 4 we show the results of small-signal conductance modulation during SET and RESET cycles for the structure with DC I-V characteristics presented in Figure 1. Prior to the pulse measurements, the structure was set to HRS. In Fig. 4(a) we can observe that structure is being set gradually, and there is a part of the steep rise of conductance and then a slower increase of conductance (first and second SET cycle). In some cycles the dependence of conductance as a function of number of pulses is linear. In the RESET cycle we observe an abrupt decrease of conductance. A single pulse is enough to turn off the structure for given measurement parameters. In Fig. 4(b) we plot the resistance as a function of the number of pulses. We can observe that the resistance in LRS is usually much below 1 kΩ, whereas in HRS it is less stable and varies between 1 and 10 kΩ. In Figure 5, we show results of another

measurement performed on the same MIS structure, showing that conductance can also exhibit steps during SET pulse trains. It can be related to the fact that conductive filament can be formed partially. From Figures 4 and 5, we can conclude that the conductive filament dissolution process is faster compared to the filament formation for a given pulse parameters.

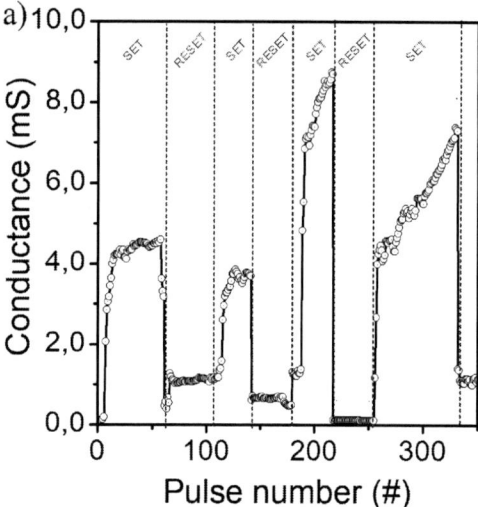

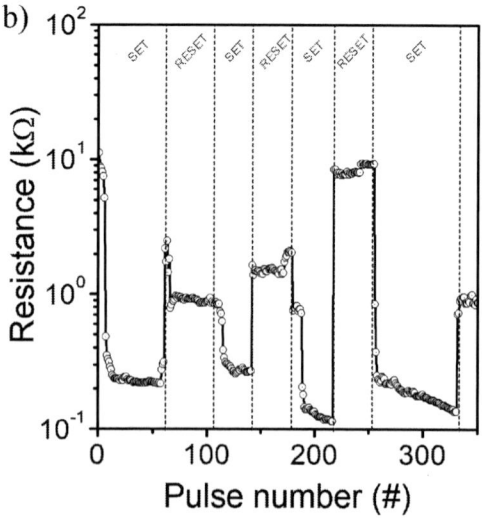

Figure 4. Conductance (a) and resistance (b) modulation during SET and RESET pulse trains for V_S/ V_R = 2 V/-2 V, t_p = 7 ms, V_{read} = 0.1 V.

978-1-6654-3746-2/21 $31.00 © 2021 IEEE

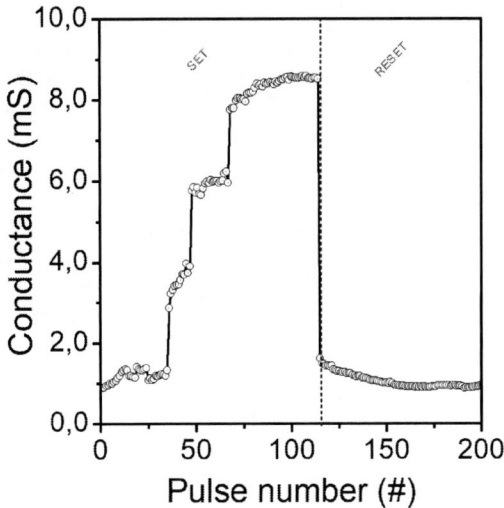

Figure 5. Conductance modulation exhibiting steps during SET pulse trains for $V_S/V_R = 2$ V/-2 V, $t_p = 7$ ms, $V_{read} = 0.1$ V.

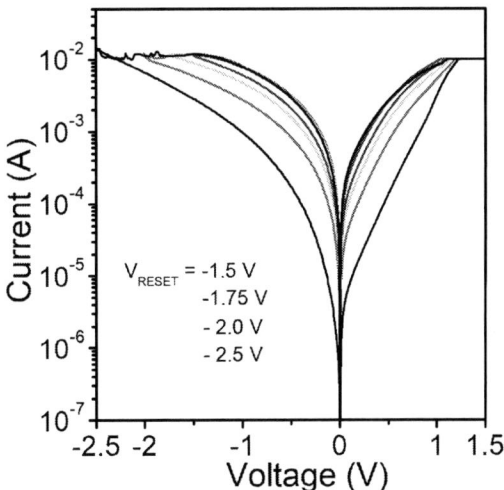

Figure 6. Current-voltage characteristics of Al/SiO₂/n++ Si device for different RESET voltage V_{RESET} values.

In Figure 6 we show the current-voltage characteristics of the MIS RS device with gate electrode diameter of 112 μm for different RESET voltage V_{RESET} values. Measurements during SET cycle were carried out with a CC of 10 mA. In the RESET cycle we stopped the measurement at different voltage values $V_{RESET} = -2.5$ V, -2.0 V, -1.75 V and -1.5 V. After every reset cycle we performed a set cycle. It resulted in a different course of current-voltage curves, which means that level of conductance/resistance at HRS can be modulated by RESET voltage. It can be observed that hysteresis loop is smaller for smaller absolute values of V_{RESET} and is almost closed for $V_{RESET} = -1.5$ V.

In Figure 7 we introduced the measurement protocol of MIS RS device for different SET voltage V_{SET} and RESET voltage V_{RESET} values. V_{SET} and V_{RESET} values were in the range 1.0 V – 1.25 V and -1.0 V – -2.0 V, respectively. Device was initially in LRS. We performed the consecutive measurements with different V_{RESET} values to reset the device. In Figure 8 we can observe the conductance modulation,

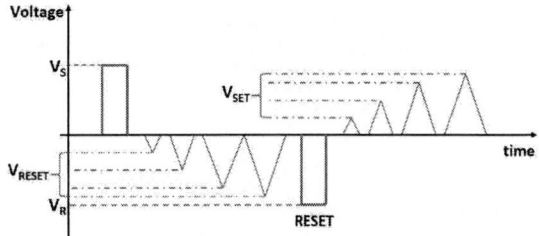

Figure 7. Measurement protocol of current-voltage characteristics of MIS RS device for different SET voltage V_{SET} and RESET voltage V_{RESET} values.

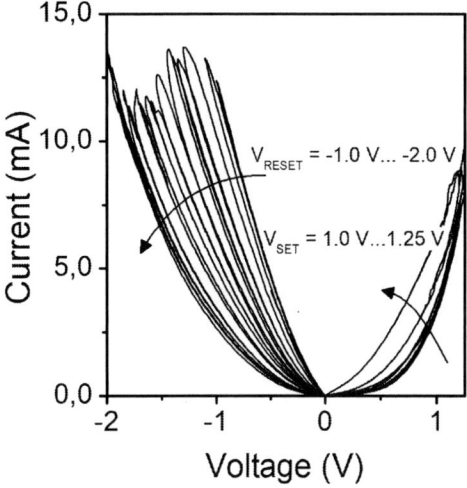

Figure 8. Current-voltage characteristics of Al/SiO₂/n++ Si device after consecutive measurements carried out for different SET voltage V_{SET} and RESET voltage V_{RESET} values.

which is related to the partial dissolution of conductive filament during consecutive measurements within the negative voltage range. Next, we carried out the consecutive measurements with different V_{SET} values to set the device in LRS. We also observe the conductance modulation as can be seen in Figure 8. It is related to the gradual conductive filament formation during consecutive measurements for positive voltage values. However, this effect is less pronounced in SET cycles than in RESET cycles. Only for larger values of V_{SET} we observe some significant change in current-voltage characteristics.

IV. CONCLUSIONS

In this work we presented the results of the measurements of Al/SiO₂/n-Si MIS resistive switching structures. We identified the current transport mechanism and show that it is possible to modulate the conductance state of the structure for a given train pulse parameters. We also show using DC measurements that conductance modulation is also possible for different SET voltage V_{SET} and RESET voltage V_{RESET} values. Further investigation of the presented devices is needed to assess their usefulness in a potential application.

ACKNOWLEDGMENT

Research was funded by POB Technologie Materiałowe of Warsaw University of Technology within the Excellence Initiative: Research University (IDUB) programme.

978-1-6654-3746-2/21 $31.00 © 2021 IEEE

REFERENCES

[1] T. C. Chang, K. C. Chang, T. M. Tsai, T. J. Chu, and S. M. Sze, "Resistance random access memory," *Materials Today*, vol. 19, no. 5. Elsevier B.V., pp. 254–264, Jun. 01, 2016, doi: 10.1016/j.mattod.2015.11.009.

[2] H. S. P. Wong *et al.*, "Metal-oxide RRAM," *Proc. IEEE*, vol. 100, no. 6, pp. 1951–1970, 2012, doi: 10.1109/JPROC.2012.2190369.

[3] I. Boybat *et al.*, "Neuromorphic computing with multi-memristive synapses," *Nat. Commun. 2018 91*, vol. 9, no. 1, pp. 1–12, Jun. 2018, doi: 10.1038/s41467-018-04933-y.

[4] B. Sarkar, B. Lee, and V. Misra, "Understanding the gradual reset in Pt/Al2O3/Ni RRAM for synaptic applications," *Semicond. Sci. Technol.*, vol. 30, no. 10, 2015, doi: 10.1088/0268-1242/30/10/105014.

[5] Z. Shen *et al.*, "Advances of RRAM devices: Resistive switching mechanisms, materials and bionic synaptic application," *Nanomaterials*, vol. 10, no. 8, pp. 1–31, 2020, doi: 10.3390/nano10081437.

[6] D. Ielmini, "Brain-inspired computing with resistive switching memory (RRAM): Devices, synapses and neural networks," *Microelectron. Eng.*, vol. 190, pp. 44–53, Apr. 2018, doi: 10.1016/j.mee.2018.01.009.

[7] Z. Jiang *et al.*, "Bidirectional Analog Conductance Modulation for RRAM-Based Neural Networks," *IEEE Trans. Electron Devices*, vol. 67, no. 11, pp. 4904–4910, 2020, doi: 10.1109/TED.2020.3025849.

[8] A. Mehonic *et al.*, "Silicon Oxide (SiOx): A Promising Material for Resistance Switching?," *Adv. Mater.*, vol. 30, no. 43, p. 1801187, Oct. 2018, doi: 10.1002/adma.201801187.

[9] E. Ambrosi, A. Bricalli, M. Laudato, and D. Ielmini, "Impact of oxide and electrode materials on the switching characteristics of oxide ReRAM devices," *Faraday Discuss.*, vol. 213, pp. 87–98, 2019, doi: 10.1039/C8FD00106E.

[10] Y.-F. Chang *et al.*, "Study of polarity effect in SiOx -based resistive switching memory," *Appl. Phys. Lett*, vol. 101, p. 52111, 2012, doi: 10.1063/1.4742894.

[11] Y.-F. Chang *et al.*, "Understanding the resistive switching characteristics and mechanism in active SiO x-based resistive switching memory," *J. Appl. Phys*, vol. 112, p. 123702, 2012, doi: 10.1063/1.4769218.

[12] Y.-F. Chang *et al.*, "Intrinsic SiO x-based unipolar resistive switching memory. I. Oxide stoichiometry effects on reversible switching and program window optimization," *J. Appl. Phys*, vol. 116, p. 43708, 2014, doi: 10.1063/1.4891242.

[13] Y.-F. Chang *et al.*, "Intrinsic SiO x-based unipolar resistive switching memory. II. Thermal effects on charge transport and characterization of multilevel programing," *J. Appl. Phys*, vol. 116, p. 43709, 2014, doi: 10.1063/1.4891244.

[14] P. Bousoulas, C. Papakonstantinopoulos, S. Kitsios, K. Moustakas, G. C. Sirakoulis, and D. Tsoukalas, "Emulating Artificial Synaptic Plasticity Characteristics from SiO2-Based Conductive Bridge Memories with Pt Nanoparticles," *Micromachines 2021, Vol. 12, Page 306*, vol. 12, no. 3, p. 306, Mar. 2021, doi: 10.3390/MI12030306.

[15] Y. Lee, C. Mahata, M. Kang, and S. Kim, "Short-term and long-term synaptic plasticity in Ag/HfO2/SiO2/Si stack by controlling conducting filament strength," *Appl. Surf. Sci.*, vol. 565, p. 150563, Nov. 2021, doi: 10.1016/J.APSUSC.2021.150563.

[16] F. C. Chiu, "A review on conduction mechanisms in dielectric films," *Adv. Mater. Sci. Eng.*, vol. 2014, 2014, doi: 10.1155/2014/578168.

[17] J. Zhu, T. Zhang, Y. Yang, and R. Huang, "A comprehensive review on emerging artificial neuromorphic devices," *Appl. Phys. Rev.*, vol. 7, no. 1, 2020, doi: 10.1063/1.5118217.

AUTHOR INDEX

Agopian, Paula G. D. 41, 162
Aguinsky, Luiz F. 81
Akhavan, N. D. 134
Allibert, Frédéric 57, 178
André, Nicolas 166
Antonov, Valentin 158, 186
Antoszewski, J. 134
Aprà, A. 114
Aragones, X. 197
Aussenac, F. 182
Ávila, Jorge 17
Bae, Jin-Hee 57
Balestra, F. 9
Balestra, Luigi 89
Barajas, E. 197
Barraud, Sylvain 142
Baumgartner, O. 5
Bawedin, M. 182
Bendra, Mario 102
Berlingard, Quentin 193
Besson, P. 182
Bosio, A. 5
Brevard, L. 182
Brugnolotto, Enrico 97
Brunetti, R. 37
Calmon, F. 49, 110
Cao, Y. X. 154
Cassé, M. 9, 114, 142, 193
Catapano, E. 114
Cathelin, A. 49, 110
Cellier, R. 49
Cepparulo, P. 174
Chalupa, Z. 178, 182
Charbon, Edoardo 93
Contamin, L. 9, 193
Crespo-Yepes, A. 197
Cretu, Bogdan 146
Cristoloveanu, S. 134
Croce, Giuseppe 61
Cuomo, O. 174
Cusano, A. 174
D'Amico, Antonio 93
De Franceschi, S. 114, 193
De Orio, Roberto L. 102
De Souza, Michelly 106
Depetro, Riccardo 61
Dhar, Rakshita P. S. 65
Dobrzynska, Jagoda 89
Doria, R. T. 106, 126
Durand, Cédric 193
Ekström, Mattias 189
El-Sayed, A.-M. 53, 69
Ender, Johannes 102
Enz, Christian 93

Esseni, David 25
Fache, Thibaud 178
Faraone, L. 134
Faynot, Olivier 142
Filipovic, Lado 33
Fiorentini, Simone 102
Flandre, Denis 166
Fonte, E. T. 126
Francis, Laurent A. 166
Gaillard, F. 114, 178, 193
Galdon, Jose 17
Galy, P. 170, 193
Gamiz, Francisco 17
Gao, S. 49, 110
García, César P. 65
Garsot, N. 197
Georgiev, Vihar P. 65, 97
Ghibaudo, G. 9, 114
Giuliano, Federico 61
Gnani, Elena 61, 89
Gnudi, Antonio 61, 89
Goes, Wolfgang 102
Golanski, D. 49, 110
Grützmacher, Detlev 57
Gu, R. 134
Hadámek, Tomáš 102
Haendler, Sebastien 170
Hagen, S. 49
Han, Hung-Chi 93
Han, Yi 57
Hartmann, J.-M. 182
Hellström, Per-Erik 189
Hoentschel, Jan 122
Hoffmann, Michael 138
Hoffmann-Eifert, Susanne 57
Hössinger, Andreas 77, 81
Hutin, Louis 178
Issartel, D. 49, 110
Jasinski, Jakub 205
Jazaeri, Farzan 93
Jech, Markus 53
Juettner, Maximilian 122
Juge, Andre 193
Kansal, Harshit 150
Khakbaz, Pedram 25
Kilchytska, Valeriya 166
Klemenschits, Xaver 33
Knoch, Joachim 57
Kosina, Hans 53, 69
Kretzschmar, Claudia 93
Kumar, Naveen 65
Lacord, J. 178, 182
Larrieu, G. 5
Le Beux, S. 5

AUTHOR INDEX

Lederer, Dimitri 130
Lee, K. 182
Lehmann, Steffen 93
Lenz, Christoph 77
Liu, W. 73
Liu, Y. N. 73, 154
Lomov, Andrey 158
Loup, V. 182
Lugo, Jose 178
Lugo-Alvarez, Jose 193
Mandorlo, F. 110
Maneux, C. 5
Maria, F. S. D. S. 9
Mariniello, Genaro 142
Marquez, Carlos 17
Martin-Martinez, J. 197
Martino, Joao A. 1, 41, 162
Matagne, Philippe 45
Mateo, D. 197
Mazurak, Andrzej 205
Medina-Bailon, Cristina 65
Medury, Aditya S. 13, 150
Mele, Leandro J. 118
Mescot, X. 182
Meunier, T. 114, 193
Miakonkikh, Andrey 158, 186
Mikolajick, T. 5
Mitrovic, I. Z. 73, 154
Morand, Yves 178
Morozzi, Arianna 138
Moulin, Maxime 178
Mroczynski, Robert 21
Mukherjee, C. 5
Mulargia, Roberto 138
Nabet, Massinissa 130
Nafria, M. 197
Nasarre, C. 197
Navarro, Carlos 17
Nyssens, Lucas 130
O'Connor, I. 5
Orobtchouk, R. 110
Otto, Michael 122
Palestri, Pierpaolo 25, 118
Palma, F. 174
Pavanello, Marcelo A. 142
Perina, Welder F. 162
Pignataro, G. 174
Pilotto, Alessandro 25
Pittet, P. 49
Plantier, Christophe 178
Poittevin, A. 5
Popov, Vladimir 158, 186
Pradhan, K. P. 85
Prucnal, Slawomir 57

Rack, Martin 130, 178
Radu, Ionut 57
Rangel, Ricardo C. 1
Rao, R. 174
Raskin, Jean-Pierre 130, 178
Recio, Maria-Isabel 17
Reggiani, Susanna 61, 89
Reiter, Tobias 33
Rodrigues, Frâncio 81
Rodrigues, Jaime C. 142
Rodriguez, R. 197
Rossetti, Mattia 61
Rudan, M. 37
Rudenko, Konstantin 158, 186
Ruvo, M. 174
Salazar, Norberto 17
Sallese, Jean-Michel 201
Sasaki, Katia R. A. 1
Sato, Shingo 29
Scharinger, Alexander 81
Schmid, Ulrich 81
Sedki, Amor 166
Seiler, Heribert 53, 69
Selberherr, Siegfried 102
Selmi, Luca 118
Seshu, Vullakula R. 85
Shaik, Rameez R. 85
Shibutani, André B. 106
Silva, Vanessa C. P. 41
Simoen, Eddy 41, 45, 146, 162
Slesazeck, Stefan 138
Smidstrup, Søren 97
Solanki, Ravi 13
Sousa, Júlia C. S. 162
Stanojevic, Z. 5
Sverdlov, Viktor 53, 69, 102
Tarkov, Sergey 158
Theodorou, C. 9
Tikhonenko, Fedor 186
Toifl, Alexander 77, 81
Tounsi, Farès 166
Trevisoli, R. 106, 126
Trommer, J. 5
Trupke, Michael 81
Umana-Membreno, G. A. 134
Van Zalinge, H. 73, 154
Vedel, Christian D. 97
Veloso, Anabela 41, 45, 146, 162
Vinet, M. 114, 142, 193
Vizioz, C. 182
Vobecký, Jan 89
Wachter, Georg 81
Wakam, F. T. 182
Waldhör, Dominic 53

AUTHOR INDEX

Wang, Q. N. 73
Weinbub, Josef 77, 81
Wisniewski, Piotr 205
Xi, Fengben..57
Yang, L. 154
Yojo, Leonardo S. 1
Zhao, C. 73, 154
Zhao, C. Z. 73, 154
Zhao, Qing-Tai 57
Zhao, Zhixing 93
Zurauskaite, Laura 189

IEEE
445 Hoes Lane
Piscataway, NJ 08854-4141

ISBN 978-1-6654-3746-2